AF358793

Forest, Foods and Nutrition

Forest, Foods and Nutrition

Editors

Alessandra Durazzo
Massimo Lucarini
Massimo Zaccardelli
Antonello Santini

MDPI • Basel • Beijing • Wuhan • Barcelona • Belgrade • Manchester • Tokyo • Cluj • Tianjin

Editors

Alessandra Durazzo
CREA—Research Centre for
Food and Nutrition
Italy

Massimo Lucarini
CREA—Research Centre for
Food and Nutrition
Italy

Massimo Zaccardelli
CREA—Research Centre for
Vegetable and Ornamental Crops
Italy

Antonello Santini
University of Napoli Federico II
Italy

Editorial Office
MDPI
St. Alban-Anlage 66
4052 Basel, Switzerland

This is a reprint of articles from the Special Issue published online in the open access journal *Forests* (ISSN 1999-4907) (available at: https://www.mdpi.com/journal/forests/special_issues/Forest_Foods).

For citation purposes, cite each article independently as indicated on the article page online and as indicated below:

LastName, A.A.; LastName, B.B.; LastName, C.C. Article Title. *Journal Name* **Year**, *Volume Number*, Page Range.

ISBN 978-3-0365-0042-3 (Hbk)
ISBN 978-3-0365-0043-0 (PDF)

Contents

About the Editors

Alessandra Durazzo was awarded her master's in Chemistry and Pharmaceutical Technology cum laude in 2003 and Ph.D. in Horticulture in 2010. Since 2005, she has been a Researcher at the CREA—Research Centre for Food and Nutrition. The core of her research is the study of chemical, nutritional, and bioactive components of food, with particular regard to the wide spectrum of substances classes and their nutraceutical features. For several years, she was involved in national and international research projects on the evaluation of several factors (agronomic practices, processing, etc.) that affect food quality, such as the levels of bioactive molecules and the total antioxidant properties as well as their possible impact on the biological role played by bioactive components in human physiology. Her research activities are also aimed toward developing, managing, and updating the Food Composition Database as well as Bioactive Compounds and Dietary Supplements databases; particular attention is given towards the harmonization of analytical procedures and classification and codification of dietary supplements.

Massimo Lucarini received his master's in Industrial Chemistry cum laude from the University of Rome "La Sapienza", Italy (1992) where he was also awarded his Ph.D. in Chemistry. His main research activities are aimed at the evaluation of nutrient content, molecules with biological and antinutrient activity in foods and diets, and stability studies of technological treatments of food products using specific process markers. Particular interest is addressed to evaluation of the nutritional quality of foods, the bioavailability of nutrients and bioactive components and their interaction with the food matrix (using in vitro models and cellular models), and applications in the nutraceutical field; recent attention has focused on the exploitation of waste from the agri-food industry, with a view toward sustainable agri-food production. In relation to the study of bioactive molecules, he has gained wide experience in this field, ranging from studies of carotenoids to phenolic substances, and from caseinophosphopeptides (CPP) to the components of dietary fiber. An integral part of the performed research is linked to institutional activity, including food composition tables, guidelines for healthy nutrition, and evaluation of fraud risk in the agri-food system. In relation to the production system, the effects of technological treatments on molecules of nutritional interest are also evaluated. He is also interested in using natural substances with strong antioxidant properties to improve the shelf-life of food products. His research activity also aims at the development of new analytical methods, the exchange of scientific information, and the acquisition of new skills both at the national and international level through training courses, participation in congresses, and seminars. The dissemination activity is carried out through the production of scientific articles, interviews released in national journals and broadcasting systems, the creation of web pages, and participation in congresses and educational and informative activities.

Massimo Zaccardelli is Research Director at CREA Research Centre for Vegetable and Ornamental Crops, located in Pontecagnano (Salerno, Italy). He obtained his Certificate of Agricultural Expert from I.T.A.S. of Ponticelli (Naples, Italy) in 1985, degree in Agricultural Science from University of Bologna (Italy) in 1991, and Ph.D. in Plant Pathology from University of Bologna (Italy) in 1998. His research topics are the use of ecocompatible agronomic techniques based on the production and use of compost and compost tea, antagonistic and PGPR microorganisms, and natural substances (saponins, polyphenols, essential oils, glucosinolates) for the control of soil-borne and aerial-borne plant pathogens of vegetable crops, optionally using biodegradable biopolymers as carriers. In recent years, he has been studying, using RNA-Seq, the effects of PGPR, compost tea, and humic acids treatments on tomato plants and, moreover, microbiological composition of suppressive compost, compost tea, and soils amended with compost using culturable methods and metagenomic approaches. Other research topics include the characterization of commercial and local varieties of grain leguminous (bean, chickpea, lentil, lupin, grasspea) and characterization and use of most efficient rhizobial strains as inoculants. He has worked on numerous nationally and internationally funded projects on the previously described topics. He has published more than 300 scientific contributions in international and national scientific magazines and congresses.

Antonello Santini Ph.D., is Professor of Food Chemistry and Food Chemistry and Analysis of Food and Nutraceuticals at the Departments of Pharmacy and Agriculture of the University of Napoli Federico II, Napoli, Italy. He is also Visiting Professor at the Albanian University of Tirana, Albania. He holds a Ph.D. in Chemical Sciences. His research areas of interest are substantiated by many international collaborations, mainly in the field of food; food chemistry, nutraceuticals, functional food; supplements; recovery of natural compounds bioactive using ecosustainable and environmentally friendly techniques from agro-food byproducts; nanocompounds; nanonutraceuticals; food risk assessment, safety and contaminants; mycotoxins and secondary metabolites; food analysis; and chemistry and food education. He is responsible for numerous funded research projects and general cultural agreements established between the University of Napoli Federico II and many Universities worldwide. His research activity is substantiated by more than 200 papers in reputed peer-reviewed international journals. He is a member of the European Food Safety Authority EFSA, ERWG, Parma, Italy; member of the Italian Authority for Food Safety (CNSA), Italian Ministry of Health, Rome Italy; member of the Managing Board, Italian Chemistry Society (SCI) Division of Teaching (DD-SCI), Rome, Italy; and expert member of Chemistry, EurSchool, European Commission, Bruxelles, Belgium.

Editorial

Forest, Foods, and Nutrition

Alessandra Durazzo [1,*], **Massimo Lucarini** [1,*], **Massimo Zaccardelli** [2,*] and **Antonello Santini** [3,*]

1   CREA-Research Centre for Food and Nutrition, Via Ardeatina 546, 00178 Roma, Italy
2   CREA-Research Centre for Vegetable and Ornamental Crops, Via Cavalleggeri 25, 84098 Pontecagnano Faiano (Salerno), Italy
3   Department of Pharmacy, University of Napoli Federico II, Via D. Montesano 49, 80131 Napoli, Italy
*   Correspondence: alessandra.durazzo@crea.gov.it (A.D.); massimo.lucarini@crea.gov.it (M.L.); massimo.zaccardelli@crea.gov.it (M.Z.); asantini@unina.it (A.S.)

Received: 30 October 2020; Accepted: 5 November 2020; Published: 9 November 2020

Abstract: Forest ecosystems are an important biodiversity environment resource for many species. Forests and trees play a key role in food production and have relevant impact also on nutrition. Plants and animals in the forests make available nutrient-rich food sources, and can give an important contributions to dietary diversity, quality, and quantity. In this context, the Special Issue, entitled "Forest, Food and Nutrition", is focused on the understanding of the intersection and linking existing between forests, food, and nutrition.

Keywords: forest; tree; edible forest products; non-edible forest products; nutritional value; biologically activecompounds; food composition databases; dedicated databases; novel food; sustainable agriculture; biodiversity

---

This Special Issue is addressed on understanding of the intersection and crosslinks existing between forests, food, and nutrition. Forest ecosystems represent relevant biodiverse environment resources of species. Forests and trees have a key role in food production and nutrition. Plants and animals in forests make available nutrient-rich food sources and can give an important contributions to dietary diversity, quality, and quantity. Moreover, forests are a relevant resource for new potentially active vegetal origin active compounds which may have a relevant impact on the diet and also contribute for functional foods, novel foods, and nutraceuticals.

Reimagining forests as an ecosystems able to support sustainable food production, allows to set a new horizon to explore. In this context, sustainable agriculture and forest vegetal resources represent a new aspect in the expansion of agricultural forest landscapes. Rediscovering the contributions of forests to food and nutrition area is leading to a relevant transition in the global food systems [1]. Firstly, the development and implementation of sustainable management of forest, as well as the optimization of yields of wild foods and fodder was here treated. As instance, the study of Kwon et al. [2] is focused on the control of fungal diseases and implementation in yields of Jujube Fruit (*Zizyphus jujuba* Miller var. inermis Rehder) orchard by means of *Lysobacter antibioticus* HS124. research on the promotion and valorization of foods from forests were discussed.

The nutritional value of forest foods has been exploited and promoted, throughout the evaluation of wild foods, to be addressed to a responsible human consumption and sustainable use of natural resources [3]. The identification, isolation, and quantification of compounds with nutritional and nutraceutical character are here outlined. The description of the main components and an assessment of their interactions, in relation particularly to factors, i.e., cultivar, weather, soil, and others have been discussed [4,5]. As instance, the geographical distribution and environmental correlation of eleutherosides and isofraxidin in *Eleutherococcus senticosus* from natural populations in the forests at Northeast of China were studied by Guo et al. [4]. The need of an updated overview, classification, and cataloguing of edible and non-edible forest products is emerging and triggering the interest of research.

Conventional and emerging procedures, with particular regards to green technologies have been reported. Innovativeg analytical techniques, i.e., multi-elemental analysis, isotopic ratio mass spectrometry, infrared spectroscopy, and nanotechnologies, joined with chemometrics, have been discussed [6,7]. In this context, it is worth mentioning the innovative research of Zhang et al. [8] on transcriptome analysis of Elm (*Ulmus pumila*) fruit in order identify genes and pathways associated phytonutrients.

The nutritional implications and the benefits of forest products have been outlined addressing the role of food forests in human nutrition. The discussion of the role of forest foods rich in compounds with nutrients and biologically active compounds to complement people's diet and the contribution of forest foods to a healthy diet has been exploited, adding information to the area of interest. The beneficial potential of medicinal plants and herbs has been investigated in different papers [9–12]. Functionally, extracts and biologically active components [13–18] from forest products are experiencing great interest for both research and potential application in nutraceutical, pharmaceutical, and cosmetic fields [19,20]. Fernández-Cervantes et al. [20] studied the essential oils of *Chamaemelum fuscatum* (Brot.) Vasc. from Spain and promoted and reinforce its ethnobotanical use. Furthermore, an application of nutraceuticals in plant defense is described throughout the case study of sage on a spontaneous Mediterranean plant to control phytopathogenic fungi and bacteria [21].

The elucidation of the role of forests for food security and nutrition was assessed, with attention to the contribution of wild and forest foods to nutrient intake among local communities. Moreover, the social and economic impact was investigated in several papers.

For instance, the study of Dejene et al. [22] attempted to provide and document Wild Edible Fruit Tree Species in Ethiopia as implementation of management strategy for sustainable utilization of natural resource. Aye et al. [23] described how mangrove forest contributes to the livelihood and dietary habits of local communities in Ayeyarwaddy Region, in Myanmar.

The ethnomycological knowledge was increased throughout semi-structured interviews with the Amhara, Agew, and Sidama ethnic groups in Ethiopia, as reported by Zeleke et al. [24]. Vlad et al. [25] studied and promoted blackberry as a traditional nutraceutical food resource from an area with high anthropogenic impact. Agúndez et al. [26] studied local preferences for production of shea nut and butter in Northern Benin. Darr et al. [27] mapped the diversity of baobab (*Adansonia digitata* L.) products in Malawi by studying the preferences of consumers and examining the major attributes on their market price.

In the food policy scenario, the work of Xie et al. [28] studied the possible constraints to the implementation of urban edible landscapes in China.

This Special Issue end points have been to contribute to the growth of this area of research, trigger research interest on forest food and its implications and impact on food security and nutrition, sustainability, novel food sources and their use, by adding information scientifically substantiated with new data.

We would like to thank all the authors and the reviewers of the papers published in this Special Issue for their great contributions and efforts. We are also grateful to the editorial board members and to the staff of the Journal for their kind support in the preparation steps of this Special Issue.

Author Contributions: All authors listed (A.D., M.L., M.Z. and A.S.) have made a substantial contribution to the work, and approved it for publication. All authors have read and agreed to the published version of the manuscript.

Funding: This research received no external funding.

Conflicts of Interest: The authors declare no conflict of interest.

References

1. Chamberlain, J.L.; Darr, D.; Meinhold, K. Rediscovering the Contributions of Forests and Trees to Transition Global Food Systems. *Forests* **2020**, *11*, 1098. [CrossRef]
2. Kwon, J.-H.; Won, S.-J.; Moon, J.-H.; Kim, C.-W.; Ahn, Y.S. Control of Fungal Diseases and Increase in Yields of a Cultivated Jujube Fruit (*Zizyphus jujuba* Miller var. *inermis* Rehder) Orchard by Employing Lysobacter antibioticus HS124. *Forests* **2019**, *10*, 1146. [CrossRef]
3. Asprilla-Perea, J.; Díaz-Puente, J.M.; Fernández, S.M. Evaluation of Wild Foods for Responsible Human Consumption and Sustainable Use of Natural Resources. *Forests* **2020**, *11*, 687. [CrossRef]
4. Guo, S.-L.; Wei, H.; Li, J.; Fan, R.; Xu, M.; Chen, X.; Wang, Z. Geographical Distribution and Environmental Correlates of Eleutherosides and Isofraxidin in *Eleutherococcus senticosus* from Natural Populations in Forests at Northeast China. *Forests* **2019**, *10*, 872. [CrossRef]
5. Cao, Y.; Fang, S.; Fu, X.; Shang, X.; Yang, W. Seasonal Variation in Phenolic Compounds and Antioxidant Activity in Leaves of *Cyclocarya paliurus* (Batal.) Iljinskaja. *Forests* **2019**, *10*, 624. [CrossRef]
6. Masek, A.; Latos-Brozio, M.; Kałużna-Czaplińska, J.; Rosiak, A.; Chrzescijanska, E. Antioxidant Properties of Green Coffee Extract. *Forests* **2020**, *11*, 557. [CrossRef]
7. Masek, A.; Latos-Brozio, M.; Chrzescijanska, E.; Podsędek, A. Polyphenolic Profile and Antioxidant Activity of *Juglans regia* L. Leaves and Husk Extracts. *Forests* **2019**, *10*, 988. [CrossRef]
8. Zhang, L.; Zhang, X.; Li, M.; Wang, N.; Qu, X.; Fan, S. Transcriptome Analysis of Elm (*Ulmus pumila*) Fruit to Identify Phytonutrients Associated Genes and Pathways. *Forests* **2019**, *10*, 738. [CrossRef]
9. Yoon, G.; Lee, M.-H.; Kwak, A.-W.; Oh, H.-N.; Cho, S.-S.; Choi, J.-S.; Liu, K.; Chae, J.-I.; Shim, J.-H. Podophyllotoxin Isolated from *Podophyllum peltatum* Induces G2/M Phase Arrest and Mitochondrial-Mediated Apoptosis in Esophageal Squamous Cell Carcinoma Cells. *Forests* **2020**, *11*, 8. [CrossRef]
10. Zhou, M.; Chen, P.; Lin, Y.; Fang, S.; Shang, X. A Comprehensive Assessment of Bioactive Metabolites, Antioxidant and Antiproliferative Activities of *Cyclocarya paliurus* (Batal.) Iljinskaja Leaves. *Forests* **2019**, *10*, 625. [CrossRef]
11. Souto, E.B.; Durazzo, A.; Nazhand, A.; Lucarini, M.; Zaccardelli, M.; Souto, S.B.; Silva, A.M.; Severino, P.; Novellino, E.; Santini, A. *Vitex agnus-castus* L.: Main Features and Nutraceutical Perspectives. *Forests* **2020**, *11*, 761. [CrossRef]
12. Nazhand, A.; Lucarini, M.; Durazzo, A.; Zaccardelli, M.; Cristarella, S.; Souto, S.B.; Silva, A.M.; Severino, P.; Souto, E.B.; Santini, A. Hawthorn (*Crataegus* spp.): An Updated Overview on Its Beneficial Properties. *Forests* **2020**, *11*, 564. [CrossRef]
13. Daliu, P.; Santini, A.; Novellino, E. A decade of nutraceutical patents: Where are we now in 2018? *Expert Opin. Ther. Patents* **2018**, *28*, 875–882. [CrossRef]
14. Santini, A.; Cammarata, S.M.; Capone, G.; Ianaro, A.; Tenore, G.C.; Pani, L.; Novellino, E. Nutraceuticals: Opening the debate for a regulatory framework. *Br. J. Clin. Pharmacol.* **2018**, *84*, 659–672. [CrossRef] [PubMed]
15. Durazzo, A.; Camilli, E.; D'Addezio, L.; Piccinelli, R.; Mantur-Vierendeel, A.; Marletta, L.; Finglas, P.; Turrini, A.; Sette, S. Development of Dietary Supplement Label Database in Italy: Focus of FoodEx2 Coding. *Nutrients* **2020**, *12*, 89. [CrossRef] [PubMed]
16. Santini, A.; Cicero, N. Development of Food Chemistry, Natural Products, and Nutrition Research: Targeting New Frontiers. *Foods* **2020**, *9*, 482. [CrossRef]
17. Durazzo, A.; Lucarini, M.; Santini, A. Nutraceuticals in Human Health. *Foods* **2020**, *9*, 370. [CrossRef] [PubMed]
18. Dini, I.; Laneri, S. Nutricosmetics: A brief overview. *Phytother. Res.* **2019**, *33*, 3054–3063. [CrossRef]
19. Farràs, A.; Cásedas, G.; Les, F.; Terrado, E.M.; Mitjans, M.; López, V.; Martínez, A.F. Evaluation of Anti-Tyrosinase and Antioxidant Properties of Four Fern Species for Potential Cosmetic Applications. *Forests* **2019**, *10*, 179. [CrossRef]
20. Fernández-Cervantes, M.; Pérez-Alonso, M.J.; Blanco-Salas, J.; Soria, A.C.; Ruiz-Téllez, T. Analysis of the Essential Oils of *Chamaemelum fuscatum* (Brot.) Vasc. from Spain as a Contribution to Reinforce Its Ethnobotanical Use. *Forests* **2019**, *10*, 539. [CrossRef]

21. Zaccardelli, M.; Pane, C.; Caputo, M.; Durazzo, A.; Lucarini, M.; Silva, A.M.; Severino, P.; Souto, E.B.; Santini, A.; De Feo, V. Sage Species Case Study on a Spontaneous Mediterranean Plant to Control Phytopathogenic Fungi and Bacteria. *Forests* **2020**, *11*, 704. [CrossRef]

22. Dejene, T.; Agamy, M.S.; Agúndez, D.; Martín-Pinto, P. Ethnobotanical Survey of Wild Edible Fruit Tree Species in Lowland Areas of Ethiopia. *Forests* **2020**, *11*, 177. [CrossRef]

23. Aye, W.N.; Wen, Y.; Marin, K.; Thapa, S.; Tun, A.W. Contribution of Mangrove Forest to the Livelihood of Local Communities in Ayeyarwaddy Region, Myanmar. *Forests* **2019**, *10*, 414. [CrossRef]

24. Zeleke, G.; Dejene, T.; Tadesse, W.; Agúndez, D.; Martín-Pinto, P. Ethnomycological Knowledge of Three Ethnic Groups in Ethiopia. *Forests* **2020**, *11*, 875. [CrossRef]

25. Vlad, I.A.; Goji, G.; Dinulică, F.; Bartha, S.; Vasilescu, M.M.; Mihăiescu, T. Consuming Blackberry as a Traditional Nutraceutical Resource from an Area with High Anthropogenic Impact. *Forests* **2019**, *10*, 246. [CrossRef]

26. Agúndez, D.; Nouhoheflin, T.; Coulibaly, O.; Soliño, M.; Alía, R. Local Preferences for Shea Nut and Butter Production in Northern Benin: Preliminary Results. *Forests* **2020**, *11*, 13. [CrossRef]

27. Darr, D.; Chopi-Msadala, C.; Namakhwa, C.D.; Meinhold, K.; Munthali, C. Processed Baobab (*Adansonia digitata* L.) Food Products in Malawi: From Poor Men's to Premium-Priced Specialty Food? *Forests* **2020**, *11*, 698. [CrossRef]

28. Xie, Q.; Yue, Y.; Hu, D. Residents' Attention and Awareness of Urban Edible Landscapes: A Case Study of Wuhan, China. *Forests* **2019**, *10*, 1142. [CrossRef]

Publisher's Note: MDPI stays neutral with regard to jurisdictional claims in published maps and institutional affiliations.

Article

Ethnomycological Knowledge of Three Ethnic Groups in Ethiopia

Gizachew Zeleke [1,2], Tatek Dejene [2], Wubalem Tadesse [2], Dolores Agúndez [3] and Pablo Martín-Pinto [2,*]

1 Sustainable Forest Management Research Institute, University of Valladolid (Palencia), Avda. Madrid 44, 34071 Palencia, Spain; gizachewzeleke@gmail.com
2 Ethiopian Environment and Forest Research Institute, Addis Ababa 30708, Ethiopia; tdejenie@yahoo.com (T.D.); wubalem16@gmail.com (W.T.)
3 INIA-CIFOR, Ecología y Genética Forestal, Carretera de la Coruña km 7.5, 28040 Madrid, Spain; agundez@inia.es
* Correspondence: pmpinto@pvs.uva.es; Tel.: +34-979-108-340; Fax: +34-979-108-440

Received: 17 July 2020; Accepted: 8 August 2020; Published: 11 August 2020

Abstract: Ethnomycological information was gathered by conducting semi-structured interviews with members of the Amhara, Agew, and Sidama ethnic groups in Ethiopia. A total of 300 individuals were involved in this study. Forest excursions were also undertaken to investigate the habitat and to identify useful wild mushroom species present in the study areas. A total of 24 useful wild mushroom species were identified. Among the three ethnic groups, the Sidama have the most extensive ethnomycological knowledge and over seven vernacular names for useful fungal species were recorded for this group. Collecting mushrooms is common practice among the Sidama and usually carried out by women and children during the main rainy season from June to September. Useful mushrooms are collected in natural forests, plantation forests, grazing areas, home gardens, and swampy areas. In terms of medicinal uses, *Lycoperdon perlatum* Pers. and *Calvatia rubroflava* (Cragin) Lloyd. are well-known treatments for wounds and skin disease. Harvest storage of wild mushroom species is unknown. Respondents in the Amhara and Agew ethnic groups were similar in terms of their use and knowledge of mushrooms. Both ethnic groups reported that although wild mushroom species were consumed by their grandparents, they do not eat mushrooms themselves, which could eventually represent a loss of mycological knowledge in these two ethnic groups. Such inconsistency between ethnic groups in terms of their knowledge may also be linked to the social valuation of mushroom resources, which could easily be mitigated by raising awareness. Thus, the baseline information obtained in this study could be useful for further investigations and documentation, and to promote ethnomycological benefits to different ethnic groups in countries with similar settings.

Keywords: mushroom; Enguday; ethnomycology; folk taxonomy; Amhara; Agew and Sidama

---

1. Introduction

One of the world's biggest challenges is to secure sufficient food for all that is healthy, safe and of high quality, and to do so in an environmentally sustainable manner [1,2]. In this context, forest resources can play an important role as a source of food [3,4], by enhancing nutritional diversity [5,6] while maintaining diversity in natural systems [7]. Thus, in recent years, there has been growing attention focused on the sustainability of foods [4,8,9] and food systems, which has highlighted the need to conserve species diversity, mainly of foods from forest systems in many parts of the world [10,11].

Wild mushroom species are vital components of the livelihoods of rural people in different parts of the world [12,13]. Many of these mushrooms are collected because they are valuable non-timber forest products (NTFPs) [14,15], enabling people to overcome vulnerability to poverty and sustain their livelihoods through a reliable source of income [12]. This has shifted ethnomycology into a discipline in different parts of the world [14]. Globally, about 140,000 important mushroom species have been reported. In various cultures, mushrooms can serve as sources of food [14], medicine [16], enzymes and various industrial compounds [17]. They serve as also important composition of food and recipes for traditional foods and recipes [18,19]. In addition, mushrooms can serve in a recreational context and in myths and beliefs [20]. Nutritionally, mushrooms are an important source of proteins, vitamins, fats, carbohydrates, amino acids, and minerals [21,22], i.e., they are a good alternative or substitute for meat and fish [22–24].

Previous ethnomycological studies have shown that local knowledge of mushrooms varies with people's cultures and beliefs [25–27]. Within local communities, conventional knowledge is passed down from one generation to the next because this is the only way of safeguarding traditional knowledge [14,25,26]. The use of questionnaires to record traditional knowledge linked to mushrooms enables mushroom "use values" to be evaluated for a specific local community to identify cultural differences between communities [25,28,29].

In Ethiopia, mushrooms are wild edible resources and important NTFPs like that of wild edible fruits, particularly in the southern and southwestern parts of the country [30–32]. Despite poor scientific knowledge, wild mushroom utilization is a common traditional practice among different ethnic groups in Ethiopia [33,34]. Mushrooms, along with other wild edible resources, are used as a coping food during periods of food shortage in localities where they are used as food [31,35]. In some local markets, mushrooms are sold by local people to provide some income to supplement the household economy [36]. However, there are few ethnomycological reports on wild mushrooms. The available reports are scanty and contain only basic information about the existence and use of mushrooms at some community levels in the country [34,36,37]. Efforts made so far have either been of a review nature and/or cross-sectional [38,39]. This lack of documentation could make local mycological knowledge vulnerable [29]. Due to a continuing exodus of people from the countryside, local communities are gradually losing an important part of their traditional knowledge, particularly about wild mushroom species [24,26,29]. There is a genuine need to record and document local traditional knowledge and perceptions about useful wild mushrooms. Furthermore, due to their economic value, efforts are needed to integrate wild mushroom species as mainstream NTFPs in Ethiopia to ensure their conservation and enhance their value as source of nutrition to improve human welfare. Thus, assessing the various uses of wild mushroom species by local people is key to the better valorization of services provided by wild useful fungi [40]. This would also enable us to better elaborate participative management and conservation plans for these forests resources. This study has, therefore, sought to assess and document ethnomycological knowledge related to wild mushroom species of three ethnic groups from three different geographical areas in Ethiopia. Thus, our specific objectives were: (i) to identify valuable wild mushroom species in three study areas; (ii) to record the use value of wild mushroom species for each of the ethnic groups; (iii) to evaluate the status of wild mushroom species within local communities; and (iv) to identify the main threats and to assess how these threats vary across the three study areas.

2. Methods

2.1. Description of the Study Area

The study was conducted in the Amhara Region and the Southern Nations, Nationalities, and Peoples' Region (SNNPR) of Ethiopia where more than half of the ethnic groups of Ethiopia are reside and from which three ethnic groups were selected: namely, the Amhara in the Fogera Woreda, the Agew in the Banja Woreda and the Sidama in the Wondo Genet Woreda (Figure 1). These three ethnic groups had been identified in this study with the assumption that they are having good knowledge of the identification, variability, and use of locally available NTFPs. The Amhara and Agew ethnic groups occupy the center of the northern highlands of the Amhara region. They live in the same agro- ecological zone and share the same cultures, and languages. They both are sedentary farmers who practice mixed agriculture, including crop production and livestock rearing. The crop varieties grown locally in these woredas include teff (*Eragrostis tef* (Zucc) Trotter), sorghum (*Sorghum bicolor* (L.) Moench), maize (*Zea mays* L.), finger millet (*Eleusine coracana* (L.) Gaertn.) and beans (*Phaseolus vulgaris* L.). The Sidama ethnic group is one of the largest ethnic groups in the southern highlands. The population density in Wondo Genet is higher than that of Fogera or Banja, with about seven inhabitants per km^2. The majority of the Sidama practice mixed agriculture, integrating cash crops, such as khat (*Catha edulis* (Vahl) Endl.) sugar cane, and Ensete (*Enset ventricosum* (Welw.) Cheesman), with livestock production, including fishing from the nearby lake.

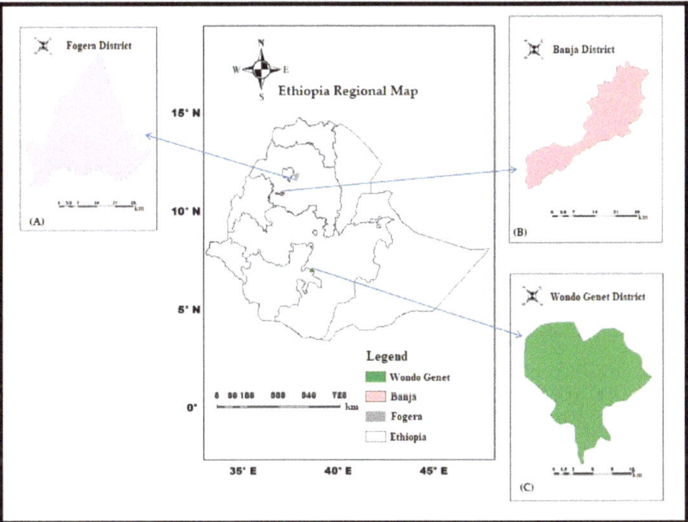

Figure 1. Location of the study woredas of the Amhara and Southern Nations, Nationalities, and Peoples' Region (SNNPR) in Ethiopia. (**A**) Fogera Woreda inhabited by the Amhara ethnic group; (**B**) Banja Woreda inhabited by the Agew ethnic group and (**C**) Wondo Genet Woreda inhabited by the Sidama ethnic group.

Wondo Genet, Banja and Fogera Woredas (Table 1) are characterized by high-altitude natural forests [41–44]. This high forest is dominated by Dry Afromontane forests [41,44], which are characterized by high humidity, a variable rainfall pattern and a prolonged dry season, making them complex and rich in biodiversity [45]. The main tree species found in these forests are *Juniperus procera* (Hochst. Ex Endl.), *Podocarpus falcatus* (Thubb.) Mirb.), *Hagenia abyssinica* (J.F.Gmel.), and *Olea africana* subsp. Cuspidata (Wall. & G.Don) Cif.) which are the main sources of timber in Ethiopia [46]. The Dry Afromontane forests also harbor various types of NTFPs [47], including edible fungi [38].

Table 1. Geographical description of the three study woredas.

Study Woredas	Ethnic Group	Geographical Location	Altitude Range (m asl)	Mean Annual Precipitation (mm)	Mean Annual Temperature (°C)	References
Wondo Genet	Sidama	7°06′ N & 38°37′ E	1760–1920	1200	19	[32]
Fogera	Amhara	11°58′ N & 37°41′ E	1780–2510	1245	20	[43]
Banja	Agew	11°10′ N & 36°15′ E	1870–2570	1300	18	[40]

2.2. Socioeconomic Data Collection

This study presents three case studies as a type of ethnomycology research, each with various forms of socioeconomic data which were collected between January and August 2019. The data were collected from the primary data sources that involved the key informant interviews, focus group discussions and household interview methods. Details of the methods used followed Mekonnen et al. [48] and are described below:

The key informant interviews required people who were relatively knowledgeable about their community, local natural resources, the culture of their community, and the use of NTFPs to share their knowledge and experience with the interviewer and, therefore, suitable participants were selected for this interview. To select these key informants, a snowball method was used in which one key informant was contacted with the assistance of local administrators and community elders. Then, he/she would inform us of a second, the second would provide the name of a third and so on until a saturation number was reached [49–51]. In total, 17 elders (10 women and 7 men aged between 41 and 77 years old) from the three study areas (three from Banja, four from Fogera and the remaining 10 from Wondo Genet) served as key informants. During the study, each key informant was visited twice to verify the reliability of the data obtained. If information conveyed about a species during the first visit was not consistent with the information provided during the second visit, the information was considered unreliable and was rejected. The second visit also helped us to gather additional information from some of the participants that were not mentioned during the first interview. However, in most cases of the participants in the key informant discussion, no new information or themes were gathered during the second visit. Thus, this was a redundancy signals and the data collection was cease [50,51].

The focus group discussions were made up of representative members of the studied communities. The discussions were made up of three independent groups per kebele, the smallest administrative level in Ethiopia, and each group were consisted of ten individuals. When forming the groups, the populations in each study keble were first split into youths, women, and elders groups. Then, the overall samples consisted of some individuals from each spited groups. In each group the members are chosen randomly [49] to obtain varied knowledge and views, giving equal chances to youths, women, and elders for being selected as member of this study. The focus group discussions assessed the participants' feelings and opinions about wild mushrooms, the perceptions, beliefs, and myths attached to wild mushrooms, their marketability, knowledge of ecological niches and phenology/calendar use, as well as their opinion regarding resource degradation and the causes of degradation at the local level. At least one key informant was included in each group to triangulate information. The information collected via the key informant interviews and the focus group discussions was used to refine and compliment the information gathered via the household interviews in each of the study areas. The collected information was qualitatively interpreted and narrated in the Results and Discussion.

Household interview were conducted using a face-to-face semi-structured questionnaire interview. The households were selected purposively based on their gender and their dependence on the forest for their livelihood [49,52]. The questionnaire was constructed to obtain information relating to the objectives of the study and had been pretested with 15 randomly selected individuals from each study woredas. Based on the results of the pretest work, the questionnaires were modified as necessary. Enumerators were recruited from the study areas and the study objectives were explained to them. The enumerators were also trained in the methods of data collection and interviewing techniques. In total, 300 households from Wondo Genet, Banja, and Fogera woredas took part in the survey. The interviews were conducted in the Amharic language. In some cases, an interpreter conducted the surveys to ensure that the meaning of the questions was not changed. Interviews were conducted in a place where the informants were most comfortable. Information regarding the gathering, preparation, use, status/abundance, etc. of wild mushroom species and their marketability was also collected. Additional discussions were conducted with the households to understand the traditional use of mushrooms for medicinal purposes.

2.3. Wild Mushroom Species Collection

Weekly wild mushroom resource assessments were undertaken in July and August in 2016 during the major rainy season in nearby remnant Dry Afromontane forests [39]. For the purpose, we established a total of nine sample plots in each study woredas, plots were established systematically about 250 m apart. Each plot covered an area of 100 m^2, with a rectangular shape (2 m × 50 m) [39]. During sampling, sample fruit bodies from encountered species in the sampling plots were collected and taken to the laboratory and dried. In the field also, specimens were photographed and their ecological characteristics were noted in order to assist and facilitate taxa identification processes. Furthermore, herbaria specimens were used for mushroom species identification. Both morphological and molecular analyses were used for taxa identification. Photographs of the wild mushroom species were used during interviews and group discussions to enable the informants involved in the study to easily recognize mushroom species in their vicinity. The relative popularity of each medicinal wild mushroom was evaluated based on the informants who independently reported its medicinal use (informant consensus) in the study area.

2.4. Data Analysis

Descriptive statistics were used to present the basic information obtained from the questionnaires. All analyses were conducted based on the number of responses from the informants. A final list of valuable wild mushroom species used by respondents was compiled from the questionnaires and from the field collection data. All fungi were identified at genus and species level whenever possible with the aid of several keys [53–60]. The statistical significance of differences between groups was obtained by performing a chi-square test. Respondents ranked the effect of different threats on a scale of 1 to 3. Thus, a Kruskal–Wallis test was performed to identify the major causes of wild mushroom degradation based on the results of the respondents' views. Clustering was used for the 24 edible species based on their use by local communities. The cluster was based on the average linkage between groups. A binary logistic regression model was used to determine the factors that influence wild mushroom use by households [61]. Data were analyzed using STATISTICA '08 edition software (StatSoft Inc., 1984–2008, Maastricht, the Netherlands).

There are many theoretical perspectives regarding food choice decision, including social behavior, social facts, and social definition theory [61]. Each offers partial insights and makes limiting assumptions that prevents food choice decision variables being fully explained. As a result, we used a food choice process model that was developed based on constructionist social definition perspectives to examine the broadest scope of factors relevant to how individuals constructed their food choice decisions. The model included components of life course, personal food systems and influences. Thus, we selected all our explanatory variables described in Table 2.

Table 2. List of variables and variable descriptions used in the binary logistic regression model.

Variables	Definition	Type of Data
Region	Administrative area within which the household head lives 1. SNNPR 2. Amhara	Categorical
Woreda	Administrative area within the region in which the household head lives expressed in years of homogeneity in terms of its ethnicity 1. Wondo Genet 2. Banja 3. Fogera	Categorical
Age	Age of the household head	Quantitative
Educational level	Educational level of the household head: 1 = illiterate 2 = can read and write 3 = primary school completed 4 = secondary school completed 5 = obtained diploma, degree or above	Categorical

Table 2. *Cont.*

Variables	Definition	Type of Data
Ethnicity	Ethnic groups to which the household head belongs—encompasses cultural variations between ethnic groups. 1. Sidama 2. Agew 3. Amhara	Categorical
Perceived value	Perception of mushroom consumption for food, income generation, and food security and as a medicine. 1. Higher 2. Medium 3. Low	Categorical
Taste experience	Consumption experience of an edible mushroom at least once in his/her lifetime 1. Yes 2. No	Categorical
Nutritional knowledge	Household head's knowledge or lack of knowledge about a mushroom's health-boosting benefits 1. I know 2. I don't know	Categorical
Indigenous identification knowledge	Household head's local knowledge of the identity of edible, poisonous and medicinal wild mushrooms 1. Yes 2. No	Categorical

3. Results

3.1. Ethnotaxa of Wild Mushrooms

During the forest survey conducted in 2016, our team collected a total of 67 wild mushroom species in the Dry Afromontane forest of the study areas. Of these, 24 species belonging to 19 genera and 9 families were classified as edible. These edible mushrooms were well recognized by the local people when shown photographic images of these species. The fungi are listed in alphabetical order by species name alongside the corresponding indigenous name used by the three ethnic groups (Table 3). The families with the greatest numbers of edible species identified were the Agaricaceae (13 species) and Psathyrellaceae (three species). These two families represented 66.67% of the identified edible wild mushroom species in the study forests, whereas the remaining 33.33% of families were represented by only a single species (Table 3).

The Amhara ethnic group appeared to have limited ethnotaxa knowledge of the wild mushroom species found in the Fogera area. They classified mushrooms as 'Enguday', which in general corresponds to 'fungi' in English. All mushrooms with caps were classified as 'Yejib-tila', which means, "Shadow of the Hyena", which may indicate their cryptic nature (Table 3).

The Sidama ethnic group from the Wondo Genet area was found to be mycophiles and to have a well-developed ethnotaxa for wild mushrooms (Table 3). Some of the names given are associated with attributes of the mushroom species. For example 'Meine' is a name given to a highly valued species that is good to eat due to its taste but scarce due to high levels of collection. 'Gadifuto', which means 'Hyena's fart', is a name given to medicinal mushroom species.

Like the Sidama ethnic group, the Agew from the Banja study area also have a well-developed folk taxonomy for wild mushrooms. 'Wagi', 'Emahoyie pinchina', 'Abahoy pinchina', 'Ye Zinjero Fes', and 'Szantila' are ethnotaxa for some of the wild mushroom species (Table 3). However, for the majority of unrelated taxa, like the Amhara ethnic group, they use the collective name 'Enguday'. Remarkably, the Agew have assigned mushrooms with specific names based on the season in which the mushrooms grow. For example, those mushrooms that grow following the first rain of the season are generally called 'Gunfane', which literally means common cold. The name 'Yejib Tila' is also used by the Agew for all cap fungi.

Table 3. Folk taxonomy of collected wild mushroom species used by the Amhara, Agew and Sidama ethnic groups in the study woredas.

Taxa Name	Family	Local Name of Mushroom Used by Ethnic Group			Edibility
		Sidama	Agew	Amhara	
Agaricus campestroides Heinem & Gooss.-Font.	Agaricaceae	Meine/Kakea	Wagi	Enguday	E
Agaricus sp$_1$. L.	Agaricaceae				
Agaricus sp$_2$. L.	Agaricaceae	-	-	-	
Agaricus sp$_3$. L.	Agaricaceae	-	-	-	
Agaricus sp$_4$. L.	Agaricaceae	-	-	-	
Agaricus sp$_5$. L.	Agaricaceae	-	-	-	
Agaricus sp$_6$. L.	Agaricaceae	-	-	-	
Agaricus subedulis Heinem.	Agaricaceae	Horoqo	Wagi	Enguday	E
Agaricus trisulphuratus Berk.	Agaricaceae	-	-	-	
Agrocybe pediades Fayod.	Strophariaceae	Shopenea	Wagi	Enguday	E
Amauroderma regulicolor Murrill.	Ganodermataceae	-	-	-	
Clitocybe elegans (Fr.) Staude	Tricholomataceae	Meine	Wagi	Enguday	E
Armillaria sp. (Fr) Staude.	Physalacriaceae	-	-	-	
Calvatia rubroflava Fr.	Agaricaceae	Gadifuto	Emahoyie pinchina	Emahoyfese	E
Collybia piperata (Beeli) Singer.	Tricholomataceae	-	-	-	E
Conocybe sp. Fayod.	Bolbitiaceae				
Coprinellus domesticus (Bolton) Vilgalys, Hopple & Jacq. Johnson.	Psathyrellaceae	Feradigamea	Wagi	Enguday	E
Coprinellus sp. P.Karst.	Psathyrellaceae	-	Wagi	Enguday	E
Coprinopsis nivea (Pers.) Redhead, Vilgalys & Moncalvo	Psathyrellaceae	Shishonea	Wagi	Enguday	E
Coprinopsis sp$_1$. P.Karst.	Psathyrellaceae	-	-	-	
Coprinopsis sp$_2$. P.Karst.	Psathyrellaceae	-	-	-	
Coprinus pseudoplicatilis Pers.	Psathyrellaceae	Shishonea	Wagi	Enguday	E
Coprinus sp. Pers.	Agaricaceae	-	-	-	
Crepidotus sp. (Fr.) Staude.	Crepidotaceae	-	-	-	
Cyptotrama asprata (Berk.) Redhead & Ginns.	Physalacriaceae	-	-	-	
Favolaschia calocera R. Heim.	Mycenaceae	-	-	-	
Ganoderma sp. (Curtis) P.Karst.	Ganodermataceae	Buki bulasa	-	-	E
Gerronema hungo (Henn.) Degreef & Eyi.	Marasmiaceae	-	-	-	E
Gymnopilus junonius (Fr.) P.D. Orton.	Cortinariaceae	-	-	-	
Gymnopilus pampeanus (Speg.) Singer.	Strophariaceae	-	-	-	E
Hygrophoropsis aurantiaca (Wulfen) Maire.	Hygrophoropsidaceae	Feradigamea	Wagi	Yejib-tila	E
Hymenagaricus fuscobrunneus Heinem.	Agaricaceae	Qochiqomalea	Szantila	Yejib-tila	E
Hymenagaricus sp. Heinem.	Agaricaceae	-	-	-	
Hypholoma fasciculare (Huds.) P. Kumm.	Strophariaceae	-	-	-	
Lepiota cristata (Bolton) P.Kumm.	Agaricaceae	-	-	-	
Leucoagaricus sp. Locq. ex Singer	Agaricaceae	Meine	Szantila	Yejib-tila	E
Leucoagaricus leucothites (Vittad.) Wasser.	Agaricaceae	Kakea	Szantila	Yejib-tila	E
Leucoagaricus rubrotinctus (Peck) Singer.	Agaricaceae	Kakea	Szantila	Yejib-tila	E
Leucoagaricus sp$_1$. Locq.ex Singer.	Agaricaceae	Adulla	Szantila	Yejib-tila	E
Leucoagaricus sp$_2$. Locq.ex Singer	Agaricaceae	Silegaga	Szantila	Yejib-tila	E
Leucocoprinus birnbaumii (Corda) Singer.	Agaricaceae	Feradigamea	Szantila	Yejib-tila	E
Leucocoprinus cepistipes (Sowerby) Pat.	Agaricaceae	Feradigamea	Szantila	Yejib-tila	E
Lycoperdon perlatum Pers.	Agaricaceae	Gadifuto	Abahoy pinchina	Abahoyfese	E
Lycoperdon sp. Pers.	Agaricaceae	-	-	-	
Marasmius buzungolo Singer.	Marasmiaceae	-	-	-	
Marasmius katangensis Singer.	Marasmiaceae	-	-	-	
Marasmius rotalis Berk & Broome.	Marasmiaceae	-	-	-	
Marasmius sp. Fr.	Marasmiaceae	-	-	-	
Microporus sp.P.Beauv.	Polyporaceae	-	-	-	E
Parasola sp$_1$.Redhead, Vilgalys & Hopple.	Psathyrellaceae	-	-	-	
Parasola sp$_2$. Redhead, Vilgalys & Hopple.	Psathyrellaceae	-	-	-	
Polyporus badius (Pers.) Schwein.	Polyporaceae	-	-	-	
Polyporus tuberaster (Jacq. ex Pers.) Fr.	Polyporaceae	-	-	-	
Psathyrella sp$_1$. Fr.ex Quél.	Psathyrellaceae	-	-	-	
Psathyrella sp$_2$. Fr.ex Quél.	Psathyrellaceae	-	-	-	
Psathyrella sp$_3$. Fr.ex Quél.	Psathyrellaceae	-	-	-	
Psathyrella sp$_4$. Fr.ex Quél.	Psathyrellaceae	-	-	-	
Psilocybe cyanescens Wakef.	Hymenogastraceae	-	-	-	
Psilocybe merdaria (Fr.) Ricken.	Hymenogastraceae	-	-	-	
Psilocybe sp. (Fr.) P.Kumm.	Hymenogastraceae	-	-	-	
Trametes versicolor (L.) Lloyd.	Polyporaceae	-	-	-	
Tremella mesenterica (Schaeff.) Retz.	Tremellaceae	-	-	-	
Un described sp$_1$.	Undescribed	-	-	-	
Un described sp$_2$.	Undescribed	-	-	-	
Un described sp$_3$.	Undescribed	-	-	-	
Un described sp$_4$.	Undescribed	-	-	-	
Xerula sp. Maire.	Physalacriaceae	-	-	-	

Note: E = edibility.

3.2. Wild Mushroom Collection and Habitat Type

Based on the respondents' answers, the three ethnic groups differed significantly in their knowledge of wild mushroom habitats (chi-square test; $p < 0.05$). Overall, six different habitat types were distinguished by respondents. However, the home garden and swampy areas are far less relevant for the Amhara and Agew ethnic groups than other habitats (Figure 2). Compared with Sidama respondents, a significantly greater proportion of respondents belonging to the Amhara and Agew ethnic groups considered the natural forest to be the main mushroom habitat (Agew–Sidama = 0.001 and Amhara–Sidama = 0.001) (Figure 2). By contrast, a significantly greater proportion of Sidama respondents considered grazing areas to be a mushroom habitat compared with the Amhara and Agew respondents ($p > 0.05$). However, the proportion of Sidama and Agew respondents that considered plantations and agricultural lands to be habitats for wild mushroom species was not significantly different (Figure 2; $p > 0.05$).

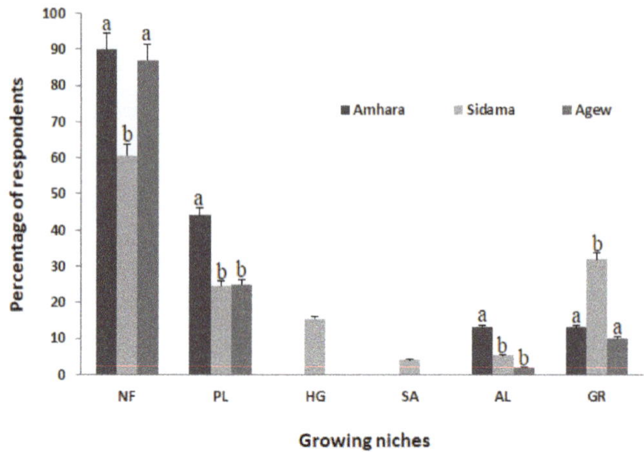

Figure 2. Perceptions of individuals in the three ethnic groups interviewed in the study areas about wild mushroom habitats. The data shown are mean results ± the standard error amount by five percent of the mean. Within each habitat type, values with the same letter above the bar are not significantly different. Abbreviations: NF, natural forest; PL, plantation forest; HG, home garden; SA, swampy area; AL, agricultural land; and GR, grazing area. Note: wild mushroom species are found in more than one habitat type; thus, the total percentage for all habitats is >100%.

We found significant differences in the awareness and use of wild mushroom species (chi-square test; $p < 0.05$) among the three ethnic groups. From the household survey, we realized that the Sidama ethnic group was more familiar with mushrooms than the Amhara and Agew ($p < 0.05$), whereas the Amhara and Agew were not significantly different from each other in terms of their awareness and use of wild mushroom species ($p > 0.05$).

The household interview revealed that the Sidama ethnic group collects wild mushroom species for food (93%) and medicinal (7%) purposes. About 63% of the respondents indicated that they usually collect wild mushrooms, 25% indicated that they occasionally collect wild mushroom species and 12% not collected mushrooms. The Sidama collect fungal species that have different nutritional modes. Saprotrophic fungi are preferred as food (90%) compared with other fungal types whereas *Ganoderma*, *Calvatia* and *Lycoperdon* (Table 4) are the three most commonly gathered genera for medicinal purposes. When dried, the spores of these mushroom species can be spread on skin to heal wounds and skin disease (Table 4).

Wild mushroom collection is common practice among the Sidama community: in most cases, children (22%) and women (70%) are the main collectors. This is because children are responsible for livestock keeping in the field and women are responsible for collecting firewood from the forests in the Wondo Genet area. Moreover, during the focus group discussions, the Sidama groups reported that the women in their communities know where and when wild mushroom species will be at their best. However, in some cases, men also collect mushrooms when they unintentionally find them as they walk to or from the forest. According to the group discussion, medicinal mushroom species are usually collected by traditional herbalists; thus herbalists in the study areas are key for determining the ethnotaxa of medicinal species.

Table 4. Traditional use and preparation of the twelve most listed wild mushroom species by the Amhara, Agew and Sidama ethnic groups in the study areas.

Species Scientific Name	Traditional Uses and Preparation			
	Amhara	Agew	Sidama	Local Preparation
Agaricus campestroides Heinem & Gooss.-Font.	NK	NK	F	Cooked with vegetables, oil and chili sauce
Agaricus subedulis Heinem.	NK	NK	F	Cooked with vegetables, oil and chili sauce
Clitocybe elegans (Fr.) Staude	NK	NK	F	Cooked with vegetables, oil and chili sauce
Calvatia rubroflava Fr.	NK	NK	F&M	Cooked with vegetables and oil. Also, spores/powder used to treat wounds
Hymenagaricus fuscobrunneus Heinem.	NK	NK	F	Cooked with vegetables, oil and chili sauce
Leucoagaricus sp. Locq. ex Singer	NK	NK	F	Cooked with vegetables and oil
Leucoagaricus leucothites (Vittad.) Wasser.	NK	NK	F	Cooked with vegetables and oil
Leucoagaricus rubrotinctus (Peck) Singer.	NK	NK	F	Cooked with vegetables and oil
Leucoagaricus sp$_1$. Locq.ex Singer.	NK	NK	F	Cooked with vegetables and oil
Leucoagaricus sp$_2$. Locq.ex Singer	NK	NK	F	Cooked with vegetables and oil
Lycoperdon perlatum Pers.	NK	F	F&M	Roasted or cooked with vegetables and oil. Also, spores/powder used to treat skin infection and wounds of both human and livestock
Ganoderma sp. (Curtis) P.Karst.	NK	NK	M	Medicinal for stomachaches and to treat wounds

Note: F, species used for food; NK, species not known as food or medicine; and M, species used for medicine.

Although none of the Amhara and Agew respondents collect wild mushrooms for food or medicinal purposes, about 57% and 72% of the Amhara and Agew respondents, respectively, indicated that they do have some knowledge of the medicinal and food use value of wild mushrooms. They obtained this information from different sources as from forefathers (Amhara, 23%; Agew, 43%), elderly people belonging to other ethnic groups (Gumze) in their vicinity (Amhara, 34%; Agew, 46%), friends (Amhara, 10%; Agew, 43%), forestry expertise (Amhara, 12%; Agew, 33%), and NGOs (Amhara, 4%; Agew, 11%).

3.3. Seasonality/Phenology of Mushrooms

The seasonality of wild mushroom appearance was not significantly different among the three study areas (chi-square test; $p > 0.05$; Figure 3). In all three cases, wild mushroom species develop during the short (peak in March) and long (peak in July) rainy seasons (Figure 3), suggesting the importance of rainfall patterns in fungal phenology.

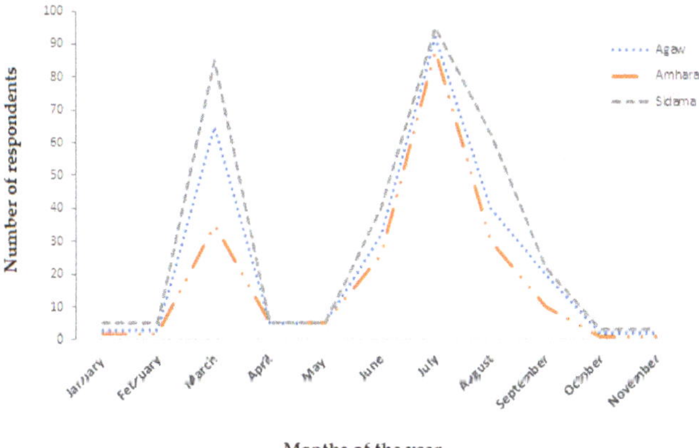

Figure 3. Phenology of mushrooms perceived by the three ethnic groups in the study area.

The Sidama collect wild mushroom species in Wondo Genet during the long rainy season (June to September). Mushroom availability in this area peaks between mid-July and the end of August (Figure 3). Interestingly, the majority of the respondents (97%) from Wondo Genet suggested that edible mushrooms grew during this season in well-known places and that the timing and pattern of their appearance was predictable. However, there were a few respondents (3%) from this study area who believed that some species could also be found during the dry season.

3.4. Wild Mushroom Use and Consumption

The identified valuable wild mushroom species clustered in three different groups and one independent species when analyzed based on their use as food, medicine, food and medicine and unknown use (Figure 4). The respondents indicated that 12 species were not known for their use by the locals. In the second group, nine wild mushroom species were identified as edible species that were consumed by locals (Figure 4). In the third group, *Calvatia rubroflava* and *Lycoperdon perlatum* were used as food and medicine. The *Ganoderma* species, which was classified as an independent species, was used for medicinal purposes only.

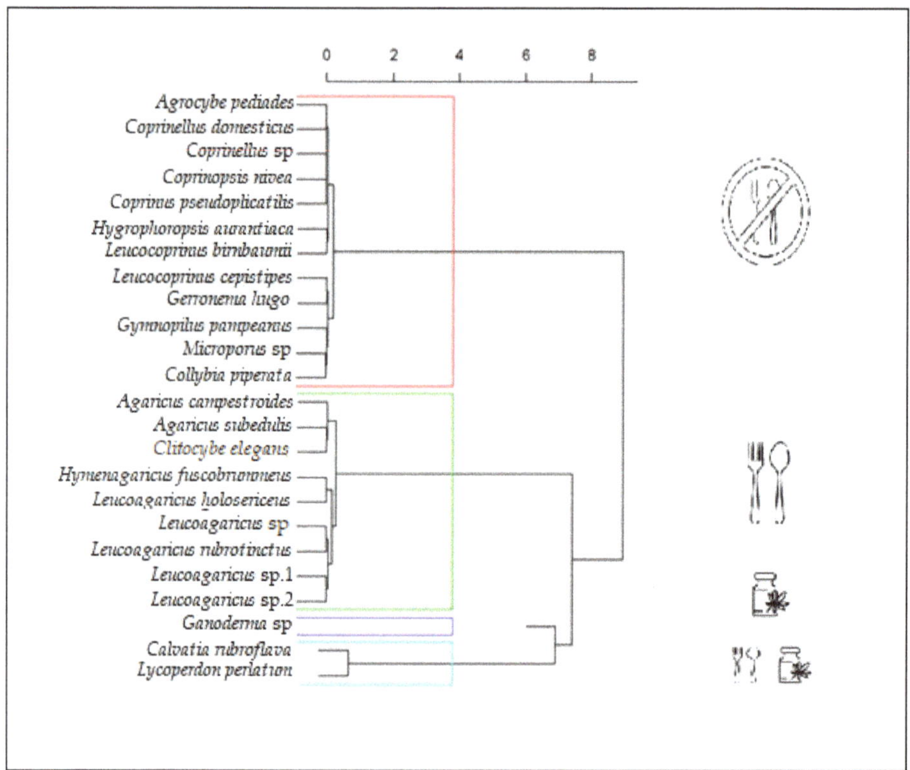

Figure 4. Dendrogram showing the classification of edible wild mushroom species based on their use by local communities. The horizontal axis represents the distance or dissimilarity between clusters and the vertical axis represents the species and clusters.

Wild mushroom use and consumption patterns varied among the three study areas (chi-square test; $p < 0.05$), with only the Sidama ethnic group using wild mushrooms. However, among the Sidama respondents, there was a significant difference in their perception of wild mushroom species (chi-square test; $p < 0.001$). About 86.7% of the interviewed individuals considered mushrooms to be a substitute for meat, whereas 21.3% considered mushrooms to be a vegetable in their daily meal. The majority of the respondents consumed mushrooms by cooking them with onions, vegetables, and oil; however, some (23%) consumed mushrooms when cooked with chili sauce or when cooked with oil and salt (8%) (Table 4).

About 71% of the Sidama that were interviewed were able to distinguish between edible and poisonous wild mushrooms. Edible species were distinguished based on aroma (like that of the soil), color (usually gray), the size of the sporocarps, habitat (leaf litter) and edibility information acquired from their forefathers. Moreover, the Sidama believe that if animals feed on a mushroom species then the mushroom must be edible. The most commonly used edible species in the Wondo Genet area included *Agaricus subedulis* Heinem., *Hymenagaricus fuscobrunneus* Heinem., *Clitocybe elegans* (Fr.) Staude., *Lycoperdon perlatum* Pers., *Calvatia rubroflava* Fr., and *Leucoagaricus leucothites* (Vittad.) Wasser. The Sidama do not preserve wild mushroom species for future use because they do not have the knowhow for mushroom preservation.

The Sidama identified poisonous mushrooms by their bright colors. In addition, if, for example, a dead insect is found on a mushroom or skin itch when the mushroom touches the body, then the mushroom is considered to be poisonous. If poisoned by ingesting a mushroom in Wondo Genet, the individuals interviewed indicated that they would drink fresh goat's blood (95%), the juice of *Kocho*, which is produced from *Ensete ventricosum* (Welw.), milk (90%), or vomit (65%). Moreover, 87% of the interviewed individuals indicated that '*Gadifuto*' (both *C. rubroflava* and *L. perlatum*) were used as a powder/spores to cure skin disease and to treat human and livestock wounds.

Although wild mushroom collection by the Amhara and the Agew ethnic groups is not common practice, the group discussions in both study areas revealed that the local people were aware that mushrooms are a food resource because their descendants and neighboring communities (mainly the Gumz) used to collect and eat mushrooms from the forest. Interestingly, some Agew individuals (3%) used to eat roasted mushrooms, and they indicated that this tradition was inherited from their forefathers. People participating in the group discussion in the Banja Woreda also confirmed that their ancestors used to eat roasted and cooked young *Lycoperdon* species with vegetables and oil. Among the interviewed individuals from both ethnic groups, 91% considered the majority of wild mushroom species to be poisonous and not good for health. However, no cases of mushroom poisoning have been recorded in the Banja or Fogera study areas. Interestingly, some members of both groups (Amhara (16%) and Agew (21%)) have some knowledge of cultivated edible mushrooms and their nutritional value (i.e., *Agaricus bisporus* (J.E.Lange) Imbach, *Lentinula edodes* (Berk.) Pegler, and *Pleurotus ostreatus* (Jacq. ex Fr.) P.Kumm) because they have received training from Woreda forestry expertise.

Wild mushroom species have not been commercialized in any of the study areas. The entire mushroom harvest in Wondo Genet is used for household consumption.

3.5. Factors Influencing Wild Mushroom Use

According to the binary logistic model, most of the evaluated factors differed significantly in their influence on wild mushroom consumption (chi-Square, $p < 0.05$; model fit, $R^2 = 0.883$). Only age, nutritional knowledge, and indigenous identification knowledge had no influence on consumption ($p > 0.05$; Table 5). The positive elasticity in the coefficients implies that a unit addition to any of the significant parameters will have a positive influence on wild mushroom consumption by the studied ethnic groups. This also holds true for the negative significant estimated coefficients. A unit reduction to any of them will have a negative influence on wild mushroom consumption by the studied ethnic groups

Table 5. Factors influencing wild mushroom consumption based on a binary logistic regression model.

Parameters	Coefficients	Standard Error	p-Value
Region	0.43	0.51	0.000 **
Woredas	5.52	0.58	0.000 **
Age	−0.03	0.02	0.889
Education level	−1.41	1.05	0.000 **
Ethnicity	9.24	1.10	0.000 **
Perceived value	−5.64	0.42	0.004 *
Taste experience	7.02	0.65	0.003 *
Nutritional knowledge	−17.83	0.49	0.078
Indigenous identification knowledge	−0.26	0.22	0.122 *
Constant	0.88	0.34	0.290

Note: ** and * indicate significance at the 0.01 and 0.05 probability level, respectively.

3.6. Perceived Status and Threats

According to the perceived view of the respondents, the status of wild mushroom species did not differ significantly among the three ethnic groups (chi-square test; $p > 0.05$). The perception of the status of wild mushroom species was decreasing in their locality was decreasing (Figure 5). The aggregate perception also indicated that 60.67% of respondents perceived that the status of wild mushroom species had decreased compared with that in previous years, 32% perceived the status to be unchanged and 7.33% perceived the status to have increased.

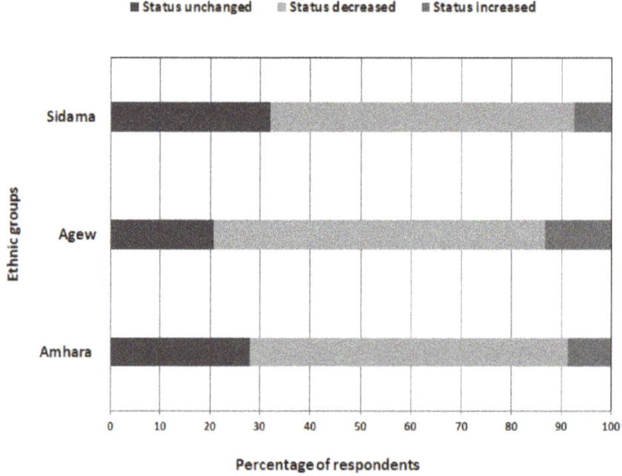

Figure 5. Perceived view of respondents in the Amhara, Agew and Sidama ethnic groups regarding the status of wild mushroom species in their locality.

Based on the results of the respondents' views, the major causes of wild mushroom degradation were perceived to be free grazing, agricultural expansion, settlements, fire incidence, climate change, and unknown reasons (Table 6). Among these, free grazing, agricultural expansion, settlements, and fire incidence were perceived to have a significant influence (Kruskal–Wallis test, $p < 0.05$) on the status of wild mushroom species in the three study areas.

Table 6. Major causes of wild mushroom degradation based on the mean risk value for each threat categorized by respondents from each ethnic group on a scale of 1 to 3. Threat: Ag, agricultural land expansion; Set, settlements; F, fire incidence in the forest; G, free grazing; CC: climate change; and UF, unknown factors.

Ethnic Group	Ag	Set	F	G	CC	UF
Amhara	2.58	1.74	2.99	2.22	1.00	1.00
Agew	1.85	1.11	2.24	1.18	1.00	1.00
Sidama	2.46	1.44	2.28	2.08	1.00	1.00
Total mean	2.34	1.43	2.45	1.90	1.00	1.00
Std. deviation	0.71	0.50	0.50	0.86	0.00	0.00
Kruskal–Wallis test (*p* value)	0.000	0.008	0.000	0.000	1.00	1.00

4. Discussion

Previous Ethiopian ethnomycological reports have usually been based on small-scale case studies [33,34,37,62]. In this ethnomycological study, we included two regions with different ethnic compositions. The Sidama ethnic group are mycophiles and have a more extensive folk taxonomy for mushrooms than the other ethnic groups [14,27,63], indicating that the Sidama ethnic group is good at sharing ethnomycological knowledge within the ethnic group [27,29,63]. Traditional healers are key informants for the identification and characterization of medicinal species because they use different species for their traditional medicinal practices [31,62]. In addition to their use as food, *Lycoperdon perlatum* Pers. and *Calvatia rubroflava* Fr. were reported to be the most useful and important medicinal wild mushroom species because they play a key role in treating wounds and skin disease. Interestingly, *Laetiporus sulphureus* (Bull.) Murrill. sensu lato has also been reported to be a common traditional medicine for lessening pain during childbirth in Kaffa areas in the southern part of Ethiopia [36], where local people preserve powder of this species for long periods in the house. Interestingly, in this study, a *Ganoderma* species was also used for medicinal purposes by the local people. Other studies have also reported that these mushrooms are used in traditional medicine by different local communities around the world to treat stress, pain, measles and lung diseases [64–66].

In our study, the surveys revealed that the Amhara and the Agew have limited ethnomycological knowledge and do not eat mushrooms even though their parents or grandparents have consumed wild mushroom species in the past and commercial mushroom species are being cultivated artificially in their locality. The non-consumption of mushrooms will eventually represent a loss of mycological knowledge in these ethnic groups [29]. Only two of the species common to their locality have been assigned a folk taxonomy by these ethnic groups (Table 3) while the common name of the majority of wild mushroom species was reported to be Enguday or Yejibtila. Therefore, rigorous work is needed in Banja and Fogera to develop a folk taxonomy [31,39]. However, this study did provide us with an opportunity to understand how different ethnic groups in Ethiopia use their ethnomycological knowledge. This sentiment echoes expressed by Tuno [34], who reported that observing the traditional ways of mushroom utilization by the Majangir ethnic group in the southwest part of Ethiopia provided a unique opportunity for studying how people belonging to traditional tribes in Africa utilize mushroom as foods [34,39]. The findings presented in [34] suggested that communities in rural part of Ethiopia are familiar with wild mushroom species growing in their locality and prize them as subsidiary food items collected in the forest [34].

In this study, communities, particularly the Sidama, were easily able to distinguish between wild mushroom species using their own traditional protocols. Broadly, they used morphological characters, smell and habitat to identify edible mushrooms. These criteria are in line with [31,34]. The oldest accounts of this practice among tribes in developing countries have been putatively reflected by different authors [67–70]. Beside these criteria, in rural areas of Ethiopia, the local people also use the presence or absence of a strong bad smell [34] to determine the edibility of mushrooms. Thus, it could be assumed that such traditional protocols are the product of ancient experimentations and possibly opportunistic discoveries by the indigenous people despite the confounded genealogy of their cultural uses [63]. Among the Sidama, women are generally involved in the collection and gathering of wild mushroom species and children are involved in the collection of well-known mushrooms. This is in line with [36], who reported that women are often responsible for this type of activity in the south and southwestern parts of the country, and also agrees with the findings of several other authors [27,34,39,63]. Because women are responsible for collecting firewood from the forests in the Wondo Genet area, they have become expert at distinguishing between edible and poisonous mushrooms [34,36] and are knowledgeable about the spatial distribution of mushrooms in terms of habitat and associated substrates in the forest and other niches [31,38].

Respondents also revealed that all mushrooms are collected during the rainy season (June to September) and used especially during food shortage periods. However, Dejene et al. [31] reported that the collection of some species, such as *Laetiporus* sp., could occur during the dry season [34,37]. Grain is the principal source of nutrition for communities in most parts of the country. In general, the availability of mushrooms in the rainy season coincides with periods of grain scarcity, suggesting that local communities may use wild mushroom species together with other edible wild foods during this period as a gap filler [31]. Common areas identified for wild mushroom collection were natural forests, farmlands, grazing lands, home gardens, and swampy areas. Most of these ecological niches are quite similar to those reported by [36,37] in the south and south western parts of the country. If communities are trained well, this culture of collecting wild mushroom species could be expanded to shift from subsistence use to income-generating small-scale business speculations through their commodification.

We have got determinant variables on wild mushroom consumption by the local communities in the three study woredas, implying that any of the significant variables has a positive effect on wild mushroom use the local people. For example, "taste experience" has a positive effect on wild mushroom use in the study area. This implies that those individuals who have tasted a wild mushroom species have better awareness of its use than those that have not tasted the species. This experience helps to understand the local livelihood context, the sources and nature of risks and the coping behavior of communities. This supports the findings of Lemenih et al. [71] who indicated that household experience is a commonly applied strategy for coping with shocks and is instrumental in poverty reduction. Thus, a local communities' experience of tasting a mushroom has implications for poverty reduction through emphasizing local available sources of the mushroom, enabling rural households to diversify their food sources. The same applied to the other positive coefficient variables, implying that a unit addition to any one of them will have a positive implication for wild mushroom consumption.

We evaluated different threats that affect wild mushroom species based on the perceived causes for degradation. Many of the threats related to habitat degradation are affecting wild mushroom species in Ethiopia in general. Free grazing and agricultural land expansion were considered to be the main threats in this study. This might be because rapidly growing human and livestock populations are driving an ever-increasing demand for crop and grazing land, which is aggravating the degradation of habitats in Ethiopia [32,72,73]. Such anthropogenic pressures are likely to also have an impact on useful known and unknown wild mushroom species, some of which could face extinction [73]. Moreover, this loss also limits the benefits that can be obtained from wild mushroom components of forest resources as well as the ethnomycological knowledge of different ethnic groups associated with mushroom use. Thus, the sustainable management of Ethiopian forest systems is mandatory in order to play major roles in the conservation and development of wild edible and medicinal mushrooms that cannot be economically cultivated, require very specific habitats, and are exceptionally difficult to reproduce in nurseries or laboratories. Furthermore, the importance of fungal resources has recently been brought to the forefront due to their ecological and economic importance. There have been many efforts to record their diversity at local and regional scales. Thus, product diversification is a fundamental strategy to integrate a model of sustainable forest exploitation and reverse the degradation of wild mushroom species through promoting ecosystem services such as biodiversity conservation.

5. Conclusions

This study has attempted to provide baseline information about the ethnomycology of three ethnic groups in Ethiopia. The Amhara and Agew ethnic groups are located in the same geographic region, are in contact with the same natural resources and have a similar knowledge of wild mushrooms. This similarity is a result of their cultural exchanges, coexistence, and shared historical events. Furthermore, parallel responses were reported in many cases, such as the consumption of wild mushroom species by their parents or grandparents even though the respondents do not eat mushrooms, which will eventually lead to the loss of mycological knowledge in these two ethnic groups. By contrast, wild mushroom consumption is an integral part of the cultural knowledge of seasonal resources among the Sidama communities in the southern part of the country. Thus, the different traditional food composition and recipes of the ethnic group should be further studied to ensure the potential value of wild mushroom species in the nutrition and food security role in the community.

Sidama women are primarily responsible for gathering and collecting valuable wild mushrooms, indicating that gender is one of the variables that influence the local knowledge of wild mushroom use and their distribution in the locality. This could also have implications for women contributing toward household food security using locally available food resources such as fungi. In other words, fungi could be a means of providing supplementary food, thereby reducing poverty and providing opportunities for women living in unfavorable areas and, hence, reducing inequality. Similarly, although most of the individuals interviewed from the Amhara and Agew ethnic groups were not familiar with the wild mushrooms, they appreciated the food value of the mushrooms when they tasted them. Thus, the experience of tasting mushrooms by local communities has implications for poverty reduction through emphasizing local available sources that enable rural households to diversify their food sources.

Although we have tried to document the traditional knowledge and uses of wild mushroom species of three ethnic groups in Ethiopia, more studies are needed to ensure that much of the potential value of wild mushroom species and the ethnomycology knowledge of local communities is not lost. Such knowledge is part of the identity of these communities: knowledge of wild mushroom uses, linguistics, and harvesting can prevent their loss as modernization proceeds due to the dominance of hegemonic culture. Documentation of this knowledge could promote a revaluation of wild mushroom species as resources and promote their conservation. The knowledge and nomenclature of useful species could also be revitalized, and their use encouraged, and in doing so, wild mushroom species could make a greater contribution to food security, especially in a country like Ethiopia where food security is a country-wide issue.

Author Contributions: Conceptualization, P.M.-P. and T.D.; methodology, D.A., P.M.-P. and T.D..; software, P.M.-P. and T.D.; validation, P.M.-P.; formal analysis, P.M.-P., G.Z. and T.D.; investigation, P.M.-P. and D.A.; data curation, G.Z.; writing—original draft preparation, T.D. and G.Z.; writing—review and editing, P.M.-P., T.D. and W.T. All authors have read and agreed to the published version of the manuscript.

Funding: This research was partially funded by the Spanish Agency for International Development Cooperation project (Sustfungi_Eth:2017/ACDE/002094).

Acknowledgments: We would like to express our gratitude to the people involved in this study from the Amhara, Agew and Sidama communities.

Conflicts of Interest: The authors declare no conflict of interest.

References

1. Godfray, H.C.; Crute, I.; Haddad, L.; Lawrence, D.; Muir, J.; Nisbett, N.; Pretty, P.; Robinson, S.; Toulmin, C.; Whiteley, R. The future of the global food system. *Philos. Trans. R. Soc. B* **2010**, *365*, 2769–2777. [CrossRef] [PubMed]
2. Pinstrup-Andersen, P. Food security: Definition and measurement. *Food Secur.* **2009**, *1*, 5–7. [CrossRef]
3. De Caluwé, E.; Halamová, K.; Van Damme, P. *Adansonia digitata* L.—A review of traditional uses, phytochemistry and pharmacology. *Afr. Focus* **2010**, *23*, 10–51. [CrossRef]

4. Termote, C.; Bwama Meyi, M.; Dhed'a Djailo, B.; Huybregts, L.; Lachat, C.; Kolsteren, P.; Van Damme, P. A biodiverse rich environment does not contribute to a better diet: A case study from DR Congo. *PLoS ONE* **2012**, *7*, e30533. [CrossRef]

5. Fentahun, M.T.; Hager, H. Exploiting locally available resources for food and nutritional security enhancement: Wild fruits diversity, potential and state of exploitation in the Amhara region of Ethiopia. *Food Secur.* **2009**, *1*, 207–219. [CrossRef]

6. Mengistu, F.; Hager, H. Wild edible fruit species cultural domain, informant species competence and preference in three districts of Amhara region, Ethiopia. *Ethnobot. Res. Appl.* **2008**, *6*, 487–502. [CrossRef]

7. DeClerck, F.; Fanzo, J.; Palm, C.; Remans, R. Ecological approaches to human nutrition. *Food Nutr. Bull.* **2011**, *32*, S41–S50. [CrossRef]

8. Powell, B.; Thilsted, S.H.; Ickowitz, A.; Termote, C.; Sunderland, T.; Herforth, A. Improving diets with wild and cultivated biodiversity from across the landscape. *Food Secur.* **2015**, *7*, 535–554. [CrossRef]

9. Rowland, D.; Blackie, R.R.; Powell, B.; Djoudi, H.; Vergles, E.; Vinceti, B.; Ickowitz, A. Direct contributions of dry forests to nutrition: A review. *Int. For. Rev.* **2015**, *17*, 45–53. [CrossRef]

10. Ickowitz, A.; Rowland, D.; Powell, B.; Salim, M.A.; Sunderland, T. Forests, trees, and micronutrient-rich food consumption in Indonesia. *PLoS ONE* **2016**, *11*, e154139. [CrossRef]

11. Johnson, K.; Jacob, A.; Brown, M. Forest cover associated with improved child health and nutrition: Evidence from the Malawi Demographic and Health Survey and satellite data. *Glob. Health Sci. Pract.* **2013**, *1*, 237–248. [CrossRef] [PubMed]

12. Cai, M.; Pettenella, D.; Vidale, E. Income generation from wild mushrooms in marginal rural areas. *For. Policy Econ.* **2011**, *13*, 221–226. [CrossRef]

13. Sarma, T.; Sarma, I.; Patiri, B. Wild edible mushrooms used by some ethnic tribes of Western Assam. *Bioscan* **2010**, *3*, 613–625. [CrossRef]

14. Boa, E. *Wild Edible Fungi: A Global Overview of Their Use and Importance to People*; Non-Wood Forest Products No. 17; FAO: Rome, Italy, 2004; p. 147.

15. Chang, Y.S.; Lee, S.S. Utilisation of macrofungi species in Malaysia. *Fungal Divers.* **2004**, *15*, 15–22.

16. Ferreira, I.C.F.R.; Vaz, J.A.; Vasconcelos, M.H.; Martins, A. Compounds from wild mushrooms with antitumor potential. *Anticancer Agents Med. Chem.* **2010**, *10*, 424–436. [CrossRef]

17. Gryzenhout, M.; Roets, F.; de Villiers, R. Fungal conservation in Africa. *Mycol. Balc.* **2010**, *7*, 43–48.

18. Marconi, S.; Durazzo, A.; Camilli, E.; Lisciani, S.; Gabrielli, P.; Aguzzi, A.; Gambelli, L.; Lucarini, M.; Marletta, L. Food composition databases: Considerations about complex food matrices. *Foods* **2018**, *7*, 2. [CrossRef]

19. Durazzo, A.; Lisciani, S.; Camilli, E.; Gabrielli, P.; Marconi, S.; Gambelli, L.; Aguzzi, A.; Lucarini, M.; Maiani, G.; Casale, G.; et al. Nutritional composition and antioxidant properties of traditional Italian dishes. *Food Chem.* **2017**, *218*, 70–77. [CrossRef]

20. Wani, B.; Bodha, R.; Wani, A. Nutritional and medicinal importance of mushrooms. *J. Med. Plants Res.* **2010**, *4*, 2598–2604.

21. Bano, Z.; Shasirekha, M.; Rajarathnam, S. Improvement of the bioconversion and biotransformation efficiencies of the oyster mushroom (*Pleurotus sajor-caju*) by supplementation of its rice straw with oil seed cakes. *Enzym. Microb. Technol.* **1993**, *15*, 985–989. [CrossRef]

22. Mattila, P.; Könkö, K.; Eurola, M.; Pihlava, J.M.; Astola, J.; Vahteristo, L.; Hietaniemi, V.; Kumpulainen, J.; Valtonen, M.; Piironen, V. Contents of vitamins, mineral elements, and some phenolic compounds in cultivated mushrooms. *J. Agric. Food Chem.* **2001**, *49*, 2343–2348. [CrossRef] [PubMed]

23. Kakon, A.; Choudhury, M.B.K.; Saha, S. Mushroom Is an Ideal Food Supplement. *J. Dhaka Natl. Med. Coll. Hosp.* **2012**, *18*, 58–62. [CrossRef]

24. Dejene, T.; Agamy, S.; Agúndez, D.; Martín-Pinto, P. Ethnobotanical survey of wild edible fruit tree species in lowland areas of Ethiopia. *Forests* **2020**, *11*, 177. [CrossRef]

25. Garibay-Orijel, R.; Caballero, J.; Estrada-Torres, A.; Cifuentes, J. Understanding cultural significance, the edible mushrooms case. *J. Ethnobiol. Ethnomed.* **2007**, *3*, 4. [CrossRef] [PubMed]

26. Härkönen, M.; Niemelä, T.; Mbindo, K.; Kotiranta, H.; Pierce, G.D. Zambian mushrooms and mycology. *Norrlinia* **2015**, *29*, 1–5.

27. Tibuhwa, D. Folk taxonomy and use of mushrooms in communities around Ngorongoro and Serengeti National Park, Tanzania. *J. Ethnobiol. Ethnomed.* **2012**, *8*, 36. [CrossRef] [PubMed]

28. Héritier, M.; Hippolyte, N.; Cephas, M.; Hoffman, B.; Gallaher, T. Importance indices in ethnobotany. *Ethnobot. Res. Appl.* **2007**, *5*, 201–218.

29. Kamalebo, H.M.; Nshimba, H.; Wa, S.; Ndabaga, C.M. Uses and importance of wild fungi: Traditional knowledge from the Tshopo province in the Democratic Republic of the Congo. *J. Ethnobiol. Ethnomed.* **2018**, *14*, 13. [CrossRef]

30. Asfaw, Z.; Tadesse, M. Prospects for sustainable use and development of wild food plants in Ethiopia. *Econ. Bot.* **2001**, *55*, 47–62. [CrossRef]

31. Dejene, T.; Oria-de-Rueda, J.A.; Martín-Pinto, P. Wild mushrooms in Ethiopia: A review and synthesis for future perspective. *For. Syst.* **2017**, *26*, eR04. [CrossRef]

32. Lulekal, E.; Asfaw, Z.; Kelbessa, E.; Van Damme, P. Wild edible plants in Ethiopia: A review on their potential to combat food insecurity. *Afr. Focus* **2011**, *24*, 71–121. [CrossRef]

33. Semwal, K.C.; Lemma, H.; Dhyani, A.; Equar, G.; Amhare, S. Mushroom: Nature's treasure in Ethiopia. *Momona Ethiop. J. Sci.* **2014**, *6*, 138–147. [CrossRef]

34. Tuno, N. Mushroom utilization by the Majangir, an Ethiopian tribe. *Mycologist* **2001**, *15*, 78–79. [CrossRef]

35. Awas, T.; Asfaw, Z.; Nordal, I.; Demissew, S. Ethnobotany of Berta and Gumuz people in western Ethiopia. *Biodiversity* **2010**, *11*, 45–53. [CrossRef]

36. Abate, D. Wild mushrooms and mushroom cultivation efforts in Ethiopia. *WSMBMP Bull.* **2014**, *11*, 1–3.

37. Muleta, D.; Woyessa, D.; Teferi, Y. Mushroom consumption habits of Wacha Kebele residents, southwestern Ethiopia. *Glob. Res. J. Agric. Biol. Sci.* **2013**, *4*, 6–16.

38. Dejene, T.; Oria-de-Rueda, J.A.; Martín-Pinto, P. Edible wild mushrooms of Ethiopia: Neglected non-timber forest products. *Rev. Fitotec. Mex.* **2017**, *40*, 391–397. [CrossRef]

39. Dejene, T.; Oria-de-Rueda, J.A.; Martín-Pinto, P. Fungal community succession and sporocarp production following fire occurrence in Dry Afromontane forests of Ethiopia. *For. Ecol. Manag.* **2017**, *398*, 37–47. [CrossRef]

40. Beza, Z.B. Challenges and prospects of community based ecotourism development in Lake Zengena and its environs, North West Ethiopia. *Afr. J. Hosp. Tour. Leis.* **2017**, *6*, 1–12.

41. Friis, B.; Demissew, S.; Breugel, P. *Atlas of the Potential Vegetation of Ethiopia*; The Royal Danish Academy of Sciences and Letters: Copenhagen, Denmark, 2010; p. 307.

42. Kebede, M.; Kanninen, M.; Yirdaw, E.; Lemenih, M. Vegetation structural characteristics and topographic factors in the remnant moist Afromontane forest of Wondo Genet, south central Ethiopia. *J. For. Res.* **2013**, *24*, 419–430. [CrossRef]

43. Takele, A. Analysis of Rice Profitability and Marketing Chain: The Case of Fogera Woreda South Gondar Zone, Amhara National Regional State, Ethiopia: ILRI-IPMS. Available online: https://agris.fao.org/agris-search/search.do?recordID=QT2016102236 (accessed on 19 December 2019).

44. Eshete, A. The Frankincense Tree of Ethiopia: Ecology, Productivity and Population Dynamics. Ph.D. Thesis, Wageningen University, Wageningen, The Netherlands, 2011. Available online: https://www.wur.nl/en/show/The-frankincense-tree-of-Ethiopia-ecology-productivity-and-population-dynamics.htm (accessed on 17 June 2020).

45. Wassie, A.; Teketay, D.; Powell, N. Church forests in North Gonder administrative zone, Northern Ethiopia. *For. Trees Livelihoods* **2005**, *15*, 349–373. [CrossRef]

46. Kassa, H.; Campbell, B.; Sandewall, M.; Kebede, M.; Tesfaye, Y.; Dessie, G.; Seifu, A.; Tadesse, M.; Garedew, E.; Sandewall, K. Building future scenarios and uncovering persisting challenges of participatory forest management in Chilimo Forest, Central Ethiopia. *J. Environ. Manag.* **2009**, *90*, 1004–1013. [CrossRef] [PubMed]

47. Shumi, G. The Structure and Regeneration Status of Tree and Shrub Species of Chilimo Forest Ecological Sustainability Indicators for Participatory Forest Management (PFM) in Oromia, Ethiopia. Available online: https://agris.fao.org/agris-search/search.do?recordID=ET2008000082 (accessed on 17 June 2020).

48. Mekonnen, Z.; Kassa, H.; Woldeamanuel, T.; Asfaw, Z. Analysis of observed and perceived climate change and variability in Arsi Negele District, Ethiopia. *Environ. Dev. Sustain.* **2018**, *20*, 1191–1212. [CrossRef]

49. Bernard, H. *Research Methods in Anthropology: Qualitative and Quantitative Approaches*, 4th ed.; AltaMira Press: New York, NY, USA, 2006; 824p.

50. Bowen, G.A. Naturalistic inquiry and the saturation concept: A research note. *Qual. Res.* **2008**, *8*, 137–152. [CrossRef]

51. Guest, G.; Bunce, A.; Johnson, L. How Many Interviews Are Enough? *Field Methods* **2006**, *18*, 59–82. [CrossRef]
52. Martin, G.J. *Ethnobotany: A Methods Manual*; Chapman & Hall: London, UK, 1995; 268p.
53. Antonín, V. *Flora of Tropical Africa, Volume 1: Monograph of Marasmius, Gloiocephala, Palaeocephala and Setulipes in Tropical Africa*; National Botanic Garden: Meise, Belgium, 2007; 177p.
54. Hama, O.; Maes, E.; Guissou, M.; Ibrahim, D.; Barrage, M.; Parra, L.; Raspe, O.; De Kesel, A. *Agaricus subsaharianus*, une nouvelle espèce comestible et consomméeau Niger, au Burkina Faso et en Tanzanie. *Crypto. Mycol.* **2010**, *31*, 221–234.
55. Heinemann, P. *Champignons Récoltés au Congo Belge*; Botanic Garden Meise: Meise, Belgium, 1956; Volume 5, pp. 99–119.
56. Morris, B. An annotated check-list of the macrofungi of Malawi. *Kirkia* **1990**, *13*, 323–364.
57. Pegler, D.; Rayner, R. A contribution to the Agaric flora of Kenya. *Kew Bull.* **1969**, *23*, 347–412. [CrossRef]
58. Rammeloo, J.; Walleyn, R. The edible fungi of Africa south of the Sahara: A literature survey. *Scr. Bot. Belg.* **1993**, *5*, 1–62.
59. Ryvarden, L.; Piearce, G.D.; Masuka, A.J. *An Introduction to the Larger Fungi of South Central Africa*; Baobab Books: Harare, Zimbabwe, 1994; Available online: https://agris.fao.org/agris-search/search.do?recordID=XF2016048164 (accessed on 17 June 2020).
60. Singer, R. Marasmius. In *Flore Iconographique des Champignons du Congo*; National Botanic Garden of Belgium: Meise, Belgium, 1965; Volume 14, pp. 253–278.
61. Gujarati, D. *Essentials of Econometrics*, 4th ed.; McGraw-Hill: New York, NY, USA, 1992; 54p.
62. Osarenkhoe, O.O.; John, O.A.; Theophilus, D.A. Ethnomycological conspectus of West African mushrooms: An awareness document. *Adv. Microbiol.* **2014**, *4*, 39–54. [CrossRef]
63. Ekandjo, L.K.; Chimwamurombe, P.M. Traditional medicinal uses and natural hosts of the genus *Ganoderma* in north-eastern parts of Namibia. *J. Pure Appl. Microbiol.* **2012**, *6*, 1139–1146.
64. Kadhila-Muandingi, N.P. The Distribution, Genetic Diversity, and Uses of Ganoderma Mushrooms in Oshana and Ohangwena Regions of Northern Namibia. Master's Thesis, University of Namibia, Windhoek, Namibia, 2010. Available online: http://repository.unam.edu.na/handle/11070/525 (accessed on 17 June 2020).
65. Sun, S.; Gao, W.; Lin, S.; Zhu, J.; Xie, B.; Lin, Z. Analysis of genetic diversity in Ganoderma population with a novel molecular marker SRAP. *Appl. Microbiol. Biotechnol.* **2006**, *72*, 537–543. [CrossRef] [PubMed]
66. Adhikari, M.K.; Devkota, S.; Tiwari, R.D. Ethnomycolgical knowledge on uses of wild mushrooms in western and central Nepal. *Our Nat.* **2005**, *3*, 13–19. [CrossRef]
67. Borah, N.; Semwal, R.L.; Garkoti, S.C. Ethnomycological knowledge of three indigenous communities of Assam, India. *Indian J. Tradit. Knowl.* **2018**, *17*, 327–335.
68. Kinge, T.R.; Apalah, N.A.; Nji, T.M.; Acha, A.N.; Mih, A.M. Species richness and traditional knowledge of macrofungi (mushrooms) in the Awing Forest Reserve and communities, Northwest Region, Cameroon. *J. Mycol.* **2017**, *2017*, 2809239.
69. Tibuhwa, D.D. Termitomyces species from Tanzania, their cultural properties and unequalled basidiospores. *J. Biol. Life Sci.* **2012**, *3*, 140–159. [CrossRef]
70. Ongoche, I.C.; Otieno, D.J.; Oluoch-Kosura, W. Assessment of factors influencing smallholder farmers' adoption of mushroom for livelihood diversification in Western Kenya. *Afr. J. Agric. Res.* **2017**, *12*, 2461–2467. [CrossRef]
71. Lemenih, M.; Olsson, M.; Karltun, E. Comparison of soil attributes under *Cupressus lusitanica* and *Eucalyptus saligna* established on abandoned farmlands with continuously cropped farmlands and natural forest in Ethiopia. *For. Ecol. Manag.* **2004**, *195*, 57–67. [CrossRef]
72. Giday, M.; Asfaw, Z.; Woldu, Z. Medicinal plants of the Meinit ethnic group of Ethiopia: An ethnobotanical study. *J. Ethnopharmacol.* **2009**, *124*, 513–521. [CrossRef]
73. Lemenih, M.; Kassa, H. Re-greening Ethiopia: History, challenges and lessons. *Forests* **2014**, *5*, 1896–1909. [CrossRef]

Article

Sage Species Case Study on a Spontaneous Mediterranean Plant to Control Phytopathogenic Fungi and Bacteria

Massimo Zaccardelli [1,*], Catello Pane [1], Michele Caputo [1], Alessandra Durazzo [2], Massimo Lucarini [2], Amélia M. Silva [3,4], Patrícia Severino [5,6,7], Eliana B. Souto [8,9], Antonello Santini [10,*] and Vincenzo De Feo [11]

1   CREA-Research Centre for Vegetable and Ornamental Crops, Via Cavalleggeri 25, 84098 Salerno, Italy; catello.pane@crea.gov.it (C.P.); caputomichele1986@libero.it (M.C.)
2   CREA-Research Centre for Food and Nutrition, Via Ardeatina 546, 00178 Roma, Italy; alessandra.durazzo@crea.gov.it (A.D.); massimo.lucarini@crea.gov.it (M.L.)
3   School of Biology and Environment, University of Trás-os-Montes e Alto Douro (UTAD), Quinta de Prados, P-5001-801 Vila Real, Portugal; amsilva@utad.pt
4   Centre for Research and Technology of Agro-Environmental and Biological Sciences (CITAB), University of Trás-os-Montes e Alto Douro (UTAD), Quinta de Prados, 5001-801 Vila Real, Portugal
5   Industrial Biotechnology Program, University of Tiradentes (UNIT), Av. Murilo Dantas 300, Aracaju 49032-490, Brazil; pattypharma@gmail.com
6   Tiradentes Institute, 150 Mt Vernon St., Dorchester, MA 02125, USA
7   Laboratory of Nanotechnology and Nanomedicine (LNMED), Institute of Technology and Research (ITP), Av. Murilo Dantas 300, Aracaju 49010-390, Brazil
8   Department of Pharmaceutical Technology, Faculty of Pharmacy, University of Coimbra, Pólo das Ciências da Saúde, Azinhaga de Santa Comba, 3000-548 Coimbra, Portugal; souto.eliana@gmail.com
9   CEB-Centre of Biological Engineering, University of Minho, Campus de Gualtar, 4710-057 Braga, Portugal
10 Department of Pharmacy, University of Napoli Federico II, Via D. Montesano 49, 80131 Napoli, Italy
11 Department of Pharmacy, University of Salerno, Via Giovanni Paolo II, 132, 84084 Salerno, Italy; defeo@unisa.it
*   Correspondence: massimo.zaccardelli@crea.gov.it (M.Z.); asantini@unina.it (A.S.); Tel.: +39-386-219 (M.Z.); +39-081-253-9317 (A.S.)

Received: 30 May 2020; Accepted: 19 June 2020; Published: 24 June 2020

Abstract: Sage species belong to the family of *Labiatae/Lamiaceae* and are diffused worldwide. More than 900 species of sage have been identified, and many of them are used for different purposes, i.e., culinary uses, traditional medicines and natural remedies and cosmetic applications. Another use of sage is the application of non-distilled sage extracts and essential oils to control phytopathogenic bacteria and fungi, for a sustainable, environmentally friendly agriculture. Biocidal propriety of non-distilled extracts and essential oils of sage are w documented. Antimicrobial effects of these sage extracts/essential oils depend on both sage species and bacteria and fungi species to control. In general, it is possible to choose some specific extracts/essential oils to control specific phytopathogenic bacteria or fungi. In this context, the use of nanotechnology techniques applied to essential oil from salvia could represent a future direction for improving the performance of eco-compatible and sustainable plant defence and represents a great challenge for the future.

Keywords: essential oils; extracts; *Salvia Africana*; *S. rutilans*; *S. munzii*; *S. mellifera*; *S. greggii*; *S. officinalis* "Icterina"; *S. officinalis*

1. Introduction

Medicinal plants are recently getting growing attention worldwide as important sources of bioactive compounds, and hence for their potential beneficial properties [1–23]. The World Health Organization (WHO) reported that about 80% of the world's population use herbal medicine for the treatment of many diseases, referring to traditional drugs only as a second choice, due to diffidence for chemical origin pharmaceuticals [24]. A lot of plants and herbs have been cultivated for their aromatic and medicinal proprieties. In particular, the interest for aromatic plants has increased not only in medicine but also in agriculture, since their essential oils are among the eco-compatible compounds potentially useful to control dangerous biotic agents of the crops, which makes them extremely interesting in view of eco-sustainable and environmentally friendly agriculture practices [25–28].

Main Features of Some Sage Species (Salvia africana, Salvia rutilans, Salvia munzii, Salvia mellifera, Salvia greggii, Salvia officinalis "Icterina", Salvia officinalis)

Salvia genus belongs to the family of *Labiatae/Lamiaceae* [29,30] and is widespread in various regions around the world, being particularly present in Mediterranean regions of Europe, South Africa, Central and South America, and South-East Asia [31,32]. It includes more than 900 species, which are used for different purposes, i.e., culinary uses, traditional medicines and remedies, and cosmetic applications. Plants of the genus *Salvia* have been reported to have a wide range of biological activities and are used for the prevention and treatment of various diseases, due to the presence of a peculiar profile of secondary metabolites/bioactive components isolated in several parts of the plants (i.e., flowers, leaves, stem), in particular, essential oils, with a large spectrum of terpenoids (i.e., α- and β-thujone, camphor, 1,8-cineole, α-humulene, β-caryophyllene and viridiflorol) as well as di- and tri-terpenes (i.e., carnosic acid, ursolic acid, carnosol and tanshinones), polyphenols, comprising an array of phenolic acids (i.e., caffeic acid and its derivatives, rosmarinic acid, salvianolic acids, sage coumarin, lithospermic acids, sagernic acid, and yunnaneic acids) and flavonoids (i.e., luteolin, apigenin, hispidulin, kaempferol and quercetin) [33].

The most common species, namely the *Salvia officinalis* L., represents an important medicinal and aromatic plant, with antioxidant, antimicrobial, anti-inflammatory and anticancer properties. Poulios et al. [34] well summarized the current advances on the extraction and identification of bioactive components of sage; further, the same authors gave a current state-of-the-art on the antioxidant activity of sage (*Salvia* spp.) and its bioactive components [35].

The study of Craft et al. [36] revealed the presence of five major chemotypes of sage (*Salvia officinalis* L.) leaf essential oils, with several subtypes: most sage oils belonged to the "typical" chemotype composition containing, in percentages (*w/w*), α-thujone > camphor > 1,8-cineole; however, the essential oil composition can vary widely and may have a profound effect on flavour and fragrance profiles, as well as on biological activities.

An update overview of pharmacological properties of *Salvia officinalis* and its components was exploited in the recent review of Ghorbani and Esmaeilizadeh [37].

S. officinalis 'Icterina' is a cultivar of *S. officinalis* characterised by yellow-green variegated leaves, as reported by Pop et al. [38]. *Salvia Africana* L., commonly known as Bruinsalie or beach sage, is distributed from Namaqualand to the Eastern Cape Province of South Africa [39]. Etsassala et al. [40] reported how the methanolic extract of *S. africana-lutea* L. is a rich source of terpenoids, especially abietane diterpenes, with strong antioxidant and anti-diabetic activities, that can be helpful to modulate the redox status of the body and could be, therefore, an excellent candidate for the prevention of diabetes.

The phytochemical composition and bioactive effects of *Salvia Africana* and *Salvia officinalis* 'Icterina', was recently reported by Afonso et al. [41]; particularly, rosmarinic acid was the dominant phenolic compound in all the extracts, yet that of *S. africana* origin was characterised by the presence of yunnaneic acid isomers, which overall accounted for about 40% of total phenolics, whereas *S. officinalis* 'Icterina' extract presented glycosidic forms of apigenin, luteolin and scuttelarein [41].

Salvia mellifera Greene, known as Black sage, is a traditional medicine of the Chumash Indians. De Martino et al. [42] reported that *Salvia mellifera* essential oils contain fifty-four monoterpenoids and several diterpenoids, such as carnosol (41%), carnosic acid (22%), salvicanol (15%) and rosmanol (9%). Adams et al. [43] reported how several chronic pain patients have reported long-term improvements in their pain after treatment with sun tea, made from the stems and leaves of the *Salvia mellifera*.

S. greggii A. Gray was studied for its terpenic compounds content [44,45], and the anti-germinative properties of its essential oils [42]. Pereira et al. [46] suggested that decoctions of *S. greggii* (*S. elegans*, *S. officinalis*) could be an effective natural antidiabetic and anti-obesity agent and help to control the glucose levels through the modulation of the α-glucosidase activity.

Salvia rutilans Carrière (synonym *S. elegans* Vahl) is native to Mexico, known as "pineapple-scented sage". The plant is used in traditional Mexican medicine for the treatment of Central Nervous System discomforts [47], and this species has been reported as a possible source of anxiolytic and antidepressant compounds [48]. The essential oil of this plant has been studied by Makino et al. [49], and its antigerminative activity was reported [50].

Salvia munzii Epling is present in xeric coastal sage scrub in Northern Baja California. It's essential oil was previously studied [50] and it was also studied for its antigerminative effects [50].

2. Sage Plants to Control Fungal and Bacterial Diseases

Sage-based phytochemicals have showed a great potential to face phytopathogens causing diseases and very important economic losses in agricultural systems worldwide. Several studies indicated the suitability of these natural substances to implement conventional crop management protocols and match the pressing sustainability challenge about safe foods and organic productions, launched by the political institutions and the large-scale distribution. In addition, the consumers' preference toward no-pesticides of chemical origin-treated crops needs to be considered. In order to reduce or eliminate the use of chemical fungicides for plant protection, research is moving toward the study and implementation of new eco-friendly non-synthetic tools able to effectively replace them in sustainable producing contests. All the plants are a source of secondary metabolites belonging to several chemical groups, that, once extracted, may be exploitable for their antifungal properties [51] alone or in integration with antagonistic microorganisms in controlling plant pathogens [52].

Aromatic plants have been extensively investigated as potential sources of natural compounds with antimicrobial activity, which can be effectively used to replace conventional environment unfriendly pesticides. This topic also includes all the bioactive sage species. Due to the natural origin, many of the sage-derived antifungal compounds could be included into the "generally recognised as safe (GRAS)" classification [53] for plant protection and food preservation. The antimicrobial compounds from sage tissues are currently extracted in raw blends, in which the activity is based on all the phytochemicals [54], such as solvent-extracts and essential oils that are among the main promising tools to totally replace or integrate chemicals for disease management.

2.1. Solvent Sage Extracts

The bioactive antimicrobial molecules are extracted from dried plant sage material to obtain their raw blends or purified compounds through water or other solvents, expression under pressure, supercritical CO_2, and solvent-free microwave methods [55,56].

Recently, Nutrizio et al. [57] explored an innovative process based on high-voltage electrical discharge for the eco-friendly recovery of bioactive compounds from Dalmatian sage. Sage extracts are characterised by antioxidant activity, high phenolic content and steroids, alkaloids and saponins to a lesser extent [58,59]. For example, caffeic acid and rosmarinic acid, reported as the most abundant phenolic compounds in sage extracts [60–62], have been associated to the control ability of Fusarium root rot in asparagus [63] and against Fusarium wilt in cyclamen [64]. The antifungal activity of *S. fruticosa* Mill. ethyl acetate extract against *B. cinerea* and *P. digitatum* has been supported by three major constituents, carnosic acid, carnosol and hispidulin [65].

As a hypothetical mechanism of action, sage extracts have been reported to affect cellular processes associated with changes in the membrane functionality involved in feeding and growth of the pathogen. A cytological study performed on *Candida albicans* yeast revealed the leakage of intracellular contents caused by membrane lipid bilayer alteration after treatment with the ethanol extract of *Salvia miltiorrhiza* [66].

A literature survey indicated a wide range of pathogens that are susceptible to the sage extract exposure (Table 1).

Table 1. Sage extracts exhibiting antimicrobial activity against phytopathogens, their extraction methods, susceptible microorganisms and bioassay to test the control efficacy.

Sage Species (*Salvia* spp.)	Extraction Methods	Target Phytopatogens	Efficacy on Bioassay	Reference
S. aegyptiaca	Ethanol and water extraction	*Phyphtora infestans*	In vitro and in vivo on tomato	[67]
S. africana-lutea	Dichloromethane: Methanol extraction	*Fusarium verticillioides* and *F. proliferatum*	In vitro determination of MICand MFC	[68]
S. cryptantha, *S. officinalis*, *S. tomentosa*	Water, ethanol and methanol extraction	*Fusarium oxysporum* f. sp. *radicis*	In vitro mycelial growth inhibition	[69]
S. cryptantha *S. officinalis*, *S. tomentosa*	Methanol and ethanol extraction	*Botrytis cinerea, Monilia laxa, Aspergillus niger*, and *Penicillum* sp.	In vivo on apple	[70]
S. fruticosa	Crude water extraction	*Alternaria solani*	In vitro mycelial growth inhibition	[71]
S. fruticosa	Extraction in semi-automated Soxlet system	*B. cinerea*	In vivo on grape	[72]
S. fruticosa	Ethanol extraction	*Sclerotinia sclerotorium*	In vitro mycelial growth inhibition	[73]
S. officinalis	Extracted in 90% methanol + 9% water + 1% acetic acid	*Alternaria alternata, A. niger*, and *A. parasiticus*	In vitro mycelial growth inhibition	[74]
S. officinalis	Metanolic extraction	*A. alternata, B. cinerea, Phytophtora cambivora,* and *F. oxysporum*	In vitro mycelial growth inhibition	[75]
S. officinalis	Ethanol and methanol extraction	*A. alternata, A. solani, Fusarium solani, F. oxysporum* f.sp. *lycopersici, P. infestans, Rhizoctonia solani, B. cinerea, Colletotrichum coccoides, Verticillum albo-atrum*	In vitro determination of MIC and MFC	[76]
S. officinalis	Ethanol extraction	*Plasmopora viticola*	In vivo on grape	[77]
S. officinalis	Water extraction	*F. oxysporum* f.sp. *asparagi*	In vitro on conidia germination	[63]
S. officinalis	Hydroethanol, ethanol and water infusion	*S. sclerotiorum*	In vitro and in vivo on lettuce	[78]
S. officinalis	Water extraction	*F. oxysporum* f.sp. *cyclaminis*	In vitro density of conidia	[64]
S. officinalis	Ethanol extraction	*Pseudoperonospora cubensis*	In vivo on cucumber	[79]
S. officinalis, *S. sclarea*	Aqueous, saline Buffer and acid extraction	*Alternaria* spp.	In vitro determination of MIC and MFC	[80]

Table 1. *Cont.*

Sage Species (*Salvia* spp.)	Extraction Methods	Target Phytopatogens	Efficacy on Bioassay	Reference
S. officinalis, S. sclarea		*Pantoea agglomerans, Erwinia chrysanthemy*	In vitro growth inhibition	[81]
S. verticillata	Methanol extraction	*F. oxysporum, A. alternata, Aureobasidium pillulans, Trichoderma harzianum, Penicillum canescens*	In vitro determination of MIC and MFC	[82]
S. virgata	Methanol and n-hexane extraction	*R. solani, A. solani, F. oxysporum* f.sp. *radicis, lycopersici, Verticillum dalhiae*	In vitro mycelial growth inhibition	[83]
S. tigrina	Ethanol extraction	*B. cinerea, F. oxysporum, A. alternata*	In vitro determination of MIC and MFC	[84]

MIC: Minimal inhibitory concentration; MFC: Minimum fungicidal concentration.

S. officinalis ethanolic extracts have proven to be able to control *Plasmopara viticola* on grapevine [77] and *Pseudoperonospora cubensis* on organic cucumber [79]. *Salvia aegyptiaca* L. ethanolic extracts have resulted as effective both in vitro and in vivo against *Phytophthora infestans*, the causal agent of late blight disease of tomato [67]. Dellavalle et al. [80] assessed the minimum inhibition concentration (MIC) and migration inhibitory factor (MIF) values of *Salvia officinalis* L. and *Salvia sclarea* L. respectively, foliar and seed acid extracts against *Alternaria* spp. quite comparable to values obtained with the conventional fungicide captants (2.5 µg mL^{-1}). In vitro inhibitory effects were also found on *Fusarium proliferatum* and *F. verticilloides* by *Salvia africana-lutea* extracts [68] and on *Aspergillus flavus, Penicillium frequentans, Botrytis cinerea, Geotrichum candidum, Fusarium oxysporum* and *Alternaria alternara* by treatments with ethanolic extracts of *Salvia tigrina* Hedge & Hub.-Mor. [84]. All these fungi are able to produce many metabolites [85–89]. Water, ethanol and methanol extracts of *S. officinalis, S. cryptantha* Montbret and Aucher and *S. tomentosa* Mill. have been found effective against *F. oxysporum* f. sp. *radicis-lycopersici* [69] and *Sclerotinia sclerotiorum, Alternaria solani, Ascochyta rabiei, Botrytis cinerea, Rhizoctonia solani, Penicillum italicum, Aspergillus niger* and *Monilia laxa* [70]. The mycelium development of *Rhizoctonia solani, Alternaria solani, Fusarium oxysporum* f. sp. *radicis lycopersici* and *Verticillium dahliae* were repressed by methanol and n-hexane extracts of *Salvia virgate* Jacq. [83], while ethanol, hexane and aqueous extracts of *S. sclarea* have shown in vitro antifungal activity against pathogenic fungi *Epicoccum nigrum* and *Colletotrichum coccodes* [90]. *Salvia fructicosa* Mill. (sage Greek) has sourced very active extracts against the mycelial growth and sclerotial formation and germination of *Sclerotinia sclerotiorum* [73] and against the early blight fungus, *Alternaria solani* [71].

The formulation of sage extracts aimed at improving their safe use and antimicrobial efficacy is a very active research field. Ghaedi et al. [91] have made available preparations based on the incorporation in Zn(OH)$_2$ nanoparticles and Hp-2-minh of extracts of *Salvia officinalis*. Chitosan-based edible coating was combined with the acetonic extract of *Salvia fruticosa*. Carrying the flavonoids hispidulin, salvigenin and cirsimaritin and the diterpenes carnosic acid, carnosol and the 12-methoxycarnosic acid as major constituents, showed encouraging efficacy against the grey mould of table grapes [72]. Salević et al. [92] recently characterised electrospun poly(ε-caprolactone) films containing a solid dispersion of *Salvia officinalis* extract for their antimicrobial potential.

2.2. Sage Essential Oils

The genus *Salvia*, as all the aromatic plant species from *Lamiaceae*, is a noble source of essential oils (EOs), obtained by the steam or hydro-distillation of different vegetative parts of the plant, including leaves, flowers, seeds and stems [93,94]. The super critical fluid-mediated extraction of bioactive EOs has also been achieved: it utilizes carbon dioxide at fluid state as the solvent in the same extractor model by merely lowering the extraction temperatures. EOs contain a wide variety of hydrophobic secondary metabolites, such as mainly terpenes, that can enhance the antimicrobial activity through synergic action [95,96] and are reported to be responsible for antiseptic properties of the phytochemical [97]. For example, Džamić et al. [98] found linalyl acetate, linalool, α-terpineol, α-pinene, 1,8-cineole, limonene, β-caryophyllene and β-terpineol as the main components of *S. sclarea* EO that affect the moderate to high antimicrobial activity against a plethora of phytopathogens. Three major constituents, α-thujone, β-thujone and myrcene, have been identified, instead, in the wide-spectrum fungistatic *S. pomifera* subsp. *calycina* EO [99,100]. The major compounds in the antimicrobial EOs of *Salvia mirzayanii* Rech. F. and Esfand were α-terpinyl acetate, eudesm-7(11)-en-4-ol, bicyclogermacrene, δ-cadinene, 1,8-cineole, germacrene D-4-ol, cis-dihyroagarofuran, linalyl acetate, α-cadinol, linalool and α-terpineol [101]. While, 1,8-cineole, α-pinene, camphene, borneol, camphor and β-pinene have been indicated among the most abundant constituents of *Salvia brachyodon* Vandas antifungal EOs [102]. The presence of cis-thujone and camphor has been associated to the fungicidal activity of *S. officinalis* EO against filamentous fungi, belonging to *Penicillium, Aspergillus, Cladosporium* and *Fusarium genera* [103].

EOs antimicrobial activity exploitable in the control of phytopatogenic diseases is due to these single constituents and their synergistic interactions [104]. Lipophilic volatile molecules of EOs penetrate the cell wall and damage cell membranes by altering permeability and seal and affect morphology, growth, reproductive functions and viability of microorganisms [105]. An injured plasma membrane may result in a leakage of cellular components, including nucleic acids and proteins, and cellular collapse until death, representing the mode of action underlying the biostatic and biocidal EOs effects [106]. EOs also may induce decreased Succinate dehydrogenase (SDH) and nicotinamide adenine dinucleotide hydride (NADH) oxidase activities and cause extreme changes in ultra-structures by penetrating and dissolving the mitochondrial membranes [107]. The exposure of *Escherichia coli* and *Staphyloccocus aureus* to *Salvia sclarea* EO caused cell plasmalemma disintegration with a massive leakage of cellular material and reduction of the intracellular ATP, as well as nuclear DNA content [108]. Among all components of the sage volatile blends, the monoterpene alcohol linalool has been shown to have the strongest antifungal activity, while 1,8-cineole was only moderate, and linalyl acetate was lower [109–111]. The antifungal in vitro activity of linalool has been attributed to leakage of intracellular material that dramatically affects radial mycelial growth and conidial production and germination of treated pathogens with severity according to dose [112]. Ultrastructural studies carried out on *Botrytis cinerea* exposed to 1,8-cineole revealed detrimental effects on cell organelles [113]. The volatile bicyclic sesquiterpene β-caryophyllene showed inhibitory effects against bacteria via suppression of DNA replication [114]. Sage EOs are assayed both in vitro and in vivo on a wide range of plant pathogens, giving encouraging indications about their fungicidal activity measured as MIC and MIF and disease control efficacy, also in comparison with synthetic molecules (Table 2).

Table 2. Sage essential oils with antimicrobial activity on phytopathogens, main active components, susceptible microorganisms and bioassay to test the control efficacy.

Sage Species (*Salvia* spp.)	Major Components	Target Phytopatogens	Efficacy on Bioassay	Reference
S. aucheeri var. aucheri		*Alternaria alternata, Penicillium italicum, Fusarium equiseti*	In vitro mycelial growth inhibition	[115]
S. cryptantha, S. officinalis, S. tomentosa	Eucalyptol, Camphor, α-pinetene, β-thujone, borneol, camphor, 3-thujonene	*Botrytis cinerea, Monilia laxa, Aspergillus niger, Penicillum* sp.	In vivo on apple	[70]
S. cryptantha, S. officinalis, S. tomentosa		*Fusarium oxysporum* f.sp. *radicis-lycopersici*	In vitro mycelial growth inhibition	[69]
S. desoleana	1,8-cineole	*A. niger, A. ochraceus, A. versicolor, A. flavus, A. terreus, A. alternata, Penicillium ochrocholon, P. funiculosum, Cladosporium cladosporoides, Trichoderma viride, Fusarium tricinctum, Phomopsis helianthi*	In vitro determination of MIC and MFC	[116]

Table 2. *Cont.*

Sage Species (*Salvia* spp.)	Major Components	Target Phytopatogens	Efficacy on Bioassay	Reference
S. fruticosa		*A. niger, A. oryza, Mucor pusillus, F. oxysporum*	In vitro mycelial growth inhibition	[117]
S. fruticosa	1,8-Cineole, Camphor	*Bipolaris/Drechslera sorociniana, Fusarium subglutinans, F. vertricilioides, F. oxysporum, F. tricinctum, F. sporotrichioides, F. equiseti, F. incarnatum, F. proliferatum, Macrophomina phaseolina*	In vitro determination of MIC and MFC	[118]
S. fruticosa	Eucalyptol, Camphor	*B. cinerea, Fusarium solani var. coeruleum, Clavibacter michiganensis* subsp. *michiganensis*	In vitro mycelial growth inhibition	[119]
S. fruticosa	Hispidulin, salvigenin, cirsimaritin, carnosic acid, carnosol, and 12-methoxycarnosic acid	*Aspergillus tubingensis, B. cinerea, P. digitatum*	In vitro determination of MIC and MFC	[65]
S. fruticosa	1,8-cineole, α-thujone, β-thujone, camphor, (E) caryophyllene	*F. oxysporum f.sp. dianthi, F. proliferatum, Rhizoctonia solani, Sclerotium sclerotiorum, F. solani f. sp. cucurbitae*	In vitro mycelial growth inhibition	[120]
S. hydrangea		*A. alternata, A. solani, Aspergillus* sp., *Botrytis* sp., *Colletotrichum* sp., *Drechslera* sp., *Fusarium acuminatum, F. chlamydosporum, F. culmorum, F. equiseti, F. graminearum, F. incarnatum, F. nivale, F. oxysporum, F. proliferatum, F. sambucinum, F. scirpi, F. semitectum, F. solani, F. tabacinum, F. verticillioides, Nigrospora* sp., *Penicillium jensenii, Phoma* sp., *Pythium ultimum, Phytophthora capsici, Rhizoctonia solani, S. sclerotiorum, Sclerotinia* sp., *Trichothecium* sp., *Verticillium albo-atrum, V. dahliae, V.tenerum*	In vitro microbial growth inhibition	[121]
S. lavenduulfolia, S. sclarea		*B. cinerea, C. gloeosporioides, F. oxysporum, P. ultimum, R. solani*	In vitro mycelial growth inhibition	[66]
S. officinalis		*F. graminearum, F. verticilloides, F. subglutinans, F. oxysporum, F. avenaceum, Diaporthe phaseolarum var. caulivora, Phomopsis viticola, Helminthosporium sativum, Colletotrichum coccodes, Thanatephorus cucumeris*	In vitro mycelial growth inhibition	[122]
S. officinalis		*A. alternata, Colletotrichum destructivum, Phytophthora parasitica*	In vitro determination of MIC and MFC	[123]
S. officinalis	Monoterpene	*Colletotrichum acutatum, B. cinerea, Clavibacter michiganensis, Xanthomonas campestris, Pseudomonas savastanoi, P. syringae pv. phaseolicola*	In vitro determination of MIC and MFC	[124]
S. officinalis	α-p-thujone, 1,8-cineole and camphor	*B. cinerea*	In vitro mycelial growth inhibition	[125]
S. officinalis		*R.solani, Streptomycetes scabies*	In vivo *on potato*	[126]

Table 2. *Cont.*

Sage Species (*Salvia* spp.)	Major Components	Target Phytopatogens	Efficacy on Bioassay	Reference
S. officinalis	Camphor, α-thujone, 1,8-cineole, viridiflorol, β-thujone, β-caryophyllene, 2,2-diphenyl-1-picrylhydrazyl radical-scavenging, linoleic acid peroxidation, ferric reducing assays	*B. cinerea, R. solani, F. oxysporum, A. alternata*	In vitro determination of MIC and MFC	[127]
S. officinalis		*F. oxysporum, B. cinerea*	In vitro mycelial growth inhibition	[128]
S. officinalis		*Verticillium fungicola var. fungicola, Cladobotryum* sp.	In vitro determination of MIC and MFC	[129]
S. officinalis		*F. oxysporum* f. sp. *cicer, Alternaria porri*	In vitro mycelial growth inhibition	[130]
S. officinalis	Camphor, camphene, eucalyptol	*Fusarium graminearum*	In vitro determination of MIC and MFC	[131]
S. officinalis	1,8-cineole, β-thujone, L-camphor	*Verticillum dalhie*	In vitro mycelial growth inhibition	[132]
S. officinalis		*Xanthomonas campestris pv. phaseoli, Clavibacter michiganensis* subsp. *michiganensis, Pseudomonas tolaasii*	In vitro determination of MIC and MFC	[133]
S. officinalis	Thujone, camphor	*Rhizopus stolonifer*	In vitro mycelial growth inhibition	[134]
S. officinalis	Camphor, α-thujone	*V. fungicola, Trichoderma harzianum*	In vitro mycelial growth inhibition	[102]
S. officinalis		*A. alternata*	In vitro and in vivo on tomato	[125]
S. officinalis		*S. sclerotiorum*	In vitro and in vivo on lettuce	[78]
S. officinalis		*C. cinerea*	In vivo on tomato	[135]
S. officinalis	Cis-thujone, camphor	*P. italicum, Aspergillus* sp., *Cladosporium cladosporioides, Fusaarium moniliforme*	In vitro determination of MIC and MFC	[103]
S. officinalis		*Phytophthora capsici*	In vitro and in vivo on zucchini	[136]
S. officinalis		*Colletotrichum nymphaeae*	In vitro and in vivo on strawberry	[137]
S. officinalis		*Aspergillus niger*	In vitro and in vivo on tomato paste	[138]
S. officinalis		*B. cinerea, P. expansum*	In vivo on apples	[139]
S. officinalis		*Monilinia laxa, B. cinerea,*	In vivo on apricot, peach and almond	[140]
S. officinalis		*F. oxysporum* f. sp. *lycopersici, R. solani, S. minor*	In vitro mycelial growth inhibition	[52]
S. officinalis		*Alternaria dauci, A. radicina, C. lindemuthianum, Ascochyta rabiei*	In vitro mycelial growth inhibition	[141]
S. officinalis, S. tomentosa		*A. rabiei*	In vitro mycelial growth inhibition	[142]

Table 2. *Cont.*

Sage Species (*Salvia* spp.)	Major Components	Target Phytopatogens	Efficacy on Bioassay	Reference
S. pomifera sp. *calycina*	α-thujone, β-thujone	*Mycogone perniciosa*	In vitro determination of MIC and MFC	[100]
S. pomifera subsp. *calycina*	α-thujone, β thujone, myrcene	*F. oxysporumf.* sp. *dianthus, F. solani f.sp. cucurbitae, F. proliferatum, S. slerotiorum, R. solani, V. dahliae, P. exspansum*	In vitro mycelial growth inhibition	[99]
S. pomifera subsp. *calycina*	α-thujone, β-thujone, myrcene	*F. oxysporum f.sp. dianthy, F. solani f.sp. cucurbitae, F. proliferatum, S. sclerotium, R. solani, V. dalhie, P. expansum*	In vitro mycelial growth inhibition	[100]
S. reflexa	Phenols, diterpenes, fatty acids, Palmitic acid, phytol (E)caryophyllene, caryophyllene oxide, hexahydrofarnezylacetone, phytol isomer, β-asarone, α-copaene, δ-cadinene	*Curvularia lunata, Helmenusathosporium maydis*	In vitro mycelial growth inhibition	[143]
S. sclarea	Linalyl acetate, linalool, α-terpineol, α-pinene, 1.8-cineole, limonene, β-caryophyllene, β-terpineol	*A. alternata, C. cladosporioides, C. fulvum, F. tricinctum, F. sporotrichoides, Phoma macdonaldii, Phomopsis helianthi, Trichoderma viride*	In vitro determination of MIC and MFC	[98]
S. sclarea	Linalyl acetate, linalool, geranyl acetate, R-terpineol,	*F. oxysporumf.* sp. *dianthi, S.sclerotiorum, Sclerotium cepivorum*	In vitro determination of MIC and MFC	[144]
S. sclarea	Linalool, linalyl acetate, geranyl acetate, β-ocimene, caryophylleneoxide	*F. oxysporum, B. cinerea, R. solani, A. solani*	In vitro determination of MIC and MFC	[109]
S. sclarea	Caryophyllene oxide, sclareol, spathulenol, 1H-naphtho, pyran, β-caryophyllene	*Epicoccum nigrum, Colletotrichum coccodes*	In vitro mycelial growth inhibition	[90]
S. officinalis, S. sclarea		*Pantoea agglomerans, Erwinia chrysanthemy*	In vitro inhibition measurement	[145]

MIC: Minimal inhibitory concentration; MFC: Minimum fungicidal concentration.

Volatility of active molecules from sage EOs benefits fumigating applications against soil-borne [120] and storage pathogens [146]. *S. officinalis* EO proved to be a potential alternative method to conventional fungicides for the control of Sclerotinia rot in lettuce [78].

Aromatic waters obtained from distillation after oil separation were also found to be active against two fungal pathogens, *Rhizoctonia solani* and *Sclerotinia minor* [147]. However, this product shaped up to be an extract.

Also, about EOs, formulation techniques have been developed especially in order to enhance the efficacy in practical application. Due to the higher hydrophobicity of compounds, innovative preparations have been developed to improve EOs aqueous dispersions, for example by absorbing them in a swelling matrix of semisynthetic copolymers [135,148], nanocapsule suspensions [149], microencapsulation within alginate [150] or solid lipid nanoparticles [151]. Kodadová et al. [152] have formulated *S. officinalis* EO monoterpenes into a chitosan-based hydrogel, while nanoemulsions containing *Salvia multicaulis* Vahl EO, have recently been pinpointed to enhance the safe control of food-borne bacteria [153].

3. Sage Essential Oils: Experimental Part of the Case Study

The case study concerns the evaluation of the bactericidal and fungicide effect of EOs extracted from seven species of sage, most of which are not well known. In detail, the sage species investigated were: *Salvia africana* L., *S. rutilans* Carrière, *S. munzii* Epling, *S. mellifera* Greene, *S. greggii* A. Gray, *S. officinalis* "Icterina" L. and *S. officinalis* L., the bacteria tested were: *Xanthomonas campestris* pv. *campestris* and *Pectobacterium carotovorum* subsp. *carotovorum*, both phytopathogens, while the fungi tested were: *Alternaria alternata, Botrytis cinerea, Sclerotinia minor, Fusarium oxysporum, F. sambucinum, F. semitectum, F. solani* and *Rhizoctonia solani*, all phytopathogenics.

Plant cultivation took place between 2005 and 2006 at "Improsta" farm, in the Campania Region, Southern Italy.

At the balsamic stage, the leaves were taken, and the essential oils were obtained by hydrodistillation. The oils extracted were stored at 4 °C until the tests for evaluation of their antibacterial and antifungal activity were run.

The EOs extracted were also analysed for their main constituents by GC and GC/MS [42].

To evaluate the ability of EOs to inhibit the growth of bacterial and fungal plant pathogens, in vitro plate tests were performed. For each essential oil, plugs (5 mm diameter) were removed from the edge of the growing mycelia or bacterial suspensions (10^8 CFU mL^{-1}) and were incubated overnight at 25 °C in sterile double distilled water (SDDW) containing 1% essential oil emulsion; for controls, plugs or bacterial suspensions were incubated in SDDW only. After incubation, plugs were transferred in the centre of potato dextrose agar (PDA) Petri plates (60 mm diameter) and incubated at 25 °C, while bacterial suspensions were incubated on nutrient agar (NA) Petri plates (60 mm diameter) at 28 °C. For each essential oil, treatments and incubation were performed in triplicate.

The diameter of the mycelia was measured daily until fungi reached the edge of the control plates, and data were expressed as % growth reduction (or, sometimes, growth increment) with respect to control; for bacteria, presence or absence of growth was evaluated after 48 h.

The EOs of the different species of sage showed different bactericidal and fungicide effects and a different composition for their main constituents. Both bacteria and phytopathogenic fungi showed different susceptibility depending on the bacterial or fungal species and depending on the essential oil used.

The bactericidal effect was generally more pronounced against *Pectobaterium carotovorum* subsp. *carotovorum*, with four out of seven essential oils, expect for *Xanthomonas campestris* pv. campestris, on which only two EOs were found to be active (Table 3).

Table 3. Antibacterial activity of essential oils of *Salvia* spp. used in this study.

Salvia **spp.**	*Xanthomonas campestris* *pv. campestris*	*Pectobacterium carotovorum* subsp. *carotovorum*
S. africana	0	-
S. rutilans	-	-
S. munzii	0	0
S. mellifera	+/−	-
S. greggii	+/−	0
S. officinalis "Icterina"	0	0
S. officinalis	-	-

Legend: 0 = no inhibition; - = total inhibition; +/- = partial inhibition.

The fungicide effect of EOs was very different, depending on the fungal species. The EOs of some species of sage have been able to completely inhibit the growth of certain fungal species, such as *S. greggi* and *S. officinalis*, that completely inhibited the development of three phytopathogenic fungi:

S. munzii, that completely inhibited two fungal species, and *S. rutilans*, that completely inhibited one of the eight species of fungi tested (see Table 4). Some EOs were found to be less active, such as *S. africana*, *S. mellifera* and *S. officinalis* "Icterina", as shown in Table 4. Some fungi were found, in general, to be very resistant to the tested EOs, such as *Fusarium sambucinum*, while other fungi were found to be more sensitive, such as the two "soil-borne" *Sclerotinia minor* and *Rhizoctonia solani* (see Table 4).

However, the EOs have also been able to increase the development of some fungal colonies, such as in the case of *S. africana* for *S. minor* and *R. solani*. It is possible to speculate that some constituents of the oils are used as a nutrient source. The different EOs were also analysed for their composition. The results are shown in Table 5. This Case Study was presented here to show how different Mediterranean species of sage may have interesting potential to provide EOs with antifungal and antibacterial properties. The research was carried out to screen new environmentally compatible suppressive means of eight fungal and two bacterial phytopathogens, which cause economically important diseases on several agrarian species worldwide. In this view, the assayed samples have displayed differential phytochemical profiles that suggest species-specific effects, likely to which the inhibitive activity could be associated. In agreement with the previous studies listed in the Table 2, the most active EOs contain thujone (cis*trans), 1,8-cineole, camphor, α-pinene and lower traces of gerianol. These compounds supported the activities of sage-sourced EOs against a plethora of fungi [118,120]. Moreover, among their major components, δ-cadinene was also found in these samples, noticed in *Salvia reflexa* essential oil producing in vitro mycelia growth inhibition of *Curvularia lunata* and *Helmenusathosporium maydis*. On the other hand, the *S. africana* essential oil profile markedly differentiated from each other with the presence of p-cymene, γ-terpinene and epizonareme among the most abundant components, which to our knowledge have not been reported as antimicrobial effector constituents of sage EOs to determine detrimental effects against plant pathogens. This circumstance may explain the reduced bioactivity shown by the sample from *S. Africana* with respect to the other.

The methodology used in this study to assess the antifungal potential of EOs by submerging fungal plugs into an EOs emulsion has been developed previously [141] and could also be applied to MIC determination if repeated at different effective concentrations. The relevance of sage material to derive suitable antifungal EOs may have a friendly impact on developing sustainable protocols for practical application in plant disease management, where the availability of alternative tools to synthetic fungicides are urgent. Furthermore, the possibility to submit the residual biomasses of the sage plants to the distillation process also opens the way to a circular economy scheme.

Table 4. Antifungal activity of essential oils of *Salvia* spp. used in this study. The numbers indicate % reduction (values with −) or increase (values with +) of the diameter of the colonies treated with essential oils with respect to not treated (control).

Salvia spp.	*Alternaria alternata*	*Botrytis cinerea*	*Sclerotinia minor*	*Fusarium oxysporum*	*Fusarium sambucinum*	*Fusarium semitectum*	*Fusarium solani*	*Rhizoctonia solani*
S. africana	0	−31.2	+50.0	+3	−5.4	−12.5	−20.0	+20.0
S. rutilans	−55.5	−50.0	−100.0	−5.6	+8.1	−21.9	−4.0	+20.0
S. munzii	0	0	+33.3	−22.7	−9.9	−100.0	−40.0	−100.0
S. mellifera	−11.1	−50.0	−33.3	−14.1	+8.1	−35.5	−40.0	−24.0
S. greggii	−100.0	−50.0	−100.0	−18.5	+3.6	−25.0	−32.0	−100.0
S. officinalis "Icterina"	−33.3	−50.0	−50.0	−7.3	−1	−21.9	+8.0	+8.0
S. officinalis	−100.0	−100.0	−100.00	0	0	−50.0	0	−50.0

Some molecules have been found in many essential oils, such as canphor and δ-cadinene, while other molecules have been found in interesting quantities only in a single species, such as p-menthadiene, isobornyl acetate, γ-terpinene, epizonareme and (Z)-β-ocimene, as shown in Table 5.

Some of these molecules are already known to have bactericidal activity, such as camphor [154]; however, biocidal effects of an essential oil can be due to a synergistic action of the many active molecules which the oil contains.

Table 5. Main constituents of essential oils (% on total oil, *w/w*) extracted from *Salvia* spp. species used in this study.

Molecules	S. mellifera	S. africana	S. rutilans	S. munzii	S. greggii	S. officinalis "Icterina"	S. officinalis
Thujone (*cis*trans*)			38.7	33.3	43.4	34.6	37.9
1,8-Cineole	38.8					4.6	4.2
Camphor	12.2		4.7	27.2	4.2	7.5	13.9
δ-Cadinene		3.8	11.5	8.9	14.0		
α-Pinene	9.2						4.4
p-Menthadiene	2.3						
Limonene	2.2						1.4
Isobornyl acetate						5.0	
Camphene						4.6	4.1
p-Cymene		17.7					1.2
γ-Terpinene		12.9					
Epizonareme		11.3					
Cubebol							
Geranyl acetate			6.9		8.7		
Geraniol		2.3	6.5	4.0	3.4		
(Z)-β-Ocimene				5.7			

4. Conclusions and Future Remarks

The use of essential oils of *Salvia* spp. to control plant disease, can be a valid eco-friendly strategy to control plant diseases, but it is very important to improve their performance by applying nanotechnologies.

Nanotechnologies represent a great challenge for the future. An emerging direction is to apply the nanotechnologies [155,156] to plant defence.

The adoption of this technology in food and agrochemical industries is emerging, addressing new nanoformulations of biopesticides [157,158]. Formulations of nanobiopesticide and nanoherbicides as a "smart delivery system" are currently studied and produced from the perspective of an eco-friendly approach, by reducing herbicide inputs and providing more effective control on where and when an active ingredient is released.

In this regard, Jampílek, and Kráľová [156] showed state-of-the-art and future opportunities for nanobiopesticides in agriculture, with particular regards to pesticide-effective organic or inorganic (poly)materials of natural origin, EOs loaded in various matrices and green-synthesised metal or metal oxide nanoparticles (NPs).

In particular, novel strategies related to both the encapsulation of vegetable oils, related methods of preparation, applications as antimicrobials and insecticide/pesticide/pest repellents, have been studied and developed [157]. Recently, de Matos et al. [158] gave an updated overview on analytical methods and challenges of essential oils in nanostructured systems. This represents a new frontier in the area of interest.

Author Contributions: M.Z., A.D. and A.S. conceived and designed the work. M.Z., C.P., A.D., M.L., E.B.S. and A.S. wrote the manuscript. M.Z., C.P., M.C., A.M.S. and P.S. validated and elaborated data information and figures. M.Z., C.P., M.C., A.D., M.L., A.M.S., P.S., E.B.S., A.S. and V.D.F. made a substantial contribution to the revision of the work and approved it for publication. All authors have read and agreed to the published version of the manuscript.

Funding: The authors acknowledge the support of the research project: Nutraceutica come supporto nutrizionale nel paziente oncologico, CUP: B83D18000140007. E.B.S. acknowledges the sponsorship of the projects M-ERA-NET-0004/2015-PAIRED and UIDB/04469/2020 (strategic fund), receiving support from the Portuguese Science and Technology Foundation, Ministry of Science and Education (FCT/MEC) through national funds, and co-financed by FEDER, under the Partnership Agreement PT2020. The authors also acknowledge Agricultural Department of Campania Region (Italy) for financial support.

Conflicts of Interest: The authors declare no conflict of interest.

References

1. Santini, A.; Novellino, E. Nutraceuticals: Beyond the diet before the drugs. *Curr. Bioact. Compd.* **2014**, *10*, 1–12. [CrossRef]

2. Durazzo, A. Extractable and non-extractable polyphenols: An overview. In *Non-Extractable Polyphenols and Carotenoids: Importance in Human Nutrition and Health*; Saura-Calixto, F., Perez-Jimenez, J., Eds.; RSC Publishing: Cambridge, UK, 2018; Volume 5, pp. 37–45.

3. Durazzo, A.; Lucarini, M.; Kiefer, J.; Mahesar, S.A. State-of-the-art infrared applications in drugs, dietary supplements, and nutraceuticals. *Hindawi J. Spectrosc.* **2020**. [CrossRef]

4. Durazzo, A.; Lucarini, M. The State of science and innovation of bioactive research and applications, health and diseases. *Front. Nutr.* **2019**, *6*, 178. [CrossRef] [PubMed]

5. Santini, A.; Novellino, E. Nutraceuticals-shedding light on the grey area between pharmaceuticals and food. *Expert Rev. Clin. Pharmacol.* **2018**, *11*, 545–547. [CrossRef]

6. Santini, A.; Novellino, E.; Armini, V.; Ritieni, A. State of the art of ready-to-use therapeutic food: A tool for nutraceuticals addition to foodstuff. *Food Chem.* **2013**, *140*, 843–849. [CrossRef] [PubMed]

7. Durazzo, A.; Lucarini, M.; Souto, E.B.; Cicala, C.; Caiazzo, E.; Izzo, A.A.; Novellino, E.; Santini, A. Polyphenols: A concise overview on the chemistry, occurrence, and human health. *Phytother. Res.* **2019**, *33*, 2221–2243. [CrossRef] [PubMed]

8. Lucarini, M.; Durazzo, A.; Kiefer, J.; Santini, A.; Lombardi-Boccia, G.; Souto, E.B.; Romani, A.; Lampe, A.; Ferrari Nicoli, S.; Gabrielli, P. Grape Seeds: Chromatographic profile of fatty acids and phenolic compounds and qualitative analysis by FTIR-ATR spectroscopy. *Foods* **2020**, *9*, 10. [CrossRef]

9. Salehi, B.; Venditti, A.; Sharifi-Rad, M.; Kręgiel, D.; Sharifi-Rad, J.; Durazzo, A.; Lucarini, M.; Santini, A.; Souto, E.B.; Novellino, E. The therapeutic potential of apigenin. *Int. J. Mol. Sci.* **2019**, *20*, 1305. [CrossRef]

10. Durazzo, A.; Lucarini, M.; Novellino, E.; Souto, E.B.; Daliu, P.; Santini, A. *Abelmoschus esculentus* (L.): Bioactive components' beneficial properties—Focused on antidiabetic role—For sustainable health applications. *Molecules* **2019**, *24*, 38. [CrossRef]

11. Abenavoli, L.; Izzo, A.A.; Milić, N.; Cicala, C.; Santini, A.; Capasso, R. Milk thistle (*Silybum marianum*): A concise overview on its chemistry, pharmacological, and nutraceutical uses in liver diseases. *Phytother. Res.* **2018**, *32*, 2202–2213. [CrossRef] [PubMed]

12. Santini, A.; Tenore, G.C.; Novellino, E. Nutraceuticals: A paradigm of proactive medicine. *Eur. J. Pharm. Sci.* **2017**, *96*, 53–61. [CrossRef]

13. Daliu, P.; Santini, A.; Novellino, E. A decade of nutraceutical patents: Where are we now in 2018? *Expert Opin. Ther. Pat.* **2018**, *28*, 875–882. [CrossRef]

14. Santini, A.; Cammarata, S.M.; Capone, G.; Ianaro, A.; Tenore, G.C.; Pani, L.; Novellino, E. Nutraceuticals: Opening the debate for a regulatory framework. *Br. J. Clin Pharmacol.* **2018**, *84*, 659–672. [CrossRef] [PubMed]

15. Bircher, J.; Hahn, E.G. Understanding the nature of health: New perspectives for medicine and public health. Improved Wellbeing at Lower Costs: New Perspectives for Medicine and Public Health: Improved Wellbeing at Lower Cost. *F1000Research* **2016**, *5*, 167. [CrossRef] [PubMed]

16. Yeung, A.W.K.; Souto, E.B.; Durazzo, A.; Lucarini, M.; Novellino, E.; Tewari, D.; Wang, D.; Atanasov, A.G.; Santini, A. Big impact of nanoparticles: Analysis of the most cited nanopharmaceuticals and nanonutraceuticals research. *Curr. Res. Biotechnol.* **2020**, *2*, 53–63. [CrossRef]

17. Daliu, P.; Santini, A.; Novellino, E. From pharmaceuticals to nutraceuticals: Bridging disease prevention and management. *Expert Rev. Clin. Pharmacol.* **2019**, *12*, 1–7. [CrossRef]

18. Durazzo, A.; D'Addezio, L.; Camilli, E.; Piccinelli, R.; Turrini, A.; Marletta, L.; Marconi, S.; Lucarini, M.; Lisciani, S.; Gabrielli, P. From plant compounds to botanicals and back: A current snapshot. *Molecules* **2018**, *23*, 1844. [CrossRef]

19. Durazzo, A.; Camilli, E.; D'Addezio, L.; Piccinelli, R.; Mantur-Vierendeel, A.; Marletta, L.; Finglas, P.; Turrini, A.; Sette, S. Development of dietary supplement label database in Italy: Focus of FoodEx2 coding. *Nutrients* **2020**, *12*, 89. [CrossRef]

20. Durazzo, A.; Lucarini, M. A current shot and re-thinking of antioxidant research strategy. *Braz. J. Anal. Chem.* **2018**, *5*, 9–11. [CrossRef]

21. Durazzo, A.; Lucarini, M. Extractable and non-extractable antioxidants. *Molecules* **2019**, *24*, 1933. [CrossRef]

22. Santini, A.; Cicero, N. Development of food chemistry, natural products, and nutrition research: Targeting new frontiers. *Foods* **2020**, *9*, 482. [CrossRef]

23. Durazzo, A.; Lucarini, M.; Santini, A. Nutraceuticals in Human Health. *Foods* **2020**, *9*, 370. [CrossRef]

24. World Health Organization (WHO). 2013. Available online: http://www.who.int/ (accessed on 18 May 2020).

25. Cimmino, A.; Andolfi, A.; Troise, C.; Zonno, M.C.; Santini, A.; Tuzi, A.; Vurro, M.; Ash, G.; Evidente, A. Phomentrioloxin: A Novel Phytotoxic Pentasubstituted Geranylcyclohexentriol Produced by *Phomopsis* sp., a Potential Mycoherbicide for *Carthamus lanathus* Biocontrol. *J. Nat. Prod.* **2012**, *75*, 1130–1137. [CrossRef]

26. Cimmino, A.; Andolfi, A.; Zonno, M.C.; Avolio, F.; Santini, A.; Tuzi, A.; Berestetskyi, A.; Vurro, M.; Evidente, A. Chenopodolin: A Phytotoxic Unrearranged ent-Pimaradiene Diterpene Produced by Phoma chenopodicola, a Fungal Pathogen for *Chenopodium album* Biocontrol. *J. Nat. Prod.* **2013**, *76*, 1291–1297. [CrossRef]

27. Mikušová, P.; Šrobárová, A.; Sulyok, M.; Santini, A. *Fusarium* fungi and associated metabolites presence on grapes from Slovakia. *Mycotoxin Res.* **2013**, *29*, 97–102. [CrossRef]

28. Salvo, A.; La Torre, G.L.; Mangano, V.; Casale, K.E.; Bartolomeo, G.; Santini, A.; Granata, T.; Dugo, G. Toxic inorganic pollutants in foods from agricultural producing areas of Southern Italy: 2 level and risk assessment. *Ecotoxicol. Environ. Saf.* **2018**, *148*, 114–124. [CrossRef]

29. Carovic-Stanko, K.; Petek, M.; Martina, G.; Pintar, J.; Bedeković, D.; Custić, M.H.; Satovic, Z. Medicinal plants of the family lamiaceae as functional foods—A review. *Czech J. Food Sci.* **2016**, *34*, 377–390. [CrossRef]

30. Salehi, B.; Armstrong, L.; Rescigno, A.; Yeskaliyeva, B.; Seitimova, G.; Beyatli, A.; Sharmeen, J.; Mahomoodally, M.F.; Sharopov, F.; Durazzo, A.; et al. Lamium plants—A comprehensive review on health benefits and biological activities. *Molecules* **2019**, *24*, 1913. [CrossRef]

31. Dweck, A.C. The folklore and cosmetic use of various Salvia species. In *SAGE-The Genus Salvia*; Kintzios, S.E., Ed.; Harwood Academic Publishers: Amsterdam, The Netherlands, 2000; pp. 1–25.

32. Walker, J.B.; Sytsma, K.J. Staminal evolution in the genus Salvia (*Lamiaceae*): Molecular phylogenetic evidence for multiple origins of the staminal lever. *Ann. Bot.* **2007**, *100*, 375–391. [CrossRef]

33. Wu, Y.B.; Ni, Z.Y.; Shi, Q.W.; Dong, M.; Kiyota, H.; Gu, Y.C.; Cong, B. Constituents from *Salvia* species and their biological activities. *Chem. Rev.* **2012**, *112*, 5967–6026. [CrossRef]

34. Poulios, E.; Giaginis, C.; Vasios, G.K. Current advances on the extraction and identification of bioactive components of sage (*Salvia* spp.). *Curr. Pharm. Biotechnol.* **2019**, *20*, 845–857. [CrossRef]

35. Poulios, E.; Giaginis, C.; Vasios, G.K. Current state of the art on the antioxidant activity of sage (*salvia* spp.) and its bioactive components. *Planta Med.* **2020**, *86*, 224–238. [CrossRef] [PubMed]

36. Craft, J.D.; Satyal, P.; Setzer, W.N. The Chemotaxonomy of common sage (*Salvia officinalis*) based on the volatile constituents. *Medicines* **2017**, *4*, 47. [CrossRef] [PubMed]

37. Ghorbani, A.; Esmaeilizadeh, M. Pharmacological properties of *Salvia officinalis* and its components. *J. Tradit. Complement. Med.* **2017**, *7*, 433–440. [CrossRef] [PubMed]

38. Pop, A.; Tofană, M.; Socaci, S.A.; Pop, C.; Rotar, A.M.; Salanţă, L. Determination of antioxidant capacity and antimicrobial activity of selected *Salvia* species. *Bull. UASVM Food Sci. Technol.* **2016**, *73*. [CrossRef]

39. Manning, J.; Goldblatt, P. *Plants of the Greater Cape Floristic Region 1: The Core Cape Flora*; South African National Biodiversity Institute: Pretoria, South Africa, 2012.

40. Etsassala, N.G.E.R.; Badmus, J.A.; Waryo, T.T.; Marnewick, J.L.; Cupido, C.N.; Hussein, A.A.; Iwuoha, E.I. Alpha-glucosidase and alpha-amylase inhibitory activities of novel abietane diterpenes from *Salvia africana-lutea*. *Antioxidants* **2019**, *8*, 421. [CrossRef]

41. Afonso, A.F.; Pereira, O.R.; Fernandes, A.; Calhelha, R.C.; Silva, A.M.S.; Ferreira, I.C.F.R.; Cardoso, S.M. Phytochemical composition and bioactive effects of *Salvia africana*, *Salvia officinalis* 'Icterina' and *Salvia mexicana* aqueous extracts. *Molecules* **2019**, *24*, 4327. [CrossRef]

42. De Martino, L.; Roscigno, G.; Mancini, E.; De Falco, E.; De Feo, V. Chemical composition and antigerminative activity of the essential oils from five Salvia species. *Molecules* **2010**, *15*, 735–746. [CrossRef]

43. Adams, J.D.; Guhr, S.; Villaseñor, E. *Salvia mellifera*-how does it alleviate chronic pain? *Medicines* **2019**, *6*, 18. [CrossRef]

44. Kawahara, N.; Inoue, M.; Kawai, K.I.; Sekita, S.; Satake, M.; Goda, Y. Diterpenoid from *Salvia greggii*. *Phytochemistry* **2003**, *63*, 859–862. [CrossRef]

45. Kawahara, N.; Tamura, T.; Inoue, M.; Hosoe, T.; Kawai, K.I.; Sekita, S.; Satake, M.; Goda, Y. Diterpenoid glucosides from *Salvia greggii*. *Phytochemistry* **2004**, *65*, 2577–2581. [CrossRef] [PubMed]

46. Pereira, O.R.; Catarino, M.D.; Afonso, A.F.; Silva, A.M.S.; Cardoso, S.M. *Salvia elegans*, *Salvia greggii* and *Salvia officinalis* decoctions: Antioxidant activities and inhibition of carbohydrate and lipid metabolic enzymes. *Molecules* **2018**, *23*, 3169. [CrossRef]

47. Mora, S.; Millán, R.; Lungenstrass, H.; Díaz-Véliz, G.; Morán, J.A.; Herrera-Ruiz, M.; Tortoriello, J. The hydroalcoholic extract of *Salvia elegans* induces anxiolytic- and antidepressant-like effects in rats. *J. Ethnopharmacol.* **2006**, *106*, 76–81. [CrossRef] [PubMed]

48. Herrera-Ruiz, M.; Garcia-Beltran, Y.; Mora, S.; Diaz-Veliz, G.; Viana Glauce, S.B.; Tortoriello, J.; Ramirez, G. Antidepressant and anxiolytic effects of hydroalcoholic extract from *Salvia elegans*. *J. Ethnopharmacol.* **2006**, *107*, 53–58. [CrossRef] [PubMed]

49. Makino, T.; Ohno, T.; Iwbuchi, H. Aroma components of pineapple sage (*Salvia elegans* Vahl). *Foods Food Ingred. J. Jpn.* **1996**, *169*, 121–124.

50. Neisess, K.R.; Scora, R.W.; Kumamoto, J. Volatile leaf oils of California salvias. *J. Nat. Prod.* **1987**, *50*, 515–517. [CrossRef]

51. Gurjar, M.; Ali, S.; Akhtar, M.; Singh, K. Efficacy of plant extracts in plant disease management. *Agric. Sci.* **2012**, *3*, 425–433. [CrossRef]

52. Pane, C.; Villecco, D.; Zaccardelli, M. Combined use of *Brassica carinata* seed meal, thyme oil and a *Bacillus amyloliquefaciens* strain for controlling three soil-borne fungal plant diseases. *J. Plant. Pathol.* **2017**, *99*, 77–84.

53. Marchev, A.; Haas, C.; Schulz, S.; Georgiev, V.; Steingroewer, J.; Bley, T.; Pavlov, A. Sage *in vitro* cultures: A promising tool for the production of bioactive terpenes and phenolic substances. *Biotechnol. Lett.* **2014**, *36*, 211–221. [CrossRef]

54. Zeng, S.L.; Duan, L.; Chen, B.Z.; Li, P.; Liu, E.H. Chemicalome and metabolome profiling of polymethoxylated flavonoids in *Citri Reticulatae Pericarpium* based on an integrated strategy combining background subtraction and modified mass defect filter in a Microsoft Excel Platform. *J. Chromatogr. A* **2017**, *1508*, 106–120. [CrossRef]

55. Kavoura, D.; Kyriakopoulou, K.; Papaefstathiou, G.; Spanidi, E.; Gardikis, K.; Louli, V.; Aligiannis, N.; Krokida, M.; Magoulas, K. Supercritical CO_2 extraction of *Salvia fruticose*. *J. Supercrit. Fluids* **2019**, *146*, 159–164. [CrossRef]

56. Šulniūtė, V.; Ragažinskienė, O.; Venskutonis, P.R. Comprehensive evaluation of antioxidant potential of 10 salvia species using high pressure methods for the isolation of lipophilic and hydrophilic plant fractions. *Plant. Foods Hum. Nutr.* **2016**, *71*, 64–71. [CrossRef]

57. Nutrizio, M.; Kljusurić, J.G.; Sabolović, M.B.; Kovačević, D.B.; Šupljika, F.; Putnik, P.; Čakić, M.S.; Dubrović, I.; Vrsaljko, D.; Maltar-Strmečki, N.; et al. Valorization of sage extracts (*Salvia officinalis* L.) obtained by high voltage electrical discharges: Process control and antioxidant properties. *Innov. Food Sci. Emerg. Technol.* **2020**, *60*, 102284. [CrossRef]

58. Abdelkaderm, M.; Ahcen, B.; Rachid, D.; Hakim, H. Phytochemical study and biological activity of sage (*Salvia officinalis* L.). *Int. J. Sch. Sci. Res. Innov.* **2014**, *8*, 1253–1287.

59. Sotiropoulou, N.S.; Megremi, S.F.; Tarantilis, P. Evaluation of antioxidant activity, toxicity, and phenolic profile of aqueous extracts of chamomile (*Matricaria chamomilla* L.) and sage (*Salvia officinalis* L.) prepared at different temperatures. *Appl. Sci.* **2020**, *10*, 2270. [CrossRef]

60. Lu, Y.; Foo, L.Y. Antioxidant activities of polyphenols from sage (*Salvia officinalis*). *Food Chem.* **2001**, *75*, 197–202. [CrossRef]

61. Salari, S.; Bakhshi, T.; Sharififar, F.; Naseri, A.; Ghasemi Nejad Almani, P. Evaluation of antifungal activity of standardized extract of Salvia rhytidea Benth. (*Lamiaceae*) against various Candida isolates. *J. Mycol. Méd.* **2016**, *26*, 323–330. [CrossRef]

62. Katanić Stanković, J.S.; Srećković, N.; Mišić, D.; Gašić, U.; Imbimbo, P.; Monti, D.M.; Mihailović, V. Bioactivity, biocompatibility and phytochemical assessment of lilac sage, *Salvia verticillata* L. (*Lamiaceae*)—A plant rich in rosmarinic acid. *Ind. Crops Prod.* **2020**, *143*, 111932. [CrossRef]

63. Ahmad, H.; Matsubara, Y. Antifungal effect of *Lamiaceae* herb water extracts against Fusarium root rot in *Asparagus*. *J. Plant. Dis. Prot.* **2020**, *127*, 229–236. [CrossRef]

64. Ahmad, H.; Matsubara, Y. Suppression of Fusarium wilt in cyclamen by using sage water extract and identification of antifungal metabolites. *Australas. Plant. Pathol.* **2020**, *49*, 213–220. [CrossRef]

65. Exarchou, V.; Kanetis, L.; Charalambous, Z.; Apers, S.L.; Pieters, V.; Gekas, V.; Goulas, V. HPLC-SPE-NMR characterization of major metabolites in *Salvia fruticosa* Mill. extract with antifungal potential: Relevance of carnosic acid, carnosol, and hispidulin. *J. Agric. Food Chem.* **2015**, *63*, 457–463. [CrossRef] [PubMed]

66. Lee, S.O.; Choi, G.J.; Jang, K.S.; Lim, H.K.; Cho, K.Y.; Kim, J.C. Antifungal activity of five plant essential oils as fumigant against postharvest and soilborne plant pathogenic fungi. *Plant. Pathol.* **2007**, *23*, 97–102. [CrossRef]

67. Baka, Z.A.M. Antifungal activity of extracts from five Egyptian wild medicinal plants against late blight disease of tomato. *Arch. Phytopathol. Plant. Prot.* **2014**, *47*, 1988–2002. [CrossRef]

68. Nkomo, M.M.; Katerere, D.D.R.; Vismer, H.H.; Cruz, T.T.; Balayssac, S.S.; Malet-Martino, M.M.; Makunga, N.N.P. Fusarium inhibition by wild populations of the medicinal plant *Salvia africana-lutea* L. linked to metabolomic profiling. *BMC Complement. Altern. Med.* **2014**, *14*, 99. [CrossRef]

69. Yilar, M.; Kadioglu, I. Antifungal activities of some *Salvia* apecies extracts on *Fusarium oxysporum* f. sp. *radicis-lycopersici* (Forl) mycelium growth *in-vitro*. *Egypt. J. Biol. Pest Control* **2016**, *26*, 115–118.

70. Yilar, M.; Izzet, K.; Telci, I. Chemical composition and antifungal activity of *Salvia officinalis* (L.), *S. cryptantha* (Montbret *et aucher ex* Benth.), *S. tomentosa* (Mill.) plant essential oils and extracts. *Fresenius Environ. Bull.* **2018**, *27*, 1695–1706.

71. Goussous, S.J.; Abu el-Samen, F.M.; Tahhan, R.A. Antifungal activity of several medicinal plants extracts against the early blight pathogen (*Alternaria solani*). *Arch. Phytopathol. Plant. Prot.* **2010**, *43*, 1745–1757. [CrossRef]

72. Kanetis, L.; Exarchou, V.; Charalambous, Z.; Goulas, V. Edible coating composed of chitosan and *Salvia fruticosa* Mill. extract for the control of grey mould of table grapes. *J. Sci. Food Agric.* **2017**, *97*, 452–460. [CrossRef]

73. Goussous, S.J.; Mas'ad, I.S.; Abu El-Samen, F.M.; Tahhan, R.A. *In vitro* inhibitory effects of rosemary and sage extracts on mycelial growth and sclerotial formation and germination of *Sclerotinia sclerotiorum*. *Arch. Phytopathol. Plant Prot.* **2013**, *46*, 890–902. [CrossRef]

74. Özcan, M.M.; AL Juhaimi, F.Y. Antioxidant and antifungal activity of some aromatic plant extracts. *J. Med. Plant. Res.* **2011**, *5*, 1361–1366.

75. Nikolova, M.; Yordanov, P.; Slavov, S.; Berkov, S. Antifungal activity of plant extracts against phytopathogenic fungi. *J. Biosci. Biotechnol.* **2017**, *6*, 155–161.

76. Pahlaviani, M.R.M.K.; Darsanaki, R.K.; Bidarigh, S. Antimicrobial activities of some plants extracts against phytopathogenic fungi and clinical isolates in Iran. *J. Med. Bacteriol.* **2018**, *7*, 5–16.

77. Dagostin, S.; Formolo, T.; Giovannini, O.; Pertot, I. *Salvia officinalis* extract can protect grapevine against *Plasmopara viticola*. *Plant. Dis.* **2010**, *94*, 575–580. [CrossRef] [PubMed]

78. Pansera, M.R.; Pauletti, M.; Fedrigo, C.P.; Camatii Sartori, V.; Da Silva Ribeiro, R.T. Utilization of essential oil and vegetable extracts of *Salvia officinalis* L. in the control of rot Sclerotinia in lettuce. *Braz. J. Appl. Technol. Agric. Sci.* **2013**, *6*, 83–88.

79. Scherf, A.; Schuster, C.; Marx, P.; Gärber, U.; Konstantinidou-Doltsinis, S.; Schmitt, A. Control of downy mildew (*Pseudoperonospora cubensis*) of greenhouse grown cucumbers with alternative biological agents. *Commun. Agric. Appl. Biol. Sci.* **2010**, *75*, 541–554. [PubMed]

80. Dellavalle, P.D.; Cabrera, A.; Alem, D.; Larrañaga, P.; Ferreira, F.; Rizza, M.D. Antifungal activity of medicinal plant extracts against phytopathogenic fungus *Alternaria* spp. *Chil. J. Agric. Res.* **2011**, *71*, 231–239. [CrossRef]

81. Chudasama, K.S.; Thaker, V.S. Biological control of phytopathogenic bacteria *Pantoea agglomerans* and *Erwinia chrysanthemy* using 100 essential oils. *Arch. Phytopathol. Plant. Protect.* **2014**, *47*, 2221–2232. [CrossRef]

82. Khosravi, D.N.; Ostad, S.N. Cytotoxic activity of the essential oil of *Salvia verticillata* L. *Res. J. Pharmacogn.* **2014**, *1*, 7–33.

83. Bayar, Y.; Yilar, M. The antifungal and phytotoxic effect of different plant extracts of *Salvia virgata* Jacq. *Fresenius Environ. Bull.* **2019**, *28*, 3492–3497.

84. Dulger, B.; Hacioglu, N. Antifungal activity of endemic *Salvia tigrina* in Turkey. *Trop. J. Pharm. Res.* **2008**, *7*, 1051–1054. [CrossRef]

85. Šrobárová, A.; Eged, S.; Teixeira Da Silva, J.; Ritieni, A.; Santini, A. The use of Bacillus subtilis for screening Fusaric Acid production by *Fusarium* spp. *Czech J. Food Sci.* **2009**, *27*, 203–209. [CrossRef]

86. Nesic, K.; Ivanovic, S.; Nesic, V. Fusarial toxins: Secondary metabolites of Fusarium fungi. *Rev. Environ. Cont. Toxicol.* **2014**, *228*, 101–120.

87. Mikušová, P.; Sulyok, M.; Samtini, A.; Šrobárová, A. *Aspergillus* spp. and their secondary metabolite production in grape berries from Slovakia. *Phytopathol. Mediterr.* **2014**, *53*, 109–114.

88. Samtini, A.; Mikušová, P.; Sulyok, M.; Krska, R.; Labuda, R.; Šrobárová, A. Penicillium strains isolated from Slovak grape berries taxonomy assessment by secondary metabolite profile. *Mycotoxin Res.* **2014**, *30*, 213–220.

89. Sakhri, A.; Chaouche, N.K.; Catania, M.R.; Ritieni, A.; Santini, A. Chemical Composition of Aspergillus creber Extract and Evaluation of its Antimicrobial and Antioxidant Activities. *Pol. J. Microbiol.* **2019**, *68*, 309–316. [CrossRef]

90. Yuce, E.; Yildirim, N.; Yildirim, N.C.; Paksoy, M.Y.; Bagci, E. Essential oil composition, antioxidant and antifungal activities of *Salvia sclarea* L. from Munzur Valley in Tunceli, Turkey. *Cell. Mol. Biol.* **2014**, *60*, 1–5.

91. Ghaedi, M.; Naghiha, R.; Jannesar, R.; Dehghanian, N.; Mirtamizdoust, B.; Pezeshkpour, V. Antibacterial and antifungal activity of flower extracts of *Urtica dioica*, *Chamaemelum nobile* and *Salvia officinalis*: Effects of Zn[OH]$^-_2$ nanoparticles and Hp-2-minh on their property. *J. Ind. Eng. Chem.* **2015**, *32*, 353–359. [CrossRef]

92. Salević, A.; Prieto, C.; Cabedo, L.; Nedović, V.; Lagaron, J.M. Physicochemical, antioxidant and antimicrobial properties of electrospun poly(ε-caprolactone) films containing a solid dispersion of sage (*Salvia officinalis* L.) extract. *Nanomaterials* **2019**, *9*, 270.

93. Soković, M.; Grubišić, D.; Ristić, M. Chemical composition and antifungal activities of essential oils from leaves, calyx and corolla of *Salvia brachyodon* Vandas. *J. Essent. Oil Res.* **2005**, *17*, 227–229. [CrossRef]

94. Karpiński, T.M. Essential oils of *Lamiaceae* family plants as antifungal. *Biomolecules* **2020**, *10*, 103. [CrossRef]

95. Abu-Darwish, M.S.; Cabral, C.; Ferreira, I.V.; Gonçalves, M.J.; Cavaleiro, C.; Cruz, M.T.; Al-Bdour, T.H.; Salgueiro, L. Essential oil of common sage (*Salvia officinalis* L.) from Jordan: Assessment of safety in mammalian cells and its antifungal and anti-inflammatory potential. *Biomed. Res. Int.* **2013**. [CrossRef] [PubMed]

96. Redondo-Blanco, S.; Fernández, J.; López-Ibáñez, S.; Miguélez, E.M.; Villar, C.J.; Lombó, F. Plant phytochemicals in food preservation: Antifungal bioactivity: A review. *J. Food Prot.* **2020**, *83*, 163–171. [CrossRef]

97. Stappen, I.; Tabanca, N.; Ali, A.; Wanner, J.; Lal, B.; Jaitak, V.; Wedge, D.E.; Kaul, V.K.; Schmidt, E.; Jirovetz, L. Antifungal and repellent activities of the essential oils from three aromatic herbs from western Himalaya. *Open Chem.* **2018**, *16*, 306–316. [CrossRef]

98. Džamić, A.; Soković, M.; Ristić, M.; Grujić-Jovanović, S.; Vukojević, J.; Marin, P.D. Chemical composition and antifungal activity of *Salvia sclarea* (*Lamiaceae*) essential oil. *Arch. Biol. Sci.* **2008**, *60*, 233–237. [CrossRef]

99. Pitarokili, D.; Tzakou, O.; Couladis, M.; Verykokidou, E. Composition and antifungal activity of the essential oil of *Salvia pomifera* subsp. *calycina* growing wild in Greece. *J. Essent. Oil Res.* **1999**, *11*, 655–659. [CrossRef]

100. Glamočlija, J.; Soković, M.; Vukojević, J.; Milenković, I.; Van Griensven, L.J.L.D. Chemical composition and antifungal activities of essential oils of *Satureja thymbra* L. and *Salvia pomifera* ssp. *calycina* (Sm.) Hayek. *J. Essent. Oil Res.* **2006**, *18*, 115–118. [CrossRef]

101. Ghasemi, E.; Sharafzadeh, S.; Amiri, B.; Alizadeh, A.; Bazrafshan, F. Variation in essential oil constituents and antimicrobial activity of the flowering aerial parts of *Salvia mirzayanii* Rech. & Esfand. Ecotypes as a folkloric herbal remedy in Southwestern Iran. *J. Essent. Oil Bear. Plants* **2020**, *23*, 51–64.

102. Soković, M.; Griensven, L.J.L.D.V. Antimicrobial activity of essential oils and their components against the three major pathogens of the cultivated button mushroom, *Agaricus bisporus*. *Eur. J. Plant. Pathol.* **2006**, *116*, 211–224. [CrossRef]

103. Pinto, E.; Salgueiro, L.R.; Cavaleiro, C.; Palmeira, A.; Gonçalves, M.J. *In vitro* susceptibility of some species of yeasts and filamentous fungi to essential oils of *Salvia officinalis*. *Ind. Crops Prod.* **2007**, *26*, 135–141. [CrossRef]

104. Grande-Tovar, C.D.; Chaves-Lopez, C.; Serio, A.; Rossi, C.; Paparella, A. Chitosan coatings enriched with essential oils: Effects on fungi involved in fruit decay and mechanisms of action. *Trends Food Sci. Technol.* **2018**, *78*, 61–71. [CrossRef]

105. Souza, D.P.; Pimentel, R.B.Q.; Santos, A.S.; Albuquerque, P.M.; Fernandes, A.V.; Junior, S.D.; Oliveira, J.T.A.; Ramos, M.V.; Rathinasabapathi, B.; Gonçalvesa, J.F.C. Fungicidal properties and insights on the mechanisms of the action of volatile oils from Amazonian Aniba trees. *Ind. Crops Prod.* **2020**, *143*, 111914. [CrossRef]

106. Lengai, G.M.W.; Muthomi, J.W.; Mbega, E.R. Phytochemical activity and role of botanical pesticides in pest management for sustainable agricultural crop production. *Sci. Afr.* **2020**, *7*, e00239. [CrossRef]

107. Chen, C.J.; Li, Q.Q.; Zeng, Z.Y.; Duan, S.S.; Wang, W.; Xu, F.R.; Cheng, Y.X.; Dong, X. Efficacy and mechanism of *Mentha haplocalyx* and *Schizonepeta tenuifolia* essential oils on the inhibition of *Panax notoginseng* pathogens. *Ind. Crops Prod.* **2020**, *145*, 112073. [CrossRef]

108. Cui, H.; Zhang, X.; Zhou, H.; Zhao, C.; Lin, L. Antimicrobial activity and mechanisms of *Salvia sclarea* essential oil. *Bot. Stud.* **2015**, *56*, 16. [CrossRef] [PubMed]

109. Fraternale, D.; Giamperi, L.; Bucchini, A.; Ricci, D.; Epifano, F.; Genovese, S.; Curini, M. Composition and antifungal activity of essential oil of *Salvia sclarea* from Italy. *Chem. Nat. Comp.* **2005**, *41*, 604–606. [CrossRef]

110. Soković, M.D.; Brkić, D.D.; Džamić, A.M.; Ristić, M.S.; Marin, P.D. Chemical composition and antifungal activity of *Salvia desoleana* Atzei & Picci essential oil and its major components. *Flavour Fragr. J.* **2009**, *24*, 83–87.

111. Morcia, C.; Malnati, M.; Terzi, V. In vitro antifungal activity of terpinen-4-ol, eugenol, carvone, 1,8-cineole (eucalyptol) and thymol against mycotoxigenic plant pathogens. *Food Addit. Contam. Part A* **2012**, *29*, 415–422.

112. Silva, K.V.S.; Lima, M.I.O.; Cardoso, G.N.; Santos, A.S.; Silva, G.S.; Pereira, F.O. Inibitory effects of linalool on fungal pathogenicity of clinical isolates of *Microsporum canis* and *Microsporum gypseum*. *Mycoses* **2017**, *60*, 387–393. [CrossRef]

113. Yu, D.; Wang, J.; Shao, X.; Xu, F.; Wang, H. Antifungal modes of action of tea tree oil and its two characteristic components against *Botrytis cinerea*. *J. Appl. Microbiol.* **2015**, *119*, 1253–1262. [CrossRef]

114. Woo, H.J.; Yang, J.Y.; Lee, M.H.; Kim, H.W.; Kwon, H.J.; Park, M.; Kim, S.K.; Park, S.Y.; Kim, S.H.; d Kim, J.B. Inhibitory effects of β-caryophyllene on *Helicobacter pylori* infection *in vitro* and *in vivo*. *Int. J. Mol. Sci.* **2020**, *21*, 1008. [CrossRef]

115. Digrak, M.; Alma, M.H.; Ilcim, A.; Sen, S. Antibacterial and antifungal effects of various commercial plant extracts. *Pharm. Biol.* **1999**, *37*, 216–220. [CrossRef]

116. Peana, A.; Moretti, M.D.; Julidano, C. Chemical composition and antimicrobial action of the essential oils of *Salvia desoleana* and *S. sclarea*. *Planta Med.* **1999**, *5*, 752–754. [CrossRef] [PubMed]

117. Ferdes, M.; Al Juhaimi, F.; Özcan, M.M.; Ghafoor, K. Inhibitory effect of some plant essential oils on growth of *Aspergillus niger, Aspergillus oryzae, Mucor pusillus* and *Fusarium oxysporum*. *S. Afr. J. Bot.* **2017**, *113*, 457–460. [CrossRef]

118. Starovic, M.; Ristic, D.; Pavlovic, S.; Ristic, M.; Stevanovic, M.; Aljuhaimi, F.; Svetlana, N.; Ozcan, M.M. Antifungal activities of different essential oils against anise seeds mycopopulations. *J. Food Saf. Food Qual.* **2016**, *67*, 61–92.

119. Daferera, D.J.; Ziogas, B.N.; Polissiou, M.G. The effectiveness of plant essential oils on the growth of *Botrytis cinerea, Fusarium* sp. and *Clavibacter michiganensis* subsp. *Michiganensis*. *Crop. Prot.* **2003**, *22*, 39–44. [CrossRef]

120. Pitarokili, D.; Tzakou, O.; Loukis, A.; Harvala, C. Volatile metabolites from *Salvia fruticosa* as antifungal agents in soilborne pathogens. *J. Agric. Food Chem.* **2003**, *51*, 3294–3301. [CrossRef] [PubMed]

121. Kotan, R.; Kordali, S.; Cakir, A.; Kesdek, M.; Kaya, Y.; Kilic, H. Antimicrobial and insecticidal activities of essential oil isolated from Turkish Salvia hydrangea DC. ex Benth. *Biochem. Syst. Ecol.* **2008**, *36*, 360–368. [CrossRef]

122. Cosic, J.; Vrandecic, K.; Postic, J.; Jurkovic, D.; Ravlic, M. In vitro antifungal activity of essential oils on growth of phytopathogenic fungi. *Polyjoprivreda* **2010**, *16*, 25–28.

123. Lu, M.; Han, Z.; Xu, Y.; Yao, L. Effects of essential oils from Chinese indigenous aromatic plants on mycelial growth and morphogenesis of three phytopathogens. *Flavour Fragr. J.* **2013**, *28*, 84–92. [CrossRef]

124. Elshafie, H.S.; Sakr, S.; Mang, S.M.; Belviso, S.; De Feo, V.; Camele, I. Antimicrobial activity and chemical composition of three essential oils extracted from mediterranean aromatic plants. *J. Med. Food* **2016**, *19*, 1–8. [CrossRef]

125. Carta, C.; Moretti, M.D.L.; Peana, A.T. Activity of the oil of *Salvia officinalis* L. against *Botrytis cinerea*. *J. Essent. Oil Res.* **1996**, *8*, 399–404. [CrossRef]

126. Arici, S.E.; Sanli, A. Effect of some essential oils against *Rhizoctonia solani* and *Streptomycetes scabies* on potato plants in field conditions. *Annu. Res. Rev. Biol.* **2014**, *4*, 2027–2036. [CrossRef]

127. Khedher, M.R.B.; Khedher, S.B.; Chaieb, I.; Tounsi, S.; Hammami, M. Chemical composition and biological activities of *Salvia officinalis* essential oil from Tunisia. *EXCLI J.* **2017**, *16*, 160–173.

128. Palfi, M.; Konjevoda, P.; Vrandečić, K.; Ćosić, J. Antifungal activity of essential oils on mycelial growth of *Fusarium oxysporum* and *Bortytis cinerea*. *Emir. J. Food Agric.* **2019**, *31*, 544–554. [CrossRef]

129. Tanovic, B.; Potocnik, I.; Delibasic, G.; Ristic, M.; Kostic, M.; Markovic, M. *In vitro* effect of essential oils from aromatic and medicinal plants on mushroom pathogens: *Verticillium fungicola* var. *fungicola*, *Mycogone perniciosa*, and *Cladobotryum* sp. *Arch. Biol. Sci.* **2006**, *61*, 231–237. [CrossRef]

130. Pawar, V.C.; Thaker, V.S. Evaluation of the anti-*Fusarium oxysporum* f. sp. *cicer* and anti-*Alternaria porri* effects of some essential oils. *World J. Microbiol. Biotechnol.* **2007**, *23*, 1099–1106. [CrossRef]

131. Tomescu, A.; Sumalan, R.M.; Pop, G.; Alexa, E.; Poiana, M.A.; Copolovici, D.M.; Mihay, C.S.S.; Negrea, M.; Galuscan, A. Chemical composition and protective antifugal activity of *Mentha Piperita* L. and *Salvia Officinalis* L. essential oils against *Fusarium Graminearum* spp. *Rev. Chem.* **2015**, *66*, 1027–1030.

132. Arsalan, M.; Dervis, S. Antifungal activity of essential oils against three vegetative compatibility groups of *Verticillium dahlia*. *World J. Microbiol. Biotechnol.* **2010**, *26*, 1813–1821. [CrossRef]

133. Todorovic, B.; Potocnik, I.; Rekanovic, E.; Stepanovic, M.; Kostic, M.; Ristic, M.; Marcic, S.M. Toxicity of twenty-two plant essential oils against pathogenic bacteria of vegetables and mushrooms. *J. Environ. Sci. Health B* **2016**, 1–8. [CrossRef] [PubMed]

134. Salteh, S.A.; Arzani, K.; Omidbeige, R.; Safaie, N. Essential oils inhibit mycelial growth of *Rhizopus stolonifer*. *Eur. J. Hort. Sci.* **2010**, *75*, 278–282.

135. Moretti, M.D.L.; Peana, A.T.; Franceschini, A.; Carta, C. *In vivo* activity of *Salvia officinalis* oil against *Botrytis cinerea*. *J. Essent. Oil Res.* **1998**, *10*, 157–160. [CrossRef]

136. Bi, Y.; Jiang, H.; Hausbeck, M.K.; Hao, J.J. Inhibitory effects of essential oils for controlling *Phytophthora capsici*. *Plant. Dis.* **2012**, *96*, 797–803. [CrossRef] [PubMed]

137. Hoseini, S.; Amini, J.; Rafei, J.N.; Khorshidi, J. Inhibitory effect of some plant essential oils against strawberry anthracnose caused by *Colletotrichum nymphaeae* under in vitro and in vivo conditions. *Eur. J. Plant. Pathol.* **2019**, *155*, 1287–1302. [CrossRef]

138. Noscirvani, N.; Fasihi, H. Control of *Aspergilus niger* in vitro and in vivo by three Iranian essential oils. *Int. Food Res. J.* **2018**, *25*, 1745–1752.

139. Lopez-Reyes, J.G.; Spadaro, D.; Gullino, M.L.; Garibaldi, A. Efficacy of plant essential oils on postharvest control of rot caused by fungi on four cultivars of apples *in vivo*. *Flavour Fragr. J.* **2010**, *25*, 171–177. [CrossRef]

140. Lopez-Reyes, J.G.; Spadaro, D.; Prelle, A.; Garibaldi, A.; Gullino, M.L. Efficacy of plant essential oils on postharvest control of rots caused by fungi on different stone fruits *in vivo*. *J. Food Prot.* **2013**, *76*, 631–639. [CrossRef] [PubMed]

141. Pane, C.; Villecco, D.; Roscigno, G.; De Falco, E.; Zaccardelli, M. Screening of plant-derived antifungal substances useful for the control of seedborne pathogens. *Arch. Phytopathol. Plant. Prot.* **2013**, *46*, 1533–1539. [CrossRef]

142. Yilar, M.; Bayar, Y. Antifungal of essential oils of *Salvia officinalis* and *Salvia Tomentosa* plants on six different isolates of *Aschochyta rabie* (PASS) Labr. *Fresenius Environ. Bull.* **2019**, *28*, 2170–2175.

143. Goswami, S.; Kanval, J.; Prakash, O.; Kumar, R.; Rawat, D.S.; Srivastava, R.M.; Pant, A.K. Chemical composition, antioxidant, antifungal and antifeedant activity of the *Salvia reflexa* Hornem essential oil. *Asian J. Appl. Sci.* **2019**, *12*, 185–191.

144. Pitarokili, D.; Couladis, M.; Panayotarou, N.K.; Tzakou, O. Composition and antifungal activity on soil-borne pathogens of the essential oil of *Salvia sclarea* from Greece. *J. Agric. Food Chem.* **2002**, *50*, 6688–6691. [CrossRef] [PubMed]

145. Jularat, U.; Apinya, P.; Peerayot, K.K.; Pitipong, T. Antifungal properties of essential oils from Thai medial plants against rice phytopathogenic fungi. *Asian J. Food Ag-Ind.* **2009**, *2*, S24–S30.

146. Bahman, H.; Alireza, E.; Seyed Mehdi, M. Fumigant toxicity of essential oil from '*Salvia leriifolia*' (Benth) against two stored product insect pests. *Aust. J. Crop. Sci.* **2013**, *7*, 855–860.

147. Zaccardelli, M.; Roscigno, G.; Pane, C.; De Falco, E. Antifungal activity of residues from aromatic waters distilled from thyme and sage. In *Multidisciplinary Approaches for Studying and Combating Microbial Pathogens*; Mendez-Vilas, A., Ed.; Universal-Publishers: Irvine, CA, USA, 2015; pp. 31–33.

148. Feng, W.; Zheng, X. Essential oils to control *Alternaria alternata* in vitro and *in vivo*. *Food Control* **2006**, *18*, 1126–1130. [CrossRef]

149. Flores, F.C.; De Lima, J.A.; Ribeiro, R.F.; Alves, S.H.; Rolim, C.M.B.; Beck, R.C.R.; Bona da Silva, C. Antifungal activity of nanocapsule suspensions containing tea tree oil on the growth of *Trichophyton rubrum*. *Mycopathologia* **2013**, *175*, 281–286. [CrossRef]

150. Soliman, E.A.; El-Moghazy, A.Y.; Mohy El-Din, M.S.; Massoud, M.A. Microencapsulation of essential oils within alginate: Formulation and in vitro evaluation of antifungal activity. *J. Encapsulation Adsorption Sci.* **2013**, *3*. [CrossRef]

151. Nasseri, M.; Golmohammadzadeh, S.; Arouiee, H.; Jaafari, M.R.; Neamati, H. Antifungal activity of *Zataria multiflora* essential oil-loaded solid lipid nanoparticles in-vitro condition. *Iran. J. Basic Med. Sci.* **2016**, *19*, 1231–1237.

152. Kodadová, A.; Vitková, Z.; Herdová, P.; Ťažký, A.; Oremusová, J.; Grančai, D.; Mikuš, P. Formulation of sage essential oil (*Salvia officinalis*, L.) monoterpenes into chitosan hydrogels and permeation study with GC-MS analysis. *Drug Dev. Ind. Pharm.* **2015**, *41*, 1080–1088. [CrossRef]

153. Gharenaghadeh, S.; Karimi, N.; Forghani, S.; Nourazarian, M.; Gharehnaghadeh, S.; Jabbari, V.; Sowtikhiabani, M.; Kafil, H.S. Application of *Salvia multicaulis* essential oil-containing nanoemulsion against food-borne pathogens. *Food Biosci.* **2017**, *19*, 128–133. [CrossRef]

154. Zaccardelli, M.; Mancini, E.; Campanile, F.; De Feo, E.; De Falco, E. Identification of bio-active coumpounds in essential oils of medicinal plants toxic for phytopathogenic fungi and bacteria. *J. Plant. Pathol.* **2007**, *89*, 3.

155. Souto, E.B.; Silva, G.F.; Dias-Ferreira, J.; Zielinska, A.; Ventura, F.; Durazzo, A.; Lucarini, M.; Novellino, E.; Santini, A. Nanopharmaceutics: Part II—Production scales and clinically compliant production methods. *Nanomater* **2020**, *10*, 455. [CrossRef] [PubMed]

156. Jampílek, J.; Kráľová, K. Nanobiopesticides in agriculture: State of the art and future opportunities. In *Nano-Biopesticides Today and Future Perspectives*; Academic Press: Cambridge, MA, USA; Elsevier: Amsterdam, The Netherlands, 2019; Chapter 17; pp. 397–447. [CrossRef]

157. Sagiri, S.S.; Anis, A.; Pal, K. Review on encapsulation of vegetable oils: Strategies, preparation methods, and applications. *Polym. Plast. Technol.* **2016**, *55*, 291–311. [CrossRef]

158. De Matos, S.P.; Lucca, L.G.; Koester, L.S. Essential oils in nanostructured systems: Challenges in preparation and analytical methods. *Talanta* **2019**, *195*, 204–214. [CrossRef]

Article

Processed Baobab (*Adansonia digitata* L.) Food Products in Malawi: From Poor Men's to Premium-Priced Specialty Food?

Dietrich Darr [1,*], Chifundo Chopi-Msadala [2,*], Collins Duke Namakhwa [3], Kathrin Meinhold [1] and Chimuleke Munthali [2]

[1] Faculty of Life Sciences, Rhine Waal University of Applied Studies, D-47533 Kleve, Germany; kathrin.meinhold@hochschule-rhein-waal.de

[2] Faculty of Environmental Sciences, Mzuzu University, Private Bag 201, Mzuzu 2, Malawi; chimuleke.munthali@gmail.com

[3] Polytechnic, Faculty of Commerce, University of Malawi, P. O. Box 278, Zomba, Malawi; cnamakhwa@gmail.com

*   Correspondence: dietrich.darr@hochschule-rhein-waal.de (D.D.); chifundochopi@gmail.com (C.C.-M.)

Received: 29 May 2020; Accepted: 21 June 2020; Published: 23 June 2020

Abstract: The baobab tree (*Adansonia digitata* L.) is an important source of non-timber forest products in sub-Saharan Africa. Its fruits contain high amounts of vitamin C, calcium, and dietary fibre. In addition, other parts of the tree are traditionally used for human consumption, particularly during lean seasons. In line with the increasing demand for natural, healthy, and nutritious food products, the baobab has great potential to contribute to human nutrition and rural livelihoods. In Malawi, where demand for baobab has substantially increased within the last decade, baobab fruits are being processed into a variety of food and non-food products, such as fruit juice, ice-lollies, sweets, and cosmetics. Yet, information on the sociodemographic background and quality preferences of baobab consumers is scanty. The current study, therefore, aimed to (1) map the diversity of baobab products available in Malawi; (2) determine consumer segments and their preferences for the most common baobab food products; and (3) examine the contribution of major attributes of processed baobab food products on their price. We employed a mixed-methods approach including the analysis of 132 baobab products and a survey of 141 consumers in formal and informal retail outlets, adopting multistage and purposive sampling. Qualitative and quantitative data were analysed using cluster analysis, cross tabulation, and hedonic regression. Results pointed to two distinct consumer segments for baobab food products, largely following the formal–informal product divide currently existing in Malawi. Both segments clearly differed with regard to preferred product attributes. We also showed that extrinsic product attributes such as packaging quality, labelling, conformity with food standards, or health claims provided distinct differentiation potential for baobab food manufacturers. In addition to providing empirical evidence for the transition of baobab food products into higher-value market segments, our results can help food processing enterprises to improve the composition and marketing of their baobab products.

Keywords: food processing industry; wild edible plants; neglected and underutilized species (NUS); Africa; urban consumers; marketing; product differentiation

---

1. Introduction

Wild edible plants collected from trees and forests are an important source of nutrition and income for communities in many parts of the world. Studies show that food products are among the most important product categories collected from forest and non-forest ecosystems, and they account for approximately 30% of the total value of products derived from the environment [1]. Large parts of rural populations in Africa regularly collect and consume wild edible plants and generate substantial parts of their income from these resources [2–4]. While food from agricultural production nevertheless remains the prime source of energy and nutrition in most cases, wild food can make an important contribution to household nutrition and well-being in times of food scarcity and economic hardship.

Given the fact that wild edible plants are typically available from public or private lands at low cost, their use and consumption has primarily been associated with poor population strata [5]. Some authors also stress the importance of traditional knowledge of the ecology for use of these plants and hence associate their consumption primarily with elder and more traditional population groups [6]. A number of studies expand on the previous literature that primarily focused on rural contexts by investigating the contribution of wild food to urban and peri-urban populations in Africa. For example, Asase and Kumordzie [7] report that indigenous vegetables cultivated or collected from the wild were highly popular among urban residents in Ghana due to their availability, affordability, and medicinal value. Catarino et al. [8] found that five species of wild and semi-cultivated leafy vegetables including the dried leaves of *Adansonia digitata* and *Bombax costatum* were available at the largest food market in the capital city of Guinea-Bissau during the dry season. A study in Kampala, Uganda, found that wild plants were commonly used and mainly collected by residents with lower income [9]. A recent study in six African cities reported that wild food and medicine remain an integral part of diets for poor urban inhabitants [10]. These and similar accounts accentuate the importance of wild indigenous food as a supplementary diet and source of income for poor population strata in light of widespread rural and urban poverty.

At the same time, there is increasing evidence that wild food products are also an integral component of the diets of wealthier consumer segments. While poor households tend to spend a higher proportion of their income on wild food, wealthy households can afford to consume more of these products in absolute terms. Likewise, the consumption of wild food considered to be luxury products such as bushmeat is often more common among higher-income population strata [11]. In addition to this, economic growth and the emergence of an urban middle class have created a group of consumers who largely purchase rather than produce or collect the food products they consume [10,12]. Given culturally rooted food preferences and eating habits, wild edible plants remain an important part of their diets. At the same time, an increasing number of urban consumers in Africa seems to appreciate the health and cultural benefits of wild indigenous food [13]. As a consequence, wild food and medicine are commonly consumed by higher-income urban populations in many African cities [10]. Aworh [14] reported of the growing acceptance of indigenous leafy vegetables among the urban elite in Kenya and elsewhere in Africa, as indicated by their increasing availability in fine dining restaurants and large urban supermarkets. In a similar vein, Garekae and Shackleton [15] reported that wild edible plants were not perceived as "food for the poor" by the majority of urban residents in two towns in South Africa. However, studies show that African traditional food products often remain poorly integrated into formal markets, supply chains, and retail outlets to date, as many of these products are mostly sold by street vendors or at fresh markets rather than by shopkeepers [16,17]. Moreover, Cloete and Idsardi [6] demonstrated that there are distinct differences between lower and higher income groups with regard to the particular wild edible plant species consumed. This underlines the ongoing differentiation and fragmentation of consumer segments with their respective tastes and product preferences, and the need for detailed and case-specific analyses.

The baobab (*Adansonia digitata* L.) is an indigenous fruit tree species widely occurring in the semi-arid parts of Africa. Due to its potential contribution to food security and household well-being it has been recommended as a priority species for domestication and commercialization [18,19]. Its fruit pulp contains high amounts of carbohydrate, dietary fibre, vitamin C, calcium, magnesium, and potassium [20,21]. Besides its fruits, various other parts of the tree are important for a diversity of traditional uses including food, fodder, medicine, cosmetics, and craft products [22,23]. Consequently, the importance of the baobab tree for household nutrition, food security, and income, particularly of the disadvantaged rural populations, has been highlighted in the literature, e.g., [24,25]. In Malawi, where demand for baobab has substantially increased within the last decade, the baobab fruits are being processed by both formal and informal enterprises to produce a variety of food and non-food products, such as fruit juice, ice-lollies, sweets, and cosmetics. Yet, little is known about the customers who purchase processed baobab food products, their preferred quality attributes, and the impact of these attributes on product price. To contribute to improved baobab commercialization, this paper aimed to (1) map the diversity of baobab products currently available in formal and informal markets in Malawi; (2) identify and characterize the main consumer segments and their preferences; and (3) determine the factors affecting the price of the two most commonly available baobab food products, i.e., baobab pulp powder and baobab juice. It is expected that our results can contribute to promoting the further commercialization and value-added processing of baobab fruit pulp in Malawi and other countries with significant resources of baobab and a developing baobab-processing sector.

2. Materials and Methods

Data were collected from nine districts, including the three major cities of Malawi, namely Mangochi, Blantyre, Mwanza, Balaka, Ntcheu, Salima, Lilongwe, Mzuzu, and Karonga (Figure 1). These districts were purposively selected as they (a) represent the centres of *Adansonia digitata* L. production and processing in Malawi, and, hence, a variety of processed products was known to be sold in these locations; (b) cover the entire north–south expansion of the country; and (c) represent a diversity of environmental and socio-economic conditions along the urban to peri-urban to rural continuum. A mixed-methods approach was employed to study consumer preferences and product characteristics. Data from 141 consumers and 132 products were collected in formal and informal retail outlets, such as supermarkets, hypermarkets, specialty stores, mini-marts, filling stations, open markets, hospital canteens, religious centres, and roadside vendors using structured questionnaires. All subjects gave their informed consent for inclusion before participating in the study. The study was conducted in accordance with the approval by the Malawi National Commission for Science and Technology (protocol number P.10/17/216). While retail outlets were purposively selected, the consumer survey was carried out in 20 retail outlets using a quota sampling procedure [26]. The samples were continuously expanded using snowballing until the target sample size was reached. To determine the consumer segments, two-step cluster analysis was used as a classification tool. Cross-tabulation and Cramer's V were used to identify relationships between demographic factors and preferred quality attributes using the SPSS software version 20. Multiple regression method was used to analyse quantitative data using a stepwise regression procedure to identify the best set of quality variables predicting the dependant variable, price.

Figure 1. Location of the study districts.

3. Results

3.1. Product Categories

Seventy-eight processed baobab products were identified in the investigated formal and informal retail outlets (Figure 2). They comprise a number of food products made of baobab pulp, as well as cosmetics products mainly made of baobab seed oil, which is not recommended for use in food products due to its content of cyclopropenoic fatty acids [27]. Sixteen different product types were found to be characterized as fruit drinks (bottled and frozen juice, i.e., ice lollies), pulp powders, sweets, and personal care products (soaps and oils). Nine of 16 product types were traded through the informal outlets, whilst formal retail outlets were selling the complete set of 16 product types. Table 1 provides a brief overview of the baobab products currently available in the Malawian market. In line with our prime interest in baobab food products, we disregarded the available cosmetics products in all further analyses.

Figure 2. Examples of baobab products from the Malawian formal and informal markets. **1A–D**: Baobab juice products; **1E**: formally processed baobab ice-lollies; **2A**: Informal baobab juice packaged in recycled plastic bottles; **2B**: Baobab sweets sold in informal outlets; **2C,D**: Informally processed baobab ice lollies; **2E**: Packaged pulp in sacks for wholesale. **3A,F**: Baobab oil; **3B**: Baobab soap; **3C**: Assorted lip balm; **3D**: Baobab Spongi soap; **3E**: Baobab body cream sold in high-end shops; **4A,C,E**: Baobab pulp powder sold in pharmacies; **4B**: Baobab coffee powder made from seeds; **4D,F**: Baobab jam; **5A**: Baobab sweets sold in supermarkets; **5B**: Baobab chocolate sold in high-end flea markets; **5C**: Baobab sweets sold in high-end markets; **5D**: Baobab pulp powder sold in high-end flea markets.

The investigated baobab products differed widely in terms of their extrinsic and intrinsic product attributes. For example, baobab juice products made by formal food processing companies were available in 250 mL, 500 mL, and 1 L polyethylene terephthalate (PET) plastic bottles of proprietary design with brand and nutrition labels (Figure 2(1A–D)) and sold widely through various retail outlets. Baobab juice made by informal food processing enterprises was sold in recycled 500 mL plastic bottles without labels (Figure 2(2A)). Likewise, baobab fruit pulp powder was sold in plastic bags or bottles with health-claim labelling in pharmacies (Figure 2(4A,C,E)) and packaged or in bulk without health-claim labelling in the informal market (Figure 2(2E,5D)).

Table 1. Characteristics of processed baobab products in Malawi.

Product Type	Number of Products Identified	Main Product Features	Estimated Baobab Content	Main Retail Outlet	Price (MKW)
Baobab fruit powder	25	Packaged or in bulk, some products with organic certification.	100%	Open markets, street vendors, supermarkets.	800–3500 (500 g)
Malambe sweets	12	Coloured sweetened pulp pieces, packaged in plastic bags (15–70 g).	90%–100%	Schools, supermarkets.	10 (15 g)
Baobab ice-lollies	11	Frozen baobab juice raw or sweetened and coloured, packaged in plastic tubes (35–50 g).	20%–30%	Street side vendors, churches, informal markets, schools.	10 (10 g)
Baobab juice, bottled	7	Packaged in 250–1000 mL PET or recycled PET bottles, with or without MBS certification.	40%–60%	Supermarkets, filling stations.	500–750 (500 mL)
Baobab coffee powder	3	Packaged in branded plastic jars (180–200 g).	100%	Supermarkets, pharmacies.	1800–3500 (200 g)
Baobab jam	3	Packaged in branded plastic jars (350–500 g).	15%–20%	Supermarkets	1900–2500 (500 g)
Baobab pure oil	3	Packaged in 100 mL glass or PET bottles.	100%	Supermarkets, pharmacies.	2000–3500 (50 mL)
Baobab lip balm	3	Packaged in wooden cases (12 g).	100%	Specialty shop, flea market.	2000–3500 (12 g)
Baobab soap	2	Packaged in plastic or branded paper wrap (110–170 g).	15%–50%	Specialty shop, high-end tourist gift shops.	500 (170 g)–6500 (110 g)
Baobab delight smoothie	2	Packaged in PET bottles (250 mL).	30%	Supermarkets.	300 (250 mL)
Baobab smoothie served in cups	2	Served in polystyrene cups (100–150 mL).	30%	Restaurants, fast-food shops.	200–2000 (300 mL)
Baobab wine	1	Packaged in 750 mL glass bottles.	10%–15%	Agriculture fair/ trade exhibits.	6000 (750 mL)
Malambe face powder	1	Plastic jar (25–40 g).	60%	Local markets.	500 (40 g)
Baobab body cream	1	Wooden jar (250 g).	30%	High-end tourist gift shops.	6500–12,000 (250 g)
Baobab chocolate	1	White chocolate bar wrapped in paper.	15%	Flea markets, tourist centres.	2000–3500 (100 g)
Baobab body lotion	1	Imported from France.	100%	Drug stores, pharmacies.	25,000 (400 mL)
Total	78				

MKW 860 = EUR 1 at the time of fieldwork. Average per-capita income in Malawi amounted to MKW 293,700 (EUR 342) in 2018 [28].

While in principle, a large variety of processed baobab food products exists in Malawi, some of these products were more widely available and hence more commonly consumed by a diversity of consumers than others. For example, baobab chocolate was only available in high-end delicacy stores in Lilongwe and Blantyre, and baobab wine was presented at an agricultural product fair. Such products represented relatively recent product innovations that were sold in very low quantities only, according to the shop owners. In contrast, baobab fruit juice was known to consumers throughout the country across various population strata and sold in significant quantities through numerous formal and informal retail channels. This may reflect the overall popularity of fruit juices, for their ease of consumption, which supply desired vitamins and nutrients in times of growing consumer interest for healthy food products.

3.2. Sociodemographics of Consumers and Preferred Attributes

For further analyses, we therefore select baobab juice (bottled and frozen) as a reference product to elicit sociodemographic consumer profiles. The results of the cluster analysis showed that there were two distinct segments of baobab juice consumers. These were primary and secondary school pupils, below 10 years of age on average with low income and concern for healthy food; and adults, 35 years on average, with more than half having completed higher education and being characterized by strong concern for healthy food (Table 2). Both segments consumed baobab juice products irregularly and in small quantities; thus, these products did not constitute an essential part of their diets. School pupils, for example, mostly consumed the frozen baobab juice as an occasional snack during school breaks, which they mostly purchased from informal channels like roadside vendors. To keep unit cost low and the product price affordable for their target group, these vendors tended to use cheap ingredients such as artificial colorants and sweeteners of questionable quality. The adult consumer segment, in contrast, purchased the product mainly through formal retail outlets like supermarkets. The majority of consumers of both segments stated that they consumed baobab juices throughout the year. As indicated by the Silhouette measure of cohesion and separation of >0.5, the analyses separated the segments with reasonable accuracy.

Table 2. Sociodemographic profiles of the baobab juice consumer segments in Malawi.

Characteristic	School Pupils	Adults
Sample (N)	54 (76.1%)	17 (23.9%)
Mean age	9.6 ± 1.89 years	35.1 ± 8.69 years
Family status [#]	Single (100%)	Married (64.7%)
Education level [#]	Primary school (98.1%)	Diploma (52.9%)
Monthly income (MWK)	≤1500 (100%)	>300,000 (29.4%)
Monthly consumption [#] (litres)	≤2 (94.4%)	≤2 (58.8%)
Concern for healthy food [#]	Weak (66.6%)	Strong (64.7%)

[#] Median categories.

3.3. Quality Attributes Preferred by Different Baobab Juice Consumer Segments

Cross-tabulation results and Chi-square tests showed that the preferred quality attributes differed significantly at $\alpha \leq 0.001$ between both customer segments, and the strength of the relationship between the preference and consumer segment variables was measured using Cramer's V (Table 3). While school pupils mainly preferred baobab juice products for their particular taste, adult consumers expressed a stronger preference for a variety of extrinsic product features, such as the availability of product information on a packaging label, the quality of the packaging material used, packaging size, and health claims made, as well as intrinsic product features such as added preservatives and nutritional properties of the baobab juice product. These preferences were strongly associated with the consumer segment variable as indicated by the Cramer's V values in Table 3.

Table 3. Quality attributes preferred by baobab juice consumers (% strong agreement).

Quality Attributes	School Pupils	Adults	Cramer's V
Product information on packaging label	0.0	52.9	0.974
Quality of packaging material	3.7	52.9	0.719
Larger package size and volume	0.0	41.2	0.619
High nutritional content	18.5	47.1	0.568
No added preservatives	0.0	35.3	0.804
High perceived health benefits	3.7	41.2	0.713
Product quality conforming to Malawi Bureau of Standards (MBS)	0.0	23.5	0.802
Sweet taste	46.3	0.0	0.316

Pearson Chi-square values showed significance at 0.1% level.

3.4. Effect of Quality Attributes on Price of Baobab Products

On average, bottled baobab juice products were sold for a price of MWK 100 to 150 per 100 mL in the various retail outlets, and baobab fruit pulp powder was sold for a price of MKW 160 to 700 per 100 g in the different markets. In order to assess how the various attributes of a baobab food product contributed to its price, we used hedonic regression analysis. We compare the two product types most commonly available in formal and informal markets, baobab pulp powder and bottled baobab juice. Identifying those attributes that increase the value of the product to consumers can help food processing enterprises further improve the composition and marketing of their products, thereby increasing value-added and profitability (Table 4).

Table 4. Determinants of baobab food product quality attributes on price.

Quality Attributes	Model 1: JuiceStandardized Coefficients	Sig.	Model 2: PowderStandardized Coefficients	Sig.
(Constant)		0.002 **		0.636
Product information on packaging label	0.276	0.000 **	0.261	0.022 *
Quality of packaging material	0.545	0.000 **		
Package size			0.744	0.000 **
High nutritional content	0.250	0.015 *		
No added preservatives			0.460	0.000 **
High perceived health benefits	0.297	0.027 *		
Product quality conforming to Malawi Bureau of Standards (MBS)	0.356	0.004 **		
Adjusted R^2	0.795		0.742	

**, * indicate significance levels of 1% and 5%, respectively.

As indicated by the adjusted R^2 values, the regression models explained between 74%–80% of the variance in the data. All coefficients had positive signs indicating that the presence of the investigated intrinsic and extrinsic product attributes increased product price. It was found that a product label, the quality of packaging materials, high nutrient content, perceived health benefits, and conformity with Malawi Bureau of Standards (MBS) quality standards significantly influenced the price of bottled juice products. Packaging quality was the variable that had the largest impact on price; the model predicts that for every increase in quality of packaging material by 1 Standard Deviation (SD), there is a 0.545 unit SD increase in price. Availability of a product label, packaging size, as well as the absence of additives significantly increased the price of baobab fruit powder, with a 1 SD increase in volume or weight of the package causing a 0.744 unit SD increase in price.

4. Discussion

Our inventory of processed baobab products in Malawi has shown that a variety of products was available through various formal and informal retail outlets, including baobab drinks, fruit powder, sweets, chocolates, oils, and soaps. This corroborates earlier studies that have reported a diversity of mostly traditional food uses of baobab in the region, e.g., [29,30]. The increasing importance of industrially processed baobab food products corresponds to the growing interest in baobab by the food processing industry, following the acceptance of baobab fruit pulp as a food ingredient by the European Union (2008/575/EC) and the US Food and Drug Administration (GRAS Notice No. GRN 000273). While the number and diversity of baobab food products currently available in Malawi demonstrates their importance for consumption and trade in line with the country's transformation into probably the largest market for baobab in Africa [31], it was lower than that reported from countries with a more developed food industry [32]. This illustrates the significant commercial opportunities in baobab products that are yet to be exploited by baobab producers and food processors in Malawi and elsewhere in Africa.

Our analysis has clearly revealed the presence of two distinct consumer segments for baobab juice products, which largely follow the formal–informal product divide currently existing in Malawi. While informal baobab juice and other informal food products like ice-lollies were largely consumed by low-income consumer groups such as school pupils, day labourers, or peasants visiting urban and peri-urban street markets, baobab juice manufactured by food processing companies in the formal sector were nearly exclusively consumed by higher-income urban consumers. In addition, evidence from interviews with shop owners and observation suggest that products such as baobab jam and chocolate sold in delicacy and specialty stores were mostly purchased by expatriates or international tourists, a consumer segment with preferences closely resembling those of consumers in developed western countries, and with purchasing power typically higher than the average Malawian urban resident. These findings conform with studies from elsewhere that found a stronger preference of better educated, higher income and well-established consumers for novel and healthy food products [33,34].

Our results also suggest that the preferences of the identified consumer segments with regard to baobab juice products differed at statistically significant levels. Our findings resemble results reported in other studies that indicated a strong association of the consumers' sociodemographic status and lifestyles with their preferences and willingness to pay for certain health-related food product attributes [35]. While low-income consumers mainly preferred baobab juice products with sweet taste, high-income consumers ranked quality attributes like product information on labels, packaging material, nutritional properties, long shelf life, health claims, and conformity with product quality standards more highly. Findings from Niger [36] also showed that the consumers' willingness to pay for baobab products correlated with their knowledge of global environmental concerns. The consumers' product preferences and choices with regard to health and food safety aspects, hence, are influenced by their distinct demographic characteristics such as education level and income [37]. This observed differentiation of baobab food products in terms of their attributes and marketing channels suggests the use of advanced product design and marketing techniques such as lifestyle marketing [38] by some of the formal baobab food product manufacturers, which address target audiences based on their particular lifestyle-related needs, interests, desires, and values.

The results of the hedonic regression analysis suggest that extrinsic product features had a strong positive association with product price for baobab fruit juice products, while the actual nutrient content was relatively less important. This was probably due to the fact that nutrient composition of various baobab juice products was relatively uniform across different brands, while extrinsic product attributes such as packaging quality, labelling, conformity with food standards, and perceived health benefits provided more distinct differentiation potential. For example, while the type and quality of packaging was partially determined by distribution channel requirements, it was also associated with the consumers' perceptions of product aesthetics [39], as well as shelf life and food safety [40]. In addition, product manufacturers may generally find it easier to differentiate their products based on extrinsic rather than intrinsic product attributes, which are generally more difficult to change. Examples show that product differentiation based on extrinsic attributes can substantially increase product value by 100% and more [35]. However, the intrinsic attributes of baobab food products also provide ample opportunities for product differentiation. For example, as noted by Cisse et al. [41], the quality of traditionally prepared baobab fruit nectars quickly deteriorates without special precautions. While the shelf life of most informal baobab juice products is therefore typically very limited, the choice of a particular technological option, such as use of artificial or natural preservatives, pasteurization, or cold stabilization using crossflow microfiltration provides the opportunity for juice processing enterprises in the formal sector to target the respective preferences held by particular consumer segments.

Past studies have shown that consumers and retailers equally place high value on food labelling [42]. Food labelling has been shown to have a positive association with price for health-conscious consumers, females, and highly educated individuals [43,44]. Studies also show that health and nutrition-related attributes had a statistically significant influence on price [33,43]; and that compliance with advanced food standards such as organic often also attracted additional price premiums in the developed world [45,46]. In contrast, baobab fruit pulp powder generally seemed to benefit less from these differentiation opportunities, giving it more the character of a commodity product. The only determinants of product price, in addition to packaging size, were product labelling and the absence of preservatives. Again, this illustrates the significant opportunities that still exist for product differentiation and additional value creation through, for example, more advanced packaging and product certification of baobab fruit pulp products in Malawi.

Notwithstanding the obvious opportunities related to the increased commercialization of baobab in Malawi, the implications of its intensified utilization on the well-being of particularly vulnerable rural populations that traditionally depended on these resources deserve further careful analysis and consideration. Circumstantial evidence suggests that increasing commercialization leads to baobab fruits being more rapidly collected right after—and sometimes even before—they mature, given that collectors compete with each other in conditions when harvesting is unregulated and property rights in baobab trees are ill-defined or poorly enforced. This might reduce the availability of these fruits for consumption by poor and vulnerable population strata; a point that has also been highlighted critically in western Africa [47]. Likewise, the higher value and price of processed baobab food products might make them less affordable and reduce the quantities consumed by poor population strata [8]. Equally important are studies that investigate the effect of increasing fruit harvesting on the structure and natural regeneration of baobab populations and wildlife.

5. Conclusions

While the overall contribution of baobab trees to the livelihoods of rural populations in large parts of sub-Saharan Africa can hardly be overestimated, our results clearly suggest that the notion of baobab serving as an emergency food for poor rural populations is a too narrow depiction of the present realities in Malawi. Along with the awakening international interest in and growing demand for baobab pulp as a nutritious natural ingredient for a variety of lifestyle food products in Europe, the US, and a number of other developed markets, and building on a long tradition of baobab use by local populations, a vivid baobab-processing sector has emerged in Malawi that supplies a diversity of food and non-food products to local consumers. While entrepreneurs running small-scale and often informal enterprises can play an important role in driving the development of an industry, innovating products, services, and business models [48], the majority of informal baobab processing enterprises in Malawi currently only offer food products of minimum quality and low price that are customized to the tastes and purchasing power of low-income population strata. This is in line with the general description of informal enterprises often being survivalist, heavily resource constrained, and of low productivity [49] and their food products being related to higher food safety and health risks [50,51]. Notwithstanding this, and given the ubiquity of informal baobab food products in Malawi during the baobab harvesting season, such products can make a certain contribution to the diversification of diets and the energy and micronutrient supply of consumers, at least during some parts of the year. However, an exact account of the contribution of these products to nutrition and food security was beyond the scope of our study.

In addition to the numerous informal baobab businesses targeting low-income consumers, growing numbers of formal enterprises have started to exploit the market opportunities emerging from the increasing differentiation and fragmentation of consumer segments and their preferences that are connected to economic development, urbanization, and the westernization of African lifestyles and societies. This empirical result somewhat expands the currently prevailing food-for-the-poor narrative for baobab fruits. Given their better access to managerial and technical know-how, financial capital, business networks, and other resources, formal companies can oftentimes more easily comply with legal or formal supply chain requirements and fulfil more advanced customer expectations in terms of the intrinsic quality or extrinsic attributes of their products. Targeting the higher-value market segments and pursuing a consequent product differentiation strategy can thus be a rational strategic choice for such enterprises. In light of contesting views in the literature [52–54], an interesting question left to future research is whether a business model targeting the bottom-of-the-pyramid consumers could be equally or even more profitable (and at the same time socially and ethically justifiable) than a model pursuing premium pricing based on product differentiation.

Although the formal baobab processing companies mostly meet the product standards required by Malawian regulators and consumers, there is room to further improve their products given a variety of anticipated yet unmet customer needs. Further investment in market research, product development and formulations, quality management, and marketing may therefore be recommended for local food processing companies to continuously develop and grow their businesses. If successful, such investments can even lead to the building of an internationally competitive baobab-processing sector in Malawi that supplies high-value processed products to international markets, an opportunity which—in light of the growing popularity of baobab products in developed markets—does not perhaps seem too far-fetched.

Author Contributions: Conceptualization, C.C.-M., D.D., K.M., and C.M.; methodology, C.C.-M. and D.D.; formal analysis, C.C.-M., C.M., C.D.N., and D.D.; investigation, C.C.-M.; data curation, C.C.-M.; writing—original draft preparation, C.C.-M.; writing—review and editing, D.D.; visualization, C.C.-M. and D.D.; supervision, D.D. and C.M.; project administration, K.M.; funding acquisition, D.D. All authors have read and agreed to the published version of the manuscript.

Funding: The study was conducted as part of the international research project "BAOFOOD—Enhancing local food security and nutrition through promoting the use of baobab (*Adansonia digitata* L.) in rural communities of Eastern Africa", which was financially supported by the German Federal Ministry of Food and Agriculture (BMEL) based on the decision of the Parliament of the Federal Republic of Germany through the Federal Office of Agriculture and Food (BLE). We gratefully acknowledge this support as well as the comments made by two anonymous reviewers that helped to further improve the manuscript.

Conflicts of Interest: The authors declare no conflict of interest.

References

1. Angelsen, A.; Jagger, P.; Babigumira, R.; Belcher, B.; Hogarth, N.J.; Bauch, S.; Börner, J.; Smith-Hall, C.; Wunder, S. Environmental income and rural livelihoods: A global-comparative analysis. *World Dev.* **2014**, *64*, S12–S28. [CrossRef]
2. Djoudi, H.; Vergles, E.; Blackie, R.R.; Koame, C.K.; Gautier, D. Dry forests, livelihoods and poverty alleviation: Understanding current trends. *Int. Forest. Rev.* **2015**, *17*, 54–69. [CrossRef]
3. Maroyi, A. Potential role of traditional vegetables in household food security: A case study from Zimbabwe. *Afr. J. Agric. Res.* **2011**, *6*. [CrossRef]
4. Zulu, D.; Ellis, R.H.; Culham, A. Collection, consumption, and sale of lusala (*Dioscorea hirtiflora*)—A wild yam—by rural households in Southern Province, Zambia. *Econ. Bot.* **2019**, *73*, 47–63. [CrossRef]
5. Nykänen, E.A.; Dunning, H.E.; Aryeetey, R.N.O.; Robertson, A.; Parlesak, A. Nutritionally optimized, culturally acceptable, cost-minimized diets for low income Ghanaian families using linear programming. *Nutrients* **2018**, *10*, 461. [CrossRef] [PubMed]
6. Cloete, P.C.; Idsardi, E.F. Consumption of indigenous and traditional food crops: Perceptions and realities from South Africa. *Agroecol. Sust. Food* **2013**, *37*, 902–914. [CrossRef]

7. Asase, A.; Kumordzie, S. Availability, cost, and popularity of African leafy vegetables in Accra markets, Ghana. *Econ. Bot.* **2018**, *72*, 450–460. [CrossRef]

8. Catarino, L.; Romeiras, M.M.; Bancessi, Q.; Duarte, D.; Faria, D.; Monteiro, F.; Moldão, M. Edible leafy vegetables from West Africa (Guinea-Bissau): Consumption, trade and food potential. *Foods* **2019**, *8*, 93. [CrossRef]

9. Mollee, E.; Pouliot, M.; McDonald, M.A. Into the urban wild: Collection of wild urban plants for food and medicine in Kampala, Uganda. *Land Use Policy* **2017**, *63*, 67–77. [CrossRef]

10. Schlesinger, J.; Drescher, A.; Shackleton, C.M. Socio-spatial dynamics in the use of wild natural resources: Evidence from six rapidly growing medium-sized cities in Africa. *Appl. Geogr.* **2015**, *56*, 107–115. [CrossRef]

11. Fargeot, C.; Drouet-Hoguet, N.; Le Bel, S. The role of bushmeat in urban household consumption: Insights from Bangui, the capital city of the Central African Republic. *Bois. Trop.* **2017**, *332*, 31–42. [CrossRef]

12. Sneyd, L. Wild food, prices, diets and development: Sustainability and food security in urban Cameroon. *Sustainability* **2013**, *5*, 4728–4759. [CrossRef]

13. Bichard, A.; Dury, S.; Schönfeldt, H.C.; Moroka, T.; Motau, F.; Bricas, N. Access to urban markets for small-scale producers of indigenous cereals: A qualitative study of consumption practices and potential demand among urban consumers in Polokwane. *Dev. South Afr.* **2005**, *22*, 125–141. [CrossRef]

14. Aworh, O.C. From lesser-known to super vegetables: The growing profile of African traditional leafy vegetables in promoting food security and wellness. *J. Sci. Food Agric.* **2018**, *98*, 3609–3613. [CrossRef] [PubMed]

15. Garekae, H.; Shackleton, C.M. Foraging wild food in urban spaces: The contribution of wild foods to urban dietary diversity in South Africa. *Sustainability* **2020**, *12*, 678. [CrossRef]

16. Mungofa, N.; Malongane, F.; Tabit, F.T. An exploration of the consumption, cultivation and trading of indigenous leafy vegetables in rural communities in the greater Tubatse local municipality, Limpopo province, South Africa. *J. Consum. Sci.* **2017**, *3*, 53–67.

17. Maseko, I.; Mabhaudhi, T.; Tesfay, S.; Araya, H.; Fezzehazion, M.; Plooy, C. African leafy vegetables: A review of status, production and utilization in South Africa. *Sustainability* **2018**, *10*, 16. [CrossRef]

18. Leakey, R.R.B.; Simons, A.J. The domestication and commercialization of indigenous trees in agroforestry for the alleviation of poverty. *Agroforest. Syst.* **1997**, *38*, 165–176. [CrossRef]

19. Akinnifesi, F.K.; Sileshi, G.; Ajayi, O.C.; Chirwa, P.W.; Mng'omba, S.; Chakeredza, S.; Nyoka, B.I. Domestication and conservation of indigenous Miombo fruit trees for improving rural livelihoods in southern Africa. *Biodiversity* **2008**, *9*, 72–74. [CrossRef]

20. Stadlmayr, B.; Charrondière, U.R.; Eisenwagen, S.; Jamnadass, R.; Kehlenbeck, K. Nutrient composition of selected indigenous fruits from sub-Saharan Africa. *J. Sci. Food Agric.* **2013**, *93*, 2627–2636. [CrossRef]

21. Chadare, F.J.; Linnemann, A.R.; Hounhouigan, J.D.; Nout, M.J.R.; van Boekel, M.A.J.S. Baobab food products: A review on their composition and nutritional value. *Crit. Rev. Food Sci. Nutr.* **2009**, *49*, 254–274. [CrossRef] [PubMed]

22. Chipurura, B.; Muchuweti, M. Post harvest treatment of *Adansonia digitata* (baobab) fruits in Zimbabwe and the potential of making yoghurt from the pulp. *IJPTI* **2013**, *3*, 392. [CrossRef]

23. Gebauer, J.; Adam, Y.O.; Sanchez, A.C.; Darr, D.; Eltahir, M.E.S.; Fadl, K.E.M.; Fernsebner, G.; Frei, M.; Habte, T.-Y.; Hammer, K.; et al. Africa's Wooden Elephant: The Baobab tree (*Adansonia digitata* L.) in Sudan and Kenya: A Review. *Genet. Resour. Crop. Evol.* **2016**, *63*, 377–399. [CrossRef]

24. Adam, Y.O.; Pretzsch, J.; Pettenella, D. Contribution of non-timber forest products livelihood strategies to rural development in drylands of Sudan: Potentials and failures. *Agric. Syst.* **2013**, *117*, 90–97. [CrossRef]

25. De Caluwé, E. Market chain analysis of baobab (*Adansonia digitata* L.) and tamarind (*Tamarindus indica* L.) products in Mali and Benin. PhD Thesis, Ghent University, Ghent, Belgium, 2011.

26. Creswell, J.W. *Research Design. Qualitative, Quantitative, and Mixed Methods Approaches*, 2nd ed.; Sage Publications: Thousand Oaks, CA, USA, 2002; ISBN 9780761924425.

27. Aitzetmüller, K. Intended use of Malvales seed oils in novel food formulations–A warning. *J. Am. Oil Chem. Soc.* **1996**, *73*, 1737–1738. [CrossRef]

28. Reserve Bank of Malawi. Monthly Economic Review. 2020. Available online: https://www.rbm.mw/Publications/EconomicReviews/Home/ (accessed on 19 June 2020).

29. Munthali, C.R.Y.; Chirwa, P.W.; Changadeya, W.J.; Akinnifesi, F.K. Genetic differentiation and diversity of Adansonia digitata L. (baobab) in Malawi using microsatellite markers. *Agroforest. Syst.* **2013**, *87*, 117–130. [CrossRef]

30. Venter, S.M.; Witkowski, E.T.F. Fruits of our labour: Contribution of commercial baobab (*Adansonia digitata* L.) fruit harvesting to the livelihoods of marginalized people in northern Venda, South Africa. *Agroforest. Syst.* **2013**, *87*, 159–172. [CrossRef]

31. Sanchez, A.C. The baobab tree in Malawi. *Fruits* **2011**, *66*, 405–416. [CrossRef]

32. Gebauer, J.; Assem, A.; Busch, E.; Hardtmann, S.; Möckel, D.; Krebs, F.; Ziegler, T.; Wichern, F.; Wiehle, M.; Kehlenbeck, K. Der Baobab (*Adansonia digitata* L.): Wildobst aus Afrika für Deutschland und Europa?! *Erwerbs-Obstbau* **2014**, *56*, 9–24. [CrossRef]

33. Verbeke, W. Consumer acceptance of functional foods: Socio-demographic, cognitive and attitudinal determinants. *Food Qual. Prefer.* **2005**, *16*, 45–57. [CrossRef]

34. Niva, M.; Mäkelä, J. Finns and functional foods: Socio-demographics, health efforts, notions of technology and the acceptability of health-promoting foods. *Int. J. Consum. Stud.* **2007**, *31*. [CrossRef]

35. Schreiner, M.; Korn, M.; Stenger, M.; Holzgreve, L.; Altmann, M. Current understanding and use of quality characteristics of horticulture products. *Sci. Hortic.* **2013**, *163*, 63–69. [CrossRef]

36. Agúndez, D.; Lawali, S.; Mahamane, A.; Alía, R.; Soliño, M. Consumer preferences for baobab products and implication for conservation and improvement policies of forest food resources in Niger (West Africa). *Econ. Bot.* **2018**, *72*, 396–410. [CrossRef]

37. Grolleau, G.; Caswell, J.A. Interaction between food attributes in markets: The case of environmental labeling. *J. Agric. Resour. Econ.* **2006**, *31*, 1–14. [CrossRef]

38. Mariotti, S.; Glackin, C. *Entrepreneurship & Small Business Management*, 1st ed.; Prentice Hall: Boston, MA, USA, 2012; ISBN 9780135030318.

39. Simms, C.; Trott, P. Packaging development: A conceptual framework for identifying new product opportunities. *Mark. Theory* **2010**, *10*, 397–415. [CrossRef]

40. Rundh, B. The role of packaging within marketing and value creation. *Brit. Food J.* **2016**, *118*, 2491–2511. [CrossRef]

41. Cisse, M.; Sakho, M.; Dornier, M.; Diop, C.M.; Reynes, M.; Sock, O. Caractérisation du fruit du baobab et étude de sa transformation en nectar. *Fruits* **2009**, *64*, 19–34. [CrossRef]

42. Steiner, B.E. Australian wines in the British wine market: A hedonic price analysis. *Agribusiness* **2004**, *20*, 287–307. [CrossRef]

43. Menrad, K. Market and marketing of functional food in Europe. *J. Food Eng.* **2003**, *56*, 181–188. [CrossRef]

44. Grunert, K.G.; Wills, J.M. A review of European research on consumer response to nutrition information on food labels. *J. Public Health* **2007**, *15*, 385–399. [CrossRef]

45. Siró, I.; Kápolna, E.; Kápolna, B.; Lugasi, A. Functional food. Product development, marketing and consumer acceptance—A review. *Appetite* **2008**, *51*, 456–467. [CrossRef] [PubMed]

46. Grunert, K.G. Food quality and safety: Consumer perception and demand. *Eur. Rev. Agric. Econ.* **2005**, *32*, 369–391. [CrossRef]

47. Buchmann, C.; Prehsler, S.; Hartl, A.; Vogl, C.R. The importance of baobab (*Adansonia digitata* L.) in rural West African subsistence–suggestion of a cautionary approach to international market export of baobab fruits. *Ecol. Food Nutr.* **2010**, *49*, 145–172. [CrossRef] [PubMed]

48. McCann, B.T.; Bahl, M. The influence of competition from informal firms on new product development. *Strat. Manag. J.* **2017**, *38*, 1518–1535. [CrossRef]

49. Bustamante, J.P.I.; Fandl, K.J. Incentivizing gray market entrepreneurs in emerging markets. *Northwest J. Int. Law Bus.* **2017**, *37*, 415–456.

50. Alexander, E.; Yach, D.; Mensah, G.A. Major multinational food and beverage companies and informal sector contributions to global food consumption: Implications for nutrition policy. *Glob. Health* **2011**, *7*, 26. [CrossRef]

51. Rock, K.T.; Mugizi, D.R.; Ståhl, K.; Magnusson, U.; Boqvist, S. The milk delivery chain and presence of Brucella spp. antibodies in bulk milk in Uganda. *Trop. Anim. Health Prod.* **2016**, *48*. [CrossRef]

52. Rao, S.; Nilakantan, R.; Iyengar, D.; Lee, K.B. On the viability of fixing leaky supply chains for the poor through benefit transfers: A call for joint distribution. *J. Bus. Logist.* **2019**, *40*, 145–160. [CrossRef]

53. Agnihotri, A. Doing good and doing business at the bottom of the pyramid. *Bus. Horiz.* **2013**, *56*, 591–599. [CrossRef]
54. Varman, R.; Skålén, P.; Belk, R.W. Conflicts at the bottom of the pyramid: Profitability, poverty alleviation, and neoliberal governmentality. *J. Public Policy Mark.* **2012**, *31*, 19–35. [CrossRef]

 *forests*

Article

Antioxidant Properties of Green Coffee Extract

Anna Masek [1,*], Malgorzata Latos-Brozio [1], Joanna Kałużna-Czaplińska [2], Angelina Rosiak [2] and Ewa Chrzescijanska [2]

[1] Faculty of Chemistry, Institute of Polymer and Dye Technology, Lodz University of Technology, Stefanowskiego 12/16, 90-924 Lodz, Poland; malgorzata.latos-brozio@dokt.p.lodz.pl

[2] Faculty of Chemistry, Institute of General and Ecological Chemistry, Lodz University of Technology, Zeromskiego 116, 90-924 Lodz, Poland; joanna.kaluzna-czaplinska@p.lodz.pl (J.K.-C.); angelina.rosiak@p.lodz.pl (A.R.); ewa.chrzescijanska@p.lodz.pl (E.C.)

* Correspondence: anna.masek@p.lodz.pl

Received: 30 March 2020; Accepted: 12 May 2020; Published: 15 May 2020

Abstract: An infusion of green coffee is a commonly consumed beverage, famous for its health-promoting properties. Green coffee owes its properties to the richness of active phytochemicals. The aim of this study was to determine the components of green coffee bean extracts and their properties. The scope of research included gas chromatography-mass spectrometry (GC-MS), Fourier transform infrared spectroscopy (FTIR) and Ultraviolet-Visible spectroscopy (UV-Vis) spectroscopy; the electrochemical determination of the behavior of green coffee extract; and the determination of antioxidant properties by colorimetric spectroscopic methods (ABTS, DPPH, FRAP and CUPRAC). Water and ethanol extracts from green coffee were characterized by significant antioxidant properties and a high capacity to reduce transition metal ions. Voltammetric tests showed that the solution has good antioxidant properties in view of it contains many polyphenolic compounds that oxidize in the potential range tested.

Keywords: green coffee; electrochemical oxidation; antioxidant activity

---

1. Introduction

The purpose of this research was to analyze the antioxidant activity of the substances present in green coffee. Coffee is a very common plant consumed by many societies. Moreover, beans of coffee are rich source of phytocompounds with noted high antioxidant activity. The phenolic compounds can be found in plant materials, and they have been testified to have biological effects, containing high antioxidant and antibacterial activity.

The main phenolic ingredients identified in green coffee beans are chlorogenic and caffeic acids, which have antimutagenic, anticancer, antibacterial and antioxidative effects. In beans of green coffee, most phenols are linked to sugars as glycosides. Phenolic acids such as chlorogenic, ferulic and caffeic acid are identified in the form of an ester bound to the cell wall, creating highly complex polysaccharide structures. Phenolics in their conjugated form limit their bioavailability due to their high molecular mass and hydrophilicity [1,2].

Polyphenolic compounds could be correlated with their antioxidant activities. To analyze the capacity of antioxidants to reduce free radicals, methods consisting of single electron transfer chemical reactions (SET) were utilized. These procedures are considered by ease of solution preparation, stability of reaction mixtures, repeatability and relatively easy determination measures. Due to the mentioned benefits, these methods are widely utilized to determine the antioxidant activity of natural extracts. SET tests include two analogous procedures, ABTS and DPPH. The ABTS technique permits the investigate of hydrophilic and hydrophobic antioxidants, while the DPPH procedure permits the determination of only hydrophobic antioxidants. Furthermore, the reduction ability of natural antioxidants is tested as the ability to reduce metals. The procedures FRAP and CUPRAC are based on the reduction of transition metal ions—iron and copper [3].

An alternative to traditional methods for determining the antioxidant activity of different extracts or plant compounds is electrochemical procedures, which in last years have attracted interest from scientists [4–6]. The benefit of electrochemical techniques is that they enable rapid, easy and inexpensive determinations that, in some cases, permit measurements in the presence of colored or masking compounds that can interfere with measurements by other techniques, e.g., spectrophotometric. Further benefits of electrochemical techniques is that it is possible to examine experimental parameters such as peak potential (E_{pa}) and peak current (i_{pa}), which are important in analyzing the antioxidant activity of the extracts tested. Low values of oxidation potential (E_{pa}) reflect the trend of a given molecule to donate electrons, and thus to show its good antioxidant activity. Cyclic voltammetry (CV) is the most usually utilized method to study the properties of electrode processes, giving data on the thermodynamics of electrode reactions and electron transfer kinetics, as well as coupled chemical reactions or adsorption processes. Other electrochemical technique applied in determining the antioxidant activity of the extracts is differential pulse voltammetry (DPV), which is typify by good detection limits and high resolution. The use of this method makes it possible to eliminate adsorbing compounds because, in this technique, they are not electroactive and, therefore, there are no visible peaks on the voltamperogram [7–14].

The aim of this study was to determine the antioxidant properties of compounds found in coffee extract by various methods, including electrochemical, using cyclic voltammetry (CV) and differential pulse voltammetry (DPV). A further objective of the work was to establish the correlation between antioxidant activities deduced from cyclic voltammograms and those previously determined by utilizing established spectrophotometric methods. Moreover, a ranking of coffee was gained, utilizing the antioxidant composite index (AC), a parameter that assigns equal weight to all the antioxidant activity assesses. Antioxidant activities applying ABTS, DPPH, FRAP and CUPRAC tests for determination of total phenolic content were performed by UV-visible spectrophotometry, FTIR spectrophotometry and gas chromatography-mass spectrometry (GC-MS). This manuscript describes different in vitro tests that characterize and compare the antioxidant activity and mechanism of action of green coffee and its bioactive substances. The study revealed a general decrease of radical scavenging capacity related to native substances of plant origin. The scientific novelty of the work is a combination of different techniques for determining the activity of green coffee extracts. So far, no such comprehensive research on green coffee extracts has been done. What is more, spectrophotometric determinations summarize the results for extracts obtained in ethanol and in water, received at different times of brewing coffee. The available literature lacks assessment of the connection among the type of solvent and antioxidant activity and reduction of transition metal ions. This publication supplements the deficiencies in the literature.

2. Materials and Methods

2.1. Reagents and Chemicals

Chemical reagents applied for testing were as follows: 2,2′-azino-di-(3-ethylbenzthiazoline sulfonic acid) (ABTS, assay ≥98%, Sigma Aldrich, Saint Louis, MO, USA), potassium peroxodisulfate (99.99%, Sigma Aldrich, Saint Louis, MO, USA), (±)-6-Hydroxy-2,5,7,8-tetramethylchromane-2-carboxylic acid (Trolox, 97%, Sigma Aldrich, Saint Louis, MO, USA), 1,1-Diphenyl-2-picrylhydrazyl radical (DPPH, ≤100%, Sigma Aldrich, Darmstadt, Germany), 2,4,6-Tris (2-pyridyl)-s-triazine (TPTZ, ≥99.0% (HPLC), Sigma Aldrich, Buchs, Switzerland), ferric chloride $FeCl_3$ (pure P.A., Chempur, Piekary Slaskie, Poland), hydrogen chloride solution HCl (40 mM, Chempur, Piekary Slaskie, Poland), sodium acetate buffer solution (0.3 M, pH 3.6, Chempur, Piekary Slaskie, Poland), 2,9-Dimethyl-1,10-phenanthroline (Neocuproine, assay ≥98%, Sigma Aldrich, Beijing, China), cupric chloride $CuCl_2$ (0.01 M, Chempur, Piekary Slaskie, Poland), ammonium acetate (NH_4Ac) buffer solution (1.0 M, pH 7.0, Chempur,

Piekary Slaskie, Poland), ethanol (pure P.A., 96%, POCH, Gliwice, Poland) and methyl cyanide (pure P.A., 99.5%, POCH, Gliwice, Poland).

Methyl cyanide, 3-(3,4-Dihydroxycinnamoyl)quinic acid, 3,4-dihydroxybenzeneacrylic acid, (+)catechin, trans-4-hydroxycinnamic acid, 4-O-Caffeoylquinic acid, (−)epicatechin, ethyl acetate, trans-4-Hydroxy-3-methoxycinnamic acid, formic acid, 3,4,5-Trihydroxybenzoic acid, hexane, ellagic acid, 5-Hydroxy-1,4-naphthoquinone, 3,4′,5,7-Tetrahydroxyflavone, methyl alcohol, 3,3′,4′,5,5′,7-Hexahydroxyflavone, trans-5-O-Caffeoylquinic acid, N,O-bis(trimethylsilyl)trifluoroacetamide, 3,3′,4′,5,6-Pentahydroxyflavone, quercetin 3-glucoside, quercetin 3-rhamnoside, quercetin-3-rutinoside hydrate, 3,5-Dimethoxy-4-hydroxycinnamic acid, 3,5-Dimethoxy-4-hydroxybenzoic acid, trimethylchlorosilane, quercetin and (+)-catechin were purchased from Sigma-Aldrich (Steinheim, Germany). Ultrapure water was prepared in the laboratory, utilizing a SimplicityTM Water Purification System (Millipore, Marlborough, MA, USA).

2.2. Method of Preparation of Green Coffee Extracts for Antioxidant Analysis

Ethanol solution: Beans of green coffee were ground in a coffee mill (organic green coffee Honduras SHG, naturally grown, 100% Arabica.) The coffee came from a specialized farm in the Santa Barbara region, grown at an altitude of 1700 m above sea level, on volcanic hills with access to clean water from nearby rivers. Plant material was extracted using a 5-fold volume of 70% ethyl alcohol under continuous mixing conditions (250 RPM—revolutions per minute, 20 °C). The extraction was performed in the dark at 20 °C for 7 days. The extracts of ground green coffee were concentrated to constant mass by utilizing a rotary evaporator under reduced pressure conditions, at 30 °C.

Water solutions: Ground coffee beans were flooded with water at 90 °C and brewed under cover for 10 min or 30 min. Then, the plant material was filtered off on filter paper, and the solutions were concentrated to constant mass by utilizing a rotary evaporator under reduced pressure conditions, at 30 °C.

2.3. Measurement Methods

2.3.1. Chromatographic Analysis

First, 5.0066 g of the powdered green coffee beans was extracted with hexane (150 mL, 4 h), in a Soxhlet apparatus (labmed, HK, Lodz, Poland), to isolate the lipid fraction. The obtained fraction was derivatized with a mixture of N,O-bis(trimethylsilyl)trifluoroacetamide and trimethylchlorosilane (100:1 *v/v*, 30 min, 75 °C) and analyzed by gas chromatograph–mass spectrometer (GC-MS, Agilent Technologies, Santa Clara, CA, USA). The solid extraction residue, after decaffeination (in ethyl acetate), was extracted with methanol in a Soxhlet apparatus (150 mL, 3 h). The obtained fraction was derivatized as described above and also analyzed by GC-MS. Analyses were carried out with an Agilent 6890N Gas Chromatograph equipped with a HP-5MS capillary column (30 m × 250 μm × 0.25 μm), coupled to Agilent 5973N Mass Selective quadrupole detector (Agilent Technologies, Santa Clara, CA, USA). The column oven was programmed for the temperature starting at 50 °C. Then, the temperature increased by 15 °C per minute, to reach 210 °C, and remained at this temperature for 1 min. Next, the temperature was increased from 210 to 230 °C, with a 5 °C change per minute, and then from 230 to 300 °C, by 15 °C change per minute. The last stage was to keep the column oven at the final temperature 300 °C for 3 min. Carrier gas (helium) flow was at a constant flow rate of 0.7 mL/min. The ion source and detector temperatures were 230 and 150 °C, respectively. The mass range was set between m/z 50 and 750.

The compounds were identified by utilizing the NIST14 and Wiley's spectra libraries. The study included compounds which had content over 0.1%. The compounds from the column (mainly siloxanes) were not included in the results.

2.3.2. FTIR (Fourier Transform Infrared) Spectroscopy and UV-Vis (Ultraviolet-Visible) Spectroscopy

The FTIR spectrum is a "fingerprint" showing the chemical compounds and their structure, which are present in extracts from green coffee. A Nicolet 670 spectrophotometer (Thermo Fisher Scientific, Waltham, MA, USA) was utilizing for the FTIR tests. Ground green coffee samples and also water or ethanolic extracts were putted at the infrared beam output. As the result of the determination, oscillating spectra were gained, the examination of which permits determination of the functional groups that the radiation interacted with.

The UV-Vis spectrum of green coffee extracts were made from a mixture of 0.2 mL of each extract and 1.8 mL of 70% ethyl alcohol or distilled water. The sample was examined at 190–1100 nm, utilizing a UV-spectrophotometer (Evolution 220, Thermo Fisher Scientific, Waltham, MA, USA) at 25 °C.

2.3.3. Cyclic and Differential Pulse Voltammetry

To assess the electrochemical oxidation mechanism and the kinetics for the flavones under investigation, cyclic voltammetry (CV) and differential pulse voltammetry (DPV) were used with an Autolab analytical unit (EcoChemie, Utrecht, Holland). The analyzer was controlled by using the Graduate Program in the Environmental Science (GPES) Program. A three-electrode system was used for the measurements consisting of a reference electrode, an auxiliary electrode (platinum wire) and a working electrode—platinum with geometric surface area of 0.5 cm^2. The potential of the working electrode was measured vs. a ferrocenium/ferrocene reference electrode (Fc$^+$/Fc) couple, as recommended by International Union of Pure and Applied Chemistry (IUPAC) [15,16]. The reference electrode was made of platinum wire immersed in a solution of ferrocene c = 1×10^{-3} mol L^{-1} in 0.1 mol L^{-1} (C$_4$H$_9$)$_4$NClO$_4$ in acetonitrile placed in a glass tube with a very tiny hole (diameter w 0.2 mm) at the bottom. The tiny hole allowed electrochemical contact between the electrolyte in the reference electrode compartment and that in the reactor compartment. The next step was coulometric oxidation to obtain an equivalent of ferrocene ion concentration ferrocenium (Fc$^+$/Fc).

Determination of antioxidants was performed by using CV and DPV techniques. CV and DPV were recorded in the potential range from 0 to 1.9 or 2.1 V. CV was recorded with various scan rates (0.01 to 1 V s^{-1}). DPV was recorded in the same potential range, with modulation amplitude of 25 mV and pulse width of 50 ms (scan rate 0.01 V s^{-1}). Before the measurements, the solutions were purged with argon in order to remove dissolved oxygen. During measurements, an argon blanket was kept over the solutions. All experiments were carried out at room temperature [17].

2.3.4. Antioxidant Activity Tested Using ABTS and DPPH Procedure

The antioxidant capacity of green coffee extracts was tested by ABTS and DPPH procedures. Green coffee extracts concentrated to a constant weight (obtained according to "Method of Preparation of green coffee extracts for antioxidant analysis") were made into solutions with concentrations of 1–4 mg/mL. The weighed concentrated green coffee extracts were dissolved in the appropriate amount of ethyl alcohol (70%)—a solvent commonly used for ABTS and DPPH procedures.

These procedures are based on reduction of radicals 2,2′-azino-bis (3-ethylbenzothiazoline-6-sulphonic acid) (ABTS) and 2,2-diphenyl-1-picrylhydrazyl (DPPH).

The ABTS$^{\bullet+}$ radical was obtained by the reaction of a 6 mM ABTS solution in water with potassium persulfate (2.45 mM) without light at 25 °C for 16 h before use. The absorbance of the ABTS$^{\bullet+}$ dilution was regulated with ethanol to 0.70 ± 0.02 at 734 nm at 25 °C. In the next step, the diluted ABTS$^{\bullet+}$ dilution (4.0 mL) was mixed with a 40 µL aliquot of all examined solution (2 mg mL^{-1}) or Trolox in ethyl alcohol. The absorbance was measured at 734 nm after 2 min at 25 °C, using a UV-spectrophotometer.

The DPPH dilution in ethanol (2.0 mL) at a concentration of 40 mg mL^{-1} (0.1 mM) was inserted to 0.5 mL of an 70% ethanol solution containing 0.02 mg mL^{-1} of green coffee-DPPH solution, which has a purple color, with a maximum absorbance at 517 nm. The progress of the reaction was spectrophotometrically monitored. Ethyl alcohol (70%) was used as a blank in both ABTS and DPPH methods.

The level of inhibition (%) of free radicals ABTS and DPPH was computed according to the following Equation (1):

$$\text{Level of inhibition (\%)} = [((A_{CS} - A_E)/A_{CS}) \times 100] \tag{1}$$

where A_{CS} is the absorbance of the control sample without extracts, and A_E is the absorbance in the presence of green coffee extract.

The level of inhibition (%) of absorbance was computed by utilizing the standard curve prepared with solution of Trolox (% inhibition level-µM Trolox). The result of green coffee extract on reduction $ABTS^{\bullet+}$ and DPPH is referred to as the Trolox equivalent antioxidant capacity (TEAC) [18].

2.3.5. Examination of Reduction of Transition Metal Ions by FRAP and CUPRAC Procedures

Green coffee extracts concentrated to a constant weight (obtained according to "Method of Preparation of green coffee extracts for antioxidant analysis") were made into solutions with concentrations of 1–4 mg/mL. The weighed concentrated green coffee extracts were dissolved in the appropriate amount of in pure water for FRAP and CUPRAC methods.

The ability of ethanolic and aqueous green coffee extracts to reduce the ferric ion (Fe^{3+}-TPTZ complex) under acidic conditions was carried out by utilizing ferric reducing antioxidant power assay (FRAP). The FRAP reagent was freshly prepared by mixing 25 mL of acetate buffer solution (0.3 M, pH 3.6), 2.25 mL of TPTZ dilution (10 mM TPTZ in 40 mM hydrogen chloride solution) and 2.25 mL of ferric chloride (20 mM) in a pure water solution. The reaction mixture was stirred and incubated at 37 °C for 20 min. In the next step, the increase in absorbance of the ferrous form with blue color (Fe^{2+}-TPTZ complex) was measured at 595 nm, using a UV-spectrophotometer.

The CUPRAC procedure is similar to the FRAP method and involves the reduction of Cu^{2+} to Cu^{1+}. Approximately 0.25 mL (0.01 M) of $CuCl_2$ was mixed with 0.25 mL of an ethanol solution of neocuproine (7.5×10^{-3} M) and 0.25 mL of buffer solution, CH_3COONH_4 (1 M), in a test tube, followed by the addition of different concentrations of green coffee extract. The total volume of samples was increased to 2 mL with pure water. After 30 min of incubation at 25 °C, the absorbance at 450 nm was measured against a reagent blank (pure water).

The reducing power of ferric (FRAP) and cupric (CUPRAC) ions was computed as follows (Equation (2)):

$$\Delta A = A_1 - A_R \tag{2}$$

where A_R is the absorbance of the reagent test, and A_1 is the absorbance after reaction [19,20].

2.4. Statistical Analysis

Computations were performed for the means and standard deviations of 3 independent extractions ($n = 3$). Statistical analysis was applied for the assessment of the means and then carried out by utilizing a Fischer LSD test (the significance level was set at $p < 0.05$).

3. Results and Discussion

3.1. Chromatographic Analysis of Green Coffee

The study began with an analysis of the chemical composition of green coffee beans. The results of the chromatographic analysis are shown below.

3.1.1. Organic Compounds in Hexane Extract of Beans of Green Coffee

The chromatogram of the hexane extract of green coffee beans is shown in Figure 1A. Table 1 shows the main organic compounds in the hexane extract of beans of green coffee. Significant amounts of saturated (palmitic, stearic and arachidic) and unsaturated (palmitelaidic, oleic and 17-octadecynoic) fatty acids and acylglycerols (1-monolinolein, glycerol monostearate) were determined. Palmitic acid and linoleic acid are the two main components of the lipid fraction. Some of the compounds determined may be potentially related to the antioxidant properties of beans of green coffee. Linoleic acid belongs to the omega-6 fatty acid group, which can act as antioxidants [21,22]. The lycopene derivative and β-tocopherol, which are also natural antioxidants [23], were determined in the sample.

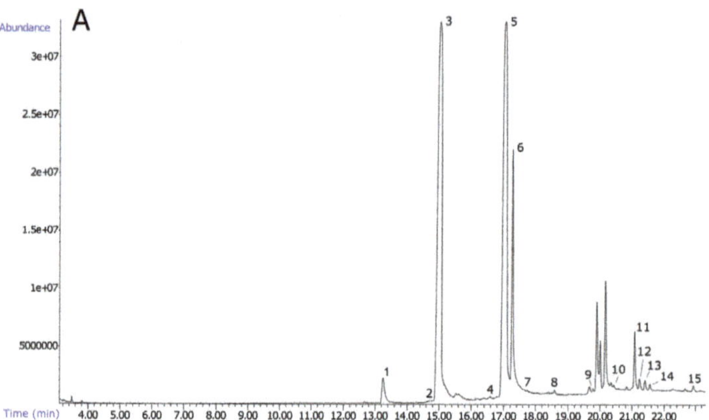

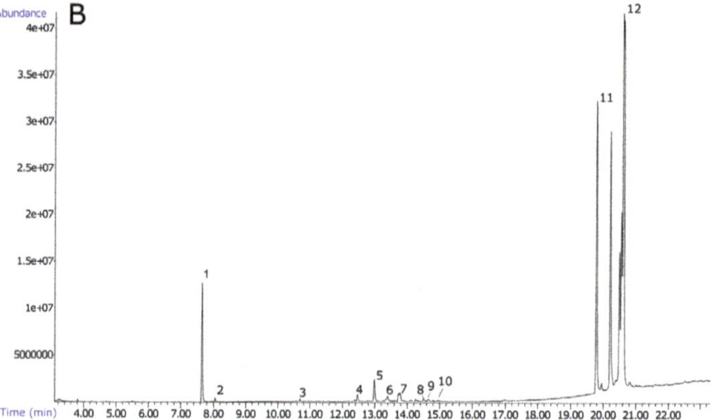

Figure 1. (**A**) The chromatogram of hexane extract of green coffee beans. The peak numbers in the figure correspond to the compounds in Table 1. (**B**) The chromatogram of methanol extract of green coffee beans. The peak numbers in the figure correspond to the compounds in Table 2.

Table 1. Main organic compounds determined in hexane extract of green coffee beans.

Peak	Retention Time (min)	Area (%)	Plant Compound
1	13.25	1.68	Caffeine
2	14.71	0.13	Palmitelaidic acid, TMS derivative
3	15.03	35.16	Palmitic Acid, TMS derivative
4	16.60	0.19	Oleic Acid, (Z)-, TMS derivative
5	17.05	32.37	Linoleic acid, TMS
6	17.27	9.96	Stearic acid, TMS derivative
7	17.61	0.47	17-Octadecynoic acid, TMS derivative
8	18.59	0.37	Arachidic acid, TMS derivative
9	19.78	0.19	2-Oleoylglycerol, 2TMS derivative
10	20.43	0.45	Methyl glycocholate, 3TMS derivative
11	21.08	1.88	1-Monolinolein, 2TMS derivative
12	21.23	0.52	Glycerol monostearate, 2TMS derivative
13	21.40	0.45	Lycopene, 1,1′,2,2′-tetrahydro-1,1′-dimethoxy-, all-trans-
14	21.55	0.26	Ethyl iso-allocholate
15	22.93	0.21	β-Tocopherol, TMS derivative

3.1.2. Organic Compounds in Methanol Extract of Beans of Green Coffee

The chromatogram of the methanol extract of green coffee beans is represented in Figure 1B. There are twelve main compounds identified in the methanol extract (Table 2). Disaccharide–sucrose is the main component present in the sample and is the most abundant carbohydrate in green coffee beans [24]. Among the compounds in the methanol extract are sugar alcohols (polyols): glycerol, D-Mannitol, Myo-Inositol and D-Pinitol, and quinic acid, which is one of the major polyphenols in green coffee beans [25].

Table 2. Main organic compounds determined in methanol extract of green coffee beans.

Peak	Retention Time (min)	Area (%)	Plant Compound
1	7.65	5.03	Glycerol, 3TMS derivative
2	8.06	0.19	Glycerol 1,2-diacetate
3	10.69	0.20	Triethanolamine, 3TMS derivative
4	12.47	0.47	D-Fructose, 5TMS derivative
5	12.99	1.44	Quinic acid
6	13.40	0.75	α-D-Glucopyranosiduronic acid, 3-(5-ethylhexahydro-2,4,6-trioxo-5-pyrimidinyl)-1,1-dimethylpropyl 2,3,4-tris-O-(trimethylsilyl)-, methyl ester
7	13.78	0.66	D-Mannitol, 6TMS derivative
8	14.51	0.29	Myo-Inositol, 6TMS derivative
9	14.66	0.24	D-Pinitol, pentakis(trimethylsilyl) ether
10	14.93	0.19	Palmitic Acid, TMS derivative
11	19.95	0.93	D-(+)-Turanose, octakis(trimethylsilyl) ether
12	20.61	29.81	Sucrose, 8TMS derivative

3.2. FTIR (Fourier Transform Infrared) Spectroscopy and UV-Vis (Ultraviolet-Visible) Spectroscopy

Figures 2A and 1B show the spectra of green coffee and its extracts. Green coffee extracts are rich in polyphenolic compounds, especially phenolic acids, such as hydroxybenzoic acid derivatives (gallic acid and vanilla acid) and also hydroxycinnamic acid derivatives (ferulic acid, p-coumaric acid and caffeic acid). Moreover, these extracts contain chlorogenic acids, which are powerful antioxidants. Moreover, (+)-Catechin, a compound from the flavonoid group, was also found in green coffee extracts [26].

Figure 2A summarizes the FTIR spectrum of ground green coffee and the spectra of ethanol and water extracts (brewing time 10 min and 30 min). The green coffee spectrum shows functional groups characteristic of cellulose, hemicellulose and lignin (3680–2950 cm^{-1}-OH stretching; 2960 2860 cm^{-1} C-H stretching; 1470–1430 cm^{-1}-O-CH$_3$; 1090–970 cm^{-1}-C-O-C). Alkyl, aliphatic and aromatic compounds are visible in the range of 1730–1700 cm^{-1}, while 900–700 cm^{-1} corresponds to C-H stretching from aromatic hydrogen compounds [27,28]. Reaction with ethyl alcohol and brewing coffee beans in water permits the extraction of active phenolic compounds from green coffee. A valuable group of plant compounds present in green coffee are polyphenols, including flavonoids. According to Heneczkowski et al. [29], the ranges 1612–1598 cm^{-1}, 1570–1560 cm^{-1} and 1488–1452 cm^{-1} are typical for the aromatic ring vibration. Another specific bands corresponding to the phenol group vibrations are C-OH deformation vibrations (1370–1308 cm^{-1}), C-OH stretching vibrations (1172–1112 cm^{-1}) and, in the case of sulfonic derivatives, vibrations are related to those substituents in ranges 1188–1180 cm^{-1} for SO$_2$ ν_{asym} and 1080–1024 cm^{-1} for SO$_2$ ν_{sym} [29]. For ethanol and aqueous extracts, the ranges of wavenumber corresponding to active compounds often overlap with the ranges corresponding to pure solvents (3390 and 3360 cm^{-1}-OH groups, 2980 cm^{-1}-CH groups, and 1090 cm^{-1} and 1050 cm^{-1}-C-O groups) [30].

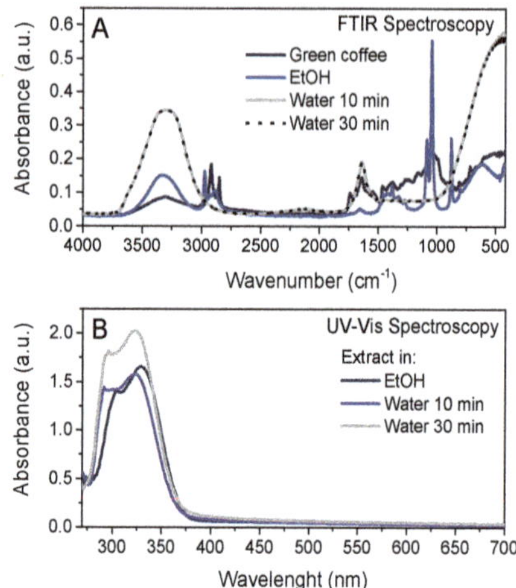

Figure 2. The FTIR spectra (**A**) and UV-Vis spectra (**B**) of extracts of green coffee.

Figure 2B shows the UV-Vis spectra of green coffee extracts. According to Song et al. and Rojas et al. [31,32], the UV-Vis spectra of phenolic acids have maxima of absorbance in the range of about 290–350 nm. Moreover, other active compounds, the flavones and related glycosides, illustration two strong absorption peaks at 240–280 nm and 300–380 nm [33]. Thus, the peaks with maxima at 337, 325, 322, 302, 296 and 288 nm on the UV-Vis spectra of green coffee ethanolic and water extracts characterize for mixture of active phenolic acids, flavones and their glycosides.

3.3. The Electrochemical Behavior of Green Coffee Extract

Voltammetric studies is used to assess the antiradicalproperties combined in various extracts of plant origin, as well as in beverages and biological systems, by many researchers [34], providing information on range and peak values. The dependence of the current on the potential of the indicator electrode was presented by the electrochemical reactions of the substances in solution on the tested electrodes. Electrochemical behavior of solutions of coffee nut extracts are contained in Figure 3. The cyclic voltammetry (CV) and differential pulse voltammetry (DPV) methods were used, because may be described by higher resolution.

On the voltamperogram for the coffee extract solution (Figure 3), four electro-oxygenation peaks are visible; the first peak (Peak I) is at 0.28 V potential, the second peak (Peak II) is at 0.96 V potential, the third poorly formed peak (Peak III) is at 1.23 V potential and the fourth peak (Peak IV) is at 1.49 V potential. In the reverse polarization cycle, one peak is visible at 0.52 V. Voltammetric tests show that the solution has good antioxidant properties because it contains many phytocompounds that oxidize in the potential field.

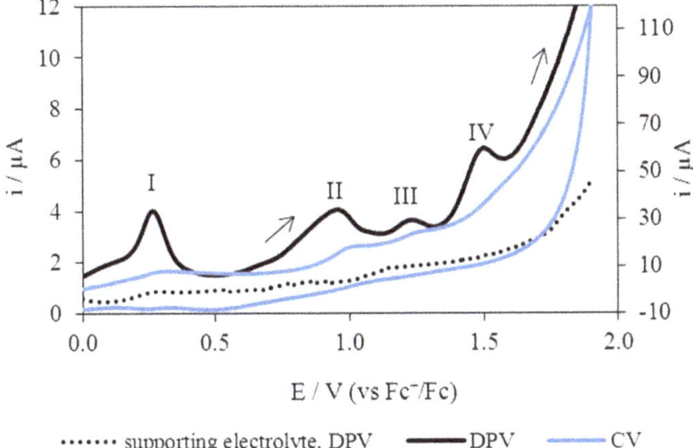

Figure 3. Cyclic voltammetry (CV) and differential pulse voltammetry (DPV) electrooxidation of extract of green coffee at Pt electrode; c = 20 mg/dm^3 in 0.1 M (C$_4$H$_9$)$_4$NClO$_4$ in acetonitrile; v = 0.05 V/s.

An electrochemical index (AC) was proposed, taking into account the main voltammetric parameters, peak potential (E_{pa}) and peak current (I_{pa}). Thus, the lower the potential (thermodynamic parameter), the higher the electron donor ability is, and the higher the peak current (kinetic parameter), the higher the amount of electroactive species is. The AC was calculated by using Equation (3):

$$AC = I_{pa1}/E_{pa1} + I_{pa2}/E_{pa2} + \ldots + I_{pan}/E_{pan} \tag{3}$$

where I_{pan} and E_{pan} correspond to current and potential values for each anodic peak observed in the CV and DPV voltammograms.

The calculated values are presented in Table 3.

Table 3. Peak potentials (Ep) and currents (Ip) determined from cyclic voltammetry (CV), differential pulse voltammetry (DPV) and antioxidant capacity (AC).

Method	Peak I		Peak II		Peak III		Peak IV		AC$_{total}$
	E$_p$ (V)	i$_p$ (μA)	E$_p$ (V)	i$_p$ (μA)	E$_p$ (V)	i$_p$ (μA)	E$_p$ (V)	i$_p$ (μA)	
CV for v = 0.05 Vs^{-1}	0.29	7.95	1.01	18.22	1.25	24.49	1.56	44.68	
DPV	0.28	3.97	0.96	4.05	1.23	3.65	1.49	6.45	
AC for CV		27.41		18.03		19.59		28.64	93.68
AC for DPV		14.17		4.21		2.96		4.32	25.69

The AC is calculated by means of the peak current (I_{pa}) and peak potential (E_{pa}), in which the first parameter would be directly proportional to the antioxidant power, while the second would be inversely so. The AC provides a useful measure of the antioxidant power of products. Since the antioxidant power is usually correlated with phenolic compounds, such products may also be evaluated by means of total phenol content.

3.4. Antioxidant Capacity of Green Coffee Extracts

The next step of the study was the determination of the antioxidant capacity of green coffee extracts by using spectrophotometric methods (Figure 4 and Table 4). Figure 4 shows the activity of ethanolic green coffee extracts and aqueous extracts obtained as a result of brewing ground coffee beans for 10 and 30 min. Antioxidant activity is shown as the capacity to reduce (% inhibition) free radicals (ABTS and DPPH) (Figure 4A,B) and as the Trolox equivalent antioxidant capacity (TEAC) (Table 3). Green coffee extracts prepared in ethanol or water, with concentrations from 1 to 4 mg/mL, were tested, and it was found that, with increasing concentration, the % inhibition of ABTS and DPPH increased. The ABTS inhibition (%) of 4 mg/mL extract of green coffee was 90.0 ± 0.20% for ethanol, 92.2 ± 0.19% for water at 10 min and 97.4 ± 0.25% for water at 30 min. Significant ability of green coffee extracts to reduce DPPH radicals was also found. The DPPH inhibition (%) of 4 mg/mL extract of green coffee was 63.9 ± 0.15% for ethanol, 69.6 ± 0.23% for water at 10 min and 81.6 ± 0.29% for water at 30 min. Analysis of antioxidant activity by ABTS and DPPH procedures showed that both the ethanol green coffee extract and aqueous solutions have good antioxidant properties. Another spectrophotometric analysis was the determination of the ability of green coffee extracts to reduce iron transition metal ions (FRAP method; Figure 4C) and copper (CUPRAC method; Figure 4D). The FRAP unit determines the capacity to reduce 1 mole of iron (III) to iron (II), and the CUPRAC unit to reduce 1 mole of copper (II) to copper (I). As in the case of ABTS and DPPH methods, it was observed that, with increasing concentration of green coffee extract, there were increases (in the range of 1–4 mg/mL) of their capacity to reduce iron and copper ions. All green coffee extracts exhibited good ability to reduce ions of transition metal; however, a slightly lower ability was found for aqueous extracts obtained by brewing coffee for 10 min. Strong antioxidant properties of green coffee extracts and the significant ability to reduce iron and copper metal ions should be combined with a high content of active compounds, such as polyphenolic compounds. The comparable activity of the ethanolic extract and aqueous solutions has proved that, thanks to the traditional brewing of green coffee, infusions with good antioxidant properties can be obtained.

All green coffee extracts were characterized by strong antioxidant activity; however, it was less than the capacity to reduce radicals ABTS and DPPH by green tea extract. Green tea extract, sold under the name Polyphenol 60, contained many polyphenolic compounds, such as gallic acid, procyanidin B1, (−)-epigallocatechin, (+)-catechin, (−)-epigallocatechingallate, (−)-epicatechin, (−)-epicatechingallate and flavonols. The extract obtained from green tea had very strong antioxidant properties: A solution with a concentration of 363 µg/mL exhibited the capacity to reduce ABTS free radicals, which was 93.6% and DPPH radicals 78.3% [18].

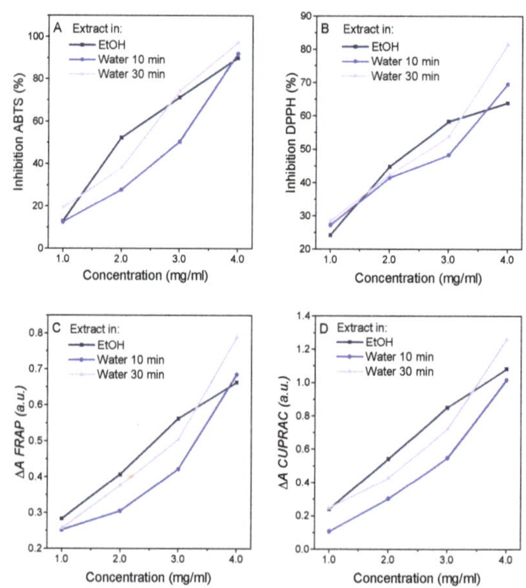

Figure 4. Antioxidant properties of green coffee extracts (ethanol and water prepared with 10 and 30 min of brewing the coffee) made by ABTS (**A**), DPPH (**B**), FRAP (**C**) and CUPRAC (**D**) techniques.

Table 4. Antioxidant capacity of green coffee extracts carried out ABTS and DPPH methods referred to as the Trolox equivalent antioxidant capacity (TEAC).

Concentration of Green Coffee Extract (mg/mL)	EtOH	Water 10 min	Water 30 min
ABTS-TEAC (mmolT/100g)			
1.00	46.6 ± 0.64	50.8 ± 0.54	73.1 ± 0.71
2.00	110.1 ± 0.94	55.2 ± 0.35	81.2 ± 0.36
3.00	94.1 ± 0.23	65.8 ± 0.15	105.5 ± 0.17
4.00	89.4 ± 0.17	91.2 ± 0.23	95.3 ± 0.23
DPPH-TEAC (mmolT/100g)			
1.00	78.5 ± 0.51	98.5 ± 0.42	96.5 ± 0.48
2.00	85.0 ± 0.26	74.2 ± 0.26	81.5 ± 0.31
3.00	67.7 ± 0.14	56.7 ± 0.29	68.9 ± 0.23
4.00	57.2 ± 0.11	61.8 ± 0.19	72.1 ± 0.22

4. Conclusions

Extensive analysis of the antioxidant properties of green coffee extract allowed for a thorough correlation of the test results by different methods. The objective of the research was extremely interesting and significant from the point of view of high consumption of this plant around the world. Studies have confirmed the strong dependence of the composition of the extract on the properties and activity of substances contained in green coffee beans.

Strong antioxidant properties of green coffee extracts and their significant ability to reduce iron and copper metal ions should be combined with a high content of active compounds, such as polyphenolic compounds. Moreover, the electroanalytical measurements and electrochemical index concept proved to be rapid to use and appropriate tools to evaluate the AC of extracts of coffee, and they can be applicable to other food sources rich in plant antioxidants with noted high antioxidant activity.

Author Contributions: Conceptualization, A.M. and M.L.-B.; methodology, A.M., E.C. and A.R.; software, M.L.-B.; validation, M.L.-B., E.C., J.K.-C. and A.R.; formal analysis, A.M.; investigation, A.M.; resources, A.M.; data curation, A.M.; writing—original draft preparation, A.M., M.L.-B. and E.C.; writing—review and editing, A.M.; visualization, M.L.-B.; supervision, A.M.; project administration, A.M.; funding acquisition, A.M. All authors have read and agreed to the published version of the manuscript.

Funding: This study was supported by the National Centre for Research and Development (NCBR) project: LIDER/32/0139/L-7/15/NCBR/2016.

Conflicts of Interest: The authors declare no conflict of interest.

References

1. Dziki, D.; Gawlik-Dziki, U.; Pecio, Ł.; Różyło, R.; Świeca, M.; Krzykowski, A.; Rudy, S. Ground green coffee beans as a functional food supplement—Preliminary study. *LWT Food Sci. Technol.* **2015**, *63*, 691–699. [CrossRef]
2. Palmieri, M.G.S.; Cruz, L.T.; Bertges, F.S.; Húngaro, H.M.; Batista, L.R.; Da Silva, S.S.; Fonseca, M.J.V.; Rodarte, M.P.; Vilela, F.M.P.; Do Amaral, M.D.P.H. Enhancement of antioxidant properties from green coffee as promising ingredient for food and cosmetic industries. *Biocatal. Agric. Biotechnol.* **2018**, *16*, 43–48. [CrossRef]
3. Masek, A.; Chrzescijanska, E.; Latos-Brozio, M.; Zaborski, M. Characteristics of juglone (5-hydroxy-1,4,-naphthoquinone) using voltammetry and spectrophotometric methods. *Food Chem.* **2019**, *301*, 125279. [CrossRef] [PubMed]
4. Chrzescijanska, E.; Wudarska, E.; Kusmierek, E.; Rynkowski, J. Study of acetylsalicylic acid electroreduction behavior at platinum electrode. *J. Electroanal. Chem.* **2014**, *713*, 17–21. [CrossRef]
5. Jara-Palacios, M.J.; Escudero-Gilete, M.L.; Hernández-Hierroa, J.M.; Herediaa, F.J.; Hernanz, D. Cyclic voltammetry to evaluate the antioxidant potential in winemaking byproducts. *Talanta* **2017**, *165*, 211–215. [CrossRef]
6. Masek, A.; Chrzescijanska, E.; Latos, M.; Zaborski, M. Influence of hydroxyl substitution on flavanone antioxidants properties. *Food Chem.* **2017**, *215*, 501–507. [CrossRef] [PubMed]

7. Barros, L.; Cabrita, L.; Boas, M.V.; Carvalho, A.M.; Ferreira, I.C.F.R. Chemical. biochemical and electrochemical assays to evaluate phytochemicals and antioxidant activity of wild plants. *Food Chem.* **2011**, *127*, 1600–1608. [CrossRef]

8. Masek, A.; Chrzescijanska, E.; Kosmalska, A.; Zaborski, M. Antioxidant activity determination in Sencha and Gun Powder green tea extracts with the application of voltammetry and UV-VIS spectrophotometry. *Comptes Rendus Chim.* **2012**, *15*, 424–427. [CrossRef]

9. Masek, A.; Chrzescijanska, E.; Kosmalska, A.; Zaborski, M. Characteristics of compounds in hops using cyclic voltammetry. UV-VIS. FTIR and GC-MS analysis. *Food Chem.* **2014**, *156*, 353–361. [CrossRef]

10. Hoyos-Arbeláez, J.; Vázquez, M.; Contreras-Calderón, J. Electrochemical methods as a tool for determining the antioxidant capacity of food and beverages: A review. *Food Chem.* **2017**, *221*, 1371–1381. [CrossRef]

11. Vicentini, F.C.; Raymundo-Pereira, P.A.; Janegitz, B.C.; Machado, S.A.S.; Fatibello-Filho, O. Nanostructured carbon black for simultaneous sensing in biologicalfluids. *Sens. Actuators B* **2016**, *227*, 610–618. [CrossRef]

12. Rebelo, M.J.; Rego, R.; Ferreira, M.; Oliveira, M.C. Comparative study of the antioxidant capacity and polyphenol content of Douro wines by chemical and electrochemical methods. *Food Chem.* **2013**, *141*, 566–573. [CrossRef]

13. Brett, C.M.A.; Brett, A.M.O. *Electrochemistry—Principles, Methods and Applications*; Oxford University Press: Oxford, UK, 1993.

14. Wudarska, E.; Chrzescijanska, E.; Kusmierek, E.; Rynkowski, J. Voltammetric study of the behaviour of N-acetyl-p-aminophenol in aqueous solutions at a platinum electrode. *Comptes Rendus Chim.* **2015**, *18*, 993–1000. [CrossRef]

15. Gritzner, G.; Kuta, J. Recommendations on reporting electrode potentials in nonaqueous solvents. *Pure Appl. Chem.* **1984**, *56*, 461–466. [CrossRef]

16. Bard, A.J.; Faulkner, L.R. *Electrochemical Methods: Fundamentals and Applications*, 2nd ed.; Wiley: New York, NY, USA, 2001.

17. Masek, A.; Chrzescijanska, E.; Zaborski, M. Electrooxidation of flavonoids at platinum electrode studied by cyclic voltammetry. *Food Chem.* **2011**, *127*, 699–704. [CrossRef]

18. Masek, A.; Chrzescijanska, E.; Latos, M.; Zaborski, M.; Podsedek, A. Antioxidant and antiradical properties of green tea extract compounds. *Int. J. Electrochem. Sci.* **2017**, *12*, 6600–6610. [CrossRef]

19. Masek, A.; Chrzescijanska, E.; Latos, M.; Kosmalska, A. Electrochemical and spectrophotometric characterization of the propolis antioxidants properties. *Int. J. Electrochem. Sci.* **2019**, *14*, 1231–1247. [CrossRef]

20. Masek, A.; Latos, M.; Chrzescijanska, E.; Zaborski, M. Antioxidant properties of rose extract (*Rosa villosa* L.) measured using electrochemical and UV/Vis spectrophotometric methods. *Int. J. Electrochem. Sci.* **2017**, *12*, 10994–11005. [CrossRef]

21. Berber, A.; Zengin, G.; Aktumsek, A.; Sanda, M.A.; Uysal, T. Antioxidant capacity and fatty acid composition of different parts of *Adenocarpus complicatus* (Fabaceae) from Turkey. *Rev. Biol. Trop.* **2014**, *62*, 337–346. [CrossRef]

22. Fagali, N.; Catalá, A. Antioxidant activity of conjugated linoleic acid isomers. linoleic acid and its methyl ester determined by photoemission and DPPH techniques. *Biophys. Chem.* **2008**, *137*, 56–62. [CrossRef]

23. Amarowicz, R. Lycopene as a natural antioxidant. *Eur. J. Lipid Sci. Technol.* **2011**, *113*, 675–677. [CrossRef]

24. Wei, F.; Tanokura, M. Chapter 17—Organic compounds in green coffee beans. In *Coffee in Health and Disease Prevention*; Preedy, V.R., Ed.; Academic Press: Cambridge, MA, USA, 2015; pp. 149–162. [CrossRef]

25. Bothiraj, K.V.; Vanitha, V. Green coffee bean seed and their role in antioxidant—A review. *Int. J. Res. Pharm. Sci.* **2020**, *11*, 233–240. [CrossRef]

26. Swieca, M.; Gawlik-Dziki, U.; Dziki, D.; Baraniak, B. Wheat bread enriched with green coffee—In vitro bioaccessibility and bioavailability of phenolics and antioxidant activity. *Food Chem.* **2017**, *221*, 1451–1457. [CrossRef]

27. Yang, H.; Yan, R.; Chen, H.; Ho Lee, D.; Zheng, C. Characteristics of hemicellulose, cellulose and lignin pyrolysis. *Fuel* **2007**, *86*, 1781–1788. [CrossRef]

28. Bolio-López, G.I.; Ross-Alcudia, R.E.; Veleva, L.; Barrios, J.A.A.; Madrigal, G.C.; Hernández-Villegas, M.M.; De la Burelo, P.; Córdova, S.S. Extraction and characterization of cellulose from agroindustrial waste of pineapple (*Ananas comosus* L. Merrill) crowns. *Chem. Sci. Rev. Lett.* **2016**, *5*, 198–204.

29. Heneczkowski, M.; Kopacz, M.; Nowak, D.; Kuźniar, A. Infrared spectrum analysis of some flavonoids. *Acta Pol. Pharm.* **2001**, *6*, 415–420.

30. Doroshenko, I.; Pogorelov, V.; Sablinskas, V. Infrared absorption spectra of monohydric alcohols. *Dataset Pap. Chem.* **2013**, *2013*, 329406. [CrossRef]

31. Song, H.; Chen, C.; Zhao, S.; Ge, F.; Liu, D.; Shi, D.; Zhang, T. Interaction of gallic acid with trypsin analyzed by spectroscopy. *J. Food Drug Anal.* **2015**, *23*, 234–242. [CrossRef]

32. Rojas, J.; Londono, C.; Ciro, Y. The health benefits of natural skin UVA photoprotective compounds found in botanical sources. *Int. J. Pharm. Pharm. Sci.* **2016**, *8*, 13–23.

33. Joshi, D.D. UV–Vis. Spectroscopy: Herbal drugs and fingerprints. In *Herbal Drugs and Fingerprints: Evidence Based Herbal Drugs*; Springer: New Delhi, India, 2012; pp. 101–120.

34. Linoa, F.M.A.; De Sá, L.Z.; Torres, I.M.S.; Rocha, M.L.; Dinis, T.C.P.; Ghedini, P.C.; Somerset, V.S.; Gil, E.S. Voltammetric and spectrometric determination of antioxidant capacity of selected wines. *Electrochim. Acta* **2014**, *128*, 25–31. [CrossRef]

 *forests*

Article

Ethnobotanical Survey of Wild Edible Fruit Tree Species in Lowland Areas of Ethiopia

Tatek Dejene [1], Mohamed Samy Agamy [2], Dolores Agúndez [3] and Pablo Martin-Pinto [2,*]

[1] Ethiopian Environment and Forest Research Institute, 30708 Addis Ababa, Ethiopia; tdejenie@yahoo.com
[2] Sustainable Forest Management Research Institute, University of Valladolid (Palencia), Avda. Madrid 44, 34071 Palencia, Spain; samym392@yahoo.com
[3] INIA-CIFOR, Ecología y Genética Forestal, Carretera de la Coruña km 7.5, 28040 Madrid, Spain; agundez@inia.es
* Correspondence: pmpinto@pvs.uva.es; Tel.: +34-979-108-340; Fax: +34-979-108-440

Received: 20 December 2019; Accepted: 31 January 2020; Published: 5 February 2020

Abstract: This study aimed to provide baseline information about wild edible tree species (WETs) through surveying of different ethnic groups in dryland areas in Ethiopia. Here the data about WETs are scant, and WETs status is unexplained under the rampant habitat degradation. Use forms, plant parts used, status, ethnobotanical knowledge, conservation needs as well as those threats affecting WETs were reviewed. The study identified 88 indigenous wild edible plants, of which 52 species were WETs. In most cases, fruits were found as the dominant use part, and they were used as raw but were occasionally cooked and preserved. Roots and bark uses are also reported from *Ximenia americana* and *Racosperma melanoxylon* respectively. June, July and August were critical periods observed for food shortage in most of the regions. However, in the Gambella region, food shortages occurred in most months of the year. The respondents in this region suggested that WETs could potentially provide them with enough food to make up for the shortage of food from conventional agricultural crops. From the respondents' perception, *Opuntia ficus-indica*, *Carissa edulis* and *Ficus vasta* were among the most difficult to locate species, and they also received the highest conservation attention. Because of the variety of WETs and existing different threats, a management strategy is required for future conservation, as WETs are vital for the livelihood of local communities and are also necessary to devise a food security strategy for Ethiopia. The lesson obtained could also be useful in other dryland parts in developing countries with similar contexts.

Keywords: wild edible tree species; biodiversity; ethnic groups; conservation; food security

---

1. Introduction

More than 700 million people are suffering from hunger worldwide [1]. In some cases, nutritional deficiencies are due to a lack of diversity in the diet [2] and an inadequate supply of micronutrients [3]. In this context, wild forest foods can play an important role as supplements to the staple diet [4–6], by enhancing the diversity of the diet of many rural people in developing countries [7], increasing the nutritional quality of rural diets [8], and supplementing other food sources [5] during drought and famine periods. Thus, in recent years, there has been increasing attention focused on the sustainability of diets [5,9,10] and food systems, which has highlighted the need to conserve species diversity, mainly of forest resources [11,12] in many parts of the world.

Wild edible tree species (WETs) are among the most widely used non-timber forest products and represent an open access source of food and medicine [13], especially for vulnerable social groups. Thus, WETs are an important source of sustenance for many people in developing countries [6,14], improving household food security under normal circumstances as well as during crop scarcity periods [15–17] in urban and rural contexts [18]. Furthermore, WETs can represent important sources of income for their users [19,20].

More than 200 tree species have been registered as WETs in Ethiopia, which have been used since antiquity by rural people [21]. The wide range of climatic and edaphic conditions in Ethiopia enables a highly diverse range of WETs to grow in this country [22,23]. However, anthropogenic factors are causing the decline of these natural resources in most habitats [24,25]. Along with these factors, multiple components of climate change are also predicted to be the main drivers of biodiversity [26], with all levels of impact on WETs.

Many of the WETs found in Ethiopia are readily available for their nutritional, medicinal and marketable use [23,27]. The use of particular WETs is also determined by culture and location, and they continue to be maintained by cultural preferences and traditional practices. Although the use of WETs in Ethiopia has been investigated in different localities [23], there are still many WETs that are inadequately characterized and neglected by research. This partly explains why the most valuable WETs remain undocumented, particularly in dry forests where there is a relatively large supply of edible products. This situation, thus, greatly undermines their conservation and sustainable utilization. Furthermore, indigenous knowledge about the use of WETs has not been sufficiently documented, leading to a cultural erosion of their uses [28,29]. Consequently, interest in documenting information about wild edible food sources such as WETs in forest systems [30] around the world has increased in recent years. Ethnobiological studies of WETs are important to record information pertaining to the diversity of species used, the relative importance of each species, their different uses by indigenous communities, seasonal availability, and the conservation needs of the identified species, together with their main threats in the dryland areas of Ethiopia.

To enhance our understanding of the management and conservation strategies required for WETs in Ethiopia, we employed mixed methods to investigate different aspects of the same phenomenon [31] and gathered data from rural communities about traditional uses of WETs. We hypothesized that different rural populations would identify different uses for WETs according to their cultural history and locality. We also expected to find that different WETs face specific threats and that the conservation status for the main WETs would vary in each location. Thus, our specific objectives were (i) to identify tree species that provide edible products that are used by rural communities during periods of food shortage across different regions of Ethiopia; (ii) to evaluate the conservation status of the most consumed WETs; and (iii) to identify the main threats to these species and to assess how these threats varied across the studied areas.

2. Materials and Methods

2.1. Characteristics of the Study Areas

The drylands of Ethiopia consist of arid, semiarid and dry subhumid regions and cover approximately 55% of the land mass [32]. They are mainly found in the north, east and central areas of the Rift valley and also in the south and southeastern parts of the country, including a very wide and diversified range of agricultural environments. The altitude ranges from −124 to 1500 m above sea level. Rainfall is low, erratic, and uneven in distribution and ranges from 200 to 700 mm annually [32]. Soils in many drylands have low organic matter content, are highly eroded and have low fertility. The two main vegetation types in the dryland areas of Ethiopia are *Acacia–Commiphora* and *Combretum–Terminalia* deciduous woodlands [32].

This study was conducted in the dry agro-ecological zones found in six administrative regions of Ethiopia: Tigray, Amhara, Oromia, Benishangul Gumuz, Gambella and the South Nations, Nationalities, and Peoples (SNNP) (Figure 1). The inhabitants of these areas are sedentary agriculturists who practice mixed agriculture, crop production and livestock rearing. They commonly practice shifting cultivation and grow sesame, cotton and sorghum. The specific sites (i.e., the Woreda and Kebele) where the surveys were conducted are listed in Tables 1 and 2.

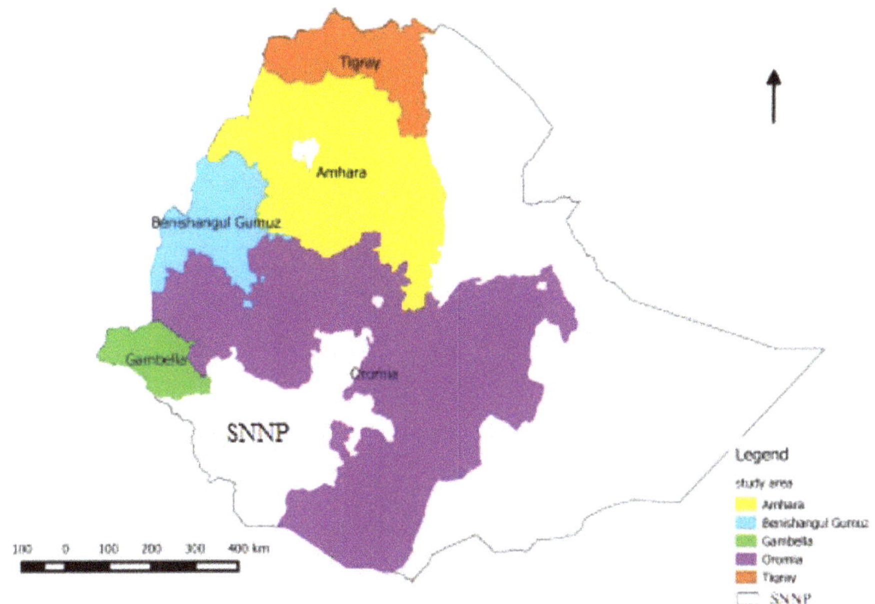

Figure 1. Map of Ethiopia showing the administrative regions that were surveyed in this study.

Table 1. Number of surveys conducted using focus groups and their location.

Region	Woreda/Kebele	Number of Surveys Conducted in the Kebele	Total Number of Surveys Conducted in the Woreda
Amhara	Kobo/Gedemeyu	3	6
	Kobo/Adis Kign	3	
Benishangul Gumuz	Bambasi/Bambisa	3	20
	Bambasi/Sonka	3	
	Debate/Debate	2	
	Debate/Parzeit	2	
	Homosha/Sherkole	3	
	Homosha/Tumet	3	
	Mandura/Duhansebeguna	2	
	Mandura/Edida	2	
Gambella	Gog/Puchala	3	12
	Gog/Gongjor	3	
	Lare/Ngour	3	
	Lare/Nip-nip	3	
Oromia	Dolo Mena/Chirri	2	2
South Nations, Nationalities and Peoples Region (SNNPR)	Hammer/Angode	3	6
	Hammer/Bita	3	
Tigray	Raya Azebo/Kara Adisho	3	6
	Raya Azebo/Hawelti	3	

Table 2. Number of surveys conducted using key informants and their location.

Region	Woreda/Kebele	Number of Surveys Conducted in the Kebele	Total Number of Surveys Conducted in the Woreda
Amhara	Kobo/Gedemeyu	4	9
	Kobo/Adis Kign	5	
Benishangul Gumuz	Bambasi/Bambisa	5	32
	Bambasi/Sonka	5	
	Debate/Debate	3	
	Debate/Parzeit	3	
	Homosha/Sherkole	5	
	Homosha/Tumet	5	
	Mandura/Duhansebeguna	3	
	Mandura/Edida	3	
Gambella	Gog/Puchala	3	18
	Gog/Gongjor	5	
	Lare/Ngour	5	
	Lare/Nip-nip	5	
Oromia	Dolo Mena/Chirri	1	3
South Nations, Nationalities, and Peoples Region (SNNPR)	Hammer/Angode	5	9
	Hammer/Bita	4	
Tigray	Raya Azebo/Kara Adisho	5	10
	Raya Azebo/Hawelti	5	

2.2. Data Collection: Sampling Technique and Sample Size

In this study, both the qualitative and quantitative data collection methods were employed. To collect data, two major primary data collection methods were used. The study was a cross-sectional survey of respondents from 19 Kebeles, the lowest administrative division in Ethiopia. This was supplemented with a total of 52 focus group discussions (Table 1) and 81 key informant surveys (Table 2). The respondents were selected using random sampling methods by ensuring the inclusion of at least 20% female in the sample. The following selection criteria were used to select the communities in the sampling: (i) a high dependence on agriculture and forestry, (ii) food shortages caused by drought, (iii) a high level of representation of the major ethnic groups and (iv) easy of accessibility of the Kebele.

2.2.1. Focus Group Discussion

Focus group discussions were conducted in the studied areas (Table 1). Each focus group consisted of 10 participants, and a total of 520 individuals were involved. These individuals were selected randomly from each study area. The purpose of the focus groups discussion was to generate information on a complete list of food trees for the study areas, which is necessary to estimate proportions of food that are obtained from wild trees and different issues including shortage period, product type collected form the forest, part of the tree used and the pattern of consumption, collection and availability time of the wild foods.

2.2.2. Key Informant Interview

The purpose of key informant interviews was to assess threats to important tree species. All key informants were familiar and knowledgeable for the area and the tree species identified in the focus group discussion. Thus, those who were known for their knowledge of the food tree products as food, traditional healers (from the area), local foresters, wood cutters, hunters, market vendors (who sell products obtained from trees) and others were included. Also, these individuals were supposed to have relatively good knowledge about their community situation, local natural resources, the culture of the community and the respective changes in the area.

A face-to-face semistructured questionnaire survey was conducted to collect primary data from the sampled key informants (Table 2). The questionnaire related to the objectives of this study was pretested with 15 randomly selected individuals in each region. Based on the results of the pretest work, necessary modifications to the questionnaires were implemented. Enumerators who were knowledgeable about the area were recruited from the study areas. Prior to performing the key informant interviews, the study objectives were explained to the enumerators, and they were trained in the methods of data collection and interviewing techniques. Finally, the survey was conducted on a total of 81 key informants. Hence, they shared their built-up knowledge and experience with the interviewer. The official language in Ethiopia is the Amharic language; however, the local people use their own languages. Thus, the surveys were translated from English to Amharic and from Amharic to local languages. In some cases, an interpreter conducted the surveys to ensure that the meaning of the questionnaires was not changed.

2.3. Data Analysis

Descriptive statistics were used to present the basic information obtained from the questionnaires. All the analyses were conducted based on the frequency of responses for each species, referring how many times the species were raised during the focus group discussion and key informant interviews. A final list of food tree species used by respondents was compiled from the questionnaires. The local names of the tree species were identified at their scientific name whenever possible following several keys [23,33–37].Clustering was used for the 30 most frequently used species based on the plant part used and how it was consumed by local communities. The cluster was based on the average linkage between groups. The statistical significance of the distance between groups was obtained by performing a chi square test using SPSS v.20. The chi square test was used for the 15 most frequently used tree species in order to analyze differences based on the following variables: period of food shortage, WETs food availability, ease of locating, conservation practices and regeneration presence. Respondents ranked the effect of different threats on each WETs on a scale of 0 to 4; therefore, Kruskal–Wallis and Mann–Whitney U tests were performed to analyze how the most frequently used species were affected. Data were analyzed using STATISTICA '08 edition software (StatSoft Inc., 1984–2008, the Netherlands).

3. Results

3.1. Diversity of Wild Edible Tree Species

A total of 88 wild edible plant species were identified by surveyed respondents as being utilized by local communities in the studied dryland areas of Ethiopia. Of these, 52 species belonging to 40 genera and 27 families were identified as WETs (Table 3). The families with the greatest numbers of edible tree species identified by survey respondents were the Malvaceae (five species) and Moraceae (four species), followed by the Boraginaceae, Anacardiaceae, Arecaceae, Rhamnaceae and Rubiaceae, which each had three edible tree species. These families represent about 46% of the registered taxa. Furthermore, approximately 31% of families were represented by more than 16 edible species, whereas the remaining 23% of families were represented by only a single species.

Table 3. List of wild edible species identified in study areas, including their plant part used and consumption form.

Species	Family	Part Used[a]	Consumption[b]	Region[c]
Adansonia digitata L.	Malvaceae	B	Rw	G
Balanites aegyptiaca Delile	Balanitaceae	L/F	Rw	T/A/B/G/SN
Balanites rotundifolia Blatt.	Balanitaceae	L/F	Rw/Ck/Pr	SN
Bauhinia thonningii Schumach.	Fabaceae	B	Rw/Pr	G/B/O
Borassus aethiopum Mart.	Arecaceae	F	Rw	G
Boscia mossambicensis Klotzsch	Capparaceae	F	Rw	SN
Carissa edulis Forssk.	Apocynaceae	L/F	Rw	T/A/B/G/O
Casimiroa edulis S.Watson	Rutaceae	F	Rw	B
Celtis africana Burm. f.	Cannabaceae	F	Rw	G
Commiphora schimperi Engl.	Burseraceae	R	Ck	SNNPR
Cordia africana Lam.	Boraginaceae	F	Rw	B

Table 3. *Cont.*

Species	Family	Part Used[a]	Consumption[b]	Region[c]
Cordia monoica Roxb.	Boraginaceae	F	Rw	T/A
Cordia sinensis Lam.	Boraginaceae	F	Rw	SN
Crateva adansonii DC.	Capparaceae	R	Ck	G
Diospyros mespiliformis Hochst.ex A.DC.	Ebenaceae	F	Rw	B/G
Dovyalis abyssinica (A. Rich.) Warb.	Flacourtiaceae	F	Rw	B
Ficus sur Forssk.	Moraceae	F	Rw	T/B/G/O/SN
Ficus sycomorus L.	Moraceae	F	Rw	B/G
Ficus vasta Forssk.	Moraceae	L/F	Rw	T/A
Flueggea virosa (Roxb. ex Willd.) Royle	Phyllanthaceae	L	Ck	G
Gardenia ternifolia Schumach. & Thonn.	Rubiaceae	F	Rw	B
Grewia bicolor Juss.	Malvaceae	F	Rw	SN
Grewia ferruginea Hochst.	Malvaceae	F	Rw	B
Grewia velutina (Forsk.) Lam.	Malvaceae	F	Rw	B
Grewia villosa Willd.	Malvaceae	L/F	Rw	T/A
Hyphaene thebaica Mart.	Arecaceae	F	Rw	G
Lannea humilis Engl.	Anacardiaceae	R	Ck	SN
Maytenus senegalensis (Lam.) Exell	Celastraceae	F	Rw	SN
Mimusops kummel Bruce ex A.DC.	Sapotaceae	F	Rw	B/G
Mitragyna inermis (Willd.) K.Schum.	Rubiaceae	F	Rw	B
Morus mesozygia Stapf	Moraceae	F	Rw	B
Nauclea latifolia Sm.	Rubiaceae	F	Rw	G
Olea capensis L.	Oleaceae	F	Rw	A/B
Oncoba spinosa Forssk.	Flacourtiaceae	F	Rw	B/G/O
Opuntia ficus-indica Mill.	Cactaceae	F	Rw	T/A
Phoenix reclinata Jacq.	Arecaceae	L/F	Rw/Ck	B
Pistacia lentiscus subsp. *emarginata* (Engl.) Al-Saghir	Anacardiaceae	F	Rw	B
Racosperma melanoxylon (R.Br.) Pedley	Fabaceae	F	Rw	G/O
Rumex nervosus Vahl	Polygonaceae	F	Rw	A
Saba comorensis (Bojer) Pichon	Apocynaceae	F	Rw	B
Searsia natalensis (Bernh. ex Krauss) F.A.Barkley	Anacardiaceae	F	Rw	T
Strychnos innocua Delile	Loganiaceae	F	Rw	B
Strychnos spinosa Lam.	Loganiaceae	F	Rw	B
Syzygium guineense DC. subsp. *guineense*	Myrtaceae	F	Rw	B/O
Tamarindus indica L.	Fabaceae	F	Rw	B/G/O/SN
Vitellaria paradoxa C.F.Gaertn.	Sapotaceae	F/B/S	Rw/Ck/Pr	G
Vitex doniana Sweet	Verbenaceae	F	Rw	B/G
Ximenia americana L.	Olacaceae	F	Rw	A/B/G/O/SN
Ximenia caffra Sond.	Olacaceae	F	Rw	B/O
Ziziphus abyssinica Hochst. ex A.Rich.	Rhamnaceae	F	Rw	G
Ziziphus mucronata Willd.	Rhamnaceae	F	Rw	SN
Ziziphus spina-christi (L.) Desf.	Rhamnaceae	F	Rw	All

[a] Part used: bark (B), fruit (F), leaf (L), root (R) and seed (S). [b] Consumed raw (Rw), cooked (Ck) or preserved (Pr). [c] Regions: All, in all regions, Gambella (G), Tigray (T), Benishangul Gumuz (B), Amhara (A), Oromia (O) and South Nations, Nationalities and Peoples (SN).

Based on the plant parts used and the method of consumption of the 52 WETs identified by respondents, the 30 most commonly reported WETs by the respondents in all regions were categorized into different groups (Figures 2 and 3). The edible plant parts that were commonly used were the fruit, leaf, bark, root and seed. Although the fruit of all 30 of the most commonly reported species was utilized by respondents, the cluster analysis categorized these WETs into two main groups and two independent species (Figure 2). The first main group comprised 18 WETs that were mainly used for their fruit. However, respondents indicated that the root of *Ximenia americana* and *Tamarindus indica* and that the bark of *Ficus sur*, *Celtis africana* and *Racosperma melanoxylon* were also used by local communities. The second main group comprised 10 species that were grouped together based on the consumption of their leaves. In addition, the bark of *Balanites aegyptiaca* was also reported as potential food. *Flueggea virosa* and *Vitellaria paradoxa*, which are used for their leaves and edible fat seed, respectively, were grouped independently of the other WETs.

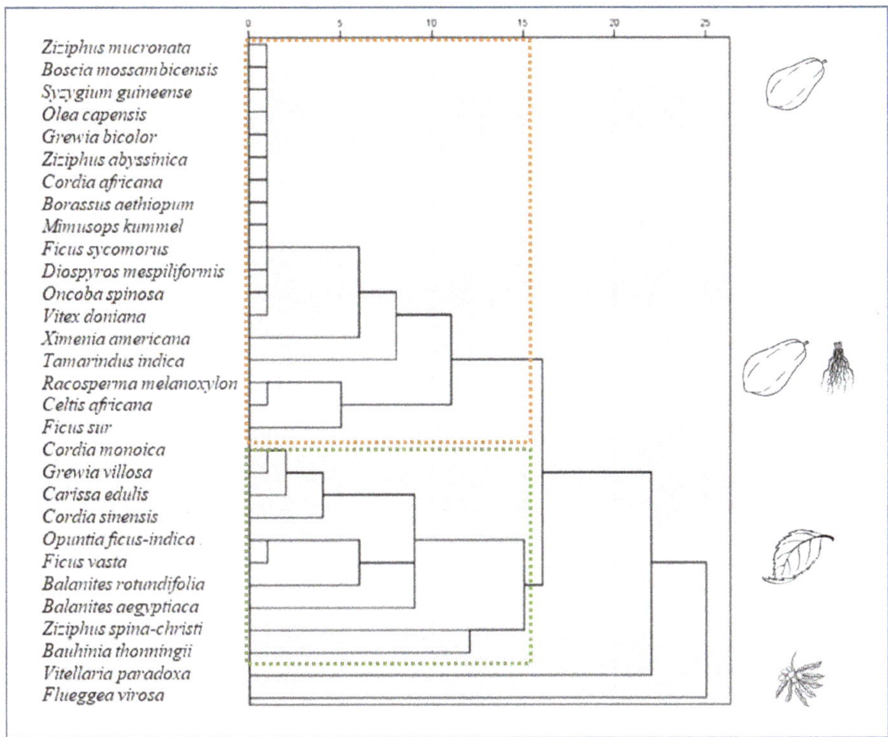

Figure 2. Dendrogram showing the classification of wild edible tree species based on the plant part used by local communities. The horizontal axis represents the distance or dissimilarity between clusters, and the vertical axis represents the species and clusters.

WETs were also clustered into three different groups and two independent species when analyzed based on the mode of consumption: raw, cooked or preserved (Figure 3). The respondents indicated that the 22 species in the first main group (Figure 3) were consumed as raw fruit. Within this group, *Tamarindus indica*, *Vitex doniana*, *Racosperma melanoxylon* and *Bauhinia thonningii* fruit were also preserved for future use. The fruit of the four species in the second main group were consumed raw and when cooked. However, *Carissa edulis* was a unique species in this group because the fruit of this species was also preserved for future use. In the third main group, *Flueggea virosa* and *Balanites rotundifolia* were consumed raw and when cooked, and they were sometimes preserved for future use. The fruit of *Balanites aegyptiaca*, which was classified as an independent species, was used raw, cooked and also preserved for future use, and *Vitellaria paradoxa* fruit was used as a form of fat.

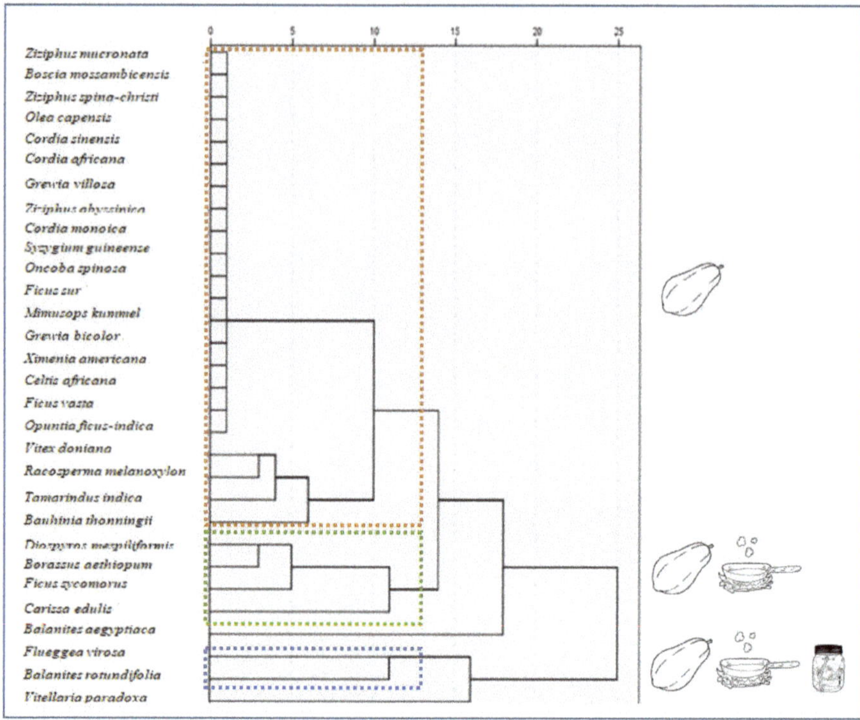

Figure 3. Dendrogram showing the classification of wild edible tree species based on the type of use or consumption by local communities. The horizontal axis represents the distance or dissimilarity between clusters, and the vertical axis represents the species and clusters. Red dashed line, fruit consumed raw; green dashed line, fruit consumed raw and cooked; blue dashed line, fruit consumed raw, cooked and preserved.

3.2. Seasonality of Wild Edible Tree Species and Shortage Periods

Periods of food shortage and the availability of food harvested from WETs strongly varied among regions (Chi square test; $p < 0.05$). The critical periods of food shortage for most of regions were May, June, July and August. However, in Gambella, food shortages occurred for most of the months over the year. Also in this region, reported by the respondents, the food shortages in March, April and May were higher than that of the other months (Figure 4a). The respondents in Gambella region suggested that WETs could potentially provide them with enough food to make up for the shortage of food from conventional agricultural crops during these periods (Figure 4b). The Oromia region had the fewest months with food shortages. Furthermore, food shortages in South Nations, Nationalities, and Peoples Region (SNNPR) were reported, by respondents, more frequently for February than for other months. Although the impact varied across regions, the data also indicated that the availability of food in all the studied regions was reduced during August, September and October, indicating that WETs did make up for this shortfall during this period (Figure 4b).

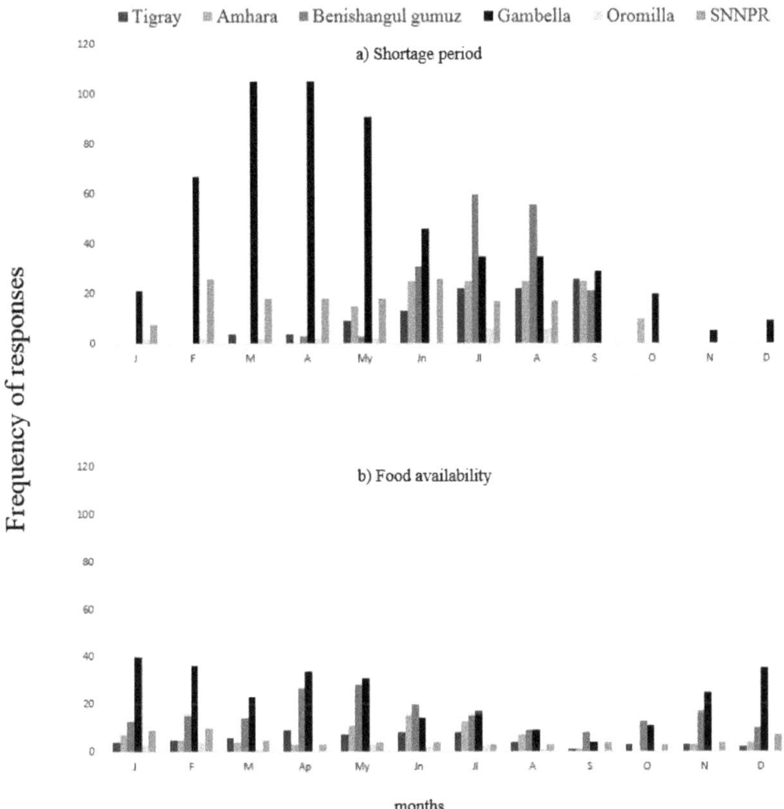

Figure 4. Seasonality of food shortage (**a**) and availability of WETs (**b**) in the study regions, Ethiopia. J = January, F = February, M = March, A = April, My = May, Jn = June, Jl = July, A = August, S = September, O = October, N = November and D = December. Frequency: frequency of positive answers from respondents regarding specific questions about periods of food shortage and the availability of food from the 15 most frequently used WETs.

Consumption patterns and plant part uses varied by regions (Chi square test; $p < 0.05$). Although the fruit of WETs were the most frequently used plant part in all the regions, far more respondents indicated that they consumed the fruit of WETs in the Gambella region than that in other regions (Figure 5a). In Tigray and SNNP regions, the number of respondents that indicated that they consumed WETs as leafy vegetables was higher than that in other regions, and more respondents in the Benishangul Gumuz region indicated that they consumed the roots of WETs than in other regions.

Although the fruit of WETs was usually consumed raw in all regions, fruit was not cooked or preserved in Oromia, Amhara and Tigray regions (Figure 5b). Root and leaf WET forms were also consumed when cooked and preserved in Benishangul Gumuz and SNNP regions, respectively. A greater number of respondents in the Gambella region indicated that they consumed wild fruit, leaves, bark and seed than respondents in other regions. People in the Gambella region consumed plant parts when raw, cooked (usually leave) or when preserved

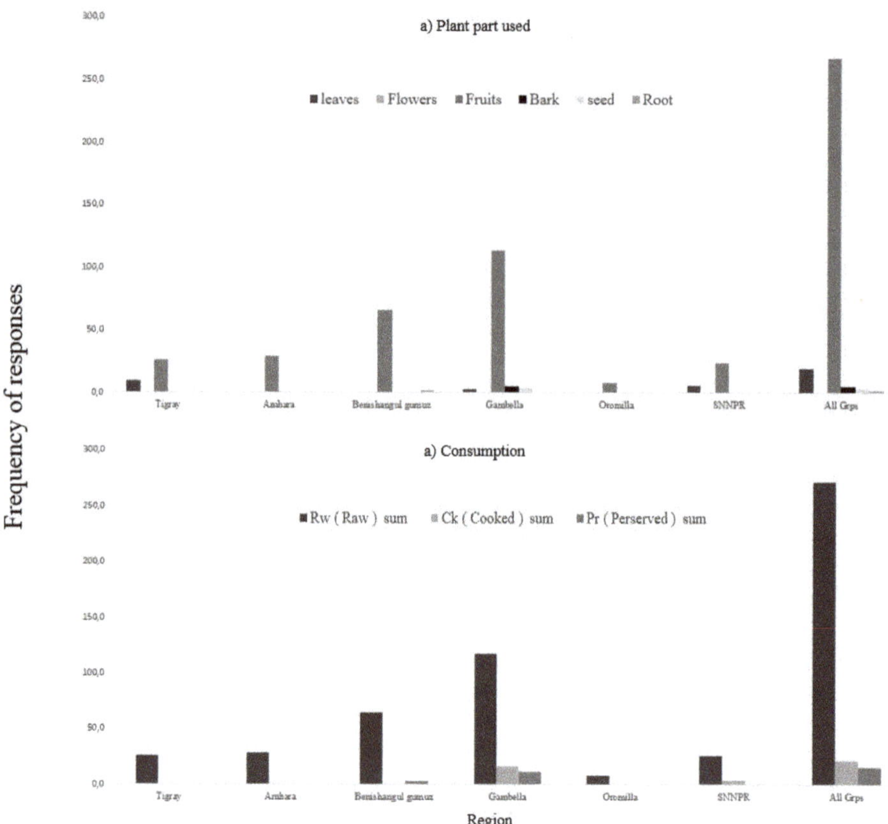

Figure 5. Plant parts used (**a**) and consumption form per region (**b**) of wild edible tree species in the study areas. Frequency: the frequency of positive answers from respondents regarding specific questions about the plant part used and their consumption of the 15 most frequently used WETs ($n = 280$; total frequency for the 15 most frequent species from focus group discussions).

3.3. Threats and Conservation Status

Based on the results from the respondents' view, most of the evaluated threats showed a significant influence (Kruskal–Wallis test, $p < 0.05$; Table 4) on WETs in the studied regions. Only grazing showed no influence on the conservation status of WETs ($p > 0.05$). The species were differently affected by the threats ($p = 0.00$). Individually, the species most threatened by all the threat factors was *Celtis africana* ($p < 0.05$). Pest and diseases, tree age and soil fertility were significant threats for *Ziziphus spina-christi*, *Balanites aegyptiaca* and *Tamarindus indica* tree species (Mann–Whitney U test; $p < 0.05$; Table 4).

Table 4. Effects of threats on the 15 most frequently used WETs based on the frequency ($n = 494$) of the responses for each threat: mean risk value for each threat categorized by respondents on a scale from 0 to 4 when considering the impact of each threat on each WETs. C: clearing of the forest, F: fire, G: grazing, T: timber harvesting, L: leaf harvesting, Fr: fruit harvesting, Fl: flower harvesting, R: root harvesting, B: bark harvesting, C: charcoal making, P: pest infestation, D: drought, AG: age of the tree, S: soil fertility condition of the site and O: others.

Name	Frequency	C	F	G	T	L	Fr	Fl	R	B	C	P	D	A	S	O
Ziziphus spina-christi	65	2.05	1.57	1.83	2.11	1.55	1.86	1.48	1.57	1.54	1.68	2.28	1.98	2.35	2.58	1.26
Tamarindus indica	54	1.96	1.80	1.78	1.52	1.30	1.76	1.37	1.33	1.30	1.31	2.65	2.11	2.15	2.28	0.43
Balanites aegyptiaca	53	2.06	1.57	1.79	1.77	1.47	1.98	1.28	1.40	1.47	2.09	2.25	1.60	2.00	2.23	0.96
Ximenia americana	45	1.82	1.60	1.73	1.40	1.27	1.87	1.07	1.24	1.24	1.38	2.13	1.62	1.47	2.29	0.67
Carissa edulis	44	1.77	1.77	1.50	1.09	1.18	1.68	1.07	1.27	1.23	1.18	1.75	1.41	1.80	1.91	0.68
Grewia villosa	29	2.21	1.66	2.07	1.86	1.79	1.93	1.62	2.03	2.00	1.79	2.69	1.83	1.83	2.79	0.93
Ficus sycomorus	31	1.86	2.07	1.43	1.68	1.32	2.11	1.50	1.43	1.46	1.32	2.18	1.64	2.00	1.68	0.75
Syzygium guineense	31	2.36	1.67	1.58	1.33	1.04	1.33	1.13	1.13	1.00	1.17	2.08	2.46	2.58	2.63	0.00
Ficus vasta	26	1.87	1.17	1.87	1.43	1.39	1.61	1.17	1.35	1.52	1.30	1.52	1.35	1.35	1.43	1.22
Opuntia ficus-indica	20	1.95	1.20	1.65	1.15	1.35	1.65	1.05	1.40	1.25	1.10	2.00	1.25	1.80	1.25	1.50
Cordia monoica	20	2.15	1.00	1.55	1.50	1.45	1.50	1.15	1.50	1.30	1.50	1.60	1.35	1.40	1.45	1.50
Bauhinia thonningii	22	2.21	1.89	1.74	1.53	1.68	2.21	1.47	1.47	1.63	2.05	2.26	2.00	2.42	2.47	1.47
Mimusops kummel	18	1.83	2.44	1.72	1.28	1.06	2.06	1.11	1.44	1.28	1.17	1.78	2.06	2.11	1.94	0.17
Diospyros nespiliformis	18	1.44	2.17	1.33	0.94	1.22	2.11	1.11	1.00	1.00	1.00	2.06	1.61	1.33	1.78	0.00
Celtis africana	18	1.83	1.61	1.89	2.11	1.83	2.28	2.33	2.67	2.67	2.28	2.39	2.44	2.78	2.44	2.06
Kruskal–Wallis (*p* value)		0.00	0.00	0.14	0.00	0.00	0.01	0.00	0.00	0.00	0.00	0.00	0.00	0.00	0.00	0.00

According to the respondents' view, the regeneration status of different species also differed significantly (Table 5; Chi square test; $p < 0.05$). *Ziziphus spina-christi, Balanites aegyptiaca, Ximenia americana* and *Tamarindus indica* showed the highest regeneration values (Figure 6). Although affected by different factors (Table 4), the perception of local people was positive regarding the regeneration of most WETs, except for *Celtis africana, Opuntia ficus-indica* and *Diospyros mespiliformis*, which were considered to have a relatively lower regeneration status respectively in their order.

Table 5. Chi square test results for regeneration status, ease of locating and conservation practices for WETs in the study areas.

	Presence of Regeneration			Ease of Locating			Conservation Practices		
	Chi Square	df	*p*	Chi Square	df	*p*	Chi Square	df	*p*
Pearson	38.13	df = 14	*p* = 0.0005	117.48	df = 28	*p* = 0.000	23.92	df = 14	*p* = 0.046
M-L	42.11	df = 14	*p* = 0.0001	130.17	df = 28	*p* = 0.000	26.96	df = 14	*p* = 0.019

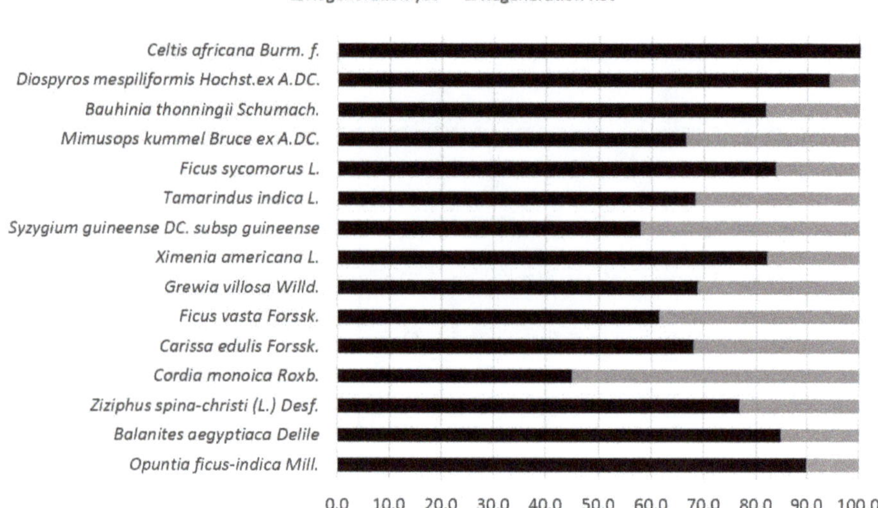

Figure 6. Regeneration status of the 15 wild edible tree species based on the responses from key informants in the study regions.

The ease of locating WETs was also significantly different ($p < 0.05$) depending on the species (Table 5). Based on the perception of the local people, *Celtis africana* was the most stable species followed by *Ficus sycomorus, Bauhinia thonningii, Diospyros mespiliformis, Tamarindus indica* and *Balanites aegyptiaca*, indicating these species were easily located in their locality. Among the key informants, 94.4% indicated that the ease of locating *C. africana* was the same as that in previous years. By contrast, *Opuntia ficus-indica* and *Cordia monoica* were perceived to be the most difficult WETs to locate compared with previous years (Figure 7a).

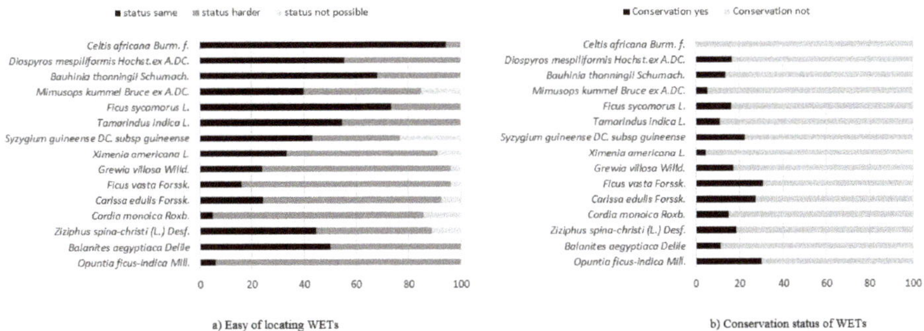

Figure 7. Values indicating the ease of locating WETs and access to these species (**a**), and their conservation status (**b**) in the studied areas. Status same: species easily located. Status harder: species harder to locate than previous years. Status not possible: not possible to locate species. Conservation: the percentage of answers indicating the existence of conservation practices by the local people for a specific WET.

The conservation status of WETs differed significantly among species (Chi square test; $p < 0.05$), with *Carissa edulis*, *Ficus vasta* and *Opuntia ficus-indica* showing higher conservation values than that of the other species (Figure 7b), indicating the conservation practice for these species by the local communities. In this case, the conservation practices considered by the local farmers included the deliberately leaving of trees on their farm land and sometimes planting of important wild edible trees in their garden.

4. Discussion

4.1. Wild Edible Tree Species

Our current knowledge of wild edible species in Ethiopia has usually been derived from small-scale case studies [23]. In this research, we included representative regions with different ethnic compositions from the dryland parts of the country. We identified a total of 88 indigenous wild edible plants (WEPs), of which 52 species (56%) were WETs, and 30 of these 52 species were considered by respondents to be commonly consumed by the community, and they were considered to be the most useful species among those listed for the studied regions. The number of WEPs reported in this study was relatively low compared with the number of species documented in previous studies carried out by [33] and [34] in southern Ethiopia. The lower number of WEPs found in the present study may be associated with differences in local traditions and customs relating to the use of wild plants in different parts of the country. Thus, this may reflect social variations in nutrition, attitudes and preferences towards wild food sources. Hence, it also explains differences in agroecology in different parts of the country.

Our results identified the existence of similar cultural practices and knowledge regarding plant parts used and the consumption pattern of WETs among different local communities in Ethiopia. Based on the social perception of respondents, fruit was the main food resource obtained from the most frequently used WETs in our survey, which agrees with findings reported by a previous study [35]. Fruit is mainly consumed raw because, in general, the storage conditions are not suitable for storing fruit for consumption during periods of food scarcity. Leaves throughout, bark, root and seed of some WETs are also commonly consumed by respondents, such as *Balanites aegyptiaca*, *Flueggea virosa* and *Vitellaria paradoxa*. Leaves, bark and roots of these species can also be preserved to complement the diet during months of food shortage in the dryland areas of Ethiopia. Even the oil from some fruit and seeds is used to prepare butter, supplementing the diet with additional calories. The role played by these kinds of WETs and WEPs was previously highlighted as supplementary and used as seasonal food sources by different communities [23,35,36] in some areas of the country as a way of combating food insecurity. Despite the importance of WETs in the study area, these tree species have generally been overlooked compared with domesticated plant food sources [23]. Our survey provided a good opportunity for ethnobotanical research of Ethiopian WETs because we were able to obtain ethnographic data directly from local communities, including observations on food culture and botany, plus field observations of different agroecological practices in rural areas where WETs are traditionally used on a daily basis.

4.2. Seasonality of Wild Edible Tree Species and Food Shortage Periods

In Ethiopia, products from WETs are generally collected for subsistence use [34,37,38] mainly because the production of edible parts, such as fruit, is seasonal and, therefore, can only be gathered for a short period of time. The fruit harvesting season and uses vary from place to place, even from species to species. This is due to climatic and intraspecific variations. In Ethiopia, seasonal food shortages are a common phenomenon and occur mainly from July to September [39] when storage bins have been emptied and the new crop is not yet ready for harvesting. Our survey revealed that WETs are commonly used during periods of food shortage, seasonally during periods of food scarcity and to add variety to the diet. The extent to which WEPs are used also varied with respect to season. For example, in most parts of the regions surveyed, local respondents reported that the food shortage period occurred in May, June, July and August. However, rural communities from Gambella reported March, April and May as the period when food shortages were most severe. In line with our results, other studies have also indicated that during these periods of food shortage, local communities depend on WETs [33,34,40,41].

The survey results also revealed that some WETs are only consumed during periods of famine, such as *Opuntia ficus-indica*, *Carissa edulis* and *Ximenia americana* in the Amhara region and *Syzygium guineense* and *Carissa edulis* in the Oromia region. This might be because these species are only used to supplement the normal diets of many rural people [42]. The respondents in this study indicated that most of the WETs have multiple edible uses. This diverse use of wild plants demonstrates that the indigenous people have a close relationship with local biological resources and that their lives are based on the use of diverse plant species [36]. Indeed, trees have been used as a source of food and medicine since time immemorial and they have become an integral part of the culture of the society throughout the country [43].

4.3. Threats and Conservation

Many threats affecting WETs are similar to those that affect other biodiversity resources in Ethiopia [44]. We evaluated threats affecting the 15 WETs reported most frequently used by local communities in the studied regions and their conservation status. The highest values/ranks were assigned to different WETs, including *Celtis africana*, *Ziziphus spina-christi*, *Balanites aegyptiaca* and *Tamarindus indica*. The respondents indicated that these plants are exploited more for their nonfood uses than for their food values. Overharvesting of these WETs to obtain fuel wood, medicine and for fencing, construction and forage purposes is aggravating the degradation status of these species in all the study areas. Moreover, the species were also affected by pests and diseases, which might have a direct influence on their degradation status. Diseases and pests start to occur when local communities change from a pastoral to an agropastoral way of life [33]. These factors can also limit the benefits that can be derived from the management and conservation of wild edible food plants in the dryland part of the country.

In general, owing to the diversity of WETs, conservation and management strategies are needed to achieve food security from the use of forest resources. Actions for the conservation and management of wild edible species include recognition of the limitations and the need to search for a way forward. Although all species with various uses deserve attention [45], the most highly valued food tree species in this study, *Celtis africana*, *Ziziphus spina-christi*, *Balanites aegyptiaca* and *Tamarindus indica*, should be prioritized when considering management strategies, their conservation and domestication. Moreover, overexploitation of plant parts (i.e., roots, leaves, bark and wood) could cause plant death or low plant productivity. Thus, unsustainable harvesting and product utilization should also be considered because these can cause the depletion of WETs in their natural habitats. Furthermore, their conservation should be encouraged and enhanced through the application of in situ and ex situ conservation programs, giving special consideration to those species currently used by local communities in different parts of the country.

5. Conclusions

This study attempts to provide baseline information that can be used as part of a management strategy for sustainable natural resource utilization in addition to documenting WETs in Ethiopia. The findings are based on the social perception of local communities who reported the actual use and demand for species found in the studied regions. From their responses, we were able to highlight the existence of valuable WETs and details relating to their utilization in the lowland parts of the country. The study also shows that WETs play a major role as a source of food and are characterized by a very high frequency of consumption by locals and contribute significantly to their livelihood through various uses. Such uses demonstrate that local people have a close relationship with their local biological resources. However, a range of factors are now affecting the WETs in Ethiopia, indicating that conservation practices should be enhanced through the application of management strategies, giving special consideration to those species currently used by local communities. Hence, there is also an urgent need for research studies on WETs to promote their use as part of a strategy to improve the food security, nutrition and livelihoods of rural communities throughout the country. Thus, the experiential knowledge of different ethnic groups should be documented because it is important for the development and conservation of important tree resources.

Author Contributions: All authors have read and agreed to the published version of the manuscript. Conceptualization, P.M.-P. and D.A.; methodology, D.A., P.M.-P. and T.D.; software, P.M.-P.; validation, P.M.-P.; formal analysis, P.M.-P.; investigation, P.M.-P. and D.A.; data curation, M.S.A.; writing—original draft preparation, M.S.A.; writing—review and editing, P.M.-P. and T.D.

Funding: This research and the APC were funded by the Spanish Agency for International Development Cooperation (AECID), PCI C/032533/10. Check carefully that the details given are accurate and use the standard spelling of funding agency names at https://search.crossref.org/funding, any errors may affect your future funding.

Acknowledgments: We would like to express our gratitude to the key informants and to the people involved in the focus groups discussion who provided the information from local communities. The research was partially funded by Spanish Agency for International Development Cooperation project (Sustfungi_Eth; 2017/ACDE/002094).

Conflicts of Interest: The authors declare no conflict of interest.

References

1. Javier, T.; Manuel, P.; Ramon, M. Ethnobotanical review of wild edible plants of Slovakia. *Bot. J. Linn. Soc.* **2006**, *152*, 27–71. [CrossRef]
2. Lachat, C.; Raneri, J.E.; Smith, K.W.; Kolsteren, P.; Van Damme, P.; Verzelen, K.; Penafiel, D.; Vanhove, W.; Kennedy, G.; Hunter, D.; et al. Dietary species richness as a measure of food biodiversity and nutritional quality of diets. *Proc. Natl. Acad. Sci. USA* **2018**, *115*, 127–132. [CrossRef] [PubMed]
3. Black, R.E.; Victora, C.G.; Walker, S.P.; Bhutta, Z.A.; Christian, P.; De Onis, M.; Ezzati, M.; Grantham-Mcgregor, S.; Katz, J.; Martorell, R.; et al. Maternal and child under nutrition and overweight in low-income and middle-income countries. *Lancet* **2013**, *382*, 427–451. [CrossRef]
4. De Caluwé, E.; Halamová, K.; Van Damme, P. *Adansonia digitata* L.-A review of traditional uses, phytochemistry and pharmacology. *Afrika Focus* **2010**, *23*, 10–51. [CrossRef]
5. Powell, B.; Thilsted, S.H.; Ickowitz, A.; Termote, C.; Sunderland, T.; Herforth, A. Improving diets with wild and cultivated biodiversity from across the landscape. *Food Secur.* **2015**, *7*, 535–554. [CrossRef]
6. Termote, C.; Meyi, M.; Ndjango, J.; Van Damme, P.; Dhed'a, D.B. Use and Socioeconomic Importance of Wild Edible Plants in Tropical Rainforest Around Kisangani District, tshopo, DR Congo. In *Systematics and Conservation of African Plants*; Van der Burgt, X., Ed.; Royal Botanic Gardens: Kew, UK, 2009; pp. 415–425.

7. Mengistu, F.; Hager, H. Wild edible fruit species cultural domain, informant species competence and preference in three districts of Amhara region, Ethiopia. *Ethnobot. Res. Appl.* **2008**, *6*, 487–502. [CrossRef]

8. Ickowitz, A.; Rowland, D.; Powell, B.; Salim, M.A.; Sunderland, T. Forests, trees, and micronutrient-rich food consumption in Indonesia. *PLoS ONE* **2016**, *11*, 1–15. [CrossRef]

9. Rowland, D.; Blackie, R.R.; Powell, B.; Djoudi, H.; Vergles, E.; Vinceti, B.; Ickowitz, A. Direct contributions of dry forests to nutrition: A review. *Int. For. Rev.* **2015**, *17*, 45–53. [CrossRef]

10. Termote, C.; Bwama Meyi, M.; Dhed'a Djailo, B.; Huybregts, L.; Lachat, C.; Kolsteren, P.; Van Damme, P. A biodiverse rich environment does not contribute to a better diet: A case study from DR Congo. *PLoS ONE* **2012**, *7*. [CrossRef]

11. Ickowitz, A.; Powell, B.; Salim, M.A.; Sunderland, T.C.H. Dietary quality and tree cover in Africa. *Glob. Environ. Chang.* **2014**, *24*, 287–294. [CrossRef]

12. Johnson, K.B.; Jacob, A.; Brown, M.E. Forest cover associated with improved child Health and nutrition: Evidence from the Malawi Demographic and Health Survey and satellite data. *Glob. Health Sci. Pract.* **2013**, *1*, 237–248. [CrossRef]

13. Beluhan, S.; Ranogajec, A. Chemical composition and non-volatile components of Croatian wild edible mushrooms. *Food Chem.* **2011**, *124*, 1076–1082. [CrossRef]

14. Vinceti, B.; Termote, C.; Thiombiano, N.; Agúndez, D.; Lamien, N. Food tree species consumed during periods of food shortage in Burkina faso and their threats. *For. Syst.* **2018**, *27*. [CrossRef]

15. Agúndez, D.; Douma, S.; Madrigal, J.; Gómez-Ramos, A.; Vinceti, B.; Alía, R.; Mahamane, A. Conservation of food tree species in Niger: Towards a participatory approach in rural communities. *For. Syst.* **2016**, *25*. [CrossRef]

16. Atato, A.; Wala, K.; Batawila, K.; Lamien, N.; Akpagana, K. Edible wild fruit highly consumed during food shortage period in Togo: State of knowledge and conservation status. *J. Life Sci.* **2011**, *5*, 1046–1057.

17. Faye, M.D.; Weber, J.C.; Abasse, T.A.; Boureima, M.; Larwanou, M.; Bationo, A.B.; Diallo, B.O.; Sigué, H.; Dakouo, J.M.; Samaké, O.; et al. Farmers' preferences for tree functions and species in the west African Sahel. *For. Trees Livelihoods* **2011**, *20*, 113–136. [CrossRef]

18. Jaenicke, H.; Hoschle-Zeledon, I. *Strategic Framework for Underutilized Plant Species Research and Development*; With special reference to Asia and the Pacific, and to Sub-Saharan Africa; ICUC, Colombo and Global Facilitation Unit for Underutilized Species: Rome, Italy, 2006; p. 33.

19. Melaku, E.; Ewnetu, Z.; Teketay, D. Non-timber forest products and household incomes in Bonga forest area, southwestern Ethiopia. *J. For. Res.* **2014**, *25*, 215–223. [CrossRef]

20. Sardeshpande, M.; Shackleton, C. Wild edible fruits: A systematic review of an under-researched multifunctional non-timber forest product. *Forests* **2019**, *10*, 467. [CrossRef]

21. Addis, G.; Urga, K.; Dikasso, D. Ethnobotanical study of edible wild plants in some selected districts of Ethiopia. *Hum. Ecol.* **2005**, *33*, 83–118. [CrossRef]

22. Friis, B.; Demissew, S.; Breugel, P. *Atlas of the Potential Vegetation of Ethiopia*; The Royal Danish Academy of Sciences and Letters: Copenhagen, Denmark, 2010; p. 307.

23. Lulekal, E.; Asfaw, Z.; Kelbessa, E.; Van Damme, P. Wild edible plants in Ethiopia: A review on their potential to combat food insecurity. *Afrika Focus* **2011**, *24*. [CrossRef]

24. Asfaw, Z. The Future of Wild Food Plants in Southern Ethiopia: Ecosystem Conservation Coupled with Enhancement of the Roles of Key Social Groups. *Acta Hortic.* **2009**, *806*, 701–708. [CrossRef]

25. Birara, E.; Mequanent, M.; Samuel, T. Assessment of Food Security Situation in Ethiopia: A Review. *Asia J. Agric. Res.* **2015**, *9*, 55–68. [CrossRef]

26. Parmesan, C. Ecological and Evolutionary Responses to Recent Climate Change. *Annu. Rev. Ecol. Evol. Syst.* **2006**, *37*, 637–699. [CrossRef]

27. Feyssa, D.H.; Njoka, J.T.; Asfaw, Z.; MM, N. Seasonal availability and consumption of wild edible plants in semiarid Ethiopia: Implications to food security and climate change adaptation. *J. Hortic. For.* **2011**, *3*, 138–149.

28. Alves, R.R.N.; Rosa, I.M.L. Biodiversity, traditional medicine and public health: Where do they meet? *J. Ethnobiol. Ethnomed.* **2007**, *3*, 1–9. [CrossRef] [PubMed]

29. Tabuti, J.R.S.; Kukunda, C.B.; Kaweesi, D.; Kasilo, O.M.J. Herbal medicine use in the districts of Nakapiripirit, Pallisa, Kanungu, and Mukono in Uganda. *J. Ethnobiol. Ethnomed.* **2012**, *8*, 1–15. [CrossRef] [PubMed]

30. Bharucha, Z.; Pretty, J. The roles and values of wild foods in agricultural systems. *Philos. Trans. R. Soc. B* **2010**, *365*, 2913–2926. [CrossRef]
31. Sarantakos, S. *Social Research*; Macmillan Press Ltd.: London, UK, 1998; p. 488.
32. Eshete, A.; Sterck, F.; Bongers, F. Diversity and production of Ethiopian dry woodlands explained by climate- and soil-stress gradients. *For. Ecol. Manag.* **2011**, *261*, 1499–1509. [CrossRef]
33. Assefa, A.; Abebe, T. Wild Edible Trees and Shrubs in the Semi-arid Lowlands of Southern Ethiopia. *J. Sci. Dev.* **2010**, *1*, 5–19.
34. Balemie, K.; Kebebew, F. Ethnobotanical study of wild edible plants in Derashe and Kucha Districts, South Ethiopia. *J. Ethnobiol. Ethnomed.* **2006**, *2*. [CrossRef]
35. Teketay, D.; Senbeta, F.; Maclachlan, M.; Bekele, M.; Barklund, P. *Edible Wild Plants in Ethiopia*; Addis Ababa University Press: Addis Ababa, Ethiopia, 2010; p. 575.
36. Awas, T.; Asfaw, Z.; Nordal, I.; Demissew, S. Ethnobotany of Berta and Gumuz people in western Ethiopia. *Biodiversity* **2010**, *11*, 45–53. [CrossRef]
37. Fentahun, M.T.; Hager, H. Exploiting locally available resources for food and nutritional security enhancement: Wild fruits diversity, potential and state of exploitation in the Amhara region of Ethiopia. *Food Secur.* **2009**, *1*, 207–219. [CrossRef]
38. Potentials and Constraints of Mushroom Production in Ethiopia; A Paper Presented at the National Mushroom Conference: Addis Ababa, Ethiopia. Available online: http://kmyb.yolasite.com/resources (accessed on 19 December 2019).
39. Getachew, O. Food Source Diversification: Potential to Ameliorate the Chronic Food Insecurity in Ethiopia. In Proceedings of the Potential of Indigenous Wild Foods, Diana, Kenya, 22–26 January 2001; Kenyatta, C., Henderson, A., Eds.; United States Agency for International Development, 2001. Available online: https://www.fsnnetwork.org/sites/default/files/indigenous_wild_foods.pdf (accessed on 19 December 2019).
40. Getahun, A. The Role of Wild Plants in the Native Diet in Ethiopia the Level of Plant Lore in Ethiopia. *AgroEcosystems* **1974**, *1*, 4556.
41. Wild-food Plants in Southern Ethiopia: Reflections on the Role of 'Famine-Foods' at a Time of Drought. Addis Ababa, Ethiopia. Available online: https://reliefweb.int/report/ethiopia (accessed on 19 December 2019).
42. The Hidden Harvest. In Seedling, the Quarterly Newsletter of Genetic Resources Action, International (GRAIN). Available online: https://www.grain.org/en/article/318-the-hidden-harvest (accessed on 19 December 2019).
43. Kassaye, K.; Amberbir, A.; Getachew, B.; Mussema, Y. A historical overview of traditional medicine practices and policy in Ethiopia. *Ethiop. J. Health Dev.* **2006**, *20*, 127–134. [CrossRef]
44. Ethiopia's Fifth National Report to the Convention on Biological Diversity. Available online: https://www.cbd.int/doc/world/et/et-nr-05-en.pdf (accessed on 17 December 2019).
45. Mamounata, B.; Moumouni, N.; Josephine, Y. Strategy of Conservation and Protection of Wild Edible Plants Diversity in Burkina Faso. *Anadolu J. AARI* **2017**, *27*, 82–90.

 *forests*

Article

Local Preferences for Shea Nut and Butter Production in Northern Benin: Preliminary Results

Dolores Agúndez [1,2,*], Théodore Nouhoheflin [3], Ousmane Coulibaly [3], Mario Soliño [4] and Ricardo Alía [1,2]

[1] INIA-CIFOR, Carretera de la Coruña km 7.5, 28040 Madrid, Spain; alia@inia.es
[2] iuFOR, Sustainable Forest Management Research Institute, University of Valladolid & INIA, Avd. Madrid s/n, 34004 Palencia, Spain
[3] LARCASS, Laboratory of Analysis and Capacity Strengthening in Social Sciences, 09 BP 932 Cotonou, Benin; tnouho@gmail.com (T.N.); o.coulibaly1995@gmail.com (O.C.)
[4] Department of Economic Analysis & ICEI, Complutense University of Madrid, 28223 Pozuelo de Alarcón, Spain; msolino@ucm.es
* Correspondence: agundez@inia.es

Received: 30 October 2019; Accepted: 14 December 2019; Published: 19 December 2019

Abstract: Shea products in Benin (West Africa) are produced in a low-developed agroindustry, but they are estimated to be the country's third largest export. The nut harvesting and quality guaranteeing in the butter process can only be achieved through improvements in the value chain, thus making it more attractive for stakeholders. The aim of this paper is to provide keys to a better product valorization, obtain a significant increase in household incomes based on shea butter marketing opportunities, and offer competitive products at the local and regional markets. Different markets were designed to catch processors and consumers' preferences for two improved shea products: butter and nuts in Northern Benin. An open-ended contingent valuation (CV) was applied, and the willingness to pay (WTP) and willingness to accept (WTA) were estimated by using a typical ordinary least squares (OLS) modelling approach. On local markets in Benin, the color, length, and weight of the nuts, as well as the color, smell, and texture of shea butter significantly influence, respectively, the processors' willingness to accept and the consumers' willingness to pay for a specific quality level. An increase in price would ensure the quality of the shea butter and would be covered by the premium to be paid by consumers. Certification design and the development of shea resources management and conservation programs should include ethnic preferences and consider gender, to avoid reducing women's profits in the shea butter local market.

Keywords: *Vitellaria paradoxa*; *Butyrospermum parkii*; agroforestry; market; non-wood forest product; contingent valuation; food; gender; Fulani

---

1. Introduction

Shea (*Vitellaria paradoxa* C.F. Gaertn, *Sapotaceae*) is one of the most important natural tree species of the Sudan–Guinean mixed open forest, in the green belt from Senegal to Ethiopia. It is present mainly in the agroforestry parks and fallows, in which various crops are grown among the trees (cotton, millet, maize, groundnut, cowpea, etc.) and in the remaining protected forests. The shea tree has traditionally been preserved alongside cultivation because of the high socioeconomic value given to shea butter, the main product resulting from the processing of its nuts that has multiple food, cosmetic, therapeutic, cultural, and religious uses. This nontimber forest product contains 45–55% of a solid cooking oil above 33 °C, and it means employment for over 18 million women collectors in Sub-Saharan Africa [1].

Nowadays, shea trees are the second most important oil crop in Africa, after the palm nut tree [2]. The trade of shea kernels in the international market, mainly for the use of their butter in cosmetics and as a fat in the production of chocolate, has been increasing since the 2000s. Western Africa represents 99.8% of total exports of shea nuts, whereas smaller quantities come from the eastern regions. Production and export estimates are that West Africa currently exports between 265,000 and 445,000 tons of shea nuts per year. The main exporters are Ghana, Burkina Faso, Benin, Côte d'Ivoire, Nigeria, Mali, and Togo [3]. Despite the potential and value of the shea tree products, there is a general regressive trend that raises high concerns about the conservation of the species and the supply of shea nuts [1,4–7].

Approximately half of the production of shea nuts has traditionally remained in the producing countries, and between 57% [3] and 41% [8] of the shea butter is consumed by domestic households. Production mainly meets the local demand, especially through the local markets, but it is not counted in the official statistics [7]. Many authors estimated that the international market potential of shea products would eventually lead to an improvement in the economies of the producing countries, their farmers, and especially the women involved [1,2,9–16]. However, the globalization of the market of this product had a weak impact on the reorganization at the regional scale. Even an increase in prices in, for example, Burkina Faso, hardly affected the quality standards and organization of the value chain governance at the local level [17,18].

The traditional production of shea butter is considered a woman's task, from the collection of the fruits in the area surrounding the villages to the sale and/or consumption of the butter at the household level, which makes women the main stakeholders in its value chain [10,19]. It is estimated that shea butter production is a source of income for hundreds of thousands of rural women [18], and that it is mainly considered as part of the household's subsistence income [20]. The majority of the processing of the kernels for their transformation into butter is still done by using traditional techniques, which are inefficient and lowers the quantity and quality of shea butter available in the market. The technical solutions required to improve the process are not, however, generally accessible to these women [21], in spite of the great importance of the shea products to their domestic economies.

The results of previous research point out the urgent need to improve the yield [10] and quality of shea butter production for the market [13]. In Benin, shea products are considered to be the country's third largest export, after cotton and cashews [22], but the sector has experienced a relatively limited development and represents only 2% of the world production [23]. The quality of the locally produced nuts speaks to Benin's untapped potential as a production country, but efforts to build a national market for shea butter remain to be made, as well as efforts to give added value to shea products, through the increased higher quality [24]. Meanwhile, the kernels are still sold at the local market or exported, and the butter produced is still consumed in the household or sold in the local or regional markets [25]. A production technique that effectively preserves the quality of the nuts and the butter would have a significant positive effect in the value chain, increase competitiveness, and, therefore, the local income [6].

Market accessibility remains a major challenge because shea processors are price takers and cannot provide the required quality consistency [3]. To solve this problem, some financial organizations have established technical support services, providing processing facilities for the producers [9]. Different initiatives, such as the Global Shea Alliance and the *Association Karité Bénin*, among others, aim at developing appropriate criteria for quality standards, provenance definitions, and processing procedures, as well as tree management and agricultural practices for nut production.

This study aims to offer preliminary key elements to more competitive shea products at the local and regional markets, as increased marketing opportunities would lead to higher domestic income. To do so, this study primarily focused on local preferences across the shea value chain in Northern Benin. An open-ended contingent valuation (CV) was applied, and the willingness to pay (WTP) and willingness to accept (WTA) were estimated by using a typical ordinary least squares (OLS) modelling approach. The main characteristics affecting the quality of the nuts and of the butter are described, as well as the socioeconomic factors influencing several steps in the value chain.

More specifically, the objective is to analyze the preferences of the local actors in regard to a future market in which a better quality of butter can be produced by (i) assessing processors' and consumers' preferences of the nut and butter quality available on the markets; (ii) analyzing the likely factors affecting better quality, and (iii) determining the marginal price to be paid/accepted when improving quality.

2. Materials and Methods

2.1. Area of Study

The area of study was selected due to the high socioeconomic value generated by the shea tree in the Bembereké parkland and more precisely in the Kalalé municipality (Figure 1), where 27% of the rural population is involved in the production and processing of shea products [26,27]. The main activity in the area is agriculture, the most important source of income for 80% of the population. Major crops grown are cotton, yams, maize, sorghums, cassavas, and groundnuts.

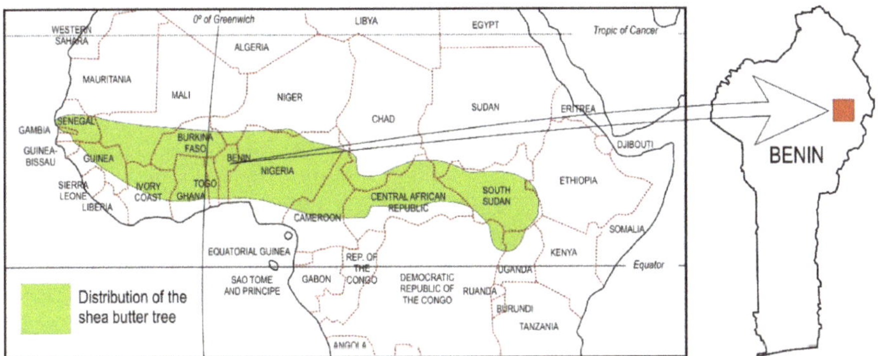

Figure 1. Greenbelt distribution of the shea tree and location of the study area in North Benin (in red).

2.2. The Sample

Three districts belonging to the Kalalé municipality were selected on the basis of their high role in shea butter production, their accessibility, and the existence of a local market. The main groups of stakeholders at the local shea value chain were selected for the interview, considering that women are the main task force. Women process nuts into butter, either through buying or self-collecting nuts, and then sell the butter. The processors' group was composed only of women, while the consumers' group was composed by both women and men. Women consumers interviewed at the local market are not involved in the production of butter and mostly buy it; men consumers interviewed at the household level are farmers with property rights on the land and/or on the shea trees. A sample of 319 individuals were contacted, and 307 were valid responses, i.e., completed surveys (Table 1).

Table 1. Survey districts in the municipality of Kalalé and total number of respondents.

Districts	Nut-into-Butter Processors (Women)	Consumers (Men at Household)	Consumers (Women at Market)	Total
Dérassi	49	25	47	
Péonga	23	15	23	
Kalalé center	51	32	42	
Total	123	72	112	307
WTP butter		72	112	184
WTA butter	123			123
WTP nuts	123			123

2.3. The Survey Design

The contingent valuation (CV) is a survey-based method. Two different questionnaires (for processors and consumers) were developed, including three different sections each. The survey was conducted with the support of independent interviewers for all the local languages (Boo, Fulani, and Baatonou) of those who filled up the questionnaires. Each interview was performed for a maximum time of 30 minutes. Women processors and men consumers were met at their homes, to answer the questionnaire, while women consumers were met at the local market, just after buying shea butter.

The first section of the questionnaire was common for shea processors and butter consumers, including 10 general questions about the respondent: ethnic group, age, level of education, household descriptions, and main activities of the respondent and his or her family.

The interviewed people belonged to different ethnic groups, including the Boo (57.5%), Fulani (15.9%), Gando (a different ethnic group that also speaks Fulani, 14.3%), and Baatombu (12.3%). The average age of respondents was 46 years old for men (min: 25; max: 80) and 40 years old for women (min: 17; max: 70). Fifty percent of the men received no formal education, 29% primary school, 17% secondary school, and 4% higher education. Ninety four percent of the women received no formal education and just 6% primary school. The household size was on average 10 people, with six persons active, involved in farm activities. The main activity was agriculture, which was also the main source of income.

Rural activities occupied two-thirds of the women, followed by processing (19%) and petty trading (16%). Women became more involved in field activities during the farming season, being, in general, more active in processing and trading activities.

The second section of the questionnaire included up to 8 questions about the factors affecting shea nut and butter production and quality: the source, period of acquisition, characteristics defining and factors affecting the quality and variation in the price for nuts and/or butter. Consumers were asked about the different usages of different, better grades of butter, the period over which the butter was mostly commercialized, and the origin and types of the butter. Processors were also asked about processing frequency and marketing of the butter. The major constraints associated with the processing of nuts into butter were also collected. This section allowed the identification of the factors affecting shea nut and butter production and quality

The third section of the questionnaire constituted the main exercise of the study and contained 5 CV-related questions. Processors were asked about the premium they were willing to pay (WTP) for a change of better-quality nuts and the price they were willing to accept (WTA) for better-quality butter. Consumers were asked about WTP for a better quality of butter (i.e., a shea butter with improved organoleptic characteristics). An open-ended format was used for the CV exercise, and the respondents were asked the following question: "How much are you willing to pay or to accept for a change dealing with a better quality of the product (nuts or butter)?" The currency used is the West African CFA Franc (XOF), which has an exchange rate that is pegged to the euro (EUR = XOF 655.957). Data were collected between November 2004 and February 2005 (see Supplementary Materials). The reference market price was XOF 1000 for approximatively 25 kg of nuts, and 100 XOF for approximatively 125 g of butter. Thus, WTP/WTA referred to these weights.

2.4. Contingent Valuation Method

The contingent valuation (CV) method [28] is based on the design of a hypothetical market in which a new product is commercialized. In the CV application, individuals were asked about their maximum willingness to pay (WTP) or minimum willingness to accept (WTA) for an improved and currently nonmarketed shea product. There are several ways to ask people about the WTP/WTA: open-ended questions, dichotomous format question, payment card, ranking, choice, etc. The choice of the elicitation format still represents an interesting academic debate [29]. In this paper, the open-ended format was used. Therefore, we directly asked people about their maximum WTP and minimum WTA, and avoided giving implicit clues to stakeholders about the value of shea products. We recognize that one of the main drawbacks of this elicitation format is that it can lead to overstating or understating

WTP values [30], being quite different attending to the public or private character of the goods (public vs. market). Different markets were designed to assess the preferences of the stakeholders for shea butter and nuts. The method was used to analyze the processors and consumers' preferences and decisions on improved characteristics of shea nuts and butter. Moreover, a descriptive analysis dealing with the variables influencing the WTP/WTA was performed thorough an ordinary least squares modelling (OLS). Previously, Pearson correlation analyses were used to select the key variables that could explain the WTP/WTA of the new hypothetical products. The OLS model is specified as follows:

$$Y_i = \beta_0 + \beta_i X_i + \varepsilon_i \tag{1}$$

where the dependent variables (Y) were the individual WTP or WTA; X is the vector of variables influencing the monetary amount that an individual is willing to pay/accept for the new shea-based products, i.e., nuts and butter of better quality than the ones currently sold at the local markets; β is a vector of parameters to be estimated; and ε is a random error term normally distributed with a zero mean and constant variance. The WTP and WTA were calculated as the average sample mean of the open responses ($\Sigma_{i=1 \dots n} Y_i/n$), where n is the sample size.

3. Results

3.1. Identification of the Factors Affecting Shea Nut and Butter Production and Quality

The main factors identified by respondents affecting the shea nuts and butter production and quality are shown in Table 2. The collection of the nuts takes place mainly during the rainy season. Nuts are picked and/or purchased without a clear grading technique. However, all the women processors recognized which nuts are better quality than others and mentioned maturity as the first criterion when selecting nuts for processing. Maturity was assessed based on physical features, such as the color, weight, diameter, and length of the nuts. A mature nut was recognized by its brownish color and by its weight: the heavier the nut, the better its quality and that of the butter.

Table 2. Identification of factors affecting nut and butter production and quality.

Shea Butter Challenges		% of Processors
Traditional processing	Low production	55.3
	Low extraction rate	44.7
Marketing constraints	Low production	77.2
	Flow	21.1
	Butter conservation	19.5
	Lack of marketing opportunities	5.7
Shea Nut Characteristics		**% of Processors**
Diameter	Large	13
	Short	87
Length	Large	40
	Short	60
Weight	Heavy	100
Color	Brown	100
Shea Butter Characteristics		**% of Consumers**
Color	White–gray	66
	Yellow	28
Odor	Slightly floral	96
Flavor	Slightly fruity	100
Texture	Solid	100

The processing techniques were described as traditional, i.e., manual, even if some form of semi-mechanization, in the form of a mill as crusher/grinder of the nuts (Figure 2), was observed at a few processing units. The remainder of the processing steps were much easier and did not require special equipment, although they were time intensive.

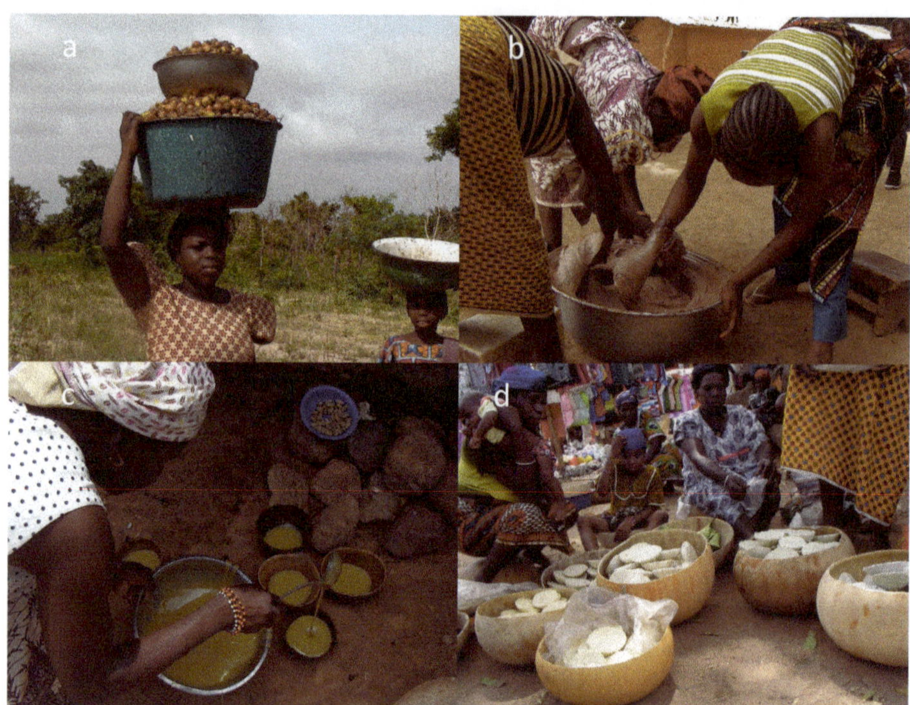

Figure 2. Pictures taken during the study in Kalalé show some of the steps in traditional process: young girls collecting shea nuts (**a**), women churning the shea paste (**b**), hot shea butter being poured into calabash molds (**c**), and selling butter at the local market (**d**).

Shea nut processors were aware about the challenges associated with the traditional process, (Table 2) such as its low production capacity and the low rate of butter extraction. The constraints women processors found related to the marketing of shea butter were said to be low production capacity, difficulties in flow, butter conservation, and lack of market opportunities.

Results showed that several categories of butter were produced. Women processors attribute these differences to the quality of the nuts used, the type of process used, and the presence of some additives. When the nuts begin to germinate, when they are not ripe, or when the drying has been poorly done, the butter obtained tends to turn to a green or yellow–green color, according to the proportion of nuts sprouted. Seventy-seven percent of the women processors report producing good butter of a white–gray color. A yellow butter color is not, however, related to the quality of the nuts used in its processing, but the result of the addition of powdered bark or root from different plants, e.g., the root of *Cochlospermum tinctorium*, which is added to the shea oil before its solidification. These additives were used by the women to change the color of their butter to satisfy their customers.

Consumers buying shea butter for household consumption searched for characteristics based on extrinsic features such as color, taste, smell, and texture. Indeed, 66% of consumers preferred a greyish-white color, while 28% chose a yellow color. The Boo ethnic group preferred white or gray–white shea butter, while the Fulani and Gando preferred butter with a yellow or white–yellow color. Indeed, Fulani and Gando processors used additives during processing to give it a yellow coloring. The Baatombu did not show any pattern of preferences for butter color. Moreover, almost all consumers (96%) were looking for butter with a slightly floral odor, fruity taste, and solid texture (100%).

3.2. Selection of Variables and Contingent Valuation

The key variables that could explain the willingness to pay/accept for the new hypothetical products are summarized in Table 3. These characteristic variables are considered to be eventual explanatory variables in the empirical model.

Table 3. Description of the variables.

Variable	Description of the Variables
KALALE	Kalalé village = 1; others = 0
FULANI	Fulani ethnic group = 1; others = 0
PASTOR	Pastoralist men Yes = 1; No = 0
ADULT	Active adults in the household (men and women, more than 15 years old) Yes = 1; Others = 0
AMAN	Active men in the household (more than 15 years old) Yes = 1; No = 0
EDU1	No education Yes = 1; No = 0
EDU4	High education Yes = 1; No = 0
COLOR	Preference for grey–brown color nuts Yes = 1; No = 0
LONG	Preference for long nuts Yes = 1; No = 0
WEIGHT	Preference for heavy nuts Yes = 1; No = 0
DIAMET	Preference for large diameter nuts Yes = 1; No = 0
BUTCOL	Butter color to be considered in the premium Yes = 1; No = 0
ODOR	Butter odor to be considered in the premium Yes = 1; No = 0
FLAVOR	Butter flavor to be considered in the premium Yes = 1; No = 0
BUTTEX	Butter texture to be considered in the premium Yes = 1; No = 0

Almost all participants were willing to contribute to ensure the quality of both nuts for processing and butter for household consumption. The models explain the factors proposed in the study that can affect this decision by 15–42% (Table 4).

All processors were willing to pay a maximum average premium of XOF 734 for nuts of a certain quality. The most important attributes influencing processors' choice were color, length, and weight. The premium processors were willing to pay for the quality of nuts is reduced when the nuts do not meet the desired quality: by XOF 358 for the length, by XOF 390.67 for the weight, and by XOF 153.85 for the color.

All of the interviewed consumers were willing to pay an average premium of 49.97 XOF (149.97 XOF) for butter with the desired quality and characteristics. The model explains in a 42% the factors that affect this decision, with a constant 68.59 XOF. The most important attributes of butter for the consumer were color and smell. If there is a variation in relation to their preferences, this premium will be reduced (24.02 XOF for color and 31.77 XOF for odor). Referring to the level of education, the premium will be reduced by 13.23 XOF if the respondent did not follow a regulated education, and, with the higher level, it will increase by 101.80 XOF. If they are from Kalalé, the WTP will be increased by 18.96 XOF; if they are Fulani, it will be increased by 16.92 XOF; and if they are men pastoralists, it will increase by 50.68 XOF. The women processors (96%) were willing to accept an average of 153.45 XOF (253.455 XOF in total) for a butter that they consider to be of better quality, i.e., with the characteristics they considered to produce a good-quality butter.

The fitness of the model is nearly 15%, and they would accept a minimum of 27.92 XOF. The most important attributes were color, smell, and texture. This premium would be increased for color (23.52 XOF) and smell (23.35 XOF), but decreased for texture (20.29 XOF).

Table 4. Ordinary least squares model.

Butter WTP Consumers	(1)				Butter WTA Processors (1)				
Variable	Coefficient		SE	t-Ratio	Variable	Coefficient		SE	t-Ratio
Constant	68.588	***	15.544	4.41	Constant	27.937	**	16.065	9.84
BUTCOL	−24.016	***	6.870	−3.50	BUTCOL	23.524	***	12.178	−3.45
ODOR	−31.771	***	9.352	−3.40	ODOR	23.351	*	17.080	4.04
FLAVOR	16.987		10.443	1.63	FLAVOR	0.048		16.880	−3.87
TEXTURE	2.782		12.192	0.23	BUTTEXT	−20.292	*	14.859	−4.54
KALALE	18.959	***	4.893	3.87	KALALE	1.257		8.209	−0.47
FULANI	16.928	**	7.647	2.21	FULANI	10.600		10.155	2.69
EDU1	−13.229	**	6.512	−2.03					
EDU4	101.802	***	18.928	5.38					
PASTOR	50.680	***	17.565	2.89					
AMAN	−9.814	*	5.029	−1.95					
ADULT	8.664		11.511	0.75					
R^2 (adjusted) = 42.21%					R^2 (adjusted) = 14.97%				
n = 184					n = 123				
Average WTP	49.97				Average WTP	53.45			

Nut WTP Processors (2)				
Variable	Coefficient		SE	t-Ratio
Constant	1055.810	***	87.687	12.04
COLOR	153.850	*	78.574	1.96
LONG	−358.045	***	81.682	−4.38
WEIGHT	−390.666	***	74.045	−5.28
DIAMET	−24.371		79.722	−0.31
KALALE	−230.535	***	83.576	−2.76
FULANI	300.250	***	104.231	2.88
R^2 (adjusted) = 35.32%				
n = 123				
Average WTP	734.55			

(1) Referred to 125 grs of butter
(2) Referred to 25 kgs of nuts

***, **, * ==> Significance at 1%, 5%, 10% level.

4. Discussion

This study assessed the factors that affect the shea butter chain and recommends strategies to improve quality in order to achieve higher household revenues. Some important social keys were also identified for consideration in the conservation and sustainable management planning of shea tree resources. Even if the data were collected between November 2004 and February 2005, some main reasons maintain the research questions and the current results: Traditional organization and the shea butter chain have slightly evolved during the last years in the area [23]; the need to improve the quality of shea butter production for the market is still a challenge [13], as well as the pending development of price premiums for the supply of improved products. This would bring the opportunity for rural producers to form tight links with buyers, institutions, or associations [3].

The income from shea products does not match Beninese local producers' expectations. Butter-making in the traditional way is a very painstaking process, involving multiple steps that can significantly affect the quality of the butter. There are particular constraints identified by women processors in the marketing of the shea butter: low production capacity, flow, butter conservation, and lack of market opportunities. Despite the efforts to promote the market for shea butter, researchers have argued that the absence of standardized measures (size, color, and weight) and the lack of market opportunities remain the main problems facing shea processors [9]. Even if the transformation and trade of quality butter is profitable, with positive gross margins average values, and net incomes and net returns on investments at the local level [31], the use of the traditional process results too often in poor-quality butter. In Ghana, women processors stated that they would be able to produce high-quality butter if they had the necessary equipment, and therefore would get competitive prices [13].

Additional means could help in improving the quality, as in the control and certification of shea products [32], although certification is costly and requires consumers to bear part of the cost. Interestingly, our study shows that the additional cost to ensure the quality of the shea butter would be covered by a premium to be paid by the consumers. Shea butter processors and consumers interviewed are, respectively, willing to accept (WTA, 28 XOF) and to pay (WTP, 69 XOF) over 100 XOF to guarantee the production and marketing of shea butter based on quality characteristics. In addition, 96% of the processors are willing to pay double (1056 XOF more over 1000 XOF) to ensure the supply of shea nuts for their processing into butter.

4.1. Shea Product Preferences by Socioeconomic Characteristics

Socioeconomic characteristics such as belonging to the Fulani ethnic group and being from Kalalé (the center of the district), significantly influence the willingness to pay of consumers of butter and nut processors. The level of education and being pastoralist also influence consumers' willingness to pay.

The preferred shea product characteristics are given more value by the Fulani ethnic group and pastoralist men. Fulani have been nomads or semi-nomadic pastoralists for centuries, but West African traditional specializations into agriculture and pastoral production systems are undergoing a process of change, affecting not only Fulani but other ethnic groups [33]. Where shea tree parklands are found, agriculturalist ethnic groups (Boo, Baatombu, and Gando, in our study) usually included shea butter for cooking the dishes. Fulani diet, in contrast, used butter from milk for cooking. Our results reflect these changes and the different behavior the Fulani show in comparison to the other ethnic groups: they show a higher WTP for both the purchase of shea butter (17 XOF) and nuts (300 XOF). To some extent, the butter from milk traditionally used in their dishes has been replaced by shea butter, while other ethnic groups have significantly reduced the food uses of shea, replacing it with other types of fat (i.e., locally produced palm or peanut oil, or other kinds of imported oil). Outside the Fulani, most women in Djougou (Benin) no longer use shea butter for cooking, replacing it with palm oil [6]. In a study in Burkina Faso, the Fulani were the group that most often used shea butter for cooking food and those that least commercialized it [34], and in Benin, Honfo [25] especially describes the Fulani as involved in the production and processing of shea butter, among others.

The fact that women are not the only shea actors in this group could be a reason why the Fulani and pastoralist men show a high WTP shea butter. Bidou et al. [6] found that, in Northern Benin, one-third of the Fulani women got help from their husbands with nut transportation. The participation of men in value chain activities traditionally reserved for women has been reported in other contexts when these activities become more remunerative [19], while the access to the land and to shea resources remains controlled by men [27]. Furthermore, in areas of high production of butter, the international marketing of the shea nuts and butter can be a source of gender conflict in the household [19]. When the value of the raw or processed shea products increases, the rights to access and the control of resources can be renegotiated and amended by the holder of the rights and other users [27]. Therefore, the future of the shea parklands and the development of shea butter market chains depends, to a large extent, on the evolution of gender relationships, which are subject to rapid cultural changes [6]. This has been the case in Ghana [15], where, in half of the households, the husband is involved in the decision-making process because of the significant income generated by the nuts. To increase women's profitability from the shea butter local market, a process supporting women in the transformation, negotiation, quality, and quantity of the production is needed [31].

4.2. Nut and Butter Characteristics for the Certification and Labeling of Shea Products

The color, length and weight of the nuts, as well as the color, smell and the texture of the butter are the factors that significantly influence the willingness to pay for a specific quality butter or nuts and to accept for the quality of the butter.

Consumers and processors agree on the most important characteristics of the butter. The variation in shea nut color would increase the premium to be paid by 154 XOF; this could be referring to the conservation status of dried nuts for storage or for being processed into butter. In addition, the size and weight of the nuts have an effect on butter performance: WTP decreases by 358 XOF for the length and 391 XOF for the weight. These results are in line with those found in a broader area of Benin, showing the degree of drying of the nuts and their appearance as the most important characteristics for the processors, and the color of the butter for the consumers [25]. Color and smell refer to the quality of the butter, which has been produced with nuts in good conditions, and the preservation of the butter itself. The preferred color varies from white to yellowish. Fulani and Gando respondents like a yellow color—resulting from adding natural products—better. A green color, together with an acid odor, mean that a high percentage of sprouted nuts have been used. The acid smell may also indicate that the butter is not from the current year's campaign and that it has not been well preserved. The processors' WTA decreases by 20 XOF for improving the texture of shea butter, which is probably either due to the current technology applied, or because they already estimate they produce high-quality butter. Furthermore, Bernard and Charlotte [35] found that the perception about the density of the product depends on the destination: light for sale by volume, heavy for sale by weight, and the price of butter paid by buyer intermediaries.

The local shea butter labeling and certification is supported by other valuation studies based on different agricultural and nontimber forest products. These studies have shown consumers' willingness to pay for certified and labeled safety or for improved products, as well as for assuring the availability of products on the market. Both young, well-educated, and wealthy consumers of certified vegetables [36], and the majority of the consumers of meat products [37], would pay price premiums to avoid health-related risks in Ghana. Consumers in Kenia place high value on the safety of leafy vegetables, suggesting the existence of a great potential for domestic market actors to improve leafy-vegetable value chains [38], as well as that of maize tested as aflatoxin-free and labeled [39] and porridge flour with improved nutritional aspects [40]. In Niger, the leaves of *Adansonia digitata* L. (African baobab) are highly important culturally as healthy ingredients in Nigerien traditional dishes. As the resource is declining, the majority of local consumers are willing to contribute to a baobab conservation program in order to ensure its availability in the market [41]. Our study offers new insights for the development of quality standards and certifications, as well as for shea tree resources sustainable management and conservation that consider ethnic and gender issues.

4.3. Consequences on Shea Tree Resources Management and Conservation

The knowledge that women have about the special features of the nuts producing butter of high quality represents a chance to ensure and to improve the supply of shea nuts. In Benin, shea tree is considered an endangered species, which is probably a consequence of the lack of appropriate protection strategies and effective regeneration methods [6], besides land management practices such as lack of fallow [42]. A great potential within the species has been found in the variation of important fruit and nut characters and yield, with a significant effect due to land use (agroforestry parklands versus protected areas) and interannual variation [43–46]. The results from the surveys in our study revealed that the nuts are picked and/or purchased without a clear grading technique, but that women could identify the factors that affect the quality and maturity of the nuts.

The women's knowledge about shea resources variation needs to be considered when planning sustainability and conservation strategies. However, they do not own the property rights of land or trees, nor can their opinions decisively influence their preservation, protection, or regeneration. In contrast, men usually do so, and they are the ones who are able to grant access to the natural resources [47]. Furthermore, landowners can have other interests, such as planting cashew orchards or cutting the shea trees to obtain wood. Merely including women in the shea nut production would not guarantee their empowerment, or the success on income development initiatives. Therefore, more equitable gender relationships should also be driven [19].

5. Conclusions

This study showed that the price to ensure the quality standards of the shea butter would be covered by a premium to be paid by the consumers. The color, smell, and texture of the butter on the one hand, and the color, length, and weight of the nuts, on the other hand, significantly influence consumers' willingness to pay. The development of the shea value chain should include the key actors in the production, collection, and processing of the shea nuts, such as the Fulani ethnic group, pastoralist men, and the women who process the nuts.

All the interviewed shea butter processors and consumers aimed at guaranteeing the quality of the shea butter production, and there was an agreement on the most important characteristics of the butter. Our study shows that a price that ensured the quality of the shea butter would be covered by a premium to be paid for by the consumers. The price premium could help to reduce the constraints that women processors find in relation to the processing and marketing of the shea butter which, in turn, are related to the traditional processing method, including production capacity, product conservation, flow, and lack of market opportunities.

This study offered key elements to increase Benin's shea products' value and the development of their potential, through the increased availability of good-quality products, and to consider labeling and certification of local shea butter. The color, smell, and texture of the butter, as well as the color, length, and weight of the nuts, significantly influence the willingness to pay for a specific quality of butter or nuts. These preferences may be considered in the design of a quality label for shea products. Other elements, such as the factors influencing the nutrition characteristics and safety, the volume or weight, and the packaging of the final product, should be considered in further studies, as the model is not completely explicative of the factors that influence the willingness to participate in the improvement of shea production quality.

New insights into shea tree sustainable management and resource conservation that consider ethnic and gender factors are offered, and the development of shea butter quality standards and certification are proposed. As the development of local shea butter production and marketing has a direct consequence on the management and conservation of the tree resources, both the Fulani ethnic group and pastoralist men should be included in the development of management programs, as well as a gender perspectives that aim at avoiding women having their profits in the shea butter local market reduced.

Moreover, the knowledge women have about shea resource variation should be considered when planning sustainable production and conservation strategies. The different existing social initiatives for shea production should invest more in capacity building, improvement of the local processing and dissemination, and developing institutional strategies.

As previously stated, the results are preliminary. This study was conceived at a period of great interest in boosting the shea product market, but the models explain 15–42% of the factors that can affect the WTP/WTA decision. Nowadays, other methodological approaches, such as auctions or choice experiments, can be considered as a superior alternative to analyze the preferences for marketed goods. Moreover, further investigations are needed on the preferences of shea butter production and their links to the conservation and sustainable management planning of shea tree resources.

Supplementary Materials: The following are available online at http://www.mdpi.com/1999-4907/11/1/13/s1.

Author Contributions: T.N., O.C., and and D.A. jointly conceived of and designed the study; T.N. implemented the study and performed the preliminary analyses; M.S. performed the formal analysis and contingent valuation modelling approach; R.A. was the supervisor of the INIA Bioversity International Joint collaboration; D.A. wrote the first draft of this article, to which the rest of the other authors provided their input. All authors have read and agreed to the published version of the manuscript.

Funding: This research was performed within the framework of the (i) "Variation of a Sub-Saharan agroforestry species. Shea butter tree (*Vitellaria paradoxa*) selection; morphological and phenological (leafing, flowering and fruiting) variation in Kalalé district (Northern Benin)" and was co-funded by AECID (International Cooperation and Development Spanish Agency), the (ii) Strengthening Regional Collaboration in Conservation and Sustainable Use of Forest Genetic Resources in Latin America and Sub-Saharan Africa. INIA-Bioversity International Joint collaboration in the SAFORGEN Programme Project financed by INIA, and the (iii) the Project AEG 17-048 established in the frame of the measure 15.2 "support to the conservation and use of forest genetic resources" and under Regulation (EU) No. 1305/2013 of the European Parliament and of the Council of 17 December 2013, on support for rural development by the European Agricultural Fund for Rural Development (EAFRD), with 75% co-financing.

Acknowledgments: We would like to thank all the people who participated in this study, mainly the respondents and the countless people who cooperated selflessly in the field work, including the local translators. We give our thanks to Oscar Eyog-Matig, former coordinator of SAFORGEN (Sub-Saharan African Forest Genetic Resources program), for his continued support in conducting this study. The authors wish to thank Patricia Grant for her editorial contribution.

Conflicts of Interest: The authors declare no conflicts of interest.

References

1. Naughton, C.C. Household food security, economic empowerment, and the social capital of women's shea butter production in Mali. *Food Secur.* **2017**, 773–784. [CrossRef]
2. Nde, B.D.; Mohammed, M.A.; César, K.; Zéphirin, M. Production zones and systems, markets, benefits and constraints of shea (*Vitellaria paradoxa* Gaertn) butter processing. *Oilseeds Fats Crop. Lipids* **2014**, *21*. [CrossRef]
3. Lovett, P.N. *The Shea Butter Value Chain: Production, Transformation and Marketing in West Africa*; West Africa Trade Hub (WATH) Technical Report; USAID: Washington, DC, USA, 2004; pp. 1–40.
4. Kafilatou, S.T.; Léonard, A.E.; Vincent, E.; Eliassou, S.H. Agro-morphological variability of shea populations (*Vitellaria paradoxa* CF Gaertn) in the Township of Bassila, Benin Republic. *J. Plant Breed. Crop Sci.* **2015**, *7*, 227–236. [CrossRef]
5. Akpona, T.J.D.; Akpona, H.A.; Djossa, B.A.; Savi, M.K.; Daïnou, K.; Ayihouenou, B.; Kakaï, R.G. Impact of land use practices on traits and production of shea butter tree (*Vitellaria paradoxa* C. F. Gaertn.) in Pendjari Biosphere Reserve in Benin. *Agrofor. Syst.* **2016**, 607–615. [CrossRef]
6. Bidou, J.-É.; KouKPéré, A.; Droy, I. *Agroforesterie et Services Écosystémiques en Zone Tropicale*; Éditions Quæ: Versailles Cedex, France, 2019; ISBN 9782759230594.
7. Seghieri, J. Shea tree (*Vitellaria paradoxa* Gaertn. f.): From local constraints to multi-scale improvement of economic, agronomic and environmental performance in an endemic Sudanian multipurpose agroforestry species. *Agrofor. Syst.* **2019**, *1*. [CrossRef]
8. Reynolds, N. *Investing in Shea in West Africa: A US Investor's Perspective*; West Africa Trade Hub (WATH) Technical Report; USAID: Washington, DC, USA, 2010; pp. 1–23.
9. Al-hassan, S. Market Access Capacity of Women Shea Processors in Ghana. *Eur. J. Bus. Manag.* **2012**, *4*, 7–18.
10. Pouliot, M. Contribution of "Women's Gold" to West African Livelihoods: The Case of Shea (*Vitellaria paradoxa*) in Burkina Faso. *Econ. Bot.* **2012**, *66*, 237–248. [CrossRef]
11. Pouliot, M.; Elias, M. Geoforum To process or not to process? Factors enabling and constraining shea butter production and income in Burkina Faso. *Geoforum* **2013**, *50*, 211–220. [CrossRef]
12. Bello-bravo, J.; Lovett, P.N.; Pittendrigh, B.R. The Evolution of Shea Butter's "Paradox of paradoxa" and the Potential Opportunity for Information and Communication Technology (ICT) to Improve Quality, Market Access and Women's Livelihoods across Rural Africa. *Sustainability* **2015**, *7*, 5752–5772. [CrossRef]
13. Kombiok, E.; Agbenyega, O. The characteristics of financing arrangements for the production and marketing of shea (*Vitellaria paradoxa*) butter in Tamale in the Northern Region of Ghana. *South. For. J. For. Sci.* **2017**, 1–8. [CrossRef]
14. Awo, M.A. A Survey-Based Qualitative Analysis of the Institutional Structures and Policy Measures in the Shea Sector of Ghana. *Res. World Econ.* **2018**, *9*, 24–37. [CrossRef]
15. Kent, R. "Helping" or "Appropriating"? Gender Relations in Shea Nut Production in Northern Ghana. *Soc. Nat. Resour.* **2018**, *31*, 367–381. [CrossRef]
16. Hammond, J.; Van Wijk, M.; Pagella, T.; Carpena, P.; Skirrow, T.; Dauncey, V. *The Climate-Smart Agriculture Papers*; Springer: Cham, Switzerland, 2019; pp. 215–226. ISBN 9783319927985. [CrossRef]
17. Rousseau, K.; Gautier, D.; Wardell, D.A. Coping with the Upheavals of Globalization in the Shea Value Chain: The Maintenance and Relevance of Upstream Shea Nut Supply Chain Organization in Western Burkina Faso. *World Dev.* **2015**, *66*, 413–427. [CrossRef]

18. Kodua, T.T.; Ankamah, J.; Addae, M. Assessing the profitability of small scale local shea butter processing: Empirical evidence from Kaleo in the Upper West region of Ghana. *Cogent Food Agric.* **2018**, *13*, 1–11. [CrossRef]

19. Elias, M.; Arora-jonsson, S. Negotiating across difference: Gendered exclusions and cooperation in the shea value chain. *Environ. Plan. D Soc. Space* **2016**, *35*, 1–19. [CrossRef]

20. Rousseau, K.; Gautier, D.; Wardell, D.A. Socio-economic differentiation and shea globalization in western Burkina Faso: Integrating gender politics and agrarian change. *J. Peasant Stud.* **2017**, *46*, 1–20. [CrossRef]

21. Kabiru, S.M. Utilisation of Modern Processing Technologies among Shea Butter Processors in Niger State, Nigeria. *J. Agric. Ext.* **2017**, 27–38. [CrossRef]

22. Ministère d'État chargé du Plan et du Développement. République du Bénin. In *Plan National de Développement 2018–2025*; BéninRévélé: Portonovo, Bénin, 2018.

23. Ahouansou, R.H.; Agbobatinkpo, P.B.; Sanya, A.; Gnonlonfin, B.; Fandohan, P. Comparative study of some physical properties of the shea kernels in the shea parks in Benin. *Int. J. Biol. Chem. Sci.* **2016**, *10*, 2151–2162. [CrossRef]

24. Association Karité Bénin. *Compte Rendu de la Conférence Annuelle du Karité*; CAK: Natitingou, Bénin, 2015.

25. Honfo, F.G. Indigenous Knowledge of Shea Processing and Quality Perception of Shea Products in Benin. *Ecol. Food Nutr.* **2012**, 37–41. [CrossRef]

26. Gnanglè, P.C.; Yabi, J.; Glèlè, K.R.; Sokpon, N. Changements climatiques: Perceptions et stratégies d'adaptations des paysans face à la gestion des parcs à karité au Centre-Bénin. In Proceedings of the SIFEComm, Niamey, Niger, 2005.

27. Agúndez, D.; Houtondji, F.; Simeni-Tchuinte, G. Land tenure, access and gender considerations in the Management and conservation of agroforestry resources in the North of Benin Republic. In *Agrarian Research for Sustainable Development: International Study Cases*; AcademicPres Publishing: Cluj-Napoca, Romania, 2014; pp. 181–202. ISBN 978-973-744-364-9.

28. Mitchell, R.C.; Carson, R.T. *Using Surveys to Value Public Goods: The Contingent Valuation Method*; Resources for the Future Press: Washington, DC, USA, 1989.

29. Carson, R.T.; Flores, N.E.; Meade, N.F. Contingent valuation: Controversies and evidence. *Environ. Resour. Econ.* **2001**, *19*, 173–210. [CrossRef]

30. Crastes dit Sourd, R.; Zawojska, E.; Mahieu, P.-A.; Louviere, J. Mitigating strategic misrepresentation of values in open-ended stated preference surveys by using negative reinforcement. *J. Choice Model.* **2018**, *28*, 153–166. [CrossRef]

31. Niculescu, N.; Badini, Z.; Diarra, M. Le beurre de karité au Burkina Faso: Entre marché domestique et filières d'exportation. *Cah. Agric.* **2009**, *18*, 369–375.

32. Garba, I.D.; Sanni, S.A.; Adebayo, C.O. Analyzing the Structure and Performance of Shea Butter Market in Bosso and Borgu Local Government Areas of Niger State, Nigeria. *Int. J. u-e-Serv. Sci. Technol.* **2015**, *8*, 321–336. [CrossRef]

33. Petit, S. Parklands with fodder trees: A Fulße response to environmental and social changes. *Appl. Geogr.* **2003**, *23*, 205–225. [CrossRef]

34. Tiétiambou, F.R.S.; Lykke, A.M.; Korbéogo, G.; Thiombiano, A.; Ouédraogo, A. Perceptions et savoirs locaux sur les espèces oléagineuses locales dans le Kénédougou, Burkina Faso. *Bois Forets des Trop* **2016**, *327*, 39–50. [CrossRef]

35. Bernard, P.B.; Charlotte, K. Des savoir-faire locaux entre IG et standardisation des produits: Exemple du beurre de karité au Burkina Faso. In Proceedings of the Actes du Colloque Localisation et Circulation des Savoir-Faire en Afrique, Aix-en-Provence, France, 19–20 March 2008; pp. 1–20.

36. Amfo, B.; Donkoh, S.A.; Gershon, I.; Ansah, K. Determinants of consumer willingness to pay for certified safe vegetables. *Int. J. Veg. Sci.* **2018**, *25*, 1–13. [CrossRef]

37. Owusu-sekyere, E.; Owusu, V.; Jordaan, H. Consumer preferences and willingness to pay for beef food safety assurance labels in the Kumasi Metropolis and Sunyani Municipality of Ghana. *Food Control.* **2014**, *46*, 152–159. [CrossRef]

38. Ngigi, M.W.; Okello, J.J.; Lagerkvist, C.J.; Karanja, N. Assessment of developing-country urban consumers'willingness to pay for quality of leafy vegetables: The case of middle and high income consumers in Urban Consumers' Willingness to Pay for Quality of Leafy Vegetables along the Value Chain: The Case o. *Int. J. Bus. Soc. Sci.* **2011**, *2*, 208–216.

39. De Groote, H.; Narrod, C.; Kimenju, S.C.; Bett, C.; Scott, R.P.B.; Tiongco, M.M.; Gitonga, Z.M. Measuring rural consumers'willingness to pay for quality labels using experimental auctions: The case of aflatoxin-free maize in Kenya. *Agric. Econ.* **2016**, *47*, 33–45. [CrossRef]

40. Chege, C.G.K.; Sibiko, K.W.; Wanyama, R.; Jager, M.; Birachi, E. Are consumers at the base of the pyramid willing to pay for nutritious foods? *Food Policy* **2019**, *87*, 101745. [CrossRef]

41. Agúndez, D.; Lawali, S.; Alía, R.; Soliño, M. Consumer Preferences for Baobab Products and Implication for Conservation and Improvement Policies of Forest Food Resources in Niger (West Africa). *Econ. Bot.* **2018**, *72*, 396–410. [CrossRef]

42. Aleza, K.; Villamor, G.B.; Nyarko, B.K.; Wala, K.; Akpagana, K. Shea (*Vitellaria paradoxa* Gaertn C. F.) fruit yield assessment and management by farm households in the Atacora district of Benin. *PLoS ONE* **2018**, *13*, e0190234. [CrossRef] [PubMed]

43. Bondé, L.; Ouédraogo, O.; Ouédraogo, I.; Thiombiano, A.; Boussim, J.I. Variability and estimating in fruiting of shea tree (*Vitellaria paradoxa* C. F. Gaertn) associated to climatic conditions in West Africa: Implications for sustainable management and development. *Plant. Prod. Sci.* **2018**, *22*, 143–158. [CrossRef]

44. Bondé, L.; Ouédraogo, O.; Traoré, S.; Thiombiano, A.; Boussim, J.I. Impact of environmental conditions on fruit production patterns of shea tree (*Vitellaria paradoxa* C. F. Gaertn) in West Africa. *Afr. J. Ecol.* **2019**, *57*, 353–362. [CrossRef]

45. Ouedraogo, S.; Bonde, L.; Ouedraogo, O.; Thiombiano, A.; Boussim, I.J.; Ouedraogo, S.; Bonde, L.; Ouedraogo, O.; Ouedraogo, A. To What Extent Do Tree Size, Climate and Land Use Influence the Fruit Production of *Balanites aegyptiaca* (L) Delile in Tropical Areas (Burkina Faso)? To What Extent Do Tree Size, Climate and Land Use Influence the Fruit Production of Balanites aegyp. *Int. J. Fruit Sci.* **2019**, 1–18. [CrossRef]

46. Gwali, S.; Bosco, J.; Okullo, L.; Eilu, G.; Nakabonge, G.; Nyeko, P.; Vuzi, P. Traditional management and conservation of shea trees (*Vitellaria paradoxa* subspecies nilotica) in Uganda. *Environ. Dev. Sustain.* **2012**, 347–363. [CrossRef]

47. Rousseau, K.; Gautier, D.; Wardell, D.A. Renegotiating Access to Shea Trees in Burkina Faso: Challenging Power Relationships Associated with Demographic Shifts and Globalized Trade. *J. Agrar. Chang.* **2016**. [CrossRef]

Article

Podophyllotoxin Isolated from *Podophyllum peltatum* Induces G2/M Phase Arrest and Mitochondrial-Mediated Apoptosis in Esophageal Squamous Cell Carcinoma Cells

Goo Yoon [1,†], Mee-Hyun Lee [2,3,†], Ah-Won Kwak [1], Ha-Na Oh [1], Seung-Sik Cho [1], Joon-Seok Choi [4], Kangdong Liu [2,3], Jung-Il Chae [5,*] and Jung-Hyun Shim [1,2,*]

[1] Department of Pharmacy, College of Pharmacy, Mokpo National University, Jeonnam 58554, Korea; gyoon@mokpo.ac.kr (G.Y.); rhkrdkdnjs12@mokpo.ac.kr (A.-W.K.); 17392303@mokpo.ac.kr (H.-N.O.); sscho@mokpo.ac.kr (S.-S.C.)

[2] The China-US (Henan) Hormel Cancer Institute, Zhengzhou 450008, China; mhlee@hci-cn.org (M.-H.L.); kangdongliu@126.com (K.L.)

[3] Basic Medical College, Zhengzhou University, Zhengzhou 450001, China

[4] College of Pharmacy, Daegu Catholic university, Havang-Ro 13-13, Havang-Eup, Gyeongsan-si, Gyeongbuk 38430, Korea; joonschoi@cu.ac.kr

[5] Department of Dental Pharmacology, School of Dentistry, BK21 Plus, Jeonbuk National University, Jeonju 54896, Korea

* Correspondence: jichae@jbnu.ac.kr (J.-I.C); s1004jh@gmail.com (J.-H.S.); Tel.: +82-63-270-4024 (J.-I.C.); +82-61-450-2684 (J.-H.S.); Fax: +82-63-270-4037 (J.-I.C.); +82-61-450-2689 (J.-H.S.)

† These authors contributed equally to this work as co-first authors.

Received: 12 November 2019; Accepted: 14 December 2019; Published: 18 December 2019

Abstract: Esophageal squamous cell carcinoma (ESCC) is one of the most common cancers in East Asia and is the seventh leading cause of cancer deaths. Podophyllotoxin (PT), a cyclolignan isolated from *podophyllum peltatum*, exhibits anti-cancer effects at the cellular level. This study investigated the underlying mechanism of anti-cancer effects induced by PT in ESCC cells. Exposure to increasing concentrations of PT led to a significant decrease in the growth and anchorage-independent colony numbers of ESCC cells. PT showed high anticancer efficacy against a panel of four types of ESCC cells, including KYSE 30, KYSE 70, KYSE 410, KYSE 450, and KYSE 510 by IC_{50} at values ranges from 0.17 to 0.3 µM. We also found that PT treatment induced G2/M phase arrest in the cell cycle and accumulation of the sub-G1 population, as well as apoptosis. Exposure to PT triggered a significant synthesis of reactive oxygen species (ROS), a loss of mitochondrial membrane potential (MMP), and activation of various caspases. Furthermore, PT increased the levels of phosphorylated c-Jun N-terminal kinase (JNK), p38, and the expression of Endoplasmic reticulum (ER) stress marker proteins via ROS generation. An increase in the level of pro-apoptotic proteins and a reduction in the anti-apoptotic protein level induced ESCC cell death via the loss of MMP. Additionally, the release of cytochrome c into the cytosol with Apaf-1 induced the activation of multi-caspases. In conclusion, our results revealed that PT resulted in apoptosis of ESCC cells by modulating ROS-mediated mitochondrial and ER stress-dependent mechanisms. Therefore, PT is a promising therapeutic candidate as an anti-cancer drug against ESCC for clinical use.

Keywords: ESCC (Esophageal squamous cell carcinoma); podophyllotoxin; ROS (reactive oxygen species); p38; JNK (c-Jun N-terminal kinase)

1. Introduction

Esophageal malignancies can be classified into adenocarcinoma and esophageal squamous cell carcinoma (ESCC). ESCC is the one of common malignant neoplasms in the worldwide, accounting for about 80% of the overall esophageal cancer-related deaths annually [1]. It originates from the esophageal epithelium, mainly the proximal and mid-esophagus [1,2]. Despite treatment of numerous cases of ESCC cancer with neoadjuvant chemoradiation or other primary treatments, almost 50% of patients suffer from recurrence [3]. The majority of therapies using HER2/neu (human epidermal growth factor receptor 2), EGFR (epidermal growth factor receptor), and VEGF (vascular endothelial growth factor) for ESCC are in clinical trials, with no remarkable findings [2]. Therefore, advances that can potentially decrease the adverse effects of conventional chemotherapy and improve the prognosis of patients are needed.

Natural products, such as plant extracts or isolates, show the benefits of less toxicity and low cost and also demonstrate anticancer, antioxidant, anti-inflammatory, and antibacterial properties [4–6]. Podophyllotoxin (PT) has been used as an herbal drug in traditional folk remedies for preventive and therapeutic applications for more than 1000 years. PT, one of the active cyclolignan compounds extracted from *Podophyllum peltatum* and *P. hexandrum*, has been reported to provide beneficial effects in patients with intestinal *Ascaris lumbicoides*-induced symptoms, venereal warts, lymphoma, and cancers [7,8]. Compelling reports have demonstrated that PT-derived anticancer drugs play a powerful role in inhibiting topoisomerase II activity and inducing apoptosis of several types of cancers, including female-related tumors, non-Hodgkin's lymphomas, and lung cancers [4,8,9]. PT also exhibits therapeutic activity in inducing cell cycle arrest (G2/M) by inhibiting the microtubule assembly [10]. Therefore, it is necessary to establish how cellular targets mediate the effects of PT to develop antitumor therapies against esophageal cancer.

Apoptosis plays an important role in eliminating damaged cells [11]. Endoplasmic reticulum (ER) stress is a cellular response produced by the accumulation of unfolded or misfolded proteins, which is related to apoptosis [12]. During sustained ER stress, the C/EBP homologous protein (CHOP), one of the death-related transcription factors, triggers the transcription of apoptosis-related genes, including Bcl-2 family proteins, death receptor (DR) 4, DR5, and Puma [13]. Anti-apoptotic proteins, such as Bcl-2 and Mcl-1, are crucial in maintaining mitochondrial functions, while pro-apoptotic proteins, such as bcl-2-like protein 4 (Bax), Bak, and Bid, promote the release of cytochrome c (cyto c) [14]. Reactive oxygen species (ROS), such as superoxide anion radicals and hydroxyl radicals, are chemically reactive molecules produced by common metabolic processes. A significant increase in ROS induces oxidative stress that can be deleterious to multiple organelles and enzymes in the cell [15]. In a recent study, ROS regulated the mitogen-activated protein kinase (MAPK) pathway [16]. MAPK comprises extracellular signal-regulated kinase (ERK), c-Jun N-terminal kinase (JNK), and p38 and plays an important role in cell proliferation and apoptosis signaling pathways [17,18].

The role of PT as an anticancer agent in ESCC has yet to be studied. The purpose of this study is to elucidate whether the anti-cancer effect of PT on esophageal cell lines (KYSE 30 and KYSE 450) is mediated via the JNK/p38 MAPK signaling pathway.

2. Materials and Methods

2.1. Reagents

Cell culture medium (RPMI-1640), fetal bovine serum (FBS), phosphate-buffered saline (PBS), penicillin, streptomycin, and trypsin were purchased from Hyclone (Logan, UT, USA) or Welgene (Daegu, Korea). Antibodies against JNK, phospho-JNK (p-JNK) (Thr183/Try185), p38, and phospho-p38 (p-p38) (Thr180/Try182) were purchased from Cell Signaling Technology (Danvers, MA, USA). The primary antibodies against β-actin, p21, p27, cell division cycle protein (cdc2), cyclin B1, CHOP, DR4, DR5, glucose response protein 78 (GRP78), Mcl-1, Bcl-2, Bax, cyto c, α-tubulin, COX4, Apaf-1, and cleaved-poly ADP-ribose polymerase (c-PARP) were purchased from Santa Cruz

Biotechnology (Santa Cruz, CA, USA). The CellTiter 96® AQueous One Solution was obtained from Promega (Madison, WL, USA). Dimethyl sulfoxide (DMSO) and PT were bought from Sigma-Aldrich (St. Louis, MO, USA).

2.2. Cell Culture

KYSE 30, KYSE 70, KYSE 410, KYSE 450, and KYSE 510 cells are human ESCC cell lines obtained by the Type Culture Collection of the Chinese Academy of Sciences (Shanghai, China). These cell lines were cultured in an RPMI 1640 medium containing 10% FBS, 100 U/mL penicillin, and streptomycin at 37 °C in a humidified incubator containing 5% CO_2.

2.3. MTS Assay

To evaluate the cytotoxicity of PT, cell viability was determined using an 3-(4,5-dimethylthiazol-2-yl)-5-(3-carboxymethoxyphenyl)-2-(4-sulfophenyl)-2H- tetrazolium inner salt (MTS) assay. Briefly, the KYSE 30, KYSE 70, KYSE 410, KYSE 450, and KYSE 510 cells were seeded into 96-well plates at different densities, as follows: 2.75×10^3/well, 10×10^3/well, 2.5×10^3/well, 3.5×10^3/well, and 5.5×10^3/well, respectively. The cells were exposed to different concentrations of PT for 24 h and 48 h and incubated with CellTiter 96® AQueous One Solution for 2 h. The absorbance was read by a spectrophotometer (Thermo Fisher Scientific, Vantaa, Finland) at 490 nm.

2.4. Soft Agar Assay

To establish colony growth by KYSE 30 and KYSE 450 cells in the soft agar, the 6-well plates were coated with 0.6% agarose in a BME medium, FBS, L-glutamine, and gentamicin with or without PT. Subsequently, the 8000 cells were layered on the agarose at the bottom comprising 0.3% agarose in a BME medium, FBS, L-glutamine, and gentamicin mixed with DMSO or PT. After two weeks, the colonies were counted, and images were observed on a phase contrast inverted microscope (Leica Microsystems, Wetzlar, Germany).

2.5. Cell Cycle Distribution Analysis

Cells were incubated with a range of PT doses (0.2 to 0.4 µM) for 48 h. Following PT exposure, the cells were washed three times with PBS and fixed with 70% ice cold ethanol for 24 h. The fixed cells were stained with Muse™ Cell cycle reagent (Merck Millipore, Billerica, MA, USA) for 30 min at 37 °C in the dark. The cell cycle distribution was measured using a Muse™ Cell Analyzer (Merck Millipore, Billerica, MA).

2.6. Annexin V/7-Aminoactinomycin D (7-AAD) Stained Cell Counting

KYSE 30 and KYSE 450 cells were seeded in 6-well plates. Cells were incubated for 48 h in the presence of PT and then harvested. Both cell lines were stained with Muse™ Annexin V and Dead Cell Reagent (Merck Millipore, Billerica, MA, USA) for 30 min at 37 °C in the dark. The stained cells were detected using a Muse™ Cell Analyzer.

2.7. Western Blots

The cells were harvested by a scraper and washed with ice-cold $1 \times$ PBS, three times. The total proteins of the KYSE 30 and KYSE 450 cell lysates were separated on 10%, 12%, and 15% SDS-PAGE gels and transferred to a polyvinylidene fluoride membrane. To block nonspecific binding, 3–5% nonfat dry milk in PBST (PBS containing 0.1% Tween 20) was added to the membrane for 2 h at room temperature (RT) followed by incubation of the membrane with the target primary antibody overnight at 4 °C. The membrane was washed with PBST for 30 min and incubated with appropriate horseradish peroxidase-conjugated secondary antibody for 2 h at RT, followed by visualization using an ECL Plus Western Blotting Detection system (Santa Cruz Biotechnology). The proteins were detected using an ImageQuant LAS 500 (GE Healthcare, Uppsala, Sweden).

2.8. ROS Assay

In order to determine the intracellular changes underlying ROS generation, a Muse™ Oxidative Stress Kit (Merck Millipore, Billerica, MA, USA) was used to measure the intracellular ROS levels in the cells of each treatment group. KYSE 30 and KYSE 450 cells were seeded on 6-well plates and treated with PT for 48 h. The cells were harvested and stained with a Muse™ Oxidative Stress Reagent in the cell incubator at 37 °C for 30 min. A Muse™ Cell Analyzer can be used to measure the intracellular ROS levels.

2.9. Measurement of Mitochondrial Membrane Potential

To evaluate the level of mitochondrial membrane potential (MMP, $\Delta\Psi$m), ESCC cells were treated with DMSO or PT (0.2 to 0.4 µM) for 48 h. The cells were washed with a 1X assay buffer and stained with Muse™ MitoPotential Dye (Merck Millipore, Billerica, MA, USA) for 20 min in the CO_2 incubator, followed by supplementation of the samples with 7-AAD (Merck Millipore, Billerica, MA, USA). The MMP measurement was conducted using a Muse™ Cell Analyzer.

2.10. Cytosolic and Mitochondrial Fractionation

Cytosolic and mitochondrial fractions were generated using digitonin-based subcellular fractionation methods reported previously [19]. The cells were resuspended in a plasma membrane extraction buffer [250 mM sucrose, 10 mM HEPES (pH 8.0), 10 mM KCl, 1.5 mM $MgC_{12}\cdot6H_2O$, 1 mM EDTA, 1 mM EGTA, 0.1 mM phenylmethylsulfonyl fluoride, 0.01 mg/mL aprotinin, and 0.01 mg/mL leupeptin] with 0.05% digitonin. After centrifugation at 13,000 rpm for 5 min at 4 °C, the cytosolic fraction was recovered. The pellet was washed with cytosolic membrane extraction buffer twice and homogenized with the same buffer containing 0.5% Triton X-100. Mitochondrial extracts were recovered by centrifugation at 13,000 rpm for 30 min at 4 °C.

2.11. Caspase Activity

The activities of caspase-1, -3, -4, -5, -6, -7, -8, and -9 were determined using a Muse™ Multi-Caspase Kit from Merck Millipore. Multi-Caspase activity was analyzed after treatment with DMSO or under different concentrations of PT for 48 h. The treated cells were resuspended in 1 × Caspase Buffer. The cells were stained with a Muse™ Multi-Caspase Reagent and a 7-AAD working solution. The multi-caspase activity was analyzed using a Muse™ Cell Analyzer.

2.12. Statistical Analysis

The data are presented as the means ± SD. Statistical analysis was performed using the Prism 5.0 statistical package. The statistical significance of differences between groups was analyzed using ANOVA. The mean values were considered statistically significant at $p < 0.05$. In the present study, the data represent experiments performed three times in triplicate.

3. Results

3.1. PT Suppresses ESCC Cell Proliferation and Colony-Forming Ability

To investigate the roles of PT in ESCC cells, we conducted an MTS assay to evaluate cell proliferation. PT (0.1, 0.2, 0.3, and 0.4 µM) treatment for 24 h and 48 h significantly decreased the cell proliferation of ESCC cells (KYSE 30, KYSE 70, KYSE 410, KYSE 450, and KYSE 510) in a concentration-dependent manner (Figure 1B–F). After 48 h of PT treatment, the half-maximal inhibitory concentrations (IC_{50}) of PT were 0.27 µM (KYSE 30), 0.3 µM (KYSE 70), 0.17 µM (KYSE 410), 0.3 µM (KYSE 450), and 0.3 µM (KYSE 510). The colony formation assay was conducted to confirm another type of anti-proliferative effect of PT on KYSE 30 and KYSE 450 cells. Both the colony number and size were counted after PT treatment for 14 days. As shown in Figure 1G,H, the colony formation in both ESCC cells was significantly suppressed under elevated concentrations of PT. These results indicated that PT inhibits the growth of ESCC cells. To identify whether the cell growth suppression with PT was triggered by apoptosis, ESCC cells, KYSE 30, and KYSE 450 treated with PT were stained with Annexin V/7-AAD and analyzed via flow cytometry. In KYSE 30 cells, the total percentages of cells stained positive with Annexin V (early and late apoptosis) were $3.10 \pm 0.38\%$, $7.14 \pm 1.49\%$, $26.77 \pm 1.58\%$, and $56.68 \pm 1.04\%$ after treatment with PT at 0, 0.2, 0.3, and 0.4 μM, respectively, for 48 h (Figure 1I). Similarly, treating KYSE 450 cells for 48 h increased the total percentage of apoptotic cells (early and late apoptosis) from $4.97 \pm 0.33\%$ at the basal level (PT 0 μM) to $6.42 \pm 0.17\%$, $10.11 \pm 0.86\%$ and $56.01 \pm 1.49\%$ with PT 0.2, 0.3, and 0.4 μM, respectively. Thus, our data indicate that treatment with PT reduced cell viability and promoted apoptosis in ESCC cells.

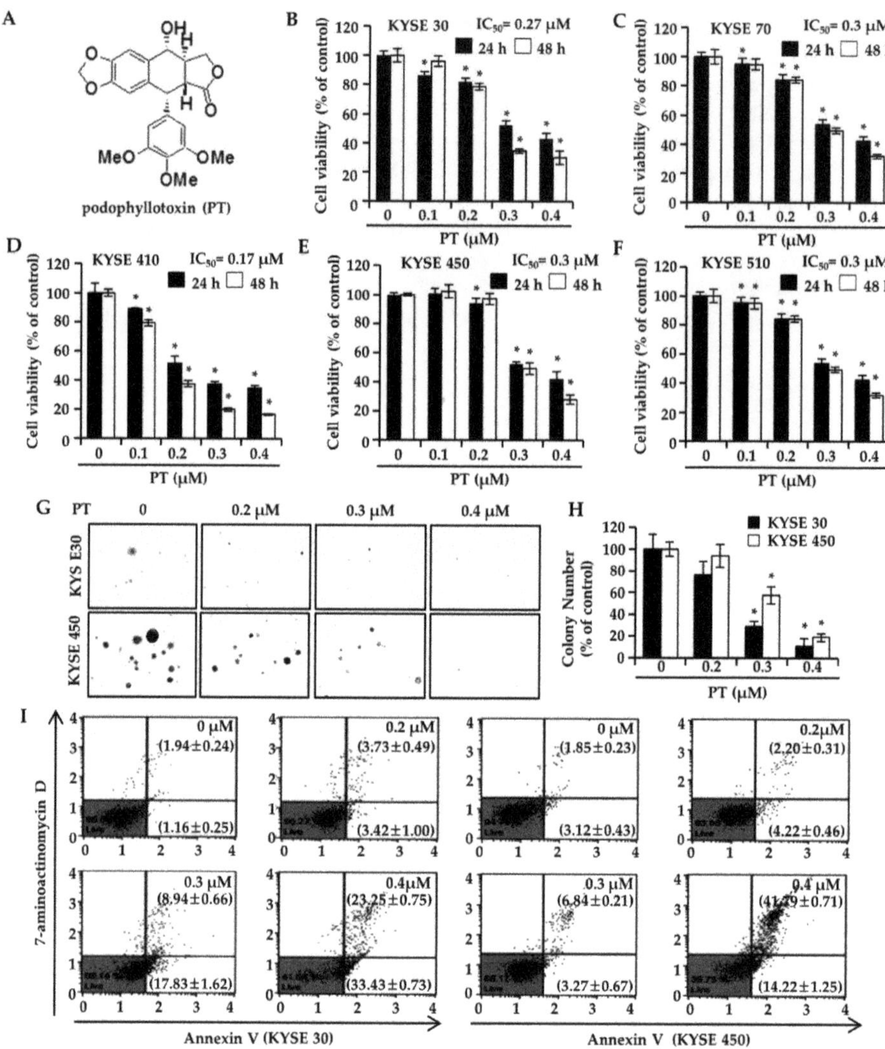

Figure 1. Inhibitory effects of podophyllotoxin (PT) on the cell growth of esophageal squamous cell carcinoma (ESCC) cell lines. (**A**) Chemical structure of PT. (**B–F**) MTS cell proliferation assays for P KYSE 30, KYSE 70, KYSE 410, KYSE 450, and KYSE 510 cells treated with increasing concentrations of the indicated PT (0.1 μM, 0.2 μM, 0.3 μM, and 0.4 μM) for 24 h and 48 h. Experiments were replicated in triplicate as the means ± SD. * $p < 0.05$. (**G**) Anchorage-independent cell growth was evaluated by a soft agar assay. KYSE 30 and KYSE 450 cells were incubated with different concentrations of PT (0 μM, 0.2 μM, 0.3 μM, and 0.4 μM) in soft agar plates for 14 days. Representative colonies formed in soft agar from ESCC cells were photographed. Each experiment was conducted in triplicate. (**H**) The colonies were counted, and the data were plotted. Data are expressed as the mean ± SD. * $p < 0.05$ versus control cells. (**I**) Representative dot plot of the Annexin V/7-AAD assay in ESCC cells. The cells were stained with Muse™ Annexin V and a dead cell reagent and analyzed using a Muse™ Cell Analyzer. The proportion of live cells is indicated in the lower left quadrant (Annexin V-/7-AAD-), the cells in early apoptosis are shown in the lower right quadrant (Annexin V+/7-AAD-), and the ones in late apoptosis are represented in the upper right quadrant (Annexin V+/7-AAD+).

3.2. PT Induces G2/M Phase Arrest of Cell Cycle in KYSE 30 and KYSE 450 Cells

To determine the effect of PT on cell cycle arrest and cell death in ESCC cells, we used flow cytometry to analyze the cell cycle distribution and quantify the sub-G1 cell fraction as a marker for cell death upon treatment of KYSE 30 and KYSE 450 cells with PT (0.2 to 0.4 µM) for 48 h. Compared with the untreated control, PT treatment affected the number of cells in the G0/G1 phase resulting in an appreciable arrest of ESCC cells at the G2/M phase of the cell cycle (Figure 2A). Exposure to PT (0.4 µM) for 48 h led to the accumulation of cells in the sub-G1 fraction, increasing by 52.87 ± 2.22% in KYSE 30 cells and 51.73 ± 2.21% in KYSE 450 cells compared with the control (4.43 ± 0.25% and 5.97 ± 0.15%), respectively (Figure 2B). To further examine the effect of PT on the G2/M phase arrest of KYSE 30 and KYSE 450 cells, we evaluated changes in the expression of checkpoint proteins using western blot. In Figure 2C, the expression of p21 and p27, which acts as a tumor suppressor protein, was significantly elevated in both cell lines following PT treatment at concentrations of 0.2, 0.3, and 0.4 µM for 48 h. In contrast, the expression of cdc2 and cyclin B1, which are major regulators of the G2-to-M phase transition, was markedly inhibited in both cell lines. Therefore, compared with the untreated controls, PT treatment induced cell cycle arrest at the G2/M phase and the sub-G1 fraction of ESCC cells.

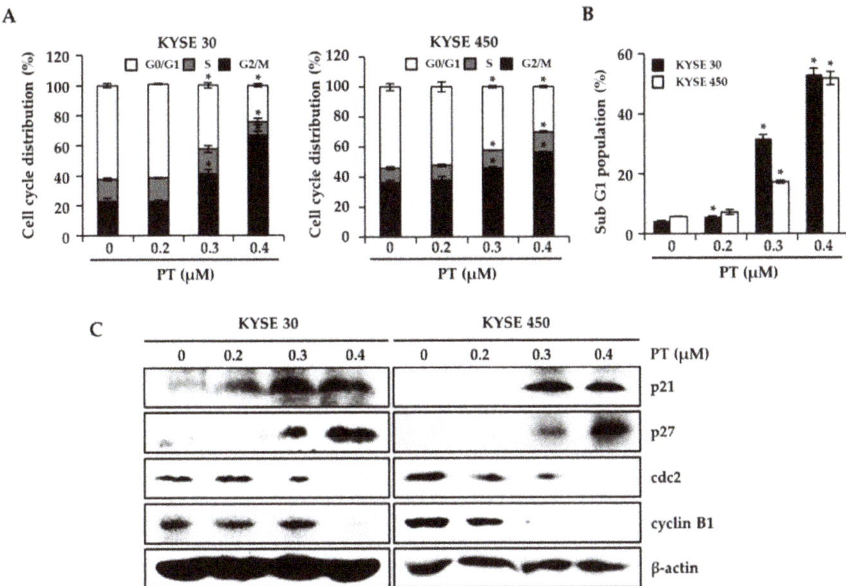

Figure 2. Effect of PT on cell cycle arrest in ESCC cells. KYSE 30 and KYSE 450 cells were cultured without and with PT (0.2, 0.3, and 0.4 µM) for 48 h. Cells were harvested, stained with Muse™ Cell cycle reagent, and analyzed to determine the cell cycle stages (**A**) and sub-G1 (**B**). Data are presented as the mean ($n = 3$) ± SD derived from three independent experiments. * $p < 0.05$ against control. (**C**) The expression of cell cycle regulators and inhibitors. Cell lysates were assessed by western blotting analysis using specific antibodies for p21, p27, cdc2, and cyclin B1. Protein loading was normalized based on β-actin. Data are representative of three independent experiments.

3.3. PT Induces Intracellular ROS Generation and Activates ROS-Dependent MAPK Pathway

To elucidate whether PT mediated the synthesis of intracellular ROS, ROS production was evaluated in ESCC cells using the Muse™ Oxidative Stress Kit after PT treatment for 48 h (Figure 3A). Intriguingly, we observed intracellular ROS accumulation when cells were treated with DMSO or PT (0.4 µM), increasing from 2.74 ± 0.64% to 31.15 ± 1.15% in KYSE 30 cells and 8.36 ± 0.90% to 43.88 ± 0.80% in KYSE 450 cells (Figure 3A). We also examined the expression of ER stress-related signaling molecules by treating ESCC cells with PT for 48 h followed by subsequent western blot analysis (Figure 3B). PT treatment dose-dependently increased the levels of CHOP, DR4, DR5, and GRP78 levels in KYSE 30 and KYSE 450 cells (Figure 3B). These data suggest that ER stress-induced proteins may be associated with apoptotic effects of PT in ESCC cells. Next, the effect of PT on the activation of the MAPK cascade in KYSE 30 and KYSE 450 cells was assessed by a western blot analysis. As shown in Figure 4, PT treatment resulted in a dose-dependent induction of p-JNK and p-p38 compared with the total JNK and p38 or β-actin levels. Our findings suggest that the PT-activated JNK and p38 signaling pathways may be regulated by ROS generation.

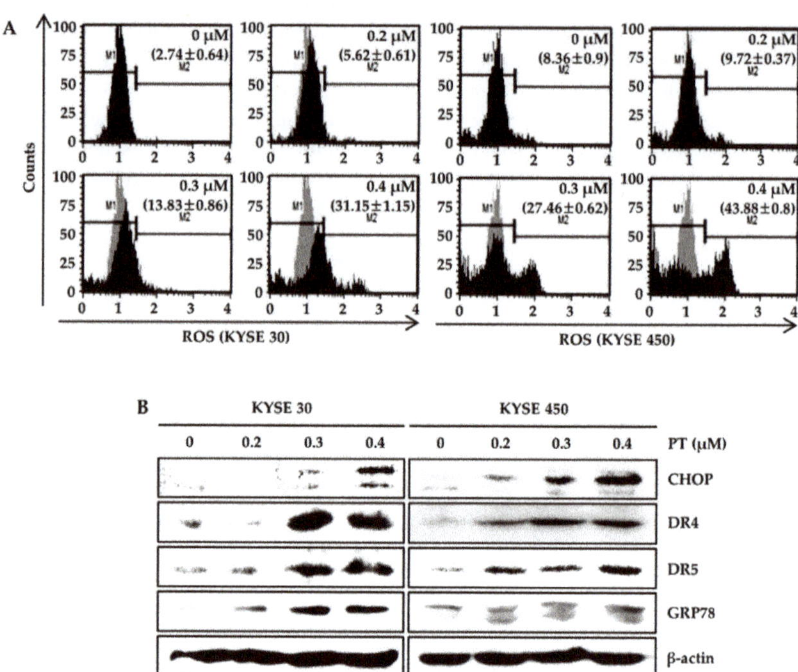

Figure 3. Effect of PT on ROS-mediated cell death. The ESCC cells were exposed to the indicated concentrations of PT for 48 h. (**A**) The amount of intracellular ROS was assessed using a Muse™ Oxidative Stress Kit and a Muse™ Cell Analyzer. "M1" and "M2" represent ROS-negative or ROS-positive cells, respectively. (**B**) The protein expression of CHOP, DR4, DR5, and GRP78 was also visualized by western blot analysis. Protein loading was normalized based on β-actin. Data are representative of three independent experiments.

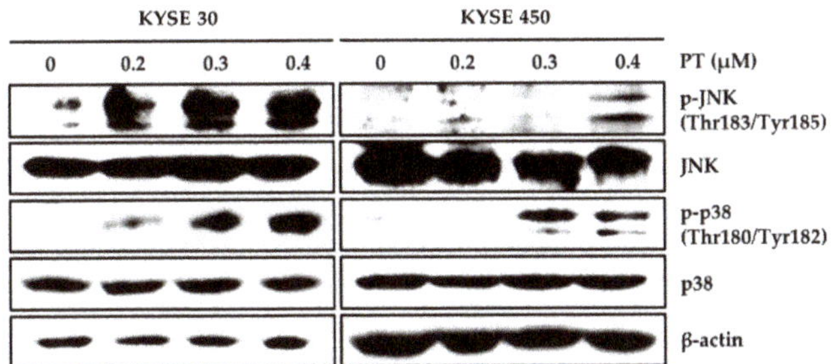

Figure 4. Effect of PT on the MAPK signaling pathway. The KYSE 30 and KYSE 450 cells were treated with various concentrations of PT and incubated for 48 h. Cell lysates were analyzed by western blot using specific antibodies, including pJNK, JNK, p-p38, p38 MAPK, and β-actin antibodies. β-actin served as the loading control. Data are representative of three independent experiments.

3.4. PT Induces Apoptosis of ESCC Cells via Reduction of Mitochondrial Membrane Potential and its Pathway

To determine whether PT-induced apoptosis was mediated via mitochondrial dysfunction, we measured MMP using a Muse™ MitoPotential Kit in PT-treated KYSE 30 and KYSE 450 cells (Figure 5A). We confirmed a concentration-dependent increase in the mean percentages of total depolarization in KYSE 30 cells, from 2.35 ± 1.96% in the untreated control to 11.77 ± 1.06%, 20.50 ± 0.71%, and 58.97 ± 6.29% in the presence of 0.2, 0.3, and 0.4 µM of PT (Figure 5A). Untreated KYSE 450 cells showed a total depolarization change of 4.97 ± 0.20%, which increased to 8.72 ± 0.24%, 19.53 ± 0.83%, and 37.43 ± 2.20% after treatment with a 0.2, 0.3, and 0.4 µM dose of PT. These results suggest that the induction of apoptosis in ESCC cells may be attributed to mitochondrial dysfunction. To further elucidate the role of PT in mitochondria-mediated apoptosis, we examined whether PT regulated the expression of pro- and anti-apoptotic proteins in ESCC cells using western blot analysis. As shown in Figure 5B, ESCC cells treated with increasing concentrations of PT exhibited augmented expression of Bax, Apaf-1, and C-PARP proteins, while the expression of Mcl-1 and Bcl-2 proteins was reduced. The results reflect a concentration-dependent increase in cytosolic cyto c after treatment with PT. Simultaneously, there was a decrease in cyto c in the mitochondrial fraction. In order to investigate the mitochondrial signals activating caspase activity following PT treatment, we analyzed the activity of multiple caspases (caspase-1, -3, -4, -5, -6, -7, -8, and -9) in KYSE 30 and KYSE 450 cells treated with different concentrations of PT for 48 h using a Muse™ Cell Analyzer (Figure 6). Flow cytometric analysis demonstrated that PT significantly induced caspase activation in a dose-dependent manner (Figure 6). The total apoptotic phenotype (multi caspase-positive/7-AAD-positive and negative) increased from 6.63 ± 0.41% to 11.51 ± 1.63%, 22.60 ± 4.73%, and 51.81 ± 0.75% after exposure to 0.2, 0.3, and 0.4 µM of PT, respectively, in KYSE 30 cells, and from 2.41 ± 0.12% in the control group to 6.52 ± 0.82%, 24.90 ± 0.64%, and 64.50 ± 0.74%, respectively, in KYSE 450 cells. Taken together, the overall results reveal that PT-induced apoptosis in ESCC cells is related to the activation of mitochondrial and caspase-dependent pathways.

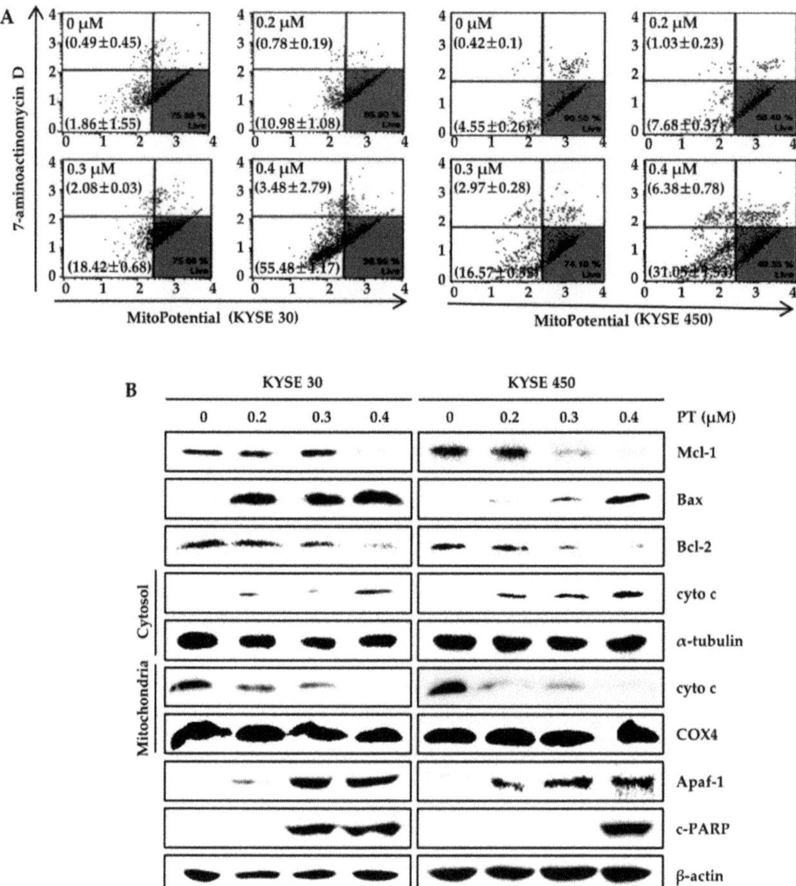

Figure 5. Effects of PT on mitochondrial dysfunction-mediated signaling in ESCC cells. (**A**) mitochondrial membrane potential (MMP) was evaluated by staining with MitoPotential dye, and 7-AAD and was analyzed using the Muse™ Cell Analyzer. The lower and upper left quadrants show mean MMP depolarization associated with early and late apoptosis, respectively. Values represent the means ± SD. Data are representative of three independent experiments; (**B**) Western blot analysis was conducted to examine the protein expression of Mcl-1, Bax, Bcl-2, Apaf-1, and c-PARP after cells were exposed to PT at different concentrations (0, 0.2, 0.3, and 0.4 µM) for 48 h. Mitochondrial and cytosolic extracts were processed via western blot using the anti-cyto c antibody. β-actin, α-tubulin, and COX4 were used as the loading controls in the whole cell, cytoplasm, or mitochondria, respectively.

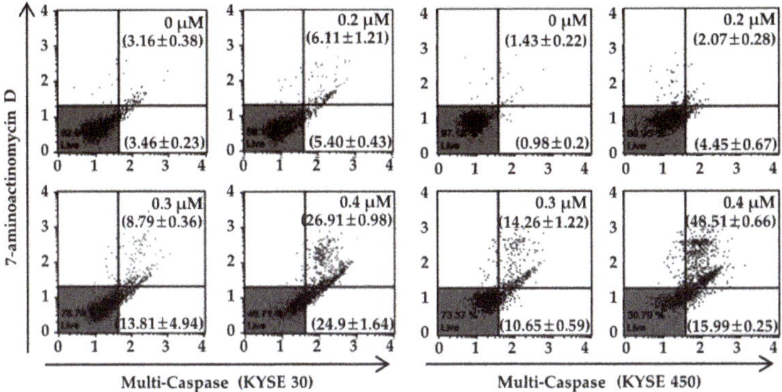

Figure 6. Effects of PT on caspase activation in ESCC cells. KYSE 30 and KYSE 450 cells were treated with different concentrations of PT (0, 0.2, 0.3, and 0.4 µM) for 48 h. Activities of multiple caspases (caspase-1, -3, -4, -5, -6, -7, -8, and -9) were measured with a Muse™ cell analyzer using a Muse™ Multi-Caspase Kit. Each quadrant shows a population of viable cells (lower-left panel), a cell population with caspase activity (lower-right panel), late stages including caspase activity/dead cells (upper-right panel), and dead cells (upper-left panel) in each experimental group. Values represent the means (caspase-positive) ± SD. Data are representative of three independent experiments.

4. Discussion

ESCC is a major health challenge worldwide, and effective treatment has yet to be identified [20]. It is imperative to explore new treatments, as existing treatments have severe side effects and poor efficacy. According to a 2018 study, weight loss and other side effects were not observed when Free PT was treated in mice xenografts to assess antitumor activity [21]. In previous reports, PT, a natural phytochemical compound, has been showed to inhibit cancers, including cervical carcinoma, breast, and prostate cancers [22,23]. Although the anti-neoplastic efficacy of PT against numerous cancers has been reported [2], its efficacy and precise pharmacological mechanisms in ESCC were not reported until now. In this study, we demonstrated that PT induced apoptosis via mitochondrial pathways in ESCC cell lines, as evidenced by an elevation in subG1 cells, phosphatidylserine externalization, ROS generation, and the loss of MMP, as well as the activation of MAPKs and caspases. Cancer cells carry defects involving the maintenance of homeostasis in cell proliferation, which is a critical hallmark of apoptosis [24,25]. In the current study, we are the first to demonstrate that PT significantly inhibits the growth of ESCC cells (Figure 1B–H). Among the five ESCC cell lines, KYSE 30 and KYSE 450 have analogical genetic backgrounds and characteristics [26] and have been selected for in-depth analysis because they show similar responses to PT treatment. The IC$_{50}$ concentrations of PT in KYSE 30 and KYSE 450 cells were 0.27 and 0.3 µM, respectively. In addition, colony formation was sharply decreased in PT-treated cells, which further indicates the growth-inhibitory action of PT (Figure 1G,H). Based on the foregoing findings, we also investigated biomarkers of apoptotic cell death, such as phosphatidylserine externalization, with annexin-V staining (Figure 1I). A loss of plasma membrane asymmetry mediated by phosphatidylserine externalization is associated with early apoptotic phenomena [27]. The significant increase in the apoptotic rate of PT-treated cells indicates that PT causes apoptosis via phosphatidylserine externalization on the cytoplasmic surface of the cell membrane (Figure 1I). A previous study showed that cellular mechanisms ensure healthy cell progression and proliferation through cell cycle arrest [28]. Cyclin B1 and cdc2 are the key regulators of the G2/M phase transition of the cell cycle, and these regulatory proteins are inhibited by p21, which serves as a cyclin-dependent kinase-inhibiting protein [28,29]. PT significantly increased the G2/M phase and sub-G1 populations in the cell cycle progression of KYSE 30 and KYSE 450 cells (Figure 2A,B). Additionally, PT induced an increase in p21 and p27 protein levels, whereas it decreased the expression of cyclin B1 and cdc2. These results suggest that PT exerts an anti-cancer effect on ESCC cells via regulation of cell cycle progression. Intracellular ROS generation stimulates ER stress, and these species enhance ER stress-mediated apoptosis [30]. In the present study, the production of cellular ROS showed that the density of ROS-positive ESCC cells is elevated by PT treatment in a dose-dependent manner (Figure 3A). It is well known that ER stress-activated JNK and p38 trigger mitochondria-dependent apoptosis in various cancer cells [31]. PT significantly induced the phosphorylation of p38 and JNK in a dose-dependent manner (Figure 4). JNK and p38 activation play an important role in the induction of apoptotic cell death in response to ER stress [31]. PT enhanced the protein levels of DR4, DR5 and ER stress markers CHOP and GRP78, indicating the pro-survival signaling of ER stress responses (Figure 3B) [32,33]. These results further suggest that PT induces ER stress-mediated apoptosis in ESCC cells. MMP is essential for the generation of ATP to modulate physiological activity, and a collapse of MMP is an indicator of cellular integrity in the early stages of apoptosis [34,35]. PT treatment increases the percentage of depolarized cells, indicating a loss of MMP in PT-induced apoptosis (Figure 5A). The levels of apoptosis-related proteins, the Bcl-2 family, correlate with the levels of MMP [36]. The loss of MMP promotes the efflux of cyto c to the cytosol and leads to the formation of a apoptosome–deoxyadenosine triphosphate-dependent complex, which consists of cyto c and Apaf-1 [37,38]. PT was shown to increase the expression of Bax, Apaf-1, and c-PARP and decrease the levels of Mcl-1 and Bcl-2 in ESCC cells. Treatment with various concentrations of PT also induced the release of cyto c into the cytosol (Figure 5B). Previous studies have reported that the release of cyto c from the mitochondria into the cytoplasm activates caspase 3, 8, and 9 and, in turn, results in single-stranded and double-stranded DNA breaks [39]. Interestingly, incubation with PT induces dose-dependent activation of multiple caspases (caspase-1, -3, -4, -5, -6, -7, -8, and -9) in KYSE 30 and KYSE 450 cells (Figure 6). It is possible that apoptosis is mediated via the mitochondrial activation of the caspase cascade in ESCC cells.

5. Conclusions

This study demonstrated for the first time that PT exerts its anti-cancer effects on ESCC cells via mitochondrial apoptotic pathways. Our findings elucidate the mechanism of PT-induced cellular apoptosis, suggesting that PT is a potential chemotherapeutic agent.

Author Contributions: Conceptualization, G.Y., M.-H.L., A.-W.K., H.-N.O., S.-S.C., J.-S.C., and K.L.; methodology, G.Y., M.-H.L., A.-W.K., H.-N.O., S.-S.C., J.-S.C., K.L., J.-I.C., and J.-H.S.; software, G.Y. and M.-H.L.; validation, G.Y., M.-H.L., A.-W.K., H.-N.O., S.-S.C., J.-S.C., and K.L.; formal analysis, G.Y., M.-H.L., A.-W.K., H.-N.O., S.-S.C., J.-S.C., and K.L.; investigation, G.Y., M.-H.L., A.-W.K., H.-N.O., S.-S.C., J.-S.C., K.L., J.-I.C., and J.-H.S.; resources, J.-I.C. and J.-H.S.; data curation, G.Y., M.-H.L., A.-W.K., H.-N.O., S.-S.C., J.-S.C., and K.L.; writing—original draft preparation, M.-H.L. and J.-S.C.; writing—review and editing, M.-H.L. and J.-S.C.; supervision, J.-I.C. and J.-H.S. All authors have read and agreed to the published version of the manuscript.

Funding: This research was funded by Basic Science Research program of National Research Foundation Korea, grant number 2019R1A2C1005899.

Acknowledgments: This research was supported by the Basic Science Research program through the National Research Foundation Korea Funded by the Ministry of Education, Science, and Technology (2019R1A2C1005899). This work was carried out by the Convergence Research Laboratory established by the Mokpo National University (MNU) Innovation Support Project.

Conflicts of Interest: The authors declare the absence of any conflict of interest.

References

1. Zhang, D.; Zhou, X.; Bao, W.; Chen, Y.; Cheng, L.; Qiu, G.; Sheng, L.; Ji, Y.; Du, X. Plasma fibrinogen levels are correlated with postoperative distant metastasis and prognosis in esophageal squamous cell carcinoma. *Oncotarget* **2015**, *6*, 38410–38420. [CrossRef] [PubMed]

2. Wang, K.; Johnson, A.; Ali, S.M.; Klempner, S.J.; Bekaii-Saab, T.; Vacirca, J.L.; Khaira, D.; Yelensky, R.; Chmielecki, J.; Elvin, J.A.; et al. Comprehensive Genomic Profiling of Advanced Esophageal Squamous Cell Carcinomas and Esophageal Adenocarcinomas Reveals Similarities and Differences. *Oncologist* **2015**, *20*, 1132–1139. [CrossRef] [PubMed]

3. Sugase, T.; Takahashi, T.; Serada, S.; Nakatsuka, R.; Fujimoto, M.; Ohkawara, T.; Hara, H.; Nishigaki, T.; Tanaka, K.; Miyazaki, Y.; et al. Suppressor of cytokine signaling-1 gene therapy induces potent antitumor effect in patient-derived esophageal squamous cell carcinoma xenograft mice. *Int. J. Cancer* **2017**, *140*, 2608–2621. [CrossRef] [PubMed]

4. Silveira, A.L.; Faheina-Martins, G.V.; Maia, R.C.; Araujo, D.A. Compound A398, a novel podophyllotoxin analogue: Cytotoxicity and induction of apoptosis in human leukemia cells. *PLoS ONE* **2014**, *9*, e107404. [CrossRef] [PubMed]

5. Zupko, I.; Jaeger, W.; Topcu, Z.; Wu, C.C. Anticancer Properties of Natural Products. *Biomed. Res. Int.* **2015**, *2015*, 242070. [CrossRef] [PubMed]

6. Blowman, K.; Magalhaes, M.; Lemos, M.F.L.; Cabral, C.; Pires, I.M. Anticancer Properties of Essential Oils and Other Natural Products. *Evid. Based Complement. Alternat. Med.* **2018**, *2018*, 3149362. [CrossRef] [PubMed]

7. Cheng, W.H.; Shang, H.; Niu, C.; Zhang, Z.H.; Zhang, L.M.; Chen, H.; Zou, Z.M. Synthesis and Evaluation of New Podophyllotoxin Derivatives with in Vitro Anticancer Activity. *Molecules* **2015**, *20*, 12266–12279. [CrossRef]

8. Liu, Y.Q.; Tian, J.; Qian, K.; Zhao, X.B.; Morris-Natschke, S.L.; Yang, L.; Nan, X.; Tian, X.; Lee, K.H. Recent progress on C-4-modified podophyllotoxin analogs as potent antitumor agents. *Med. Res. Rev.* **2015**, *35*, 1–62. [CrossRef]

9. Zhang, L.; Zhang, Z.; Chen, F.; Chen, Y.; Lin, Y.; Wang, J. Aromatic heterocyclic esters of podophyllotoxin exert anti-MDR activity in human leukemia K562/ADR cells via ROS/MAPK signaling pathways. *Eur. J. Med. Chem.* **2016**, *123*, 226–235. [CrossRef]

10. Singh, A.; Yashavarddhan, M.H.; Kalita, B.; Ranjan, R.; Bajaj, S.; Prakash, H.; Gupta, M.L. Podophyllotoxin and Rutin Modulates Ionizing Radiation-Induced Oxidative Stress and Apoptotic Cell Death in Mice Bone Marrow and Spleen. *Front. Immunol.* **2017**, *8*, 183. [CrossRef]

11. Jiang, Z.; Song, F.; Li, Y.; Xue, D.; Zhao, N.; Zhang, J.; Deng, G.; Li, M.; Liu, X.; Wang, Y. Capsular Polysaccharide of Mycoplasma ovipneumoniae Induces Sheep Airway Epithelial Cell Apoptosis via ROS-Dependent JNK/P38 MAPK Pathways. *Oxid. Med. Cell Longev.* **2017**, *2017*, 6175841. [CrossRef]

12. Lin, C.L.; Lee, C.H.; Chen, C.M.; Cheng, C.W.; Chen, P.N.; Ying, T.H.; Hsieh, Y.H. Protodioscin Induces Apoptosis Through ROS-Mediated Endoplasmic Reticulum Stress via the JNK/p38 Activation Pathways in Human Cervical Cancer Cells. *Cell Physiol. Biochem.* **2018**, *46*, 322–334. [CrossRef] [PubMed]

13. Li, J.; He, J.; Fu, Y.; Hu, X.; Sun, L.Q.; Huang, Y.; Fan, X. Hepatitis B virus X protein inhibits apoptosis by modulating endoplasmic reticulum stress response. *Oncotarget* **2017**, *8*, 96027–96034. [CrossRef] [PubMed]

14. Onyeagucha, B.; Subbarayalu, P.; Abdelfattah, N.; Rajamanickam, S.; Timilsina, S.; Guzman, R.; Zeballos, C.; Eedunuri, V.; Bansal, S.; Mohammad, T.; et al. Novel post-transcriptional and post-translational regulation of pro-apoptotic protein BOK and anti-apoptotic protein Mcl-1 determine the fate of breast cancer cells to survive or die. *Oncotarget* **2017**, *8*, 85984–85996. [CrossRef] [PubMed]

15. Jiang, Y.; Wang, X.; Hu, D. Furanodienone induces G0/G1 arrest and causes apoptosis via the ROS/MAPKs-mediated caspase-dependent pathway in human colorectal cancer cells: A study in vitro and in vivo. *Cell Death Dis.* **2017**, *8*, e2815. [CrossRef]

16. Zhu, J.; Yu, W.; Liu, B.; Wang, Y.; Shao, J.; Wang, J.; Xia, K.; Liang, C.; Fang, W.; Zhou, C.; et al. Escin induces caspase-dependent apoptosis and autophagy through the ROS/p38 MAPK signalling pathway in human osteosarcoma cells in vitro and in vivo. *Cell Death Dis.* **2017**, *8*, e3113. [CrossRef]

17. Li, H.Y.; Zhang, J.; Sun, L.L.; Li, B.H.; Gao, H.L.; Xie, T.; Zhang, N.; Ye, Z.M. Celastrol induces apoptosis and autophagy via the ROS/JNK signaling pathway in human osteosarcoma cells: An in vitro and in vivo study. *Cell Death Dis.* **2015**, *6*, e1604. [CrossRef]

18. Zhang, J.; Chen, Q.; Wang, S.; Li, T.; Xiao, Z.; Lan, W.; Huang, G.; Cai, X. alpha-Mangostin, A Natural Xanthone, Induces Apoptosis and ROS Accumulation in Human Rheumatoid Fibroblast-Like Synoviocyte MH7A Cells. *Curr. Mol. Med.* **2017**, *17*, 375–380. [CrossRef]

19. Oh, H.N.; Oh, K.B.; Lee, M.H.; Seo, J.H.; Kim, E.; Yoon, G.; Cho, S.S.; Cho, Y.S.; Choi, H.W.; Chae, J.I.; et al. JAK2 regulation by licochalcone H inhibits the cell growth and induces apoptosis in oral squamous cell carcinoma. *Phytomedicine* **2019**, *52*, 60–69. [CrossRef]

20. Han, G.; Wu, Z.; Zhao, N.; Zhou, L.; Liu, F.; Niu, F.; Xu, Y.; Zhao, X. Overexpression of stathmin plays a pivotal role in the metastasis of esophageal squamous cell carcinoma. *Oncotarget* **2017**, *8*, 61742–61760. [CrossRef]

21. Zhou, H.; Lv, S.; Zhang, D.; Deng, M.; Zhang, X.; Tang, Z.; Chen, X. A polypeptide based podophyllotoxin conjugate for the treatment of multi drug resistant breast cancer with enhanced efficiency and minimal toxicity. *Acta Biomater.* **2018**, *73*, 388–399. [CrossRef] [PubMed]

22. Hu, L.L.; Zhou, X.; Zhang, H.L.; Wu, L.L.; Tang, L.S.; Chen, L.L.; Duan, J.L. Exposure to podophyllotoxin inhibits oocyte meiosis by disturbing meiotic spindle formation. *Sci. Rep.* **2018**, *8*, 10145. [CrossRef] [PubMed]

23. Xie, S.; Li, G.; Qu, L.; Zhong, R.; Chen, P.; Lu, Z.; Zhou, J.; Guo, X.; Li, Z.; Ma, A.; et al. Podophyllotoxin Extracted from Juniperus sabina Fruit Inhibits Rat Sperm Maturation and Fertility by Promoting Epididymal Epithelial Cell Apoptosis. *Evid. Based Complement. Alternat. Med.* **2017**, *2017*, 6958982. [CrossRef] [PubMed]

24. Hanahan, D.; Weinberg, R.A. Hallmarks of cancer: The next generation. *Cell* **2011**, *144*, 646–674. [CrossRef] [PubMed]

25. Tower, J. Programmed cell death in aging. *Ageing Res. Rev.* **2015**, *23*, 90–100. [CrossRef] [PubMed]

26. Hao, J.J.; Shi, Z.Z.; Zhao, Z.X.; Zhang, Y.; Gong, T.; Li, C.X.; Zhan, T.; Cai, Y.; Dong, J.T.; Fu, S.B.; et al. Characterization of genetic rearrangements in esophageal squamous carcinoma cell lines by a combination of M-FISH and array-CGH: Further confirmation of some split genomic regions in primary tumors. *BMC Cancer* **2012**, *12*, 367. [CrossRef]

27. Siedlecka-Kroplewska, K.; Wronska, A.; Stasilojc, G.; Kmiec, Z. The Designer Drug 3-Fluoromethcathinone Induces Oxidative Stress and Activates Autophagy in HT22 Neuronal Cells. *Neurotox. Res.* **2018**. [CrossRef]

28. Yu, H.; Yin, S.; Zhou, S.; Shao, Y.; Sun, J.; Pang, X.; Han, L.; Zhang, Y.; Gao, X.; Jin, C.; et al. Magnolin promotes autophagy and cell cycle arrest via blocking LIF/Stat3/Mcl-1 axis in human colorectal cancers. *Cell Death Dis.* **2018**, *9*, 702. [CrossRef]

29. Chang, C.C.; Hung, C.M.; Yang, Y.R.; Lee, M.J.; Hsu, Y.C. Sulforaphane induced cell cycle arrest in the G2/M phase via the blockade of cyclin B1/CDC2 in human ovarian cancer cells. *J. Ovarian Res.* **2013**, *6*, 41. [CrossRef]

30. Rouault-Pierre, K.; Lopez-Onieva, L.; Foster, K.; Anjos-Afonso, F.; Lamrissi-Garcia, I.; Serrano-Sanchez, M.; Mitter, R.; Ivanovic, Z.; de Verneuil, H.; Gribben, J.; et al. HIF-2alpha protects human hematopoietic stem/progenitors and acute myeloid leukemic cells from apoptosis induced by endoplasmic reticulum stress. *Cell Stem Cell* **2013**, *13*, 549–563. [CrossRef]

31. Chen, W.Y.; Hsieh, Y.A.; Tsai, C.I.; Kang, Y.F.; Chang, F.R.; Wu, Y.C.; Wu, C.C. Protoapigenone, a natural derivative of apigenin, induces mitogen-activated protein kinase-dependent apoptosis in human breast cancer cells associated with induction of oxidative stress and inhibition of glutathione S-transferase pi. *Investig. New Drugs* **2011**, *29*, 1347–1359. [CrossRef] [PubMed]

32. Gu, L.L.; Shen, Z.L.; Li, Y.L.; Bao, Y.Q.; Lu, H. Oxymatrine Causes Hepatotoxicity by Promoting the Phosphorylation of JNK and Induction of Endoplasmic Reticulum Stress Mediated by ROS in LO2 Cells. *Mol. Cells* **2018**, *41*, 401–412. [CrossRef] [PubMed]

33. Shen, M.; Wang, L.; Wang, B.; Wang, T.; Yang, G.; Shen, L.; Wang, T.; Guo, X.; Liu, Y.; Xia, Y.; et al. Activation of volume-sensitive outwardly rectifying chloride channel by ROS contributes to ER stress and cardiac contractile dysfunction: Involvement of CHOP through Wnt. *Cell Death Dis.* **2014**, *5*, e1528. [CrossRef] [PubMed]

34. Ansari, S.S.; Sharma, A.K.; Soni, H.; Ali, D.M.; Tews, B.; Konig, R.; Eibl, H.; Berger, M.R. Induction of ER and mitochondrial stress by the alkylphosphocholine erufosine in oral squamous cell carcinoma cells. *Cell Death Dis.* **2018**, *9*, 296. [CrossRef] [PubMed]

35. Zhou, X.; Wang, H.Y.; Wu, B.; Cheng, C.Y.; Xiao, W.; Wang, Z.Z.; Yang, Y.Y.; Li, P.; Yang, H. Ginkgolide K attenuates neuronal injury after ischemic stroke by inhibiting mitochondrial fission and GSK-3beta-dependent increases in mitochondrial membrane permeability. *Oncotarget* **2017**, *8*, 44682–44693. [CrossRef] [PubMed]

36. Chou, W.H.; Liu, K.L.; Shih, Y.L.; Chuang, Y.Y.; Chou, J.; Lu, H.F.; Jair, H.W.; Lee, M.Z.; Au, M.K.; Chung, J.G. Ouabain Induces Apoptotic Cell Death Through Caspase- and Mitochondria-dependent Pathways in Human Osteosarcoma U-2 OS Cells. *Anticancer Res.* **2018**, *38*, 169–178. [CrossRef]
37. Su, L.Y.; Shi, Y.X.; Yan, M.R.; Xi, Y.; Su, X.L. Anticancer bioactive peptides suppress human colorectal tumor cell growth and induce apoptosis via modulating the PARP-p53-Mcl-1 signaling pathway. *Acta Pharmacol. Sin.* **2015**, *36*, 1514–1519. [CrossRef]
38. Wang, R.; Ma, L.; Weng, D.; Yao, J.; Liu, X.; Jin, F. Gallic acid induces apoptosis and enhances the anticancer effects of cisplatin in human small cell lung cancer H446 cell line via the ROS-dependent mitochondrial apoptotic pathway. *Oncol. Rep.* **2016**, *35*, 3075–3083. [CrossRef]
39. Huang, X.; Zou, L.; Yu, X.; Chen, M.; Guo, R.; Cai, H.; Yao, D.; Xu, X.; Chen, Y.; Ding, C.; et al. Salidroside attenuates chronic hypoxia-induced pulmonary hypertension via adenosine A2a receptor related mitochondria-dependent apoptosis pathway. *J. Mol. Cell Cardiol.* **2015**, *82*, 153–166. [CrossRef] [PubMed]

Article

Control of Fungal Diseases and Increase in Yields of a Cultivated Jujube Fruit (*Zizyphus jujuba* Miller var. *inermis* Rehder) Orchard by Employing *Lysobacter antibioticus* HS124

Jun-Hyeok Kwon [1], Sang-Jae Won [1], Jae-Hyun Moon [1], Chul-Woo Kim [2] and Young-Sang Ahn [1,*]

[1] Department of Forest Resources, College of Agriculture and Life Sciences, Chonnam National University, Gwangju 61186, Korea; wg6102@naver.com (J.-H.K.); lazyno@naver.com (S.-J.W.); mjh132577@naver.com (J.-H.M.)

[2] Division of Special-purpose Trees, National Institute of Forest Science, Suwon 16631, Korea; futuretree@korea.kr

* Correspondence: ysahn@jnu.ac.kr; Tel.: +82-62-530-2081

Received: 7 November 2019; Accepted: 13 December 2019; Published: 15 December 2019

Abstract: The objective of this study is to investigate the inhibitory effects of *Lysobacter antibioticus* HS124 on fungal phytopathogens causing gray mold rot, stem rot, and anthracnose. Another objective of this study is to promote the yield of fruit in jujube farms. *L. antibioticus* HS124 produces chitinase, a lytic enzyme with the potential to reduce mycelial growth of fungal phytopathogens involving hyphal alterations with swelling and bulbous structures, by 20.6 to 27.3%. Inoculation with *L. antibioticus* HS124 decreased the appearance of fungal diseases in jujube farms and increased the fruit yield by decreasing fruit wilting and dropping. In addition, *L. antibioticus* HS124 produced the phytohormone auxin to promote vegetative growth, thereby increasing the fruit size. The yield of jujube fruits after *L. antibioticus* HS124 inoculation was increased by 6284.67 g/branch, which was 2.9-fold higher than that of the control. Auxin also stimulated fine root development and nutrient uptake in jujube trees. The concentrations of minerals, such as K, Ca, Mg, and P in jujube fruits after *L. antibioticus* HS124 inoculation were significantly increased (1.4- to 2.0-fold greater than the concentrations in the control). These results revealed that *L. antibioticus* HS124 could not only control fungal diseases but also promote fruit yield in jujube farms.

Keywords: auxin; biocontrol; chitinase; fruit; fungal pathogen; jujube; *Lysobacter antibioticus* HS124; mineral concentration; production; Rhamnaceae

---

1. Introduction

Jujube (*Zizyphus jujuba* Miller var. *inermis* Rehder) is an elliptical or spherical-shaped fruit [1]. Approximately 170 species of jujube have been cultivated in China over the past 5000 years [1,2]. They are mainly cultivated in central and southern China and southern and eastern Europe where the climate is warm [3]. Jujube fruit is mostly consumed fresh. Its dried form is also consumed in various food products, such as bread, cake, candy, powder, and juice [4]. Jujube fruit is rich in fiber, minerals, phenolic compounds, and vitamins [1,5]. It can strengthen the cardiovascular system with its antioxidant capacity [6,7]. Eating one jujube fruit per day provides an adult with the dose of vitamins B and C recommended by the Food and Agriculture Organization of the United Nations (FAO)/World Health Organization (WHO) [1,8]. In addition, jujube seeds are known to contain saponins with pharmacological properties, such as antibacterial, insecticidal, anti-inflammatory, and immunity-enhancing activities [9,10]. In ancient China and Korea, jujube fruit was commonly used for health promotion and as a medicinal ingredient [10–13].

A high incidence of fruit fungal pathogens has been reported in intensive cultivation of jujube because of its high sugar content of approximately 5.4 to 10.5 g/100 g and its high moisture content of over 80% [1,14]. Fungal diseases in jujube fruits and trees include gray mold rot, stem rot, and anthracnose [15–18], which can reduce jujube production. Fungi and spores are mainly present in the soil and atmosphere. Rain, increases the atmospheric humidity, resulting in the cracking of jujube fruits. Thus, the interior part of the jujube is exposed to the outer environment where fungi and spores are present. Generally, it is difficult to control fungal diseases and fungicides can be used to control the fungal diseases in jujube farms [19,20]. However, the continuous use of fungicides results in an increased number of fungi with acquired resistance. Fungicide-resistant fungi require higher doses and increased fungicidal application frequency, eventually necessitating the development of new fungicides [20]. Although intensive farming methods ensure high yields and quality, they also require the use of excessive chemicals, such as fungicides, pesticides, and fertilizers [19]. Excessive use of chemicals can lead to environmental pollution, such as the eutrophication of water quality and salt accumulation in the soil [21], the destruction of soil microorganisms, and the inhibition of plant growth [22,23]. Recently, increased focus has been directed toward environmentally friendly practices in fruit cultivation systems. However, the cultivation of fruit tree orchards still suffers from fungal diseases because of the lack of knowledge regarding biological control methods.

Recent changes in awareness about the environment and changes in consumer trends favoring environmentally friendly crop products have prompted a demand to transform the existing farming methods [24,25]. Among various other eco-friendly cultivation methods, environmentally friendly cultivation methods that employ biological control by using microorganisms can promote crop production. Thus, they are gaining popularity because of their high potential value [25–28]. Plant growth promoting rhizobacteria (PGPR) play a significant role in reduction of infection by plant pathogenic fungi leading to be a promising alternative to control plant diseases. PGPR are known to produce lytic enzymes, preventing infection by fungal diseases [27–30]. However, the biocontrol efficacy of PGPR on diseases associated with jujube fruit is not clearly investigated. PGPR provide plants with phytohormones [24,31], especially auxin, which can loosen the cell wall of root tissue cells, allowing cells to absorb additional nutrients, thereby promoting the growth of root hair and lateral roots [24,25,27,28,31,32]. In addition, auxin helps establish symbiosis with rhizobia or mycorrhiza and facilitates the uptake of nutrients from soils, promoting plant production [24,25,27,28,31,32]. In particular, after the flower is fertilized, auxin acts on the division and expansion of cells, promoting vegetative growth and increasing fruit size and yield [31,33,34]. *Lysobacter* species are isolated from rhizosphere soil. Genus *Lysobacter* is famous for their positive effects on plant health [30,35]. *Lysobacter antibioticus* strains can produce a variety of bioactive compounds, including lytic enzymes and antimicrobial compounds, that can effectively inhibit the growth of phytopathogenic fungi [30]. However, the effects of bioenhancers derived from *L. antibioticus* strains on the production and nutrient uptake in fruit trees remain poorly understood.

Fruit trees have been studied with respect to many bacterial species that can act as PGPR. These bacterial species have been reported to successfully provide biological control of diseases and can also improve fruit production. However, there are only a few studies on the simultaneous use of PGPR for the biological control of disease and the promotion of fruit production [26,36]. Currently, the increasing demand for fruit tree production, along with a significant reduction in the use of synthetic chemical fertilizers and fungicides, is a major challenge. Despite fungicide treatment for field survey areas of cultivated jujube orchards, several other flowers and fruits have been found to be contaminated with fungal diseases, including gray mold rot, stem rot, and anthracnose (Figure 1c). In Korea, fungal diseases caused by *Botrytis cinerea*, *Botryosphaeria dothidea*, and *Colletotrichum gloeosporioides* are major diseases in jujube fruit. These causative agents are potential fungal pathogens in jujube farms [15–18]. To improve the production of jujube fruit, it is crucial to consider the capacity of beneficial microbes on fungal diseases and the production of jujube fruits in cultivated orchards. Therefore, the aim of this study is to observe the biocontrol capacities of fungal plant pathogens such as *B. cinerea*,

B. dothidea, and *C. gloeosporioides*. Another objective of this study was to promote fruit yield in cultivated jujube orchards through the performance of antagonistic bacteria, such as *L. antibioticus* HS124.

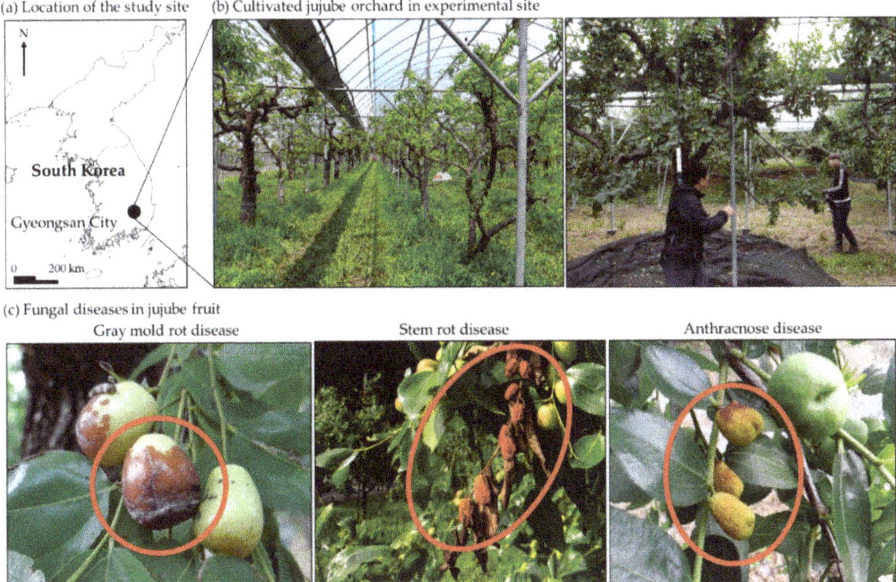

Figure 1. Locations of the study sites (**a**). Cultivated jujube orchard at the experimental site (**b**). gray mold rot by *B. cinerea* (left), stem rot by *B. dothidea* (center), and anthracnose by *C. gloeosporioides* (right) (**c**).

2. Materials and Methods

2.1. Bacterial Culture

The antagonistic bacterial strain *L. antibioticus* HS124 was isolated from field rhizosphere soil in Naju City, Korea [30]. It was subcultured in casein-yeast (CY) agar medium (Pancreatic digest of casein (Neogen, Lansing, MI, USA), 3 g/L; $CaCl_2 \cdot 2H_2O$ (Daejung chemicals, Siheung, Korea), 1.36 g/L; yeast extract (Daejung chemicals, Siheung, Korea), 1 g/L; agar (Daejung chemicals, Siheung, Korea), 20 g/L; and distilled water, 1 L). The CY medium composition and preparation for inoculation of *L. antibioticus* HS124 followed instructions of the Korean Agriculture Culture Collection (KACC, Suwon, Korea). It was incubated at 30 °C for three days. A single colony from fresh culture medium was pre-inoculated into CY broth (Pancreatic digest of casein, 3 g/L; $CaCl_2 \cdot 2H_2O$, 1.36 g/L; yeast extract, 1 g/L; and distilled water, 1 L) and cultured for three days. The pre-inoculated broth culture of strain HS124 was also checked with an ultraviolet (UV) spectrometer (Shimadzu, Kyoto, Japan) at 600 nm. Its optical density (OD) value was 1.71 [37]. Then, 100 μL of the pre-inoculated *L. antibioticus* HS124 culture (10^7 CFU/mL) was inoculated into 100 mL of CY broth and incubated at 30 °C for 10 days with shaking (130 rpm). Samples were collected on each inoculation day and spread onto CY agar medium using the serial dilution technique. The viable bacterial cells were counted as colony forming unit (CFU) on each inoculation day and the growth pattern of *L. antibioticus* HS124 was examined [30].

2.2. Chitinase Activity

To examine the chitinase activity based on incubation time, *L. antibioticus* HS124 was cultured in chitin–potato-dextrose (CPD) broth (Colloidal chitin [38], 10%; yeast extract, 0.05%; and potato dextrose broth (Daejung chemicals, Siheung, Korea), 1.2%) at 30 °C for 10 days in a shaking incubator at 130 rpm [39]. Samples were taken at 2-day intervals and centrifuged at 12,000 rpm for 10 min. The resulting supernatant was used to investigate chitinase activity on each sampling day.

Chitinase activity was measured by the Lingappa and Lockwood method [40]. Briefly, a reaction mixture containing 50 µL of *L. antibioticus* HS124 supernatant, 450 µL of 0.2 M sodium acetate buffer (pH 5.0), and 500 µL of 0.5% colloidal chitin was incubated at 37 °C. Subsequently, 200 µL of 1 N NaOH (Yakuri pure chemicals, Kyoto, Japan) was added to terminate the reaction. The mixture was centrifuged at 12,000 rpm for 10 min at 4 °C. Next, 750 µL of the supernatant was mixed with 1 mL of Schales' reagent and 250 µL of distilled water and incubated at 100 °C boiling water for 15 min. The amount of reducing sugar was then quantitatively analyzed at 420 nm using a UV-spectrometer. The assay was repeated three times. One unit of chitinase enzyme activity was determined as the quantity of enzyme releasing 1 µmol of *N*-acetyl-glucosamine per hour at 37 °C.

2.3. Antifungal Activity of L. antibioticus HS124 toward Fungal Phytopathogens

The phytopathogenic fungi used in this study, *B. cinerea* (KACC 41008), *B. dothidea* (KACC 45481), and *C. gloeosporioides* (KACC 40897), were provided by the KACC. These three phytopathogenic fungi were cultured in potato dextrose agar (PDA) medium at 25 °C for seven days. The antagonistic activities of *L. antibioticus* HS124 against three phytopathogenic fungi were determined by the dual culture method. One-loopful of *L. antibioticus* HS124 colonies was streaked on one side of a CY agar plate and incubated at 30 °C for three days [30]. Then, a 5-mm plug of each phytopathogenic fungus from a 7-day-old culture plate was placed on the other side of the same plate at a distance of 4 cm and the plates were incubated at 25 °C. Depending on the growth rate of each phytopathogenic fungus, the incubation days were different: *B. cinerea*, four days; *B. dothidea*, five days; and *C. gloeosporioides*, eight days. A plate without inoculation of *L. antibioticus* HS124 was used as a control. The experiment was repeated three times with three replications. The inhibition of fungal growth by the HS124 strain was determined with the following formula: inhibition (%) = [(α − β)/α] × 100, where α was the radial growth of phytopathogenic fungus on the control plate and β was the radial growth of phytopathogenic fungus on the dual culture plate [27,28,30].

A small piece of mycelium at the boundary of the fungal colony inhabited by *L. antibioticus* HS124 was taken and observed for hyphal deformation and degradation caused by *L. antibioticus* HS124 under a light microscope at 200× magnification (Olympus BX41TF, Tokyo, Japan).

2.4. Indole-3-Acetic Acid (IAA) Production by L. antibioticus HS124

Quantitative analyses of IAA production by *L. antibioticus* HS124 were performed using a UV spectrometric method. Briefly, *L. antibioticus* HS124 was cultured in a medium containing 0.1 g/L crab shell powder (Purne, Jangseong, Korea), 0.2 g/L Na_2HPO_4 (Daejung chemicals, Siheung, Korea), 0.1 g/L KH_2PO_4 (Daejung chemicals, Siheung, Korea), 0.5 g/L NaCl (Daejung chemicals, Siheung, Korea), 0.1 g/L NH_4Cl (Yakuri pure chemicals, Kyoto, Japan), 0.05 g/L $MgSO_4$ $7H_2O$ (Shimakyu's pure chemicals, Osaka, Japan), 0.05 g/L $CaCl_2$ $2H_2O$, 0.01 g/L yeast extract, and 0.1 g/L L-tryptophan (Junsei chemical, Tokyo, Japan). The culture was incubated at 30 °C in a shaking incubator (140 rpm). Samples were taken every two days five times from the day of inoculation. Quantitative measurement of the HS124 strain was performed according to Salkowski's method [41]. Briefly, the samples were centrifuged at 12,000 rpm for 10 min at 4 °C and 1 mL of the resulting supernatant was mixed with 2 mL of Salkowski's reagent. Subsequently, the reaction mixture was incubated at room temperature under dark conditions for 25 min. The IAA concentration of each sample was measured at 530 nm using a UV-spectrometer.

2.5. Study Area and Field Experimental Conditions

The experimental sites of a cultivated jujube orchard were located (35°84′69″ N, 128°80′40″ E) in Gyeongsan City, Gyeongbuk Province, Korea (Figure 1a). The soils at the study sites were fluvial deposits. The major soil type of the study area was sandy loam. The experimental sites in Gyeongsan

City had a temperate climate with a mean temperature of 14.6 °C in 2016. The precipitation on-site was 1227 mm, approximately 57% of which fell between July 2016 and September 2016.

The jujube trees in the study area were planted in the 1970s (Figure 1b). In 2016, approximately 46-year-old trees with a height of approximately 4 m and a diameter of approximately 15 cm were distributed in the study areas. In addition, a rain shelter with transparent vinyl was installed above the tree crown to prevent the fall of fruits because of rain (Figure 1b). Jujube field experiment was arranged using a rectangular plot design measuring 8 m wide × 24 m long with a distribution of 10 trees (Figure 1b). The following two treatment groups were used in the field experiment, each with three replicates: (1) control without *L. antibioticus* HS124 inoculation and (2) *L. antibioticus* HS124 inoculation.

The *L. antibioticus* HS124 cultures were prepared in CY medium. The liquid form of microbial product containing 10^{10} CFU/mL of *L. antibioticus* HS124 (GCM+, Purne, Jangseong, Korea) was used for large scale cultivation of *L. antibioticus* HS124. Typically, 150 mL of *L. antibioticus* HS124 liquid product was inoculated into 500 L of CY medium and cultured at 30 °C for seven days using a fermenter. The *L. antibioticus* HS124 cultures were diluted with tap water (1:2 v/v) and poured onto the soil adjacent to the tree roots at approximately 2-week intervals 14 times from April to September 2016.

2.6. Mineral Concentration, Fruit Characteristics, and Yield

To determine the fruit characteristics (length and diameter) and fruit yield after treatments, six jujube trees in each plot were selected and the fruits of branches with a similar length and height and facing south compared to a standard were harvested in October 2016. The weight of fresh fruit from a standard branch in each sample tree was recorded to determine the yield. In addition, fruit length and diameter were calculated with respect to 10 fruit in each sample tree.

Undamaged jujube fruits were selected and washed thoroughly with distilled water to remove dirt and air-dried. The fruit was finely chopped into small pieces using a sharp knife and the seeds were separated from the pulp. The prepared sample was pretreated using a dry decomposition method [42]. For dry decomposition, 0.7 g of the sample was ashed at 550 °C, then 10 mL of a diluted solution (HCl:distilled water = 1:1) was added to the ash, followed by decomposition at 25 °C for six hours. Thereafter, a funnel was inserted into a 50-mL volumetric flask, the ash solution was filtered through filter paper, massed with distilled water, and used as a test solution. The mineral compositions were determined using a procedure described by the AOAC (Association of Official Analytical Chemists) [42]. Each mineral (potassium (K), calcium (Ca), magnesium (Mg), and phosphorus (P)) was detected with an inductively coupled plasma optical emission spectrometer (PerkinElmer, Waltham, MA, USA).

2.7. Statistical Analysis

All statistical calculations were performed using the Statistical Package for the Social Sciences (SPSS) software, version 23 (Armonk, New York, NY, USA). The results are reported as the mean ± standard deviation. The data were evaluated by *t*-tests with significance considered at $p < 0.05$.

3. Results

3.1. Inhibitory Effect of L. antibioticus HS124 on Growth of Fungal Pathogens

3.1.1. Chitinase Activity

The growth of *L. antibioticus* HS124 was low until five days post-inoculation (Figure 2) and exhibited a rapid increase six days post-inoculation. The maximum growth (5.67×10^7 CFU/mL) was found seven days post-inoculation. Subsequently, the growth moderately decreased until the end of the inoculation period (Figure 2).

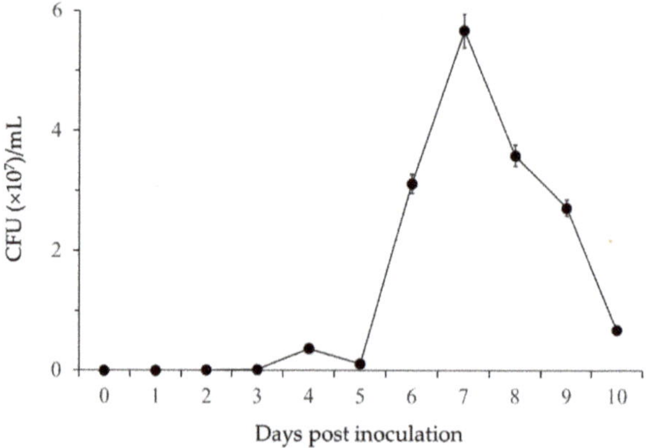

Figure 2. Cell growth curve of *L. antibioticus* HS124 in CY medium at 30 °C for 10 days. Error bars represent the standard deviation of three replications. CFU: colony forming unit.

The chitinase activity of *L. antibioticus* HS124 increased after two days and eventually reached a maximum value of 81.1 unit/mL over a period of four days (Figure 3). Thereafter, chitinase activity remained steady until eight days post-inoculation (Figure 3).

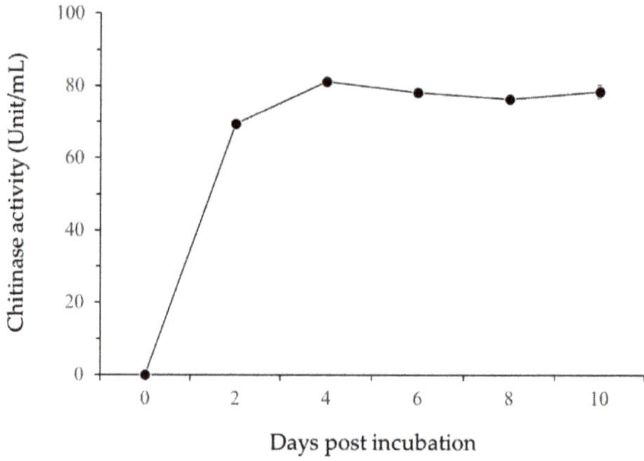

Figure 3. Chitinase activity of *L. antibioticus* HS124 cultured in CPD broth at 30 °C for 10 days. Quantitative measurement of chitinase enzymes produced by strain HS124 was done using a UV-spectrophotometer at 420 nm.

3.1.2. Growth Inhibition of Phytopathogenic Fungi by *L. antibioticus* HS124

The antifungal activities of *L. antibioticus* HS124 against different phytopathogenic fungi, including *B. cinerea*, *B. dothidea*, and *C. gloeosporioides*, were assayed in CY agar medium using the dual culture test (Figure 4). *L. antibioticus* HS124 showed the highest inhibition (27.25%) against *B. dothidea* and the lowest inhibition (20.56%) against *C. gloeosporioides* (Figure 4). Moreover, it showed 22.03% mycelial growth inhibition against *B. cinerea* (Figure 4).

The hyphal morphologies by *L. antibioticus* HS124 were abnormal, showing degradation, deformation, and lysis, compared to controls not inoculated with *L. antibioticus* HS124. The controls showed normal hyphal structures (Figure 5).

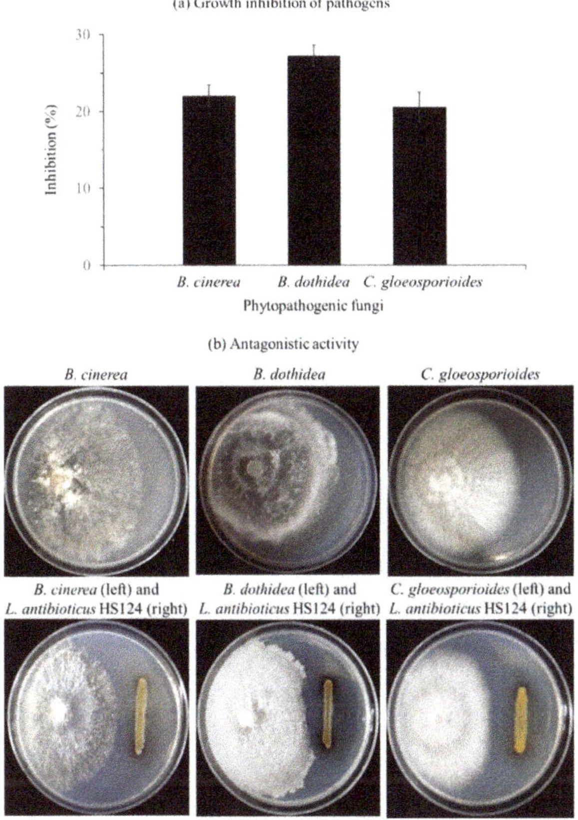

Figure 4. Inhibitory effect of *L. antibioticus* HS124 on mycelial growth of *B. cinerea*, *B. dothidea*, and *C. gloeosporioides* (**a**), and antagonistic activity of *L. antibioticus* HS124 against fungal pathogens (**b**) by the dual culture method.

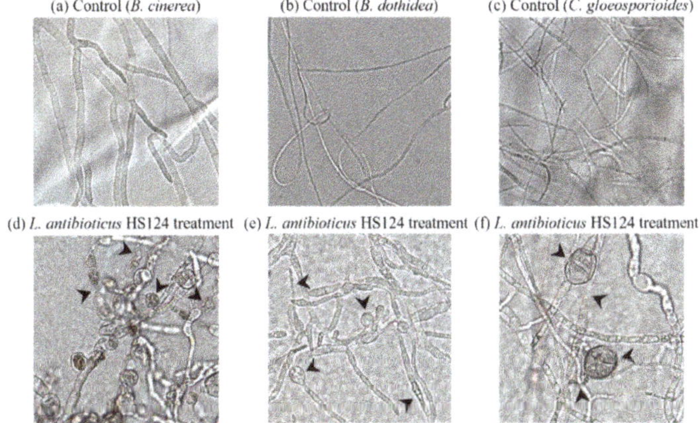

Figure 5. Deformed hyphal morphologies of *B. cinerea* (**d**); *B. dothidea* (**e**); and *C. gloeosporioides* (**f**) affected by *L. antibioticus* HS124 compared to *B. cinerea* control (**a**); *B. dothidea* control (**b**); and *C. gloeosporioides* control (**c**) under a light microscope. Arrows indicate hyphal alterations with swelling and bulbous structures caused by *L. antibioticus* HS124.

3.2. Effect of L. antibioticus HS124 on Fruit Yield

3.2.1. IAA Production

L. antibioticus HS124 produced auxin (Figure 6). The IAA concentration steadily increased for six days, eventually reaching a maximum value of 9.3 mg/L. Thereafter, the IAA concentration decreased rapidly (Figure 6).

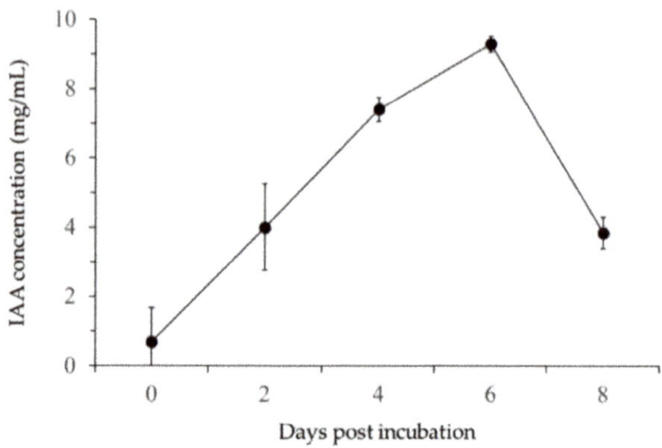

Figure 6. IAA (indole-3-acetic acid) production by *L. antibioticus* HS124.

3.2.2. Mineral Concentration, Characteristics, and Yield of Jujube Fruits

Potassium (K) was the most abundant mineral (2.12 g/kg) in the fruit, followed by P (0.21 g/kg) (Table 1). The fruit contained relatively low concentrations of Ca and Mg. A significant increase in the concentrations of K, Ca, Mg, and P was observed in the jujube fruits inoculated with *L. antibioticus* HS124 when compared to that of the control group (Table 1). The K and Mg concentrations in fruits with bacterial inoculation were 1.8- and 2.0-fold more than that of control group, respectively (Table 1). Moreover, the P and Ca concentrations in fruits that were inoculated with bacteria were 1.4- and 1.6-fold more than that of control group, respectively (Table 1).

Table 1. Mineral concentration, characteristics, and yields of jujube fruits in control and treatment groups with *L. antibioticus* HS124 inoculation in cultivated orchards.

Treatment	Fruit Mineral (g/kg)				Fruit Characteristics (mm)		Fruit Yield (g)
	K	Ca	Mg	P	Length	Diameter	
Control	1.18 ± 0.18 *	0.05 ± 0.01 *	0.04 ± 0.01 *	0.15 ± 0.02 *	29.66 ± 1.67 *	23.77 ± 1.24 *	2165.78 ± 221.19 *
Bacterial inoculation	2.12 ± 0.17 *	0.08 ± 0.01 *	0.08 ± 0.01 *	0.21 ± 0.03 *	38.37 ± 1.41 *	28.99 ± 1.53 *	6284.67 ± 1207.23 *

* A significant difference between treatments was observed by *t*-test at $p < 0.05$.

The length of the jujube fruits ranged from 35.39 to 40.85 mm with *L. antibioticus* HS124 inoculation and from 26.93 to 33.70 mm in the control without *L. antibioticus* HS124 inoculation (Table 1). The length of the fruits inoculated with bacteria was significantly higher than that of the control (Table 1). The average diameter of the jujube fruits ranged from 23.77 mm in the control to 28.99 mm with bacterial inoculation (Table 1). Regarding fruit characteristics of jujube cultivated in the orchard, the length and diameter of fruits inoculated with bacteria were significantly higher than those of the control fruits (Table 1).

The average yield of jujube fruit ranged from 2165.78 g/branch in the control to 6284.67 g/branch in trees inoculated with *L. antibioticus* HS124 (Table 1). A remarkable improvement in fruit yield was observed in treatment with *L. antibioticus* HS124 as compared to that of control (Table 1).

4. Discussion

Cell wall-degrading enzymes such as chitinase produced by antagonistic bacteria are known to play key roles in the suppression of phytopathogenic fungi [25–30]. In the present study, results of quantitative chitinase assay indicated that strain HS124 produced high levels of chitinase in colloidal containing medium (Figure 3). Therefore, the action of chitinase could be involved in antagonism against these fungal pathogens, including *B. cinerea*, *B. dothidea*, and *C. gloeosporioides* (Figures 4 and 5). In the presence of *L. antibioticus* HS124, the hyphae of phytopathogenic fungi showed abnormal morphologies, including swelling and bulb formation (Figure 5). Our results are in agreement with previous report of Won et al. [27,28], showing that chitinase enzymes produced by *Bacillus licheniformis* MH48 were involved in the control of diseases caused by fungi such as *Fusarium oxysporum*, *B. cinerea*, *Glomerella cingulate, Pestalotia diospyri, and Pestalotiopsis karstenii*. In a similar experiment, Brzezinska and Jankiewicz [43] showed that a chitinase secreted by *Aspergillus niger* LOCK62 had antifungal activities against various other fungi such as *Fusarium solani*, *Fusaium culmorum*, and *B. cinerea*. According to Velusamy and Kim [44], a chitinase secreted by *Enterobacter* sp. could decompose mycelia of *C. gloeosporioides* and *B. cinerea* and inhibit their growth. The antagonistic activity of *L. antibioticus* HS124 against phytopathogenic fungi resulted in a significant increase in the production of jujube fruits in cultivated orchards (a 2.9-fold increase) compared to the control (Table 1). Control jujube fruit and trees were easily infected by fungal diseases (Figure 1c), resulting in wilting and dropping of fruit during production. Fungal pathogen-infected fruits were discarded during harvesting (Figure 1b), thus decreasing the fruit production (Table 1). In particular, our field observations frequently found anthracnose disease which led to a reduction in the production of jujube fruit because of the increased flower wilting and fruit dropping. In addition, *L. antibioticus* HS124 produced auxin (Figure 6). Auxin acts at the time of cell division after fertilization of flowers [31,33,34], thus increasing the size of the fruit and improving the yield of trees inoculated with *L. antibioticus* HS124 (Table 1). PGPR also provides nutrients to the soil by fixing nitrogen from the atmosphere and solubilizing phosphorus in the soil [26,45]. Plants with increased absorption of nitrogen and phosphorus can promote photosynthesis and protein biosynthesis, thus increasing the size and yield of the fruit [46–49]. In cultivated orchard experiments, the size and yield of jujube fruits from trees inoculated with *L. antibioticus* HS124 were significantly increased (Table 1).

Well development in root system of trees are the most essential attribute for enhancing minerals and nutrient uptake [25,27,28,49]. Auxin produced by *L. antibioticus* HS124 inoculation might promote root development and stimulate the formation of absorbent root hairs and lateral roots [24,25,27,28,31,32]. This finding indicates that jujube trees inoculated with *L. antibioticus* HS124 could absorb more mineral elements such as K, Ca, Mg, and P in fruits (Table 1). Because PGPR including *Lysobacter* sp. can secrete organic acids and mineralize mineral elements such as K, Ca, Mg, Fe, and P from soil and enrich them in the soil [26,45,47,49], roots developed by auxin can promote the absorption of minerals and nutrient elements [24,25,27,32,48,50]. In addition, the photosynthesis of plants in the nitrogen-enhanced soils caused by nitrogen fixation from PGPR is enhanced, resulting in increased growth and mineral absorption in plants [26,45,48–52]. According to Ipek et al. [45] and Pirlak et al. [51] apples and strawberries with PGPR such as *Pseudomonas* sp., *Bacillus* sp., *Alcaligenes* sp., *Staphylococcus* sp., and *Agrobacterium* sp. can promote the photosynthesis caused by nitrogen fixation and stimulate the growth and contents of mineral elements. Results of the present study indicate that *L. antibioticus* HS124 is not only an effective biocontrol agent for phytopathogenic fungi, but also a beneficial agent for increasing the fruit mineral content and yield in jujube farms.

5. Conclusions

L. antibioticus HS124 could suppress fungal diseases caused by *B. cinerea*, *B. dothidea*, and *C. gloeosporioides* by producing lytic enzymes, including chitinase (Figures 3–5), thereby decreasing the fruit wilting and dropping during the production and harvest periods of cultivated jujube orchards (Figure 1c). The improvement in size and yield of jujube fruits in the present study could be associated with the secretion of auxin like IAA by *L. antibioticus* HS124 (Table 1). Therefore, inoculating *L. antibioticus* HS124 to jujube trees should be considered as a potential way to control fungal diseases and increase the production of jujube fruits in a sustainable and ecological cultivation system of orchards.

Author Contributions: Conceptualization, funding acquisition and project administration, Y.-S.A.; investigation and experiments, J.-H.K., S.-J.W., and J.-H.M.; data analysis, J.-H.K. and S.-J.W.; resources, C.-W.K.; writing—original draft preparation, J.-H.K.; writing—review and editing, Y.-S.A.

Funding: This study was supported by the R&D program for Forest Science & Technology Projects (No. 2018122B10-1820-AB01) funded by the Korea Forest Service (Korea Forestry Promotion Institute). Additionally, this research was supported by a grant (No. 2018R1D1A1B07050052) of the National Research Foundation (NRF) of Korea under the Basic Science Research Program.

Acknowledgments: The authors would like to thank Kil-Yong Kim at the Chonnam National University for the technical assistance in the field and laboratory. The senior author also thanks Yun-Serk Park, CEO of Purne for their skillful assistance in analyzing the bacteria.

Conflicts of Interest: The authors declare no conflicts of interest.

References

1. Pareek, S. Nutritional composition of jujube fruit. *Emir. J. Food Agric.* **2013**, *25*, 463–470. [CrossRef]
2. Liu, M.J.; Cheng, C.Y. A taxonomic study on the genus *Ziziphus*. *Acta Hortic.* **1995**, *390*, 161–165. [CrossRef]
3. Hernández, F.; Legua, P.; Melgarejo, P.; Martinez, R.; Martinez, J.J. Phenological growth stages of jujube tree (*Ziziphus jujube*): Codification and description according to the BBCH scale. *Ann. Appl. Biol.* **2015**, *166*, 136–142. [CrossRef]
4. Krša, B.; Mishra, S. Sensory evaluation of different products of *Ziziphus jujuba* Mill. *Acta Hortic.* **2009**, *840*, 557–562. [CrossRef]
5. Gao, Q.H.; Wu, C.S.; Wang, M.; Xu, B.N.; Du, L.J. Effect of drying of jujubes (*Ziziphus jujuba* Mill.) on the contents of sugars, organic acids, α-tocopherol, β-carotene, and phenolic compounds. *J. Agric. Food Chem.* **2012**, *60*, 9642–9648. [CrossRef]
6. Wojdyło, A.; Ángel, A.C.B.; Legua, P.; Hernández, F. Phenolic composition, ascorbic acid content, and antioxidant capacity of Spanish jujube (*Ziziphus jujube* Mill.) fruit. *Food Chem.* **2016**, *201*, 307–314. [CrossRef]
7. Shahrajabian, M.H.; Khoshkharam, M.; Zandi, P.; Sun, W.; Cheng, Q. Jujube, a super-fruit in traditional Chinese medicine, heading for modern pharmacological science. *J. Med. Plants Stud.* **2019**, *7*, 173–178.
8. Godbole, S. Overview of traditional knowledge and scientific research on medicinal uses of leaves and flowers offered during festivals. *World J. Pharmaceut. Res.* **2018**, *7*, 523–546. [CrossRef]
9. Goyal, M.; Nagori, B.P.; Sasmal, D. Review on ethnomedicinal uses, pharmacological activity and phytochemical constituents of *Ziziphus mauritiana* (*Z. jujube* Lam., non Mill). *Spatula DD* **2012**, *2*, 107–116. [CrossRef]
10. Marrelli, M.; Conforti, F.; Araniti, F.; Statti, G.A. Effects of saponins on lipid metabolism: A review of potential health benefits in the treatment of obesity. *Molecules* **2016**, *21*, 1404. [CrossRef]
11. Hernández, F.; Luis, N.A.; Burló, F.; Wojdyło, A.; Ángel, A.C.B.; Legua, P. Physico-chemical, nutritional, and volatile composition and sensory profile of Spanish jujube (*Ziziphus jujube* Mill.) fruits. *J. Sci. Food Agric.* **2016**, *96*, 2682–2691. [CrossRef] [PubMed]
12. Guo, S.; Duan, J.A.; Tang, Y.; Su, S.; Shang, E.; Ni, S.; Qian, D. High-performance liquid chromatography–wo wavelength detection of triterpenoid acids from the fruits of *Ziziphus jujuba* containing various cultivars in different regions and classification using chemometric analysis. *J. Pharmaceut. Biomed.* **2009**, *49*, 1296–1302. [CrossRef] [PubMed]
13. Zhang, H.; Jiang, L.; Ye, S.; Ye, Y.; Ren, F. Systematic evaluation of antioxidant capacities of the ethanolic extract of different tissues of jujube (*Ziziphus jujube* Mill.) from China. *Food Chem. Toxicol.* **2010**, *48*, 1461–1465. [CrossRef] [PubMed]
14. Morton, J.F. *Fruits of Warm Climates*; Purdue University: Miami, FL, USA, 1987; pp. 272–275.
15. Williamson, B.; Tudzynski, B.; Tudzynski, P.; Kan, J.A.L.V. *Botrytis cinerea*: The cause of grey mould disease. *Mol. Plant Pathol.* **2007**, *8*, 561–580. [CrossRef] [PubMed]

16. Phoulivong, S.; Cai, L.; Chen, H.; McKenzie, E.H.C.; Abdelsalam, K.; Chukeatirote, E.; Hyde, K.D. *Colletotrichum gloeosporioides* is not a common pathogen on tropical fruits. *Fungal Divers.* **2010**, *44*, 33–43. [CrossRef]

17. Zhu, H.Y.; Tian, C.M.; Fan, X.L. Studies of botryosphaerialean fungi associated with canker and dieback of tree hosts in Dongling Mountain of China. *Phytotaxa* **2018**, *348*, 63–76. [CrossRef]

18. He, C.; Zhang, Z.; Li, B.; Xu, Y.; Tian, S. Effect of natamycin on *Botrytis cinerea* and *Penicillium expansum*—Postharvest pathogens of grape berries and jujube fruit. *Postharvest Biol. Technol.* **2019**, *151*, 134–141. [CrossRef]

19. O'connell, P.F. Sustainable agriculture—A valid alternative. *Outlook Agric.* **1992**, *21*, 5–12. [CrossRef]

20. Gisi, U.; Sierotzki, H.; Cook, A.; McCaffery, A. Mechanisms influencing the evolution of resistance to Qo inhibitor fungicides. *Pest Manag. Sci.* **2002**, *58*, 859–867. [CrossRef]

21. Savci, S. An agricultural pollutant: Chemical fertilizer. *Int. J. Environ. Sci. Dev.* **2012**, *3*, 77–80. [CrossRef]

22. Juntunen, M.L.; Hammar, T.; Rikala, R. Leaching of nitrogen and phosphorus during production of forest seedlings in containers. *J. Environ. Qual.* **2002**, *31*, 1868–1874. [CrossRef] [PubMed]

23. Youssef, M.M.A.; Eissa, M.F.M. Biofertilizers and their role in management of plant parasitic nematodes. A review. *E3 J. Biotechnol. Pharm. Res.* **2014**, *5*, 1–6.

24. Glick, B.R.; Penrose, D.M.; Li, J. A Model for the lowering of plant ethylene concentrations by plant growth-promoting bacteria. *J. Theor. Biol.* **1998**, *190*, 63–68. [CrossRef] [PubMed]

25. Glick, B.R. Plant growth-promoting bacteria: Mechanisms and applications. *Scientifica* **2012**, *2012*, 963401. [CrossRef] [PubMed]

26. Orhan, E.; Esitken, A.; Ercisli, S.; Turan, M.; Sahin, F. Effects of plant growth promoting rhizobacteria (PGPR) on yield, growth and nutrient contents in organically growing raspberry. *Sci. Hortic.* **2006**, *111*, 38–43. [CrossRef]

27. Won, S.-J.; Choub, V.; Kwon, J.-H.; Kim, D.-H.; Ahn, Y.-S. The control of fusarium root rot and development of coastal pine (*Pinus thunbergii* Parl.) seedlings in a container nursery by use of *Bacillus licheniformis* MH48. *Forests* **2019**, *10*, 6. [CrossRef]

28. Won, S.-J.; Kwon, J.-H.; Kim, D.-H.; Ahn, Y.-S. The effect of *Bacillus licheniformis* MH48 on control of foliar fungal diseases and growth promotion of *Camellia oleifera* seedling in the coastal reclaimed land of Korea. *Patohgens* **2019**, *8*, 6. [CrossRef]

29. Glick, B.R. The enhancement of plant growth by free-living bacteria. *Can. J. Microbiol.* **1995**, *41*, 109–117. [CrossRef]

30. Ko, H.-S.; Jin, R.-D.; Krishnan, H.B.; Lee, S.-B.; Kim, K.-Y. Biocontrol ability of *Lysobacter antibioticus* HS124 against *Phytophthora* blight is mediated by the production of 4-Hydroxyphenylacetic acid and several lytic enzymes. *Curr. Microbiol.* **2009**, *59*, 608–615. [CrossRef]

31. Gravel, V.; Antoun, H.; Tweddell, R.J. Growth stimulation and fruit yield improvement of greenhouse tomato plants by inoculation with *Pseudomonas putida* or *Trichoderma atroviride*: Possible role of indole acetic acid (IAA). *Soil Biol. Biolchem.* **2007**, *39*, 1968–1977. [CrossRef]

32. Goswami, D.; Thakker, J.N.; Dhandhukia, P.C. Portraying mechanics of plant growth promoting rhizobacteria (PGPR): A review. *Cogent Food Agric.* **2016**, *2*. [CrossRef]

33. Srivastava, A.; Handa, A.K. Hormonal regulation of tomato fruit development: A molecular perspective. *J. Plant Growth Regul.* **2005**, *24*, 67–82. [CrossRef]

34. Liu, D.J.; Chen, J.Y.; Lu, W.J. Expression and regulation of the early auxin-responsive *Aux/IAA* genes during strawberry fruit development. *Mol. Biol. Rep.* **2011**, *38*, 1187–1193. [CrossRef] [PubMed]

35. Han, T.; Cho, M.-Y.; Lee, Y.-S.; Park, Y.-S.; Park, R.-D.; Nam, Y.; Kim, K.-Y. Biocontrol of pepper diseases by *Lysobacter enzymogenes* LE429 and Neem Oil. *Korean J. Soil Sci. Fert.* **2010**, *43*, 490–497.

36. Esitken, A.; Ercisli, S.; Karlidag, H.; Sahin, F. Potential use of plant growth promoting rhizobacteria (PGPR) in organic apricot production. In Proceedings of the International Scientific Conference, Polli, Estonia, 7–9 September 2005; Libek, A., Kaufmane, E., Sasnauskas, A., Eds.; Tartu University Press: Tartu, Estonia, 2005; pp. 90–97.

37. Widdel, F. Theory and measurement of bacterial growth. *Grund. Mikrobiol.* **2007**, *4*, 1–11.

38. Berger, L.R.; Reynolds, D.M. The chitinase system of a strain of *Streptomyces griseus*. *Biochim. Biophys. Acta* **1958**, *29*, 522–534. [CrossRef]

39. Kim, Y.-T.; Monkhung, S.; Lee, Y.S.; Kim, K.Y. Effects of *Lysobacter antibioticus* HS124, an effective biocontrol agent against *Fusarium graminearum*, on crown rot disease and growth promotion of wheat. *Can. J. Microbiol.* **2019**, *65*, 904–912. [CrossRef]

40. Lingappa, Y.; Lockwood, J. Chitin media for selective isolation and culture of Actinomycetes. *Phytopathology* **1962**, *52*, 317–323.

41. Rahman, A.; Sitepu, I.R.; Tang, S.Y.; Hashidoko, Y. Salkowski's reagent test as a primary screening index for functionalities of rhizobacteria isolated from wild dipterocarp saplings growing naturally on medium-strongly acidic tropical peat soil. *Biosci. Biotechnol. Biochem.* **2010**, *74*, 2202–2208. [CrossRef]

42. Association of Analytical Communities. *Official Methods of Analysis of the Association of Official Analytical Chemists*, 15th ed.; Helich, K., Ed.; Association of Official Analytical Chemists, Inc.: Arlington, VA, USA, 1990; Volume 1, pp. 40–58.

43. Brzezinska, M.S.; Jankiewicz, U. Production of antifungal chitinase by *Aspergillus niger* LOCK 62 and its potential role in the biological control. *Curr. Microbiol.* **2012**, *65*, 666–672. [CrossRef]

44. Velusamy, P.; Kim, K.Y. Chitinolytic activity of *Enterobacter* sp. KB3 antagonistic to *Rhizoctonia solani* and its role in the degradation of living fungal hyphae. *Int. Res. J. Microbiol.* **2011**, *2*, 206–214.

45. Ipek, M.; Pirlak, L.; Esitken, A.; Donmez, M.F.; Turan, M.; Sahin, F. Plant growth-promoting rhizobacteria (PGPR) increase yield, growth and nutrition of strawberry under high-calcareous soil conditions. *J. Plant Nutr.* **2014**, *37*, 990–1001. [CrossRef]

46. Brewster, J.L. Onions and other vegetable alliums. *Sci. Hortic.* **1994**, *62*, 145–149. [CrossRef]

47. Osman, A.G.; Elaziz, F.I.A.; Elhassan, G.A. Effects of biological and mineral fertilization on yield, chemical composition and physical characteristics of faba bean (*Vicia faba* L.) cultivar seleim. *Pak. J. Nutr.* **2010**, *9*, 703–708. [CrossRef]

48. Rizk, F.A.; Shaheen, A.M.; Elsamad, E.H.A.; Sawan, O.M. Effect of different nitrogen plus phosphorus and Sulphur fertilizer levels on growth, yield and quality of onion (*Allium cepa* L.). *J. Appl. Sci. Res.* **2012**, *8*, 3353–3361.

49. Raklami, A.; Bechtaoul, N.; Tahiri, A.; Anli, M.; Meddich, A.; Oufdou, K. Use of rhizobacteria and mycorrhizae consortium in the open field as a strategy for improving crop nutrition, productivity and soil fertility. *Front. Microbiol.* **2019**, *10*, 1106. [CrossRef]

50. Yolcu, H.; Gunes, A.; Gullap, M.K.; Cakmakci, R. Effects of plant growth-promoting rhizobacteria on some morphologic characteristics, yield and quality contents of Hungarian vetch. *Turk. J. Field Crop.* **2012**, *17*, 208–214.

51. Pirlak, L.; Turan, M.; Sahin, F.; Esitken, A. Floral and foliar application of plant growth promoting rhizobacteria (PGPR) to apples increase yield, growth, and nutrient element contents of leaves. *J. Sustain. Agric.* **2007**, *30*, 145–155. [CrossRef]

52. Park, H.-G.; Lee, Y.-S.; Kim, K.-Y.; Park, Y.-S.; Park, K.-H.; Han, T.-H.; Park, C.-M.; Ahn, Y.S. Inoculation with *Bacillus licheniformis* MH48 promotes nutrient uptake in seedling of the ornamental plant *Camellia japonica* grown in Korea reclaimed coastal lands. *Hortic. Sci. Technol.* **2017**, *35*, 11–20. [CrossRef]

Article

Residents' Attention and Awareness of Urban Edible Landscapes: A Case Study of Wuhan, China

Qijiao Xie [1,2], Yang Yue [1] and Daohua Hu [1,*

[1]   School of Resources and Environmental Science, Hubei University, Wuhan 430062, China;
    xieqijiao@126.com (Q.X.); yueyang428@163.com (Y.Y.)
[2]   Hubei Key Laboratory of Regional Development and Environmental Response (Hubei University),
    Wuhan 430062, China
*   Correspondence: hdh@hubu.edu.cn; Tel.: +86-027-8866-1699

Received: 1 November 2019; Accepted: 11 December 2019; Published: 13 December 2019

Abstract: More and more urban residents in China have suffered from food insecurity and failed to meet the national recommendation of daily fruit and vegetable consumption due to rapid urbanization in recent years. Introducing edible landscapes to urban greening systems represents an opportunity for improving urban food supply and security. However, residents' opinion on urban edible landscapes has rarely been discussed. In this study, questionnaire surveys were performed in eight sample communities in Wuhan, China, to collect the information on residents' attention and awareness of urban edible landscapes. Results indicated that nearly one-third of the respondents were unaware of edible landscapes before the interview. Most residents thought that an edible landscape could promote efficient land use (57.26%) and express special ornamental effects (54.64%), but quite a few didn't believe that growing edible plants in urban public spaces could increase food output (37.10%) and improve food quality (40.12%). Overall, 45.65% and 32.73% of the growers performed their cultivation behavior in private and semiprivate spaces, respectively. Lack of public areas for agriculture use was regarded as the main barrier restricting the development of urban horticulture by 55.86% of growers and 59.51% of non-growers. The residents were also worried about their property manager's opposition, possible conflicts, and complex relationships with their neighbors. Food policies and infrastructure support from local governments and official institutions were needed to ensure the successful implementation of edible landscapes in urban areas.

Keywords: edible landscape; food security; urban horticulture; community garden

1. Introduction

Rapid urbanization leads to the increase of urban population and the replacement of agricultural lands by construction lands. With the increasing demand for food, city dwellers have to rely on processed food or that transported over long distances, which results in nutrient loss and high food prices. To some extent, the urbanization process reduces food availability and food accessibility. Food security, that is "when all people, at all times, have physical and economic access to sufficient safe and nutritious food that meets their dietary needs and food preferences for an active and healthy life (FAO 2008)" [1], thus will be greatly challenged. To solve this problem, some related research has been conducted. Subsistence farming in urban areas was regarded as a strategy for improving food and nutrition security due to the easy access to food, especially in developing countries [2–4]. Edible elements such as vegetables, fruits, herbs, and crops were introduced to urban public spaces in many countries [5–7]. This created a new landscape, characterizing both an aesthetic effect and food production, defined as edible landscape. Besides the traditional benefits of carbon sequestration, rainwater retention, and heat island mitigation [8,9], it also has irreplaceable advantages in alleviating food safety crisis, strengthening social cohesion, undertaking science popularization education, and enriching urban species diversity [10–12].

Orsinia et al. reported that the annual output of fruits and vegetables per m^2 could be as high as 50 kg [13]. Effective implementation of urban edible landscape strategy can meet 15%–20% of the global food demand [14]. By being engaged in food planting, urban residents can save transportation costs, reduce food miles, and obtain safety food. Urban horticulture is the main source of daily food and nutrition, especially for low-income residents in some developing countries [15,16]. It provides a way for retired people and housewives to integrate into society [17,18], which is of great significance for easing social conflicts and stabilizing social relations [19]. Residents feel close to nature by participating in cultivation activities, which deepens their understanding of the urban ecosystem [20]. The integration of edible plants with ornamental ones increases species diversity in urban areas. This helps to maintain the stability of the urban ecosystem [21] and promote sustainable urban development [22].

The relationships between urban and rural, human and nature, and industry and agriculture have gradually become out of balance. As a part of the urban green infrastructure, the edible landscape plays an important role in improving the environment, maintaining self-supporting systems, and enriching green-space functions. More than 80% of community gardens in New York were arranged for cultivating edible plants [9]. In New Zealand, 71% of early education schools and 52.9% of primary and secondary schools have edible gardens [23,24]. Germany built Andrnach, the first "edible city" in history [25]. In addition, the edible landscape has created considerable ecological, economic, and social value for cities. Kulak et al. reported that edible landscapes could reduce the emission of greenhouse gas in the London Sutton district by 34 tons per hm^2 in a year, which was higher than the carbon sequestration rate of traditional green lands [26]. Smith et al. confirmed that the average productivity of community gardens in the United States in 2010 was $15.19 per m^2 [27]. Meanwhile, Orsiniab et al. concluded that 77% of Bologna's agricultural output came from urban edible landscapes [28]. Growing edible plants in Brazilian cities reduced the food insecurity index from 30.2% in 2009 to 22.6% in 2013, so as to guarantee the health of urban residents [12].

For developing countries, food security is the guarantee to reduce social conflicts and maintain social stability [29–31]. In recent years, much attention has been paid to urban food insecurity in China. More and more Chinese experts and urban planners have realized the significance of developing edible landscapes in cities. Activities of planting food species in urban areas require residents' negotiation and collaboration, since every dweller acts as a participant and beneficiary in this process. However, what the Chinese residents' opinion is on edible landscapes has rarely been discussed [32–34]. This study, therefore, aimed to understand city dwellers' perception of edible landscapes and what factors contribute to their development. The specific objectives were as follows: (1) to find out urban residents' awareness of edible landscapes; (2) to investigate residents' experience of growing food species in their daily life; (3) to identify the possible constraints to the implementation of urban edible landscapes in China; and (4) to discuss the key points of policies related to edible landscapes in the future. The study can provide references for the construction of edible landscapes in emerging communities or public spaces in China or other developing countries undergoing rapid urbanization.

2. Methods

2.1. Study Area

Wuhan (113°41′–115°05′ E, 29°58′–31°22′ N), the capital city of Hubei Province in China, is located in the east of the Jianghan plain and at the confluence of the Yangtze River and the Hanjiang River (Figure 1a). It belongs to the north subtropical monsoon climate, with four distinct seasons and abundant rainfall, which is suitable for crop growth. As a result of its strong economic power and early modernization, Wuhan is more urbanized and has a population of over 11 million. As an ecological garden city, Wuhan has been committed to improving the urban ecology and living environment through landscape greening, and the green area has reached 1510 km^2. In recent years, Wuhan has been experiencing rapid urbanization. Urban construction extent sprawls outward around the city center, forming a typical ring-shaped pattern. This study was mainly carried out inside of the 3rd ring-road of Wuhan (Figure 1b), which covers an area of 628 km^2.

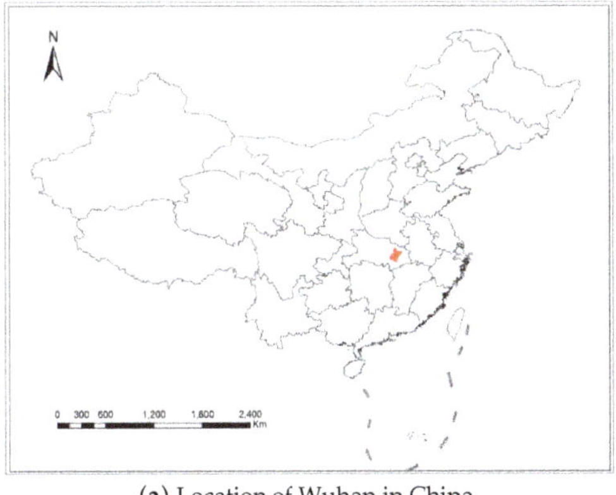
(**a**) Location of Wuhan in China.

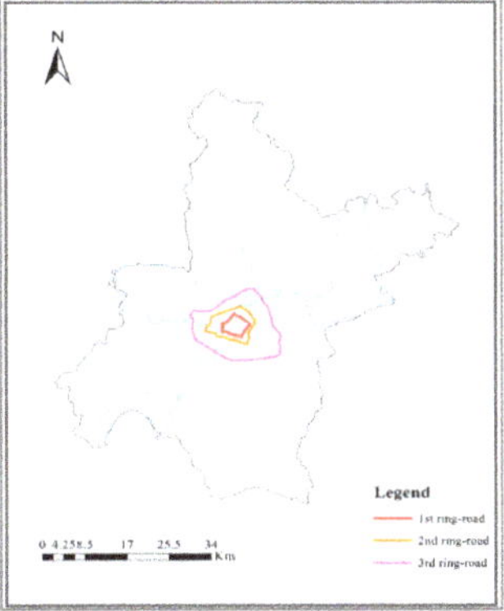
(**b**) Location of the study area in Wuhan.

Figure 1. Location of the study area.

2.2. Questionnaire

The semi-closed questionnaire was designed with three main parts: declaration, personal information, and questions about the residents' attitudes toward edible landscapes. To ensure the respondents' right to know about the investigation, some information on edible landscapes and the objective of this survey were stated at the beginning of the questionnaire. In addition, the anonymity of the questionnaire and the security of the personal information of the interviewees were also declared. Respondents' characteristics, such as their gender, age, and educational status, were recorded. We then investigated the knowledge and perception of urban residents of edible landscape, including the cognitive channel, awareness of edible landscapes, and their experience shown on edible species and planting space. Barriers that restricted the implementation of the Edible Landscape Project from the perspective of growers and non-growers were also covered. All activities were approved by the Academic Committee and Professor Committee of the School of Resources and Environmental Science, Hubei University (Approval No. ZH20180301). A small range of presurvey research was conducted in order to observe the respondents' reactions, and the standard time for complete responses was recorded. Questions that respondents were hesitant about were adjusted, aiming to make the information more direct and concise.

2.3. Data Collection

Taking into consideration the uniformity and representativeness of the interviewees, as well as the geographical location, construction age, and the green coverage of the communities, eight typical residential areas in Wuhan were selected as the main implementation area of the questionnaire survey, as shown in Table 1. These eight sample communities were evenly distributed in Wuhan urbanized area, as far as possible, from the inner city to the 3rd ring-road. The areas of the selected communities varied from 60,000 to 5,500,000 m^2, and the green coverage was 20%~40%. The communities were built 6~40 years ago, with 8000~180,000 residents.

The formal survey was conducted from April to June in 2018. To keep the same inquiry tone and interpretation, unified training was conducted for investigators before the interview. Residents of different ages were interviewed face-to-face in public spaces, such as garden tours, vegetable markets, or shopping malls near the study communities. At the same time, an online questionnaire survey was also conducted, as a useful supplementary in order to improve the response rate and to attract younger participants. The online survey was distributed to the residents through instant-messaging apps, such as QQ groups or WeChat groups, with the assistance of the community property owner committees or the neighborhood committees of the eight residential areas.

Table 1. Basic information on the eight investigated communities.

Community	Location	Years	Area (m^2)	Population	Green Coverage (%)
Tiandiyujiang	Inside 1st ring-road	8	125,000	8000	30.0
Xujiapeng Changlun	Inside 1st ring-road	40	60,000	11,000	20.0
Parrot Community	1st—2nd ring-road	19	120,000	7000	35.0
Tongxin Community	1st—2nd ring-road	14	160,000	8200	26.7
Baibuting Community	2nd—3rd ring-road	24	5,500,000	180,000	36.0
Changqing Community	2nd—3rd ring-road	22	4,500,000	120,000	40.0
Golden Harbor	Outside 3rd ring-road	19	250,000	15,000	38.0
Gezhouba Sun City	Outside 3rd ring-road	6	130,000	3000	20.0

2.4. Data Analysis

The collected questionnaire results were preliminarily distinguished according to the consistency of contextual logic and the response time. Questionnaires with incomplete or illogical answers were eliminated. For example, those who chose "F" as the answer for Question 7 could only choose "F" for answering Question 8 instead of other options (shown in Appendix A). Otherwise, it was out of logical consistency and invalid. The reasonable response time was determined in the phase of pre-investigation, with the value of 72 s. The questionnaires with the answer time of far less than 72 s were defined invalid and excluded. Then the response data were entered into an Excel database and counted separately, according to different issues.

3. Results

3.1. Characteristics of Interviewees

A total of 496 residents were interviewed through questionnaires. Their basic information of gender, age, education level, and length of residence in Wuhan is presented in Figure 2. Because not all residents were cooperative, those people who were willing to be interviewed expressed their attention to the edible landscape to a certain extent. Female residents paid more attention to edible landscapes than male ones, with a higher proportion (59.88%) than that of the males (40.12%). Of 496 respondents, 341 were between 20 and 40 years old, accounting for the largest proportion, which was 68.75%. Then there were respondents aged 40–60 years, which accounted for 25%. Those who were over 60 years old and less than 20 years old were uncooperative with the interview. The younger respondents and the elder ones only accounted for 2.62% and 3.63%, respectively.

For the education level, 60.96% of the respondents had received an undergraduate education. Those who had postgraduate education and senior high school education levels accounted for 18.35% and 15.12%, respectively. Only 5.85% of the respondents had education experience in junior high school and below. In total, 66.13% of respondents settled in Wuhan for more than 10 years. The interviewees who lived in Wuhan for 2–5 years and 5–10 years made up a similar proportion, with the values of 12.90% and 12.70%, respectively. Only 8.27% of the respondents lived in Wuhan for less than two years.

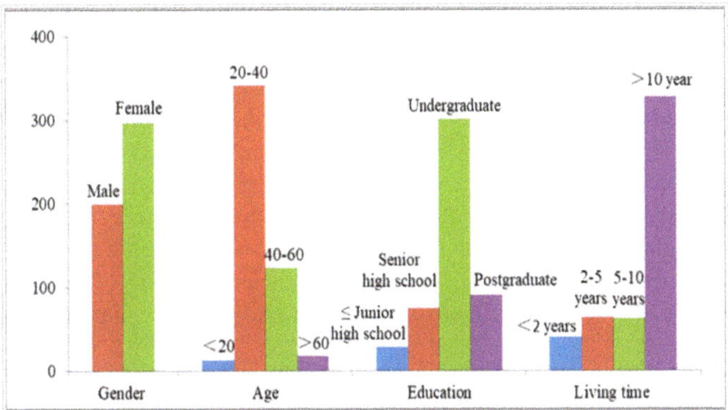

Figure 2. Characteristics of the respondents (n = 496).

3.2. Perception of Edible Landscape

Community residents' awareness and perception of edible landscapes are of great significance to the implementation of food planting in cities, especially in urban public spaces. For most Chinese urban residents, the edible landscape is new to community gardens. As shown in Table 2, of the 496 respondents, 32.06% were unaware of edible landscapes before this interview, 23.79% learned about edible landscapes and the related information from internet, and 20.97% were introduced to edible landscapes by their relatives or neighbors. A minority of the interviewees were informed about the edible landscape through magazines (5.44%), television programs (6.65%), and other ways (11.09%).

In this study, six typical benefits of edible landscapes were displayed, such as providing food production, high food quality, natural education, physical exercise, efficient land use, and distinct visual effects. As Table 3 shows, 284 respondents believed that food planting in community gardens could promote efficient land use, accounting for the largest proportion of 57.26%, and 54.64% accepted the distinct visual effect of edible landscape to the traditional one in public areas. Half of the respondents said that planting edible species in public areas provides the opportunity for natural education for urban residents, especially for children, and 49.19% thought that cultivating behavior could enhance their physical exercise and help them relax in their spare time. Only 37.10% and 40.12% believed that food planting in community gardens could provide food products and improve food quality.

Table 2. Ways to learn about the edible landscape (n = 496).

Cognitive Channels	Number	Proportion
By reading newspapers and magazines	27	5.44%
By watching TV	33	6.65%
By surfing on Internet	118	23.79%
By the introduction of relatives and neighbors	104	20.97%
By other ways	55	11.09%
Be unaware of it	159	32.06%

Table 3. Residents' awareness on the benefits of edible landscapes (n = 496).

Benefits of Edible Landscape	Number	Proportion
Forming distinct visual effect with traditional landscape	271	54.64%
Improving food quality	199	40.12%
Educating children on the science of nature	248	50%
Enhancing physical exercise and being relaxed	244	49.19%
Promoting efficient land use	284	57.26%
Providing food products	184	37.1%
Be unaware of it	159	32.06%

3.3. Growing Experience

Residents' growing experience of edible plants reflected their preference for cultivating location and species. Figure 3 records five common types of planting sites in urban areas. Of the 496 respondents, 163 had no experience growing edible plants in their daily life, accounting for 32.86% and 30.65% planted vegetables and small fruits on the balcony, followed by 16.53% in the courtyard, 8.87% in other locations, 5.65% in community vacant land, and 5.44% on the roof. Among the 333 respondents who had experienced food planting, 152 (45.65%) performed their cultivation behavior on the balcony, a private place. A total of 82 (24.62%) and 27 (8.11%) respondents planted vegetables and fruits in semiprivate places, namely the courtyard and the roof, respectively. Only 28 (8.41%) interviewees said their planting experience took place in public spaces, such as roadsides, the waterfront, or community gardens.

Figure 4 displays the preference of different growers for planting edible species. The majority of the growers preferred to plant vegetables, such as cucumbers, tomatoes, peppers, eggplants, and towel gourds, with the largest proportion of 71.47%. Fifty-eight respondents planted fruit trees, such as grapes, loquats, peaches and oranges, accounting for 17.42% of the growers. There were a few people who had planted crops (6.61%) and herbs (3.90%). Only two respondents planted mushrooms, accounting for 0.60%.

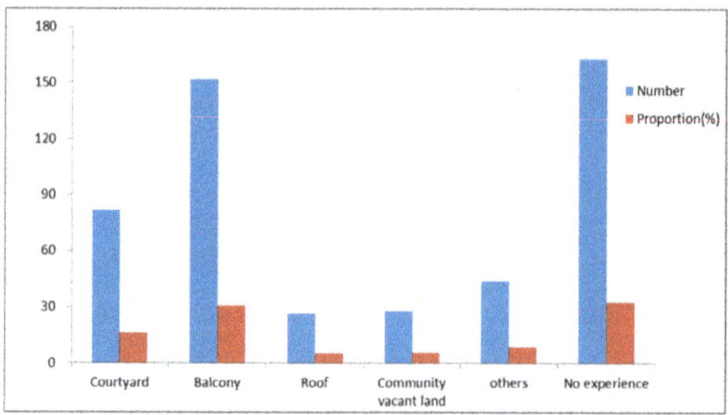

Figure 3. Locations where residents grow edible plants ($n = 496$).

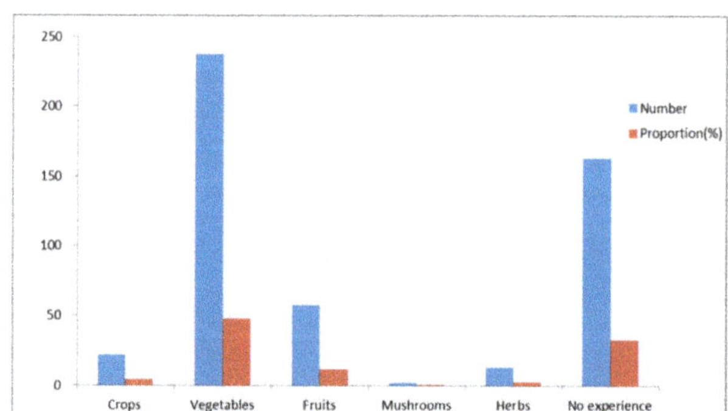

Figure 4. Edible elements urban residents cultivate ($n = 496$).

3.4. Barriers

To make it clear what possibly restricted the residents from implementing the Edible Landscape Project, the barriers were discussed. Considering the different understanding of the obstacles in implementing edible landscapes between growers and non-growers, they were discussed separately.

3.4.1. Barriers for Growers

Figure 5 counts the perceptions of 333 growers on the barriers influencing their planting. A total of 186 growers said that the largest trouble was lack of planting spaces, with the proportion of 55.86%. In total, 35.14% thought that the possible conflicts between growers and non-growers, as well as between residents and property managers, made it difficult to successfully manage the cultivating process; 33.03% of growers didn't think that they had enough free time for food planting; and 25.83% and 24.32% of growers also complained about the difficulty of obtaining plant seeds and fertilizer and lack of help in planting, respectively. In addition, their property manager's opposition, high planting costs, and food process difficulties were regarded as barriers by 19.82%, 14.11%, and 16.52% of growers, respectively.

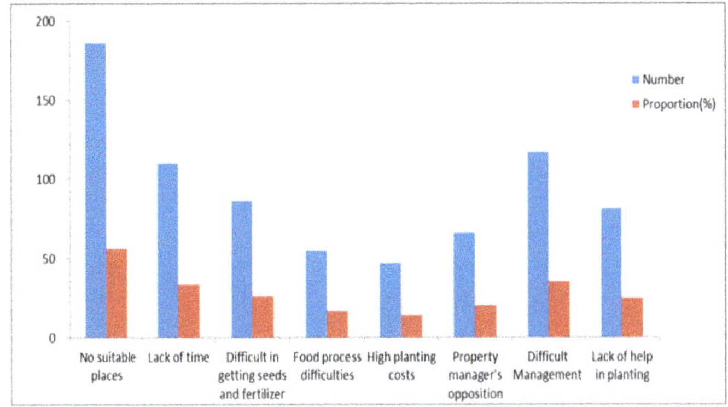

Figure 5. Barriers influencing growers' cultivation activities ($n = 333$).

3.4.2. Barriers of Non-Growers

Figure 6 expresses the understanding of 163 non-growers on the barriers constraining their planting. A total of 59.51% of people thought they had no suitable places to grow edible plants; 47.85% complained they were too busy to cultivate; 39.88% of non-growers didn't think they could finish the whole cultivation process only by themselves; and 33.13% of them gave up planting partly because they were unwilling to face the complex relationships with their neighbors. Other possible barriers, such as a property manager's objections, high planting costs, and difficulties in getting seeds and processing food products also troubled 21.47%, 20.25%, 19.02%, and 12.27% of non-growers, respectively, and then prevented them from planting.

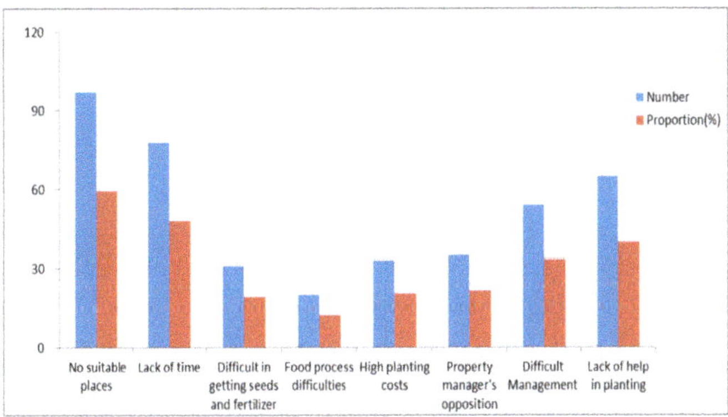

Figure 6. Barriers constraining non-growers' cultivation activities ($n = 163$).

4. Discussion

4.1. Lack of Awareness of the Benefit of Edible Landscapes in Improving Food Security

As a kind of urban landscape, the edible landscape has special ornamental and aesthetic effects compared with the traditional ones. This benefit was well recognized by most respondents in our investigation. Instead, the primary service of food production was not reasonably evaluated by the interviewees. Only about 40% of the respondents believed that growing edible plants in urban areas can actually alleviate the urban food crisis (Table 3). This significantly differed from that of the other countries. In most developing countries, edible landscapes (formerly known as urban horticulture) have been encouraged and promoted in both urban public and private spaces, for they play a vital role in providing food products and improving food quality for urban dwellers [35]. Food production with the yield potential of 50 kg/m^2 vegetables and fruits per year can almost meet the household fresh-food needs of urban residents [13]. The benefit of food producing is regarded as one of the most important services of edible landscapes by the majority of urban residents [36].

Lack of awareness about the benefit of edible landscapes in producing fresh food seriously blocked the residents' enthusiasm for growing food plants. On the one hand, residents don't fully believe that engagement in farming can allow them and their family to consume better and more nutritious diets from fresh vegetables and fruits, so they don't have a strong incentive to grow edible plants. On the other hand, the public does not understand that the implementation of edible landscapes in urban areas can effectively reduce the amount of food import, thus promoting urban food security and stability of urban food systems. They can't accept other residents planting edible species in public areas, such as city parks and community gardens. Correspondingly, the respondents in this study could grow food plants mainly in private or semiprivate areas, such as their balcony and courtyard (shown in Figure 3). Only 5.65% of the respondents had planting experience in public vacant spaces which were not planned for farming. Improving the urban residents' cognition of edible landscapes is of great significance for the successful implementation of the edible-landscape strategy.

4.2. Production Potential of Vegetables and Fruits

To some extent, displaying data is the most intuitive and persuasive way to change residents' cognition. The community gardens in our eight sample plots were designed to meet residents' needs for leisure and entertainment activities. Community residents are used to chatting, jogging, exercising, and playing cards there, so it is hard for them to imagine that these lands could produce fresh and nutritious vegetables and fruits, to meet their daily needs, even though this has already been achieved in many other countries. To provide solid evidence with clear data, the food-production potential of investigated community gardens was calculated. Table 1 lists the detailed information of the sample communities about the community area, greening coverage, and population. If 44% of the community garden area was used for fruit and vegetable planting [37], with the annual potential yields of 50 kg per m^2 [13], the supply of fruits and vegetables for each sample community was obtained. According to the Dietary Guidelines for Chinese Resident (2016) [38], one Chinese resident needs at least 500–850 g of vegetables and fruits every day, to ensure a healthy diet. The minimum intake of 500 g per day was used as a reference to estimate the fresh-food demand.

Table 4 displays the supply and demand of fresh vegetables and fruits for each sample community. The annual production of vegetables and fruits had a wide range of 264 to 43,560 t among different communities, depending on the areas of the community gardens. Residents' annual demand for vegetables and fruits varied from 547.5 to 32,850 t, which were currently imported into the cities by long-distance transportation from the nearby suburbs. With the self–production function of community gardens, 14%–180% of the demand for fruits and vegetables can be satisfied. For big communities such as Changqing Community and Baibuting Community, fresh food supply is much more than the demand, with the supply and demand ratio of 1.8 and 1.3, respectively. The total supply-and-demand ratio of fruits and vegetables in estimated areas reached 7. This meant that there were still a large number of fruits and vegetables surplus on the premise of meeting the residents' daily demand. The remaining fruits and vegetables can feed some residents of other communities within this city, which can help to reduce the food import and to improve the city's self-sufficiency.

Table 4. Potential production assessment of eight sample communities.

Community	Tiandi-yujiang	Xujiapeng Changlun	Parrot Community	Tongxin Community	Baibuting Community	Changqing Community	Golden Harbor	Gezhouba Sun City	Total
Supply (t)	825	264	924	939.84	43,560	39,600	2090	572	88,774.8
Demand (t)	1460	2007.5	1277.5	1496.5	32,850	21,900	2737.5	547.5	64,276.5
Ratio (%)	56.5	13.2	72.3	62.8	132.6	180.8	76.3	104.5	699.0

4.3. Significance of Government's Plan and Management

The potential benefits of community gardens and other urban open spaces have encouraged many governments to plan specific spaces for food planting [39,40]. Edible parks, community farms, school gardens, and rooftop gardens are the main growing areas in most countries [9,23,24]. In China, there were no public spaces planned for agriculture use in urban areas. As shown in Figure 3, 261 of 333 growers in this study conducted their cultivation behaviors in private and semiprivate spaces, such as their balcony and roof garden. Only vegetables and small fruit trees can be planted in the narrow private spaces (shown in Figure 4). Such small-scale planting can't provide enough food products for the residents. They need the local government's systematic plan for the urban food system. The first step is to plan some open or semi-open spaces for urban horticulture [41], which was expected by most of the respondents (see Figures 5 and 6).

Governments play a vital role in the successful integration of different benefits of urban open spaces, such as urban parks, community gardens, and school yards [42–44]. For example, the Canadian national government pursued the People's Food Policy, a national food policy in 2011. In this project, a recommendation to increase local food production through urban agriculture activities was put forward. At the White House, former First Lady Michelle Obama established a vegetable garden and advocated healthy eating and community gardening through her Let's Move initiative. Some local governments in other countries, such as Australia [45,46], Japan [47], and Uruguay [48] have planned to protect urban agriculture lands and support household food production. Urban horticulture plans and the government's political backing can ensure the legitimacy of food planting in public areas and give the needed infrastructure support [49,50]. From our investigation, most of the barriers restraining urban residents' planting behaviors, such as lack of cultivation spaces, their property manager's opposition, and possible conflicts and complex relationships (Figure 5) with their neighbors, were associated with the related policies, which could be completely or partly eliminated through the government's guidance and management. Therefore, more and more official institutions and specific policies backed by the government must be established to promote the edible landscape projects.

4.4. Limitations and Prospects

Some limitations still existed in this study. First of all, the limited number of respondents and the age proportion of different groups might influence the results. This study was carried out in Wuhan, due to the extensive transportation system and mixed people from all over China. Our investigation was conducted in eight typical communities, taking the geographical location, community age, and resident characteristics into consideration. To some extent, the collected information from 496 respondents could represent the residents' awareness and perception of edible landscapes. Though, due to lack of participation, young respondents below 20 and old ones over 60 accounted for only 6.25% of the total (Figure 2). Their responses did not fully express the residents' attitudes of the corresponding age groups. Secondly, an online questionnaire was used as an important supplement to the on-site interview in this study. It attracted more residents to participate in the survey and efficiently expanded the sample size. However, if the online respondents had any questions during the questionnaire process, they couldn't be answered in time. This probably led to some small deviations in the results.

Thirdly, although the Edible Landscape Project has attracted more and more attention from Chinese ecologists and planners, it is still a newcomer to most urban residents. Their understanding of questions in the questionnaire differed from each other, depending on their age, education level, and living background. The answers collected from the questionnaires may not fully reflect their real opinions. Finally, the data in this study were obtained mainly by questionnaire survey. The survey provided enough information on how urban residents responded to the Edible Landscape Project, but it did not provide information about why they had those responses. A combined method of questionnaire survey with in-depth interview is needed, which can be improved in the future.

China has been experiencing rapid urbanization in recent years with expanding urban sprawl. Urban residents have suffered from serious food insecurity and have failed to meet the national recommendation of daily fruit and vegetable consumption [51]. Recently, this issue has attracted great attention from many Chinese scholars and designers [52,53], but there is still a big gap between the development of urban edible landscapes in China and other countries. Table 5 summarizes the residents' awareness of edible landscapes in this study and some other typical studies; this information provides guidance for the future direction in regard to the implementation of edible landscapes in China. In most developed countries, edible elements such as vegetables, fruits, and herbs were introduced into community gardens, school gardens, and home gardens [4,6,43,45,54]. This mainly provided the social benefit of natural education for the public [3,23,24], which attracts most urban residents' participation [37,43]. Compared with developed countries, most developing countries such as Argentina, Brazil, and Nepal implemented edible landscape projects to provide high-quality food supply [3,16,36]. Many housewives were engaged in community and family farming to get crops and fresh food [6,12,13,55], so they expected the horticulture infrastructures to be improved in order to enhance food production [3,43]. However, in China, though suffering from food insecurity, urban residents generally believed that growing edible plants in cities serves only to promote land-use efficiency and visual effect and couldn't fully accept the benefit of food production (Table 3); thus, they only planted some vegetables and small fruit trees in their courtyard and on their balcony, as no public places were planned and provided (Figure 5). By comparing our research with previous studies, we find that it is urgent to raise the residents' awareness about the benefits of local food in improving food access and quality. Good communication and cooperation between the government and community residents is the premise that helps guarantee the implementation of an urban food strategy. Food education through official publicity, systematic training, and planting practice in some pilot gardens is encouraged. Government-guided food-system plans and policies are needed to legitimize and institutionalize food planting in urban public spaces.

Table 5. Residents' awareness of the urban edible landscape in different studies.

	Developed Countries		Other Developing Countries		Wuhan, China
	Overall Perspectives	Typical Countries	Overall Perspectives	Typical Countries	Perspectives
Main purpose	Social and education benefits [3,23,24]	New Zealand	Improving food security [3,16,36]	Argentina	Promoting land-use efficiency
Participants	Most residents [37,43]	United States	Female [6,16,17]	South Africa	Female
Growing sites	Community and home gardens [4,6,45]	United States	Community and home gardens [3,12,13]	Brazil	Courtyard and balcony
Edible plants	Vegetables, fruits, and herbs [43,54]	France	Vegetables, fruits, crops, and herbs [3,55]	Nepal	Vegetables and small fruits
Main barriers	Insecurity of future land access [7,43]	Canada	Insufficient infrastructure and services [3,43]	Cuba	No suitable places

5. Conclusions

In this study, a questionnaire survey was conducted to collect information about the residents' attention and awareness of urban edible landscapes in eight sample communities in Wuhan, China. In total, 496 urban residents were interviewed, most of them aged 20~40 (68.75%). Female residents

(59.88%) paid more attention to edible landscapes than male ones (40.12%). Those with an undergraduate education background (60.96%) showed more interest in edible landscapes. Our investigation indicated that nearly one-third of the respondents were unaware of edible landscapes. The respondents learned about edible landscape and the related information mainly from the internet (23.79%) and their relatives or neighbors (20.97%). Most residents thought that edible landscapes could promote efficient land use (57.26%) and express special ornamental effects (54.64%), but quite a few didn't believe that growing edible plants in urban public spaces could increase food output (37.10%) and improve food quality (40.12%). Only 67.14% of the respondents had experience in growing edible plants. They usually performed their cultivation behavior in private or semiprivate spaces, such as their balcony (30.65%) and courtyard (16.53%), in their spare time. The most frequent edible plants they grew were vegetables (47.98%) and fruits (11.69%). Few public spaces were privately used for food planting, because the government did not plan open spaces for agriculture use, and this was regarded as the main barrier restricting the development of urban horticulture in China. For growers, the three biggest troubles in their planting were lack of planting spaces (55.86%), possible conflicts with other residents (35.14%), and lack of time (33.03%). Meanwhile, for non-growers, the three biggest barriers constraining their planting were lack of suitable places (59.51%), lack of time (47.85%), and lack of confidence in their abilities (39.88%). These conclusions provide scientific references for the policy decisions and construction of edible landscapes in public spaces in China or other developing countries. This study is still at the preliminary stage for implementing and promoting edible landscapes and has some limitations which need to be counteracted by more research in the future.

Author Contributions: Q.X. designed the research and wrote the manuscript; Y.Y. performed the investigation and data analysis; D.H. provided some effective suggestions and revised the manuscript.

Funding: This research was sponsored by MOE (Ministry of Education in China) Project of Humanities and Social Sciences (19YJCZH195) and the Natural Science Foundation of Hubei Province of China (2019CFB538).

Conflicts of Interest: The authors declare no conflicts of interest.

Appendix A

Survey on Wuhan Residents' Attention and Awareness of Edible Landscapes

Dear Sir/madam:

Hello! We are from Hubei University. A survey on the attention and awareness of Wuhan residents on edible landscapes is being conducted. This survey is anonymous, and the relevant personal information is confidential. All the information you provide is only for our research, so please feel free to fill in. Thank you for the cooperation!

Note: the edible landscape is a new type of landscape. The main elements are food plants, including cultivated vegetables (such as radish, cabbage, etc.), food crops (such as rice, peanuts, etc.), fruits (such as bananas, grapes, etc.), herbs (such as honeysuckle, mint, etc.), and mushrooms (such as lentinus edodes, enoki mushroom, etc.).

1. Your gender is ( ).
A. Male
B. Female
2. Your age is ( ).
A. <20
B. 20–40
C. 40–60
D. >60
3. Your education background is ( ).
A. Junior high school and below
B. Senior high school
C. Undergraduate
D. Postgraduate

4. You have lived in Wuhan for ( ).
A. <2 years
B. 2–5 years
C. 5–10 years
D. >10 years
5. How did you learn about edible landscape? ( )
A. By reading newspapers and magazines
B. By watching TV
C. By surfing on Internet
D. By the introduction of relatives and neighbors
E. By other ways
F. Unaware of it

6. What do you think are the main advantages of edible landscape compared with traditional one (multiple choices)? ( )
A. Forming distinct view effect with traditional landscape
B. Improving food quality
C. Educating children on the science of nature
D. Enhancing physical exercise and being relaxed
E. Promoting efficient land use
F. Providing food products

G. Nothing

7. Where do you and your family perform your cultivation activities? ( )
A. Courtyard
B. Balcony
C. Roof
D. Community vacant land
E. Others
F. No experience

8. Which specie do you and your family usually grow? ( )

A. Crops
B. Vegetables
C. Fruits
D. Mushrooms
E. Herbs
F. No experience

9. Which are the main barriers in growing edible plants (multiple choices)? ( )

A. No suitable place
B. Lack of time
C. Difficult in getting seeds and fertilizer
D. Food process difficulties
E. High planting costs
F. Property manager's opposition
G. Difficult management
H. Lack of help in planting

References

1. FAO. *Climate Change and Food Security: A framework Document*; Food and Agriculture Organization of the United Nations: Rome, Italy, 2008; Available online: http://www.fao.org/forestry/15538-079b31d45081fe9c3dbc6ff34de4807e4.pdf (accessed on 10 December 2019).

2. Obeng-Odoom, F. Underwriting food security the urban way: Lessons from African countries. *Agroecol. Sustain. Food Syst.* **2013**, *37*, 614–628. [CrossRef]

3. Eigenbrod, C.; Gruda, N. Urban vegetable for food security in cities. A review. *Agron. Sustain. Dev.* **2014**, *35*, 483–498. [CrossRef]

4. Mok, H.F.; Williamson, V.G.; Grove, J.R.; Burry, K.; Barker, S.F.; Hamilton, A.J. Strawberry fields forever? Urban agriculture in developed countries: A review. *Agron. Sustain. Dev.* **2014**, *34*, 21–43. [CrossRef]

5. Björklund, J.; Eksvärd, K.; Schaffer, C. Exploring the potential of edible forest gardens: Experiences from a participatory action research project in Sweden. *Agrofor. Syst.* **2019**, *93*, 1107–1118. [CrossRef]

6. Taylor, J.R.; Lovell, S.T.; Wortman, S.E. Urban home food gardens in the Global North: Research traditions and future directions. *Agric. Hum. Values* **2014**, *31*, 285–305. [CrossRef]

7. Drake, L.; Lawson, L.J. Results of a US and Canada community garden survey: Shared challenges in garden management amid diverse geographical and organizational contexts. *Agric. Hum. Values* **2015**, *32*, 241–254. [CrossRef]

8. Mattsson, E.; Ostwald, M.; Nissanka, S.P.; Pushpakumara, D.K.N.G. Quantification of carbon stock and tree diversity of homegardens in a dry zone area of Moneragala district, Sri Lanka. *Agrofor. Syst.* **2015**, *89*, 435–445. [CrossRef]

9. Gittleman, M.; Farmer, C.J.Q.; Kremer, P.; McPhearson, T. Estimating stormwater runoff for community gardens in New York City. *Urban. Ecosyst.* **2017**, *20*, 129–139. [CrossRef]

10. Goldstein, B.; Hauschild, M.; Fernández, J.; Birkved, M. Testing the environmental performance of urban agriculture as a food supply in northern climates. *J. Clean. Prod.* **2016**, *135*, 984–994. [CrossRef]

11. Sima, R.M.; Micu, I.; Maniutiu, D.; Sima, N.F. Edible landscaping-integration of vegetable garden in the landscape of a private property. *Horticulture* **2010**, *1*, 278–283.

12. Medeiros, N.A.; Carmo, D.L.; Priore, S.E.; Santos, R.H.S.; Pinto, C.A. Food security and edible plant cultivation in the urban gardens of socially disadvantaged families in the municipality of Viçosa, Minas Gerais, Brazil. *Env. Dev. Sustain.* **2019**, *21*, 1171–1184. [CrossRef]

13. Orsini, F.; Kahane, R.; Nono-Womdi, R.; Gianquinto, G. Urban agriculture in the developing world: A review. *Agron. Sustain. Dev.* **2013**, *33*, 695–720. [CrossRef]

14. Abdulkadir, A.; Dossa, L.H.; Lompo, J.P. Characterization of urban and peri-urban agroecosystems in three West African cities. *Int. J. Agric. Sustain.* **2012**, *10*, 289–314. [CrossRef]

15. Specht, K.; Siebert, R.; Hartmann, I.; Freisinger, U.B.; Sawicka, M.; Werner, A.; Thomaier, S.; Henckel, D.; Walk, H.; Dierich, A. Urban agriculture of the future: An overview of sustainability aspects of food production in and on buildings. *Agric. Hum. Values* **2014**, *31*, 33–51. [CrossRef]

16. Hamilton, A.J.; Burry, K.; Mok, H.F.; Barker, S.F.; Grove, J.R.; Williamson, V.G. Give peas a chance? Urban agriculture in developing countries. A review. *Agron. Sustain. Dev.* **2014**, *34*, 45–73. [CrossRef]

17. Olivier, D.W.; Heinecken, L. Beyond food security: women's experiences of urban agriculture in Cape Town. *Agric. Hum. Values* **2017**, *34*, 743–755. [CrossRef]

18. Tei, F.; Benincasa, P.; Farneselli, M.; Caprai, M. Allotment Gardens for Senior Citizens in Italy: Current status and technical proposals. *Acta Hortic.* **2010**, *881*, 91–96. [CrossRef]

19. He, B.J.; Zhu, J. Constructing community gardens? Residents' attitude and behaviour towards edible landscapes in emerging urban communities of China. *Urban For. Urban Green.* **2018**, *34*, 156–165. [CrossRef]

20. Fischer, L.K.; Brinkmeyer, D.; Karle, S.J.; Cremer, K.; Huttner, E.; Seebauer, M.; Nowikow, U.; Schütze, B.; Voigt, P. Biodiverse edible schools: Linking healthy food, school gardens and local urban biodiversity. *Urban For. Urban Green.* **2018**, *40*, 35–43. [CrossRef]

21. Klepacki, P.; Kujawska, M. Urban Allotment Gardens in Poland: Implications for Botanical and Landscape Diversity. *J. Ethnobiol.* **2018**, *38*, 123–137. [CrossRef]

22. Clarke, L.W.; Jenerette, G.D. Biodiversity and direct ecosystem service regulation in the community gardens of Los Angeles, CA. *Landsc. Ecol.* **2015**, *30*, 637–653. [CrossRef]

23. Dawson, A.; Richards, R.; Collins, C.; Reeder, A.I.; Gray, A. Edible gardens in early childhood education settings in Aotearoa, New Zealand. *Health Promot. J. Aust.* **2013**, *24*, 214–218. [CrossRef] [PubMed]

24. Collins, C.; Richards, R.; Reeder, A.I.; Gray, A.R. Food for thought: Edible gardens in New Zealand primary and secondary schools. *Health Promot. J. Aust.* **2015**, *26*, 70–73. [CrossRef] [PubMed]

25. Kosack, L. Die Essbare Stadt Andernach. Urbane Landwirtschaft im öffentlichen Raum. *Standort* **2016**, *40*, 138–144. [CrossRef]

26. Kulak, M.; Graves, A.; Chatterton, J. Reducing greenhouse gas emissions with urban agriculture: A Life Cycle Assessment perspective. *Landsc. Urban. Plan.* **2013**, *111*, 68–78. [CrossRef]

27. Smith, V.M.; Harrington, J.A. Community Food Production as Food Security: Resource and Economic Valuation in Madison, Wisconsin (USA). *J. Agric. Food Syst. Community Dev.* **2014**, *4*, 61–80. [CrossRef]

28. Orsini, F.; Gasperi, D.; Marchetti, L.; Piovene, C.; Draghetti, S.; Ramazzotti, S.; Bazzocchi, G.; Gianquinto, G. Exploring the production capacity of rooftop gardens (RTGs) in urban agriculture: The potential impact on food and nutrition security, biodiversity and other ecosystem services in the city of Bologna. *Food Secur.* **2014**, *6*, 781–792. [CrossRef]

29. Crush, J.; Caesar, M. City Without Choice: Urban Food Insecurity in Msunduzi, South Africa. *Urban. Forum* **2014**, *25*, 165–175. [CrossRef]

30. Goedele, V.D.B.; Maertens, M. Horticultural exports and food security in developing countries. *Glob. Food Secur.* **2016**, *10*, 11–20.

31. Gido, E.O.; Ayuya, O.I.; Owuor, G.; Bokelmann, W. Consumer Acceptance of Leafy African Indigenous Vegetables: Comparison Between Rural and Urban Dwellers. *Agric. Food Econ.* **2017**, *5*, 346–361. [CrossRef]

32. Luan, B.; Wang, X.; Huang, S.H.; Shao, W.W.; Chen, J.X. Design exploration of edible landscape in Chinese. *Landsc. Archit.* **2017**, *9*, 36–42.

33. Zheng, G.W.; Wen, J.Y. Feasibility analysis of edible landscape in space landscape planning of nursing home. *Archit. Cult.* **2019**, *6*, 190–191.

34. Liu, Y.L.; Xu, J.L.; Yin, K.L. Participatory construction of high density urban community public space-taking community garden as an example. *Landsc. Archit.* **2019**, *26*, 13–17.

35. Weinberger, K.; Lumpkin, T.A. Diversification into Horticulture and Poverty Reduction: A Research Agenda. *World Dev.* **2007**, *35*, 1464–1480. [CrossRef]

36. Zezza, A.; Tasciotti, L. Urban agriculture, poverty, and food security: Empirical evidence from a sample of developing countries. *Food Policy* **2010**, *35*, 265–273. [CrossRef]

37. Gregory, M.M.; Leslie, T.W.; Drinkwater, L.E. Agroecological and social characteristics of New York city community gardens: Contributions to urban food security, ecosystem services, and environmental education. *Urban. Ecosyst.* **2016**, *19*, 763–794. [CrossRef]

38. Dietary Guidelines for Chinese Residents. 2016. Available online: http://dg.cnsoc.org/article/2016b.html (accessed on 10 December 2019).

39. McClintock, N.; Cooper, J.; Khandeshi, S. Assessing the potential contribution of vacant land to urban vegetable production and consumption in Oakland, California. *Landsc. Urban Plan.* **2013**, *111*, 46–58. [CrossRef]

40. Al-Delaimy, W.K.; Webb, M. Community gardens as environmental health interventions: Benefits versus potential risks. *Curr. Env. Health Rep.* **2017**, *4*, 252–265. [CrossRef]

41. Clarke, L.W.; Li, L.; Jenerette, G.D.; Yu, Z. Drivers of plant biodiversity and ecosystem service production in home gardens across the Beijing Municipality of China. *Urban Ecosyst.* **2014**, *17*, 741–760. [CrossRef]

42. Teig, E.; Amulya, J.; Bardwell, L.; Buchenau, M.; Marshall, J.A.; Litt, J.S. Collective Efficacy in Denver, Colorado: Strengthening Neighborhoods and Health through Community Gardens. *Health Place* **2009**, *15*, 1115–1122. [CrossRef]

43. Guitart, D.; Pickering, C.; Byrne, J. Past results and future directions in urban community gardens research. *Urban. Urban. Green.* **2012**, *11*, 364–373. [CrossRef]

44. Leuven, J.R.F.W.; Rutenfrans, A.H.M.; Dolfing, A.G.; Leuven, R.S.E.W. School gardening increases knowledge of primary school children on edible plants and preference for vegetables. *Food Sci. Nutr.* **2018**, *6*, 1960–1967. [CrossRef] [PubMed]

45. Kingsley, J.; Townsend, M. 'Dig in' to social capital: Community gardens as mechanisms for growing urban social connectedness. *Urban Policy Res.* **2006**, *24*, 525–537. [CrossRef]

46. Millar, J.; Roots, J. Changes in Australian agriculture and land use: Implications for future food security. *Int. J. Agric. Sustain.* **2012**, *10*, 25–39. [CrossRef]

47. Kimura, A.H.; Nishiyama, M. The chisan-chisho movement: Japanese local food movement and its challenges. *Agric. Hum. Values* **2008**, *25*, 49–64. [CrossRef]

48. Santandreu, A.; Gome, P.A.; Terrile, R.; Ponce, M. Urban agriculture in Montevideo and Rosario: A response to crisis or a stable component of the urban landscape? *Urban. Agric. Mag.* **2009**, *22*, 12–13.

49. Clark, K.H.; Nicholas, K.A. Introducing urban food forestry: A multifunctional approach to increase food security and provide ecosystem services. *Landsc. Ecol.* **2013**, *28*, 1649–1669. [CrossRef]

50. Castro, J.; Ostoić, S.K.; Cariñanos, P.; Fini, A.; Sitzia, T. "Edible" urban forests as part of inclusive, sustainable cities. *Unasylva* **2018**, *69*, 59–65.

51. Hu, T.; Ju, Z.S.; Zhou, W. Regional pattern of grain supply and demand in China. *Acta Geogr. Sin.* **2016**, *71*, 1372–1383.

52. He, W.; Li, H. Study on space carrier, design concept and technology of edible landscape in community. *Landsc. Arch.* **2017**, *9*, 43–49.

53. Chen, C.R. Edible landscape-A new approach to the construction of beautiful rural landscape. *Chin. Hortic. Abstr.* **2017**, *33*, 156–158.

54. Pourias, J.; Aubry, C.; Duchemin, E. Is food a motivation for urban gardeners? Multifunctionality and the relative importance of the food function in urban collective gardens of Paris and Montreal. *Agric. Hum. Values* **2015**, *33*, 257–273. [CrossRef]

55. Sunwar, S.; Thornstrom, C.-G.; Subedi, A.; Bystrom, M. Home gardens in western Nepal: Opportunities and challenges for on-farm management of agrobiodiversity. *Biodivers. Conserv.* **2006**, *15*, 4211–4238. [CrossRef]

Article

Polyphenolic Profile and Antioxidant Activity of *Juglans regia* L. Leaves and Husk Extracts

Anna Masek [1,*, Malgorzata Latos-Brozio [1], Ewa Chrzescijanska [2] and Anna Podsedek [3]**

[1] Institute of Polymer and Dye Technology, Faculty of Chemistry, Lodz University of Technology, Stefanowskiego 12/16, 90-924 Lodz, Poland; malgorzata.latos-brozio@edu.p.lodz.pl

[2] Institute of General and Ecological Chemistry, Faculty of Chemistry, Lodz University of Technology, Zeromskiego 116, 90-924 Lodz, Poland; ewa.chrzescijanska@p.lodz.pl

[3] Institute of Molecular and Industrial Biotechnology, Faculty of Biotechnology and Food Sciences, Lodz University of Technology, Stefanowskiego 4/10, 90-924 Lodz, Poland; anna.podsedek@p.lodz.pl

* Correspondence: anna.masek@p.lodz.pl; Tel.: +48-42-631-32-93

Received: 8 October 2019; Accepted: 30 October 2019; Published: 6 November 2019

Abstract: The aim of this study is to characterize the antioxidant capacity and establish the profile of polyphenolic compounds in walnut extracts (different extracts prepared from walnut leaf and green husks). The correlation between bioingredients of the product tested and their ability to scavenge free radicals and reduce them by chelating various metal ions were examined. Research technology combining TG (thermogravimetry), FTIR (Fourier-transform infrared spectroscopy), high-performance liquid chromatography system (HPLC) with electrochemical methods (cyclic and differential pulse voltammetry) and spectrophotometric methods (ABTS, FRAP, and DPPH assays) was used to rate the potential oxidation-reduction components of walnut extracts. A high affinity for scavenging free radicals ABTS and DPPH was found for natural substances present in leaves and green husks. The walnut is beneficial to health as it contains alpha-linolenic acid in its lipid fraction and, as demonstrated in this study, its husks are rich in polyphenolics with high antioxidant capacity.

Keywords: walnut (*Juglans regia* L.); electrochemical oxidation; UV-VIS; antioxidant; ABTS; DPPH

---

1. Introduction

Natural polyphenols are a large group of secondary plant metabolites, arising from phenylalanine or shikimic acid. These plant compounds perform an important function in counteracting different types of stress factors, such as ultraviolet irradiation, aggression by pathogens, parasites, and plant predators. Moreover, natural polyphenols affect the organoleptic properties of plants and foods of plant origin [1]. In addition, these compounds are well-known for their advantageous effects on human health. The literature describes their antioxidant, cardioprotective, cytotoxic, anti-inflammatory, as well as antibacterial properties [2,3]. As they are widely found in plants, these substances are a beneficial alternatives to antibiotics and additives of chemical origin. The application of plant origin compounds as natural preservatives in the foodstuffs production because of their advantage to human health has been intensively tested in recent years. Walnut leaves and husks are among the sources studied. The deciduous tree *Juglans regia* L. is commonly called walnut and belongs to the Juglandaceae family. Plant materials (leaves, bark, unripe nuts) obtained from this tree was commonly used in folk medicine for its strong antioxidant capacity, antidiabetic, antibacterial, anti inflammatory, anti-atherogenic, and liver-protective properties [4,5]. The main active ingredients in raw material are tannins (gallotannins and ellagitannins) and 1,4,5-trihydroxynaphthalene-4-beta-D-glucoside, which is converted into a naphthoquinone derivative-juglon (5-hydroxy-1,4-naphthoquinone) during drying of the leaves and green husks. Moreover, the walnut leaf contains flavonols (derivatives of quercetin and kaempferol), phenolic acids (caffeic, p-coumaric), and oils [6–10]. Furthermore,

Oliveira et al. (2008) [11] have shown that aqueous extracts of green husk or walnut leaves have significant antioxidant properties. The phenolic compounds contained in the raw material are responsible for the antioxidant effects. Walnuts are a rich source of linolenic, palmitic, oleic, and stearic acid, phytosterols, nonsodium minerals, g-tocopherol, melatonin, vitamin E, and polyphenols. The most important polyphenols in plant materials obtained from walnuts, are ellagitannins, which are metabolized to urolithins, compounds with antioxidant, anti-inflammatory, cytotoxic as well as prebiotic effects [12]. Anderson et al. and Jahanban-Esfahlan et al. [13,14] described that walnut extract contains gallic acid, ellagic acid, and flavonoids, tannins, folate, proteins, melatonin and sterols and minerals.

An alternative to traditional methods to determine the antioxidant effect of different extracts of plant origin or natural substances are electrochemical methods. In recent years, these methods have aroused great interest of scientists [15–17]. Electrochemical methods have many advantages, such as quick, simple, and low price. In addition, these methods allow measurements in the presence of colored or other masking compounds that may interfere with measurements by other methods, e.g., spectrophotometric. The experimental parameters, which are useful in examination the antioxidant properties of the analyzed compounds, such as peak potential (E_{pa}) and peak current (i_{pa}) can be determined by using electrochemical methods. Low values of oxidation potentials (E_{pa}) indicate the tendency of a given molecule to donate electrons, therefore to demonstrate its strong antioxidant effect. Cyclic voltammetry (CV) is the method used to test the properties of electrode processes, giving data on the thermodynamics of electrode reactions and kinetics of electron transfer, as well as and also coupled chemical reactions or adsorption processes. One more electrochemical method applied to examine the antioxidant effect of the extracts of plant origin under study is differential pulse voltammetry (DPV). This method is characterized by a good detection limit and high resolution [18–24]. Evaluation of redox behavior by means of electrochemical characterizes provide an antioxidant profile. Furthermore, it is possible to calculate the electrochemical index value, which is a reliable measure associated with the antioxidant potential (EI) [15,25]. The aim of this work was to determine the antioxidant capacity of leaf and walnut green husk extracts by using cyclic voltammetry (CV) and differential pulse voltammetry (DPV). The methods are known for their suitability for food control and monitoring the levels of antioxidant activity in samples of biological origin. The advantage of the CV and DPV methods compared to traditional in vitro methods is that they are cheap methods, giving very precise results and at the same time correlate with traditional methods, e.g., DPPH, ABTS.

In the last decade, FTIR spectroscopy has been shown to be a powerful method for the analysis of natural molecules and of complex biological systems such as tissues and cells. An advantage of FTIR spectroscopy is that this method can be applied to powdered, dehydrated, or aqueous samples. The FTIR method is simple, selective, validated, and ecofriendly [26,27].

Scheme (Figure S1) of Polyphenolic Profile and Antioxidant Activity of *Juglans regia* L. Leaves and Husk Extracts has been included in Supplementary Materials.

2. Materials and Methods

2.1. Reagents and Chemicals

Samples of leaves and green walnut husks of walnut were obtained from trees growing on the farm in central Poland. Plant materials were taken from seven different walnut trees, variety: non-grafted, age 16-years. The species was confirmed by a specialist in the field of horticulture. Plant materials were harvested in October. The trees grow on a sandy clay loam, in temperate climate, southern exposure to sun. The trees are bred in accordance with the principles of ecological horticulture.

The following reagents were used for the tests: 2,2′-azino-bis(3-ethylbenzothiazoline-6-sulfonic acid) diammonium salt (ABTS, purity ≥98% (HPLC), Sigma Aldrich, Saint Louis, MO, USA), potassium persulfate (99.99%, Sigma Aldrich, Saint Louis, MO, USA), 2,2-Diphenyl-1-picrylhydrazyl (DPPH, ≤100%, Sigma Aldrich, Darmstadt, Germany), 2,4,6-Tris (2-pyridyl)-s-triazine (TPTZ, ≥99.0% (HPLC),

Sigma Aldrich, Buchs, Switzerland), Iron (III) chloride $FeCl_3$ (pure P.A., Chempur, Piekary Slaskie, Poland), hydrochloric acid HCl (Standard solution 40 mmol·L^{-1}, Chempur, Piekary Slaskie, Poland), acetate buffer solution (0.3 mol·L^{-1}, pH 3.6, Chempur, Piekary Slaskie, Poland), 2,9-Dimethyl-1, 10-phenanthroline (Neocuproine, purity ≥98%, Sigma Aldrich, Beijing, China), copper (II) chloride $CuCl_2$ (standard solution 0.01 mol·L^{-1}, Chempur, Piekary Slaskie, Poland), ammonium acetate (NH_4Ac) buffer solution (1.0 mol·L^{-1}, pH 7.0, Chempur, Piekary Slaskie, Poland), ethyl alcohol (pure P.A., 96%, POCH, Gliwice, Poland), acetonitrile (pure P.A., 99.5%, POCH, Gliwice, Poland).

Acetonitrile, chlorogenic acid, caffeic acid, (+)catechin, coumaric acid, cryptochlorogenic acid, (−)epicatechin, ferulic acid, formic acid, gallic acid, ellagic acid, juglone, kaempferol, myricetin, neochlorogenic acid, quercetin, quercetin 3-glucoside, quercetin 3-rhamnoside, rutin, sinapic acid, syringic acid, quercetin and (+)-catechin were bought from Sigma-Aldrich (Steinheim, Germany). Ultra purity water was done in the laboratory using a SimplicityTM Water Purification System (Millipore, Marlborough, MA, USA).

2.2. Preparation of the Extracts

Material from the seven different walnut trees was collected for testing: leaves and green husks. From the husks and leaves, collected from each of the seven trees, three leaves and husks extracts from one tree were prepared. Concentration of all extracts were 50 mg/mL.

Leaves and husks of the walnut were cut into pieces and then ground in a ball mill. The particle size of the shredded plant materials was less than 1 mm. Plant materials were extracted using a five-fold volume of 70% ethanol under continuous mixing conditions (200 RPM, 25 °C). The extraction was carried out at 25 °C and at dark for 7 days. The final extracts of leaves and husks of the walnut were concentrated to constant weight using a rotary evaporator under reduced pressure conditions at 30 °C (Scheme 1).

Scheme 1. Preparation of extracts from walnut leaves and husks.

In all considered analysis, the tests were repeated three times on samples of husks and leaf extracts obtained from seven trees. The average results are presented in the manuscript.

2.3. Measurement Methods

2.3.1. Thermal Decomposition

The thermogravimetric (TG) analysis of leaves and husks of the walnut was determined using a Mettler Toledo Thermobalance (TA Instruments, Greifensee, Switzerland). The samples of 5 mg were inserted in aluminum pans and heated from 25 °C to 800 °C under a dynamic nitrogen flow (50 mL/min). A heating rate of 5 °C/min was used.

2.3.2. FTIR and UV-VIS Spectra

Using infrared spectroscopy (FTIR) and ultraviolet visible (UV-VIS) spectroscopy, the presence of active plant substances in raw walnut leaves and its extract was investigated. Samples of leaves and its extract were placed in the infrared beam output of a Nicoled 670 spectrometer (Thermo Fisher Scientific, Waltham, MA, USA). Analysis of the oscillatory spectra obtained allows examine of the functional groups with which the radiation interacted. The UV-VIS spectra of the walnut leaf extract solution were recorded from a mixture of 0.2 mL of the extract and 1.8 mL of 70% ethanol. 70% ethanol was used as a blank. The mixture was scanned at 190–1100 nm using a UV-spectrometer (Evolution 220, Thermo Fisher Scientific, Waltham, MA, USA).

2.3.3. HPLC-PDA Analysis of Phenolic Compounds

Phenolic profiles were performed using a high-performance liquid chromatography system (Waters, Milford, MA) that consisted of a gradient pump (1525), photodiode array detector (2998), auto-injector (2707), and Breeze 2 system controller equipped with a 250 × 4.6 mm i.d, 5 μm Symmetry C18 column (Waters). The mobile phase was a binary gradient with A, water/formic acid (90:10, *v/v*), and B, water/acetonitrile/formic acid (40:50:10, *v/v/v*), with a flow rate of 1 mL/min [28]. The binary gradient was as follows: 100–12% B (0 min), 12–30% B (0–26 min), 30–100% B (26–40 min): 100% B (40–43 min), 12% B (43–48 min), and 12% B (48–50 min). Detector was set at 280 nm for hydroxybenzoic acid derivatives and flavanols, 320 nm for hydroxycinnamic acid derivatives and 360 nm for flavonols. Phenolic compounds were identified by comparison of its retention time and absorption spectra (240–500nm) with standards and literature [29–32]. The identified compounds were quantified according to the peak area measurements, which were reported in calibration curves of the corresponding standards. Unidentified hydroxybenzoic acids, hydroxycinnamic acids, and flavonols were quantified as gallic acid, chlorogenic acid, and quercetin 3-galactoside, respectively. Data are reported as means ± standard deviations of two independent analyses.

2.3.4. Cyclic and Differential Pulse Voltammetry

Determination and testing of antioxidant capacity of compounds found in solutions obtained from walnut leaf and husks extracts were made by electroanalysis using cyclic voltammetry (CV) and differential pulse voltammetry (DPV). The tests were performed using an Autolab electroanalytical unit (EcoChemie, Holland). The solution was placed in an electrolytic cell with three electrodes, the indicator electrode being platinum with a geometric surface of 1 cm^2, whose potential was measured against the ferricinium/ferrocene reference electrode (Fc$^+$/Fc). The third electrode was platinum as the auxiliary electrode, and the current dependence of the potential of the indicator electrode in the potential range from 0 V to 2 V was recorded. CV were recorded for a polarization rate (v) of 0.1 V · s^{-1}. DPV were recorded in the same potential range with modulation amplitude 25 mV, pulse width 50 ms (scan rate (v) 0.01 V · s^{-1}).

Before the measurements, oxygen dissolved in the solution was displaced with argon, and an argon cushion was maintained on the test solution during measurement. The tests were determined at 25 °C.

2.3.5. Antioxidant Activity Properties Measured by ABTS and DPPH Methods

The antioxidant capacity of walnut leaves were carried out by ABTS and DPPH assay. The methods are based on the reduction of 2,2'-azino-*bis*(3-ethylbenzothiazoline-6-sulphonic acid) ABTS and 2,2-diphenyl-1-picrylhydrazyl DPPH.

ABTS (6 mmol·L^{-1}) and potassium persulfate (2.45 mmol·L^{-1}) solution were mixed in ethanol, in volume ratio 7:1:2 *v/v/v*, then the mixture was allowed to stand for 16 h to generate reactive radicals. The ABTS/radical solution was diluted with ethanol to an absorbance of 0.70 at 734 nm. Then, 2 mL of diluted ABTS solution was added to 50 μL of walnut leaves extract. Absorbance was recorded with a UV-spectrometer (Evolution 220, Thermo Fisher Scientific, Waltham, MA, USA) at 734 nm.

The ethanol solution of the DPPH (2.0 mL) with a concentration 0.1 mmol·L^{-1} was added to 0.5 mL of alcohol solution (70% ethanol) that contained 50 μL of walnut leaf extract. Then, after 10 min of mixing, the absorbance of the solutions was determined at 517 nm.

The inhibition level (%) of the ABTS or DPPH radical (A%) was determined using the Equation (1):

$$\text{Inhibition (A\%)} = (((A_0 - A_1)/A_0) \times 100) \tag{1}$$

where A_0 is the absorbance of the control (reagent mixture without plant extract), and A_1 is the absorbance in the presence of the extract.

2.3.6. Determination of Ion Reduction—Iron by FRAP Method and Copper by CUPRAC Method

The FRAP method is based on reduction of ferric ion (Fe^{3+}-TPTZ complex) under acidic conditions. The solution of the oxidant in the FRAP method was obtained by the addition of 25 mL of acetate buffer (0.3 mol·L^{-1}, pH 3.6), 2.25 mL of TPTZ solution (10 mmol·L^{-1} TPTZ in 40 mmol·L^{-1} HCl) and 2.25 mL of $FeCl_3$ (20 mmol·L^{-1} in water solution). Walnut leaf extract (0.2 mL) was added to the oxidant solution and, after 4 min, absorbance at 595 nm was measured. As a blank, the reagent mixture without walnut leaf extract was used.

The CUPRAC method is analogous to the FRAP method and involves the reduction of copper ions. $CuCl_2$ (0.01 mol·L^{-1}, 0.25 mL), neokuproin ethanol solution (7.5 × 10^{-3} mol·L^{-1}, 0.25 mL) and CH_3COONH_4 buffer solution (pH 7.0, 1 mol·L^{-1}, 0.25 mL) were mixed in a test tube, and then walnut leaf extract (0.2 mL) was added. Absorbance at 450 nm was measured against blank reagent (reagent mixture without walnut leaf extract) after 30 minutes incubation at 25 °C.

The ferric and cupric ions reducing power was calculated as Equation (2):

$$\Delta A = A_{AR} - A_0 \tag{2}$$

where: A_0—absorbance of the reagent test, A_{AR}—absorbance of sample after reaction.

3. Results and Discussion

3.1. Characteristics of Thermal Decomposition of Plant Materials

In order to assess the thermal stability of walnut leaves and husks obtained, thermogravimetry was performed by means of thermic analysis (Figure 1, Table 1). Decomposition temperature and mass loss of plant materials were determined. The distribution of samples is in three-stage. The plant substances in the leaf extract are slightly more stable than those found in the husks. This is evidenced by the fact that 50% weight loss for the leaf extracts is recorded at 267 °C while for the husks 50% loss is already achieved at 240 °C.

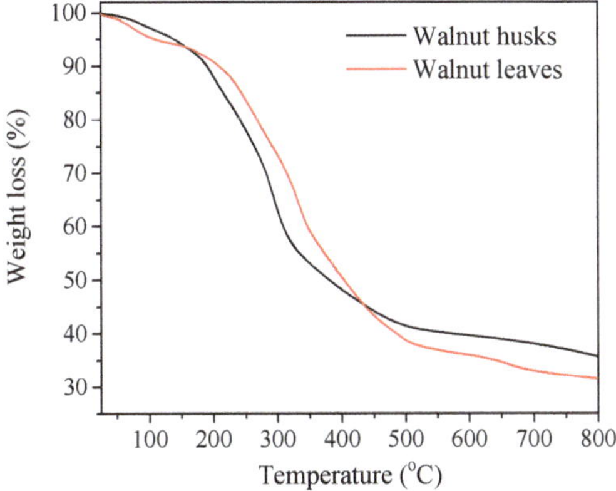

Figure 1. Thermogravimetry analysis (TGA) curves of plant materials: walnut husks and leaves.

Table 1. T_{20}, T_{50}, and T_{65} of plant materials: walnut husks and leaves.

Sample	T_{20}	T_{50}	T_{65}
Walnut husks	240	380	799
Walnut leaves	267	403	640

3.2. Analysis of Polyphenolic Profile of Extracts

The FTIR and UV-VIS analyzes were performed on a sample of raw plant leaf material as well as on leaf extracts. To FTIR and UV-VIS analyzes, leaves and walnut leaf extract were selected as samples representing the composition of the material obtained from the walnut tree. The FTIR and UV-VIS spectroscopy of walnut husk and walnut husk extract has been described in another authors' publication [33].

Analysis of the composition of the extracts was made on the basis of FTIR and UV-VIS analyses (Figure 2). The FTIR spectrum reflects the composition of the material. From the spectra, the presence of phenol functionalities was confirmed. Absorption peaks present in the spectrum at about 3200 cm^{-1} are responsible for the phenol O-H stretching groups, while 1603 cm^{-1} is from the resonance groups of the aromatic C=C, and C-H aliphatic is between 2850 and 3000 cm^{-1}. In the region of 1530 cm^{-1}, a band from in-plane bending of phenyl C-H bonds was noted. Absorptions at 1442 and 1325 cm^{-1} were assigned to C-H deformations and in plane O-H bending. Peaks identified from 1200 to 1030 cm^{-1} were corresponded to C-O stretching and-OH deformation vibrations in secondary alcohols and phenols, and also to C-O-C glycosidic linkage vibrations [34–37].

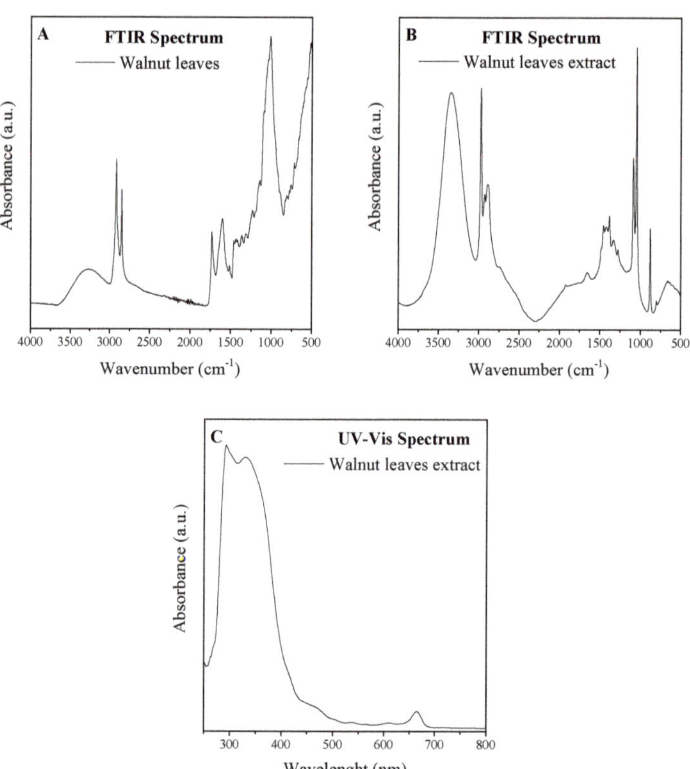

Figure 2. Fourier-transform infrared (FTIR) spectrum of leaves of walnut (**a**) and extract of leaves of walnut (**b**) and ultraviolet visible (UV-VIS) spectrum of extract of leaves of walnut (**c**).

Sarai Agustin-Salaza [38] made a thorough analysis of the composition of nutshell extract by means of FIA-ESI-IT-MS/MS method. The data obtained allowed description of the composition of the extract examined, which contained, among others, (epi)gallocatechin, anthocyanidin, taxifolin, ellagic acid, (epi)catechin, trans-resveratrol, gallic acid, coumaric acid, protocatechuic acid, trans-cinnamic acid, and protocatechualdehyde.

The UV-VIS spectra of flavones and related glycosides exhibit two strong absorption peaks at 300–380 nm and 240–280 nm [39]. The maximum absorbance of flavonol was determined at 350 nm [39]. Additionally, the peaks of natural phenolic acids, e.g., gallic acid, ferulic acid, *p*-coumaric acids, and vanillic acid, have absorbance maxima in the range of approximately 290–350 nm [40–43]. Chlorophylls a and b exhibited maximum absorbance in the ranges 400–500 nm and 600–700 nm [44].

The results of qualitative and quantitative analyses of phenolic compounds by HPLC method allowed to conclude that walnut extracts tested had the different phenolic profile (Table 2, Figure 3). The contents of phenolic compounds was expressed in mg/g of extract as a mean value ± standard deviation (Table 2). Among the tested walnut extracts, higher content of total phenolics was found in leaf extract. It contained about five times more phenolic compounds than walnut husk extract.

Table 2. Phenolic compounds content in walnut leaf extract and walnut husk extract.

Compounds	Content (mg/g of Extract)	
	Leaf Extract	**Husk Extract**
Gallic acid	0.59 ± 0.03	6.06 ± 0.18
Syringic acid	-	1.75 ± 0.16
Other hydroxybenzoic acids [1]	9.11 ± 1.64	4.68 ± 0.06
Ellagic acid	-	0.44 ± 0.01
Neochlorogenic acid	11.20 ± 0.25	3.82 ± 0.14
Chlorogenic acid	3.56 ± 0.17	-
Caffeic acid	0.65 ± 0.03	-
Sinapic acid	0.12 ± 0.01	0.73 ± 0.04
Ferulic acid	0.09 ± 0.00	-
p-Coumaric acid	0.07 ± 0.01	-
Other hydroxycinnamic acids [2]	50.25 ± 6.15	17.68 ± 1.47
Myricetin	0.45 ± 0.04	-
Quercetin	1.85 ± 0.31	-
Kaempferol	1.03 ± 0.08	-
Quercetin 3-rhamnoside	8.04 ± 0.52	0.94 ± 0.04
Quercetin 3-galactoside	18.91 ± 0.03	0.43 ± 0.04
Rutin	0.23 ± 0.03	-
Other flavonols [3]	79.89 ± 1.12	2.47 ± 0.10
Total	186.04 ± 5.30	39.00 ± 0.65

Mean ± SD, *n* = 2. [1] as gallic acid equivalents; [2] as chlorogenic acid equivalents; [3] as quercetin-3-galactoside equivalents.

The UV spectra of the compounds obtained by HPLC-PDA analysis revealed that flavonols (59.34% of total phenolics) were the main groups of phenolic compounds in leaf extract while hydroxycinnamic acids (57.00% of total phenolics) dominated in the husk extract. Among the identified phenolics, quercetin 3-galactoside and gallic acid were the main compounds of walnut leaf and husk extracts, respectively. Quercetin 3-galactoside was the main phenolic compound in the walnut leaves studied by Pereira et al. [31]. Nour at al. [30] reported ellagic acid as the dominating phenolic acid of fresh walnut leaves. Unfortunately, in the presented studies this compound was present only in the walnut husk extract. The reasons for the lack of this substance in the tested extract is certainly how the plant was collected, stored, and prepared. Literature reports that ellagic acid is present only in fresh plant. Yield of extraction of walnut appears to depend on temperature. Higher temperature reduces the efficiency of extraction. Juglone degrades in certain solvents and aquatic conditions that include acetonitrile, methanol, acidic solutions, alkaline solutions, and saline water [45,46], but juglon is stable in acidic conditions. According to the above-cited authors catechin hydrate and myricetin were the main flavonoids in walnut leaves. In contrast, our extracts did not contain catechins, while the content of myricetin in the leaf extract was lower than other flavonol aglycons (quercetin and kaempferol). In addition, in both extracts, no juglone (5-hydroxy-1-4-naphthoquinone) was found. According to literature, juglone is present in considerable amounts in all green and growing parts of the tree but because of polymerization phenomena, juglone only occurs in dry leaves at vestigial amounts [31,47].

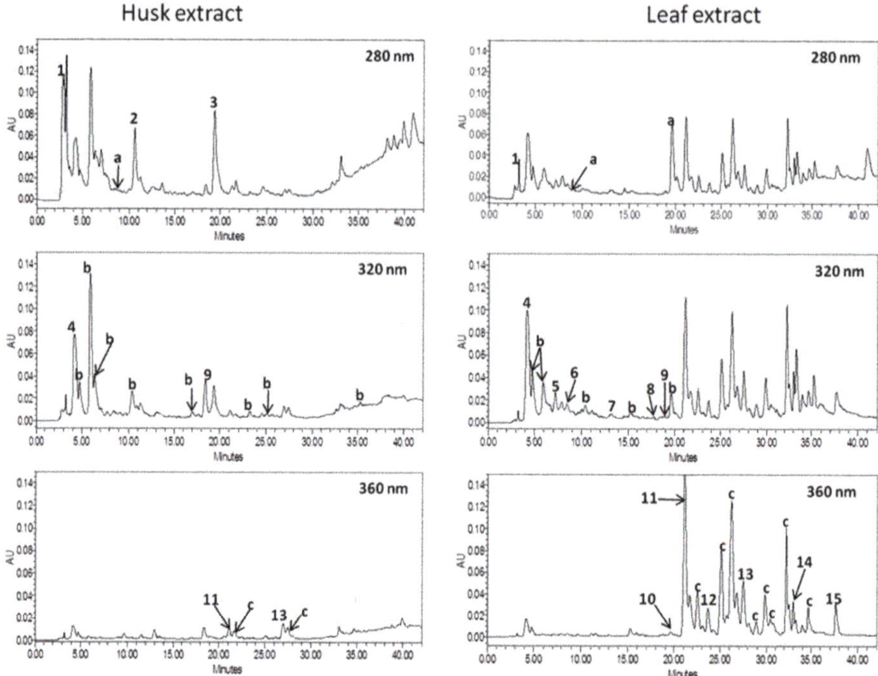

Figure 3. Chromatograms of walnut leaf and husk extracts at λ = 280 nm, 320 nm, and 360 nm. Peaks: 1, gallic acid; 2, syringic acid; 3, ellagic acid; 4, neochlorogenic acid; 5, chlorogenic acid; 6, caffeic acid; 7, *p*-coumaric acid; 8, ferulic acid; 9, sinapic acid; 10, rutin; 11, Quercetin 3-galactoside; 12, myricetin; 13, quercetin 3-rhamnoside; 14, quercetin; 15, kaempferol; a, unidentified hydroxybenzoic acids; b, unidentified hydroxycynnamic acids; c, unidentified flavonols.

3.3. Analysis of Properties of Extracts

The Electrochemical Behavior of Walnut Leaves and Husk at the Pt Electrode

Voltammetric analysis is used to evaluate the antioxidant properties of compounds found in various plant extracts by many researchers [18–20], as well as in beverages [21] and biological fluids [22], providing information on the value of peak potential and current [23]. The dependence of the current on the potential of the indicator electrode is characterized by the electrochemical reactions of compounds in solution on the tested electrodes. Electrochemical behavior of walnut extract solutions obtained from leaf and husks is presented in Figures 4 and 5. The cyclic voltammetry (CV) method and the differential pulse voltammetry (DPV) method, which is characterized by higher resolution, were used. Typical cyclic voltammograms for the test extracts (first cycle) are shown in Figure 4. A voltammograms was also recorded for the supporting electrolyte (0.1 mol·L^{-1} (C$_4$H$_9$)$_4$NClO$_4$ in acetonitrile) which does not show peaks in the range of potentials studied.

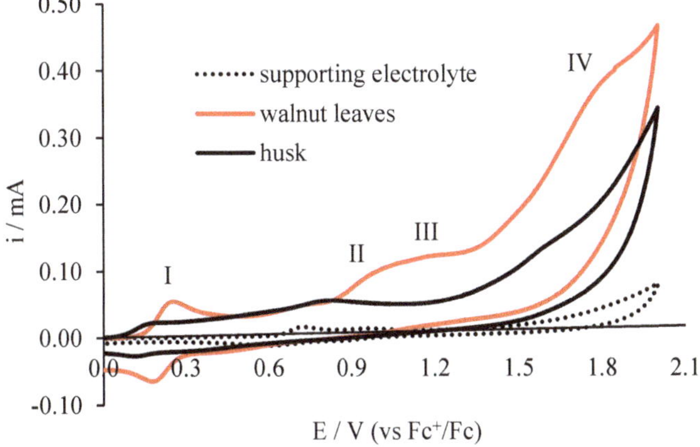

Figure 4. Cyclic voltammetry (CV) electrooxidation of leaf extracts and walnut husks on the electrode Pt; c = 20 mg/dm^3 in 0.1 M $(C_4H_9)_4NClO_4$ in acetonitrile, v = 0.1 V/s.

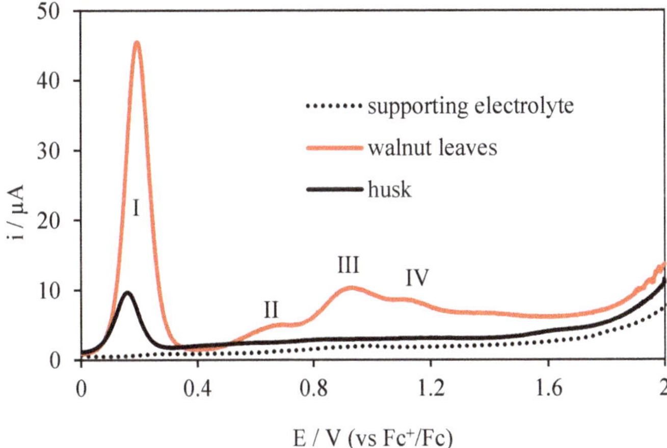

Figure 5. Differential pulse voltammograms (DPV) electrooxidation of extracts from walnut leaf and husks on the Pt electrode; c = 20 mg/dm^3 in 0.1 mol·L^{-1} $(C_4H_9)_4NClO_4$ in acetonitrile.

On the voltammogram for the walnut leaf extract solution, four electrooxidation peaks are visible. The first peak (I peak) is at a potential of 0.254 V, the second peak at 0.968 V, the third poorly developed peak at 1.184 V potential and the fourth peak at 1.777 V potential. In the reverse polarization cycle on the voltammogram, one peak is visible at a potential of 0.184 V, whereas on the voltammogram made for walnut husk extract, three weakly formed peaks are observed, characterized by lower currents than for the walnut leaf extract: first peak at 0.159 V potential, second peak at 0.764 V potential, and third peak at 1.576 V potential. Peak currents for walnut leaf extract are higher than for walnut husk extract.

Because of the shape of the voltammograms (their profile), it can be stated that the extracts tested are characterized by different qualitative and quantitative composition of compounds that oxidize (different composition). The purpose of this work is to determine the antioxidant capacity of these extracts. On the basis of voltammetric tests, it can be concluded that the antioxidant properties of the extract of walnut leaves are enhanced, because it contains more polyphenolic compounds oxidizing in the range of potentials tested than the walnut husks extracts.

The use of the second DPV method is associated with its higher resolution and the elimination of absorbable compounds, because they are not electroactive in this technique and, for this reason, there are no visible peaks on the voltammogram [24]. The differential pulse voltammograms (DPV) for the walnut extracts and basic electrolyte studied are shown in Figure 5. Four peaks are visible for the walnut leaf extract: first peak at 0.194 V potential, second peak at 0.648 V potential, third peak at 0.905 V potential, and fourth peak at 1.118 V potential. Regarding CV, a new peak at a potential of 0.648 V appeared on the DPV, but there is no peak at a potential 1.777 V. This indicates better DPV resolution and the adsorptive character of the electrooxidation of the compound at a potential of 1.777 V.

The DPV voltammogram obtained for walnut husks extract shows one peak at 0.144 V potential and a second very poorly developed peak at 1.564 V potential. However, there is no peak visible on the CV at 0.764 V potential. This indicates the adsorptive nature of the electrode reaction taking place at this potential. Electrochemical tests performed using the DPV method and the potentials and peak currents determined, confirm that the extract from walnut leaves has better antioxidant properties. Electrochemical studies have shown better antioxidant properties of leaf extract, therefore further spectrophotometric studies were performed for this extract only.

The electrochemical index (EI) was calculated taking into account the main voltammetric parameters, peak potential (E_{pa}), and peak current (I_{pa}). Based on the fact that the lower the E_{pa} (thermodynamic parameter), the higher is the electron donor ability, and the higher the I_{pa} (kinetic parameter), the higher is the amount of electroactive species, EI was calculated using the following equation [15,48]

$$EI = I_{pa1}/E_{pa1} + I_{pa2}/E_{pa2} + \cdots + I_{pan}/E_{pan}. \tag{3}$$

The determined parameters E_{pa}, I_{pa} and the calculated EI values for the examined extracts are included in the Table 3.

Table 3. Peak potentials (E_p) and currents (I_p) determined from CV i DPV, antioxidant capacity (EI).

Method	Peak I		Peak II		Peak III		Peak IV		EI$_{total}$
	E_p (V)	i_p (μA)	E_p (V)	i_p (μA)	E_p (V)	i_p (μA)	E_p (V)	i_p (μA)	
CV for husk	0.15	21.00	0.80	56.00	1.58	134.00			
CV for walnut leaves	0.23	51.00	0.94	87.00	1.17	122.00	1.75	357.00	
DPV for husk	10.15	9.07							
DPV for walnut leaves	1.90	45.00	0.65	4.77	0.89	9.96	1.11	8.64	
EI for husk (CV)		138.16		70.00		84.81			292.97
EI for walnut leaves (CV)		221.74		92.55		104.27			418.57
EI for husk (DPV)		0.89							0.89
EI for walnut leaves (DPV)		23.68		7.34		11.19		7.79	50.00

From the values presented in Table 3, it follows that walnut husks have better antioxidant properties.

The literature [38] shows that nutshells contain a number of substances with antioxidant properties. Another very interesting aspect is the presence of substances with bactericidal and fungicidal properties. This last property is mainly due to the presence of juglone. As part of the publication, the antiradical and reducing properties of the mixture of polyphenols present in the extract tested were examined. Currently, in both medicine and other fields, researchers are looking for substances with strong stabilizing properties. An additional advantage is their natural and renewable origin. Polyphenols play a very important role in the absorption or neutralization of free radicals. According to the tests carried out on the SET mechanism (transfer of a single electron), the walnut leaf extract showed some antioxidant activity (Table 4). The inhibition of the subject examined by the reduction of free radicals according to the ABTS and DPPH method extends to about 6.71%–6.94%, compared to the standard herbal extract which has high antioxidant capacity, ranging from 8%–39%. A similar effect was noted for the FRAP and CUPRAC methods. The power to reduce iron ions is 0.82 and copper 0.38 a.u., in comparison with from 0.2 to 2 a.u. for a green tea extract [49]. From this, it can certainly be said that substances present in the plant materials tested have strong antioxidant capacity. This work has been attempted to correlate the key chemical parameters of biosubstances of extract including the radical scavenging, chelating, reducing capacity with their electrochemical properties. It is surprising that confirmation of one study in another was obtained. Both the electrochemical and spectrophotometric parts are confirmed by the high antioxidant potential of substances present in the tested extract.

Table 4. Antioxidant capacity and ability to reduce iron and cooper ions of extracts of walnut leaves.

Method		Walnut Leaves
Antioxidant capacity		
ABTS	inhibition [%]	6.94 ± 0.35
DPPH	inhibition [%]	6.72 ± 0.34
Reduction of iron and cooper ions		
FRAP	$Fe^{3+} \rightarrow Fe^{2+}$ (ΔA, a.u.)	0.82 ± 0.04
CUPRAC	$Cu^{2+} \rightarrow Cu^{1+}$ (ΔA, a.u.)	0.38 ± 0.02

4. Conclusions

CV and DPV electrochemical methods were applied to determine the antioxidant capacity of leaf extracts and walnut husks. They showed correlation with other methods such as spectrophotometric analysis ABTS, DPPH, FRAP, and CUPRAC. Four qualitative peaks were designated on the basis of the electrical tests. The first peak (I peak) is at a potential of 0.254 V, the second peak at 0.968 V, the third peak at 1.184 V potential, and the fourth peak at 1.777 V potential. Substances present in the extract with antioxidant capacity have a high affinity for scavenging free radicals, and less ability to reduce transition metal ions. HPLC analysis shows the great advantage of the phenolic compounds in the leaves relative to the husk. Unfortunately, there is no juglone, which is very unstable and quickly degrades. Thermogravimetric analysis allowed determination of the thermal stability of the plant materials. It was found that plant materials (leaf and husks) derived from walnut can be processed up to 240 °C. It should be said that walnut extracts are very rich in materials that inhibit oxidation processes. These bioactive compounds present in walnut extracts are suggested as an interesting economical source of antioxidants for use in the food and nutraceutical [50–52] industries. The use of electrochemical techniques in combination with spectrophotometric methods has allowed an in-depth study of the antioxidant potential of bioactive substances present in the tested extract. The research proposed and the results obtained would be of great interest to understand the relationship between content and scavenging mechanism of all the nutrients in walnuts.

Supplementary Materials: The following are available online at http://www.mdpi.com/1999-4907/10/11/988/s1, Figure S1: Scheme of Polyphenolic Profile and Antioxidant Activity of Juglans regia L. Leaves and Husk Extracts.

Author Contributions: A.M. conceived, designed, and performed the experiments, analysed and interpreted the data, and wrote the paper; A.P., E.C. and M.L.-B. performed the experiments and analysed and interpreted the data.

Funding: This study was supported by the National Centre for Research and Development (NCBR) project: LIDER/32/0139/L-7/15/NCBR/2016.

Conflicts of Interest: The authors declare no conflict of interest.

References

1. Di Mauro, M.D.; Giardina, R.C.; Fava, G.; Mirabella, E.F.; Acquaviva, R.; Renis, M.; D'Antona, N. Polyphenolic profile and antioxidant activity of olive mill wastewater from two Sicilian olive cultivars: Cerasuola and Nocellara etnea. *Eur. Food Res. Technol.* **2017**, *243*, 1895–1903. [CrossRef]
2. Mauro, M.D.D.; Fava, G.; Spampinato, M.; Aleo, D.; Melilli, B.; Saita, M.G.; Centonze, G.; Maggiore, R.; Antona, D.; Di Mauro, M.D.; et al. Polyphenolic Fraction from Olive Mill Wastewater: Scale-Up and in Vitro Studies for Ophthalmic Nutraceutical Applications. *Antioxidants* **2019**, *8*, 462. [CrossRef] [PubMed]

3. Fraga, C.G.; Croft, K.D.; Kennedy, D.O.; Tomás-Barberán, F.A. The effects of polyphenols and other bioactives on human health. *Food Funct.* **2019**, *10*, 514–528. [CrossRef] [PubMed]

4. Acquaviva, R.; D'Angeli, F.; Malfa, G.A.; Ronsisvalle, S.; Garozzo, A.; Stivala, A.; Ragusa, S.; Nicolosi, D.; Salmeri, M.; Genovese, C. Antibacterial and anti-biofilm activities of walnut pellicle extract (*Juglans regia* L.) against coagulase-negative staphylococci. *Nat. Prod. Res.* **2019**, *2019*, 1–6. [CrossRef]

5. Masek, A.; Chrzescijanska, E.; Latos-Brozio, M.; Zaborski, M. Characteristics of juglone (5-hydroxy-1,4, -naphthoquinone) using voltammetry and spectrophotometric methods. *Food Chem.* **2019**, *301*, 125279. [CrossRef]

6. Salejda, A.M.; Janiewicz, U.; Korzeniowska, M.; Kolniak-Ostek, J.; Krasnowska, G. Effect of walnut green husk addition on some quality properties of cooked sausages. *LWT* **2016**, *65*, 751–757. [CrossRef]

7. Amaral, J.S.; Valentão, P.; Andrade, P.B.; Martins, R.C.; Seabra, R.M. Do Cultivar, Geographical Location and Crop Season Influence Phenolic Profile of Walnut Leaves? *Molecules* **2008**, *13*, 1321–1332. [CrossRef]

8. Zhou, Y.; Yang, B.; Jiang, Y.; Liu, Z.; Liu, Y.; Wang, X.; Kuang, H. Studies on Cytotoxic Activity against HepG-2 Cells of Naphthoquinones from Green Walnut Husks of Juglans mandshurica Maxim. *Molecules* **2015**, *20*, 15572–15588. [CrossRef]

9. Rusu, M.E.; Gheldiu, A.-M.; Mocan, A.; Moldovan, C.; Popa, D.-S.; Tomuta, I.; Vlase, L. Process Optimization for Improved Phenolic Compounds Recovery from Walnut (*Juglans regia* L.) Septum: Phytochemical Profile and Biological Activities. *Molecules* **2018**, *23*, 2814. [CrossRef]

10. Liu, R.; Wu, L.; Du, Q.; Ren, J.-W.; Chen, Q.-H.; Li, D.; Mao, R.-X.; Liu, X.-R.; Li, Y. Small Molecule Oligopeptides Isolated from Walnut (*Juglans regia* L.) and Their Anti-Fatigue Effects in Mice. *Molecules* **2019**, *24*, 45. [CrossRef]

11. Oliveira, I.; Sousa, A.; Ferreira, I.C.; Bento, A.A.; Estevinho, L.; Pereira, J.A.; Estevinho, M.L.M.F. Total phenols, antioxidant potential and antimicrobial activity of walnut (*Juglans regia* L.) green husks. *Food Chem. Toxicol.* **2008**, *46*, 2326–2331. [CrossRef] [PubMed]

12. Ros, E.; Izquierdo-Pulido, M.; Sala-Vila, A. Beneficial effects of walnut consumption on human health role of micronutrients. *Curr. Opin. Clin. Nutr. Metab. Care* **2018**, *21*, 498–504. [CrossRef] [PubMed]

13. Jahanban-Esfahlan, A.; Ostadrahimi, A.; Tabibiazar, M.; Amarowicz, R. Comparative Review on the Extraction, Antioxidant Content and Antioxidant Potential of Diferent Parts of Walnut (*Juglans regia* L.) Fruit and Tree. *Molecules* **2019**, *24*, 2133. [CrossRef] [PubMed]

14. Anderson, K.J.; Teuber, S.S.; Gobeille, A.; Cremin, P.; Waterhouse, A.L.; Steinberg, F.M. Walnut Polyphenolics Inhibit In Vitro Human Plasma and LDL Oxidation. *J. Nutr.* **2001**, *131*, 2837–2842. [CrossRef] [PubMed]

15. Lino, F.; De Sá, L.; Torres, I.; Rocha, M.; Dinis, T.; Ghedini, P.; Somerset, V.; Gil, E.; Rocha, M. Voltammetric and spectrometric determination of antioxidant capacity of selected wines. *Electrochim. Acta* **2014**, *128*, 25–31. [CrossRef]

16. Jara-Palacios, M.J.; Escudero-Gilete, M.L.; Hernández-Hierro, J.M.; Heredia, F.J.; Hernanz, D. Cyclic voltammetry to evaluate the antioxidant potential in winemaking by-products. *Talanta* **2017**, *165*, 211–215. [CrossRef]

17. Masek, A.; Chrzescijanska, E.; Latos, M.; Zaborski, M. Influence of hydroxyl substitution on flavanone antioxidants properties. *Food Chem.* **2017**, *215*, 501–507. [CrossRef]

18. Barros, L.; Cabrita, L.; Boas, M.V.; Carvalho, A.M.; Ferreira, I.C. Chemical, biochemical and electrochemical assays to evaluate phytochemicals and antioxidant activity of wild plants. *Food Chem.* **2011**, *127*, 1600–1608. [CrossRef]

19. Masek, A.; Chrzescijanska, E.; Kosmalska, A.; Zaborski, M. Antioxidant activity determination in Sencha and Gun Powder green tea extracts with the application of voltammetry and UV-VIS spectrophotometry. *Comptes Rendus Chime* **2012**, *15*, 424–427. [CrossRef]

20. Masek, A.; Chrzescijanska, E.; Kosmalska, A.; Zaborski, M. Characteristics of compounds in hops using cyclic voltammetry, UV–VIS, FTIR and GC–MS analysis. *Food Chem.* **2014**, *156*, 353–361. [CrossRef]

21. Hoyos-Arbeláez, J.; Vázquez, M.; Contreras-Calderón, J. Electrochemical methods as a tool for determining the antioxidant capacity of food and beverages: A review. *Food Chem.* **2017**, *221*, 1371–1381. [CrossRef] [PubMed]

22. Vicentini, F.C.; Raymundo-Pereira, P.A.; Janegitz, B.C.; Machado, S.A.; Fatibello-Filho, O. Nanostructured carbon black for simultaneous sensing in biological fluids. *Sens. Actuators B Chem.* **2016**, *227*, 610–618. [CrossRef]

23. Rebelo, M.; Rego, R.; Ferreira, M.; Oliveira, M.C. Comparative study of the antioxidant capacity and polyphenol content of Douro wines by chemical and electrochemical methods. *Food Chem.* **2013**, *141*, 566–573. [CrossRef] [PubMed]

24. Brett, C.M.A.; Brett, A.M.O. *Electrochemistry—Principles, Methods and Applications*; Oxford University Press: Oxford, UK, 1993.

25. Escarpa, A. Food electroanalysis: Sense and simplicity. *Chem. Rec.* **2012**, *12*, 72–91. [CrossRef]
26. Duygu, D.; Baykal, T.; Acikgoz, D.; Yildiz, K. Fourier Transform Infrared (FT-IR) Spectroscopy for Biological Studies. *J. Sci.* **2009**, *22*, 117–121.
27. Ashokkumar, R.; Ramaswamy, M. Phytochemical screening by FTIR spectroscopic analysis of leaf extracts of selected Indian Medicinal plants. *Int. J. Curr. Microbiol. App. Sci.* **2014**, *3*, 395–406.
28. Dyrby, M.; Westergaard, N.; Stapelfeldt, H. Light and heat sensitivity of red cabbage extract in soft drink model systems. *Food Chem.* **2001**, *72*, 431–437. [CrossRef]
29. Amaral, J.S.; Seabra, R.M.; Andrade, P.B.; Valentão, P.; Pereira, J.A.; Ferreres, F. Phenolic profile in the quality control of walnut (*Juglans regia* L.) leaves. *Food Chem.* **2004**, *88*, 373–379. [CrossRef]
30. Nour, V.; Trandafir, I.; Cosmulescu, S. HPLC Determination of phenolic acids, flavonoids and juglone in walnut leaves. *J. Chromatogr. Sci.* **2013**, *51*, 883–890. [CrossRef]
31. Pereira, J.A.; Oliveira, I.; Sousa, A.; Valentão, P.; Andrade, P.B.; Ferreira, I.C.; Ferreres, F.; Bento, A.A.; Seabra, R.; Estevinho, L. Walnut (*Juglans regia* L.) leaves: Phenolic compounds, antibacterial activity and antioxidant potential of different cultivars. *Food Chem. Toxicol.* **2007**, *45*, 2287–2295. [CrossRef]
32. Sipkina, N.Y.; Skorik, Y.A. Detection and determination of some phenolic and cinnamic acids in plant extracts. *J. Anal. Chem.* **2015**, *70*, 1406–1411. [CrossRef]
33. Latos-Brozio, M.; Masek, A. Effect of Impregnation of Biodegradable Polyesters with Polyphenols from Cistus Linnaeus and *Juglans regia* Linnaeus Walnut Green Husk. *Polymers* **2019**, *11*, 669. [CrossRef] [PubMed]
34. Dos Santos, R.C.; Amorim, A.D.G.N.; Thomasi, S.S.; Figueiredo, F.C.; Carneiro, C.S.; Da Silva, P.R.P.; Neto, W.R.D.V.; Ferreira, A.G.; Júnior, J.R.D.S.; Leite, J.R.D.S.D.A. Development of an electrolytic method to obtain antioxidant for biodiesel from cashew nut shell liquid. *Fuel* **2015**, *144*, 415–422. [CrossRef]
35. Foo, L.Y. Proanthocyanidins: Gross Chemical Red Spectra Structures by Infra-Red Spectra. *Phytochemistry* **1981**, *20*, 1397–1402. [CrossRef]
36. Kacuráková, M. FT-IR study of plant cell wall model compounds: Pectic polysaccharides and hemicelluloses. *Carbohydr. Polym.* **2000**, *43*, 195–203. [CrossRef]
37. McGhie, T.K.; Rowan, D.R.; Edwards, P.J. Structural Identification of Two Major Anthocyanin Components of Boysenberry by NMR Spectroscopy. *J. Agric. Food Chem.* **2006**, *54*, 8756–8761. [CrossRef]
38. Agustin-Salazar, S.; Gamez-Meza, N.; Medina-Juárez, L.Á.; Malinconico, M.; Cerruti, P. Stabilization of Polylactic Acid and Polyethylene with Nutshell Extract: Efficiency Assessment and Economic Evaluation. *ACS Sustain. Chem. Eng.* **2017**, *5*, 4607–4618. [CrossRef]
39. Joshi, D.D. UV–Vis. Spectroscopy: Herbal Drugs and Fingerprints. In *Herbal Drugs and Fingerprints: Evidence Based Herbal Drugs*; Springer: Delhi, India, 2012; pp. 101–120. [CrossRef]
40. Engida, A.M.; Faika, S.; Nguyen-Thi, B.T.; Ju, Y.-H. Analysis of major antioxidants from extracts of Myrmecodia pendans by UV/visible spectrophotometer, liquid chromatography/tandem mass spectrometry, and high-performance liquid chromatography/UV techniques. *J. Food Drug Anal.* **2015**, *23*, 303–309. [CrossRef]
41. Song, H.; Chen, C.; Zhao, S.; Ge, F.; Liu, D.; Shi, D.; Zhang, T. Interaction of gallic acid with trypsin analyzed by spectroscopy. *J. Food Drug Anal.* **2015**, *23*, 234–242. [CrossRef]
42. Rojas, J.; Londono, C.; Ciro, Y. The health benefits of natural skin UVA photoprotective compounds found in botanical sources. *Int. J. Pharm. Pharm. Sci.* **2016**, *8*, 13–23.
43. Ebrahimia, I.; Gashtib, M.P. Extraction of juglone from Pterocarya fraxinifolia leaves for dyeing, anti-fungal finishing, and solar UV protection of wool. *Colora Technol.* **2015**, *131*, 451–457. [CrossRef]
44. Butnariu, M.; Coradini, C.Z. Evaluation of Biologically Active Compounds from Calendula officinalis Flowers using Spectrophotometry. *Chem. Central J.* **2012**, *6*, 35. [CrossRef] [PubMed]
45. Hadjmohammadi, M.R.; Kamel, K. Determination of juglone (5-hydroxy 1,4,-naphthoquinone) in Pterocarya flaxinifolia by RP-HPLC. *Iran. J. Chem. Chem. Eng.* **2006**, *25*, 73–76.
46. Wright, D.; Mitchelmore, C.; Dawson, R.; Cutler, H. The Influence of Water Quality on the Toxicity and Degradation of Juglone (5 Hydroxy 1,4 Naphthoquinone). *Environ. Technol.* **2007**, *28*, 1091–1101. [CrossRef] [PubMed]
47. Solar, A.; Colaric, M.; Usenik, V.; Stampar, F. Seasonal variations of selected flavonoids, phenolic acids and quinines in annual ahoots of common walnut (*Juglans regia* L.). *Plant Sci. J.* **2006**, *170*, 453–461. [CrossRef]
48. De Macêdo, I.Y.L.; Garcia, L.F.; Neto, J.R.O.; de Siqueira Leite, K.C.; Ferreira, V.S.; Ghedini, P.C.; de Souza Gil, E. Electroanalytical tools for antioxidant evaluation of red fruits dry extracts. *Food Chem.* **2017**, *217*, 326–331. [CrossRef]

49. Masek, A. Antioxidant and Antiradical Properties of Green Tea Extract Compounds. *Int. J. Electrochem. Sci.* **2017**, *12*, 6600–6610. [CrossRef]

50. Daliu, P.; Santini, A.; Novellino, E. A decade of nutraceutical patents: Where are we now in 2018? *Expert Opin. Ther. Patents* **2018**, *28*, 875–882. [CrossRef]

51. Santini, A.; Novellino, E. Nutraceuticals-shedding light on the grey area between pharmaceuticals and food. *Expert Rev. Clin. Pharmacol.* **2018**, *11*, 545–547. [CrossRef]

52. Daliu, P.; Santini, A.; Novellino, E. From pharmaceuticals to nutraceuticals: Bridging disease prevention and management. *Expert Rev. Clin. Pharmacol.* **2019**, *12*, 1–7. [CrossRef]

Article

Geographical Distribution and Environmental Correlates of Eleutherosides and Isofraxidin in *Eleutherococcus senticosus* from Natural Populations in Forests at Northeast China

Shenglei Guo [1,†], Hongxu Wei [2,†], Junping Li [1], Ruifeng Fan [1], Mingyuan Xu [1], Xin Chen [2] and Zhenyue Wang [1,*]

[1] College of Pharmacy, Heilongjiang University of Chinese Medicine, Harbin 150040, China; guoshenglei@163.com (S.G.); 15645015705@163.com (J.P.); ruifeng-fan@163.com (R.F.); xumingyuan2000@163.com (M.X.)
[2] Northeast Institute of Geography and Agroecology, Chinese Academy of Sciences, Changchun 130120, China; weihongxu@iga.ac.cn (H.W.); chenxin_iga@sina.com (X.C.)
*   Correspondence: wangzhen_yue@163.com; Tel.: +86-451-8219-5025
†   These authors contributed equally.

Received: 31 July 2019; Accepted: 27 September 2019; Published: 4 October 2019

Abstract: Non-wood forest products (NWFPs) derived from understory plants are attracting attention about sustainable forestry development. Geographical distribution and climate correlates of bioactive compounds are important to the regional management for the natural reserves of medical plants in forests. In this study, we collected *Eleutherococcus senticosus* individuals from 27 plots to map the special distribution of concentrations of eleutheroside B, eleutheroside E, and isofraxidin in forests of Northeast China. Compound concentrations in both aerial and underground organs were further detected for relationships with the average of 20-year records of temperature, precipitation, and relative humidity (RH). We found higher shoot eleutheroside B concentration in populations in northern and low-temperature regions ($R = -0.4394$; $P = 0.0218$) and in eastern and high-RH montane forests ($R = 0.5003$; $P = 0.0079$). The maximum-likelihood regression indicated that both RH ($Pr >$ Chi-square, 0.0201) and longitude ($Pr >$ Chi-square, 0.0026) had positive contributions to eleutheroside B concentration in roots, but precipitation had strongly negative contributions to the concentrations of eleutheroside E ($Pr >$ Chi-square, 0.0309) and isofraxidin ($Pr >$ Chi-square, 0.0014) in roots. Both geography and climate factors had effects on the special distribution of medical compounds in *E. senticosus* plants in natural populations in Northeast China. The management of NWFP plants at the regional scale should consider effects from climatic geography.

Keywords: ciwujia; Siberian ginseng; *Acanthopanax*; secondary metabolite; Changbai Mountains; Khingan Mountains

---

1. Introduction

Climate change is expected to drive the shift in structure and composition of forest ecosystems [1]. Forest vegetation is a critical player in land–atmosphere interactions by regulating the energy, water, and carbon cycles [2]. Over the last decades, it was found that growth of temperate forest vegetation is heavily dependent on rising temperature [3,4]. In contrast, the rising (CO_2) consequence of this growth contributed to the increase of daily maximum temperature, which can reinforce the mean temperature increase in temperate forests [2]. Currently, both empirical models and field investigations have intensively detected the change of general vegetation in temperate forest ecosystems in the scenario with increasing temperature. Although plants in the understory plants are sensitive and adaptable to local environment conditions [5–8], their relationship with climatic factors has been less understood.

Many understory plants are of significant values in edible and medicine manufacture, which can be classified into the category of non-wood forest products (NWFPs). NWFPs are defined in this context as all the products other than woods derived from forests, shrubs, tree plantations, and trees outside forests [9,10]. NWFPs can be taken as a potential asset in forest management system generating socioeconomic benefits along the entire value-chain spanning rural development [11]. Increasing needs to touch nature and to consume forest products promote the international trade of NWFPs through countries and interest in NWFPs is growing in Europe and East Asia [9,11–13]. Because of the cover of tree canopy, it is unlikely to directly detect the relationship between NWFP plants and regional climate factors using the model with data about dominant trees. Therefore, it is of strong necessity to make field investigation for functional traits of NWFP plants to establish their responsive dataset. This consumes significant workload and time and much less is known about the response of NWFP plants to regional climate change.

Traditional culture of original stocks to produce NWFPs was conducted on farmland using cloth-shading at high density [14–16]. However, practices are accumulating to cultivate NWFP plants in forests at the understory layer where trees provide necessary shade and soils supply sufficient rhizosphere condition [13,15]. The forest-culture mode can further contribute to the conservation of natural forest ecosystem by fully exploiting the resource of the understory space. Mapping the distribution of natural populations can supply essential information for ecologically sustainable management of NWFP plants [17]. Environmental factors varied with geographical patterns which drive the spatial distribution of phenotype of local plants [5,18]. Therefore, to map the geographical distribution of traits of NWFP plants would benefit the illustration of adaption to location environment and further analyze the response to climate change.

Eleutherococcus senticosus is one of the NWFP species with thorny stem and branch which is also one of the best-known medical plants from the Araliaceae family. *E. senticosus* is also referred to as *Acanthopanax senticosus* as well with so-called names of Siberian ginseng in Europe and America, ciwujia in China, and gasiogapi in Korea [19]. *E. senticosus* is commonly used as an ingredient of folk medicines which have been documented in both Chinese and European pharmacopoeia [20]. Abundant content of multiple bioactive ingredients contributes to the medical value of *E. senticosus*. Eleutheroside B is the main effect compound due to antioxidant, immunomodulatory, and anti-inflammatory activities [21,22]. Eleutheroside E is another bioactive ingredient that has stress-protective activities. The structures of eleutherosides B and E are elucidated as lignans of syringin and acanthoside D, respectively (Figure 1). Isofraxidin (7-hydroxy-6,8-dimethoxycoumarin; Figure 1) has similar effect as eleutherosides and performs better in antifatigue and antistress [21]. Organs of stem, branch, and roots all had records to be detectable for these bioactive compounds [21,22]. Studies of environmental factors on metabolites are scattered and placed by limited attention. It was indicated that temperature in the range of 12–18 °C resulted in the accumulation of total phenolics, flavonoids, and eleutheroside E in *E. senticosus* [3,23]. These results suggest that the trait of metabolites in *E. senticosus* may be distributed with some geographical pattern by the driving of regional environmental factors. However, to the best of our knowledge, relevant evidence is quite scarce for either a geographical map of metabolites or clear relationship with environmental factors.

Forests in Northeast China reserve abundant amount of natural *E. senticosus* populations, which were chosen as the objectives of this study. We conducted a field investigation to collect aerial and below-ground organs to detect contents of eleutheroside B, eleutheroside E, and isofraxidin. The aim of the study was to map the special distribution of these bioactive compounds in forests of Northeast China and further detect the environmental correlates. With regard to the experimental results that eleutherosides accumulate in cooler temperature [3,23], it was hypothesized that metabolite content was higher in the northern part of the research region or in montane areas at higher elevations.

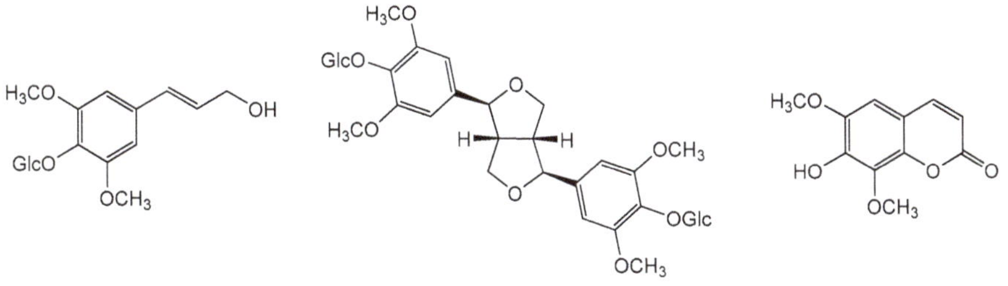

Eleutheroside B   Eleutheroside E   Isofraxidin

Figure 1. Chemical structures of eleutherosides B and E and isofraxidin [24,25].

2. Materials and Methods

2.1. Study Area

Forests in Northeast China were chosen as the study area. Northeast China is referred to including the eastern region of Inner Mongolia Province, Heilongjiang Province, Jilin Province, and Liaoning Province. According to the topography, mountains in Northeast China include Changbai Mountains (eastern montane regions across Jilin and Liaoning Provinces), the lesser Khingan Mountains (mid- to mid-eastern regions of Heilongjiang Provinces), and the Greater Khingan Mountains (northeastern regions in Heilongjiang Province and northwestern regions of Inner Mongolia Province) [26,27]. The regional climate in mountains of Northeast China is characterized by a long and chilling winter, a dry and windy spring and autumn, and a rainy summer with mild temperature. In eastern mountains in Heilongjiang and Jilin Provinces, the mean annual temperature varies between −4.7 and 10.7 °C with an annual precipitation of 866 mm. Regional relative humidity (RH) was averaged to be about 65%. The targeted forests that harbored natural *E. senticosus* populations were mainly dominated by trees of *Abies nephrolepis*, *Alnus mandshurica*, *Betula platyphylla*, *Larix gmelinii*, *Pinus koraiensis*, *Populus davidiana*, and *Quercus mongolica* [28–30].

2.2. Data Collection

Environmental factors were studied with the focus on temperature, precipitation, and relative humidity (RH) [26]. Data about these three parameters were averaged for the historical record of annual amounts during 1996 and 2016 for 27 meteorological stations. Specific locations and topographic information for these stations are listed in Table 1. These stations were chosen because at least one natural *E. senticosus* population was found nearby. During the time from July to late August in 2017, three sub-plots were randomly set around the coordinate of the meteorological station. Each sub-plot had an area of 400 m^2 in a 20 m × 20 m spacing stand, wherein a number of 10 *E. senticosus* individuals were randomly investigated across the stand for height and root-collar diameter (RCD) (Table 1). Thereafter, both aerial and below-ground parts of investigated individuals were harvested. Roots were carefully shoveled from soils and whisked rhizosphere soils. Harvested individuals were transported to the laboratory on ice (0–2 °C) where roots were further carefully rinsed to clean the roots free from soils. Both aerial shoots and underground roots were rinsed by distilled water and prepared for further determination.

Table 1. Plot characteristics of natural *Acanthopanax senticosus* populations in forests in Northeast China.

Plot Number	Latitude	Longitude	Elevation (m)	Slope (°)	Forest Type	CD [1] (%)	Shoot Height (cm)	RCD [2] (cm)
1	49°28′41″	126°46′29″	535	5	Broadleaf-conifer	0.5	161.67 ± 10.41	1.13 ± 0.06
2	49°16′33″	128°53′30″	198	15	Mongolian oak [3]	0.9	168.33 ± 7.64	1.33 ± 0.15
3	48°47′24″	129°25′29″	317	10	Broadleaf-conifer	0.4	193.33 ± 36.17	1.57 ± 0.38
4	48°3′24″	128°59′16″	346	8	Broadleaf-conifer	0.65	203.33 ± 27.54	1.87 ± 0.23
5	47°23′57″	129°32′30″	345	10	Secondary forest	0.2	158.33 ± 16.07	1.77 ± 0.15
6	45°49′58″	130°14′24″	301	15	Secondary forest	0.5	166.67 ± 55.08	1.20 ± 0.35
7	45°43′23″	129°38′38″	164	0	Larch [4]	0.3	178.33 ± 25.17	1.30 ± 0.20
8	47°36′31″	128°30′11″	322	5	Broadleaf-conifer	0.65	196.00 ± 20.81	1.33 ± 0.21
9	47°12′35″	129°41′21″	412	4	Secondary forest	0.45	195.00 ± 21.79	1.43 ± 0.32
10	46°55′57″	128°54′37″	439	10	Broadleaf-conifer	0.7	170.00 ± 5.00	1.20 ± 0.09
11	46°38′35″	128°51′35″	327	13	Broadleaf-conifer	0.7	170.00 ± 10.00	1.68 ± 0.16
12	46°9′10″	128°40′2″	168	5	Secondary forest	0.55	168.33 ± 10.41	1.41 ± 0.12
13	45°22′45″	127°36′18″	363	13	Broadleaf-conifer	0.5	168.33 ± 7.64	1.36 ± 0.21
14	44°49′52″	129°5′47″	674	5	Fir [5]	0.7	197.33 ± 23.69	1.33 ± 0.20
15	44°52′22″	129°8′16″	540	10	Broadleaf-conifer	0.67	150.00 ± 10.00	1.09 ± 0.08
16	43°2′1″	127°59′42″	714	5	Broadleaf-conifer	0.7	176.67 ± 11.55	1.50 ± 0.20
17	42°48′58″	127°54′19″	586	15	Secondary forest	0.7	165.00 ± 21.79	1.33 ± 0.25
18	42°35′48″	127°42′32″	791	3	Secondary forest	0.5	173.33 ± 5.77	1.47 ± 0.29
19	42°2′41″	127°30′44″	839	3	Secondary forest	0.5	160.00 ± 17.32	7.40 ± 5.39
20	42°2′24″	126°43′45″	723	8	Broadleaf-conifer	0.7	186.67 ± 11.55	1.43 ± 0.15
21	41°39′48″	126°28′47″	416	10	Secondary forest	0.7	140.00 ± 0.01	0.87 ± 0.06
22	41°16′35″	126°5′16″	726	15	Secondary forest	0.7	173.33 ± 20.82	1.23 ± 0.23
23	43°52′48″	126°54′16″	305	33	Secondary forest	0.7	196.67 ± 15.28	1.13 ± 0.15
24	44°38′48″	127°27′13″	266	20	Secondary forest	0.7	146.67 ± 5.77	1.27± 0.16
25	45°52′49″	132°08′11″	245	20	Broadleaf forest [6]	0.6	176.00 ± 14.42	1.57 ± 0.15
26	46°53′47″	133°48′57″	398	30	Broadleaf forest	0.6	165.00 ± 15.00	1.60 ± 0.36
27	45°17′22″	129°54′28″	392	25	Broadleaf-conifer	0.6	171.67 ± 12.58	1.43 ± 0.35

[1] CD, canopy density; [2] RCD, root-collar diameter; [3] Dominated by *Quercus mongolica* Fisch.; [4] Dominated by *Larix gmelinii*; [5] Dominated by *Abies nephrolepis* (Trautv.) Maxim.; [6] Dominated by *Populus davidiana, Betula platyphylla, Q. mongolica* Fisch., and *Acer palmatum*.

2.3. Bioactive Compound Analysis

The methodology of bioactive compound extraction was modified from that of Wu et al. [20]. Samples were oven-dried at 45 °C for 72 h and ground to pass a 60-mesh sieve (0.25 mm). Approximately 1 g of dry powdered plant material was extracted with 25 mL of methanol (70%; v/v) by supersonic for 75 min. The extract was moved to a 2 mL tube for centrifugation at 4000 rpm at 4 °C for 10 min. The supernatant was filtered to pass the 0.45 μm millipore-filter and reserved at 0–2 °C until the analysis using high-performance liquid chromatography (HPLC) (Waters 2695 Separations Module Waters 2998 PDA Detector, Waters Inc., Milford, MA, USA).

Measurement of the bioactive compounds were carried out with the HPLC system using a Welchrom C18 (250 mm × 4.6 mm, 5 μm) chromatographic column. The elution gradient of the mobile phase consisted of chromatographically pure acetonitrile (J&K Scientific Ltd. Headquarters, Beijing, China) and phosphoric acid (0.1%; w/w) at the rate of 1 mL min^{-1}: 10%–15% acetonitrile at 0–12 min, 15–20% acetonitrile at 12–35 min. The column temperature was maintained at 35 °C and the injection volume was 10 μL. The determination wavelength was used at 206 nm.

A weight of 2.502, 2.510, and 2.501 mg of reference substances for eleutheroside B (HPLC ≥ 98%, batch number P14J7F8937), eleutheroside E (HPLC ≥ 98%, batch number W11J7K8869), and isofraxidin (HPLC ≥ 99%, batch number Y12J7S17732) (Yuanye Bio. S&T. Inc., Shanghai, China) was prepared and placed in 10 mL volumetric flasks, respectively. Flasks were added to using methanol to the volume of 10 mL to obtain the references of three compounds. References were used to make the prepared model by regressing known concentrations in a given range to the peak area (Table 2). The chromatogram is shown in Figure 2.

Table 2. Prepared model of linear regression with ranged of known concentrations as independent and the peak area as dependent.

Compound	Regression Model	R^2	Independent Range (μg)
Eleutheroside B	Y = 3,970,604.45x − 1316.93	0.9997	0.0390–2.502
Eleutheroside E	Y = 675,578.96x − 23,980.50	0.9997	0.0390–2.510
Isofraxidin	Y = 6,040,711.41x − 148,307.39	0.9996	0.0195–2.507

Using this regression model, the references were extracted six times in 10 μL for three compounds to calculate the relative standard deviation (RSD). As a result, RSD for the peak area of 0.62%, 0.40%, and 0.32% for eleutheroside B, eleutheroside E, and isofraxidin, respectively, was obtained indicating the precision of the determination can be acceptable. RSD of determination using samples at 0, 4, 6, 8, 10, and 12 h after preparation for eleutheroside B, eleutheroside E, and isofraxidin were 1.19%, 0.88%, and 0.79%, respectively, suggesting samples were stable for12 h. The six repeated analyses of each parameter resulted in RSD ranging from 0.78% to 1.20%, which indicated an acceptable precision of technically repeated measures.

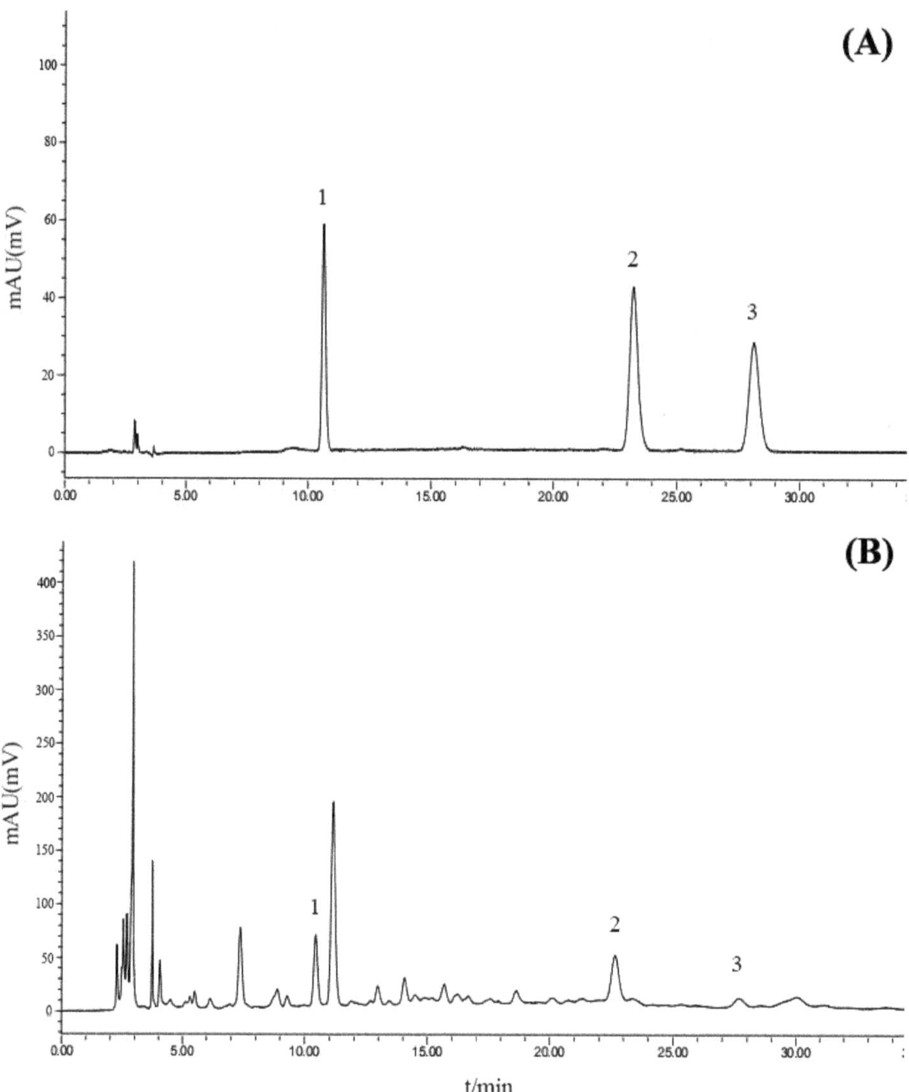

Figure 2. Chromatogram for peaks of eleutheroside B, eleutheroside E, and isofraxidin in *Eleutherococcus senticosus*. (**A**) Peaks of eleutheroside B (1), eleutheroside E (2), and isofraxidin (3) in the reference. (**B**) Peaks of eleutheroside B (1), eleutheroside E (2), and isofraxidin (3) in the tested samples.

2.4. Statistical Analysis

Values of both parameters about regional environment (temperature, precipitation, and RH) and bioactive compounds (eleutheroside B, eleutheroside E, and isofraxidin) were interpolated by kriging using ArcGIS software (V-9.3, Esri China Information Technology Ltd., Beijing, China) to map their spatial distributions. Data about bioactive compounds for one plot of *E. senticosus* population were averaged for the mean from three sub-plots. Data about compound concentrations were calculated and mapped separately in aerial and underground organs.

Statistical analyses were finished using SAS software (ver. 9.4 64-bit, SAS Institute, Cary, NC, USA). All data were checked for the normal distribution and necessary transformation was made to pass the normality test. Pearson correlation was employed twice. Firstly, it was performed to detect the relationship between data about regional climate (temperature, precipitation, and RH) and topographical factors (longitude, latitude, elevation, and slope). Subsequently, the linear correlation was employed again to detect the relationship between any of the unique abiotic factors and bioactive compound concentrations. Thereafter, the stepwise regression was employed to identify the contribution of multiple abiotic variables to concentrations of the three compounds. When the linear regression failed to test the multiple variables, raw data were used for the maximum-likelihood regression. The significance of probability for both Pearson correlation and multiple-variable-regression was accepted at the 0.05 level.

3. Results

3.1. Spatial Distribution of Climatic Factors

The average temperature from 1996 to 2016 presented a geographically descending gradient from the south to the north (Figure 3A). Precipitation was also found to be higher in the southern regions in the study area (Figure 3B). However, RH was lower in the central part of the research area than in the surrounding regions especially the eastern edge and the western and southwestern orientations (Figure 3C).

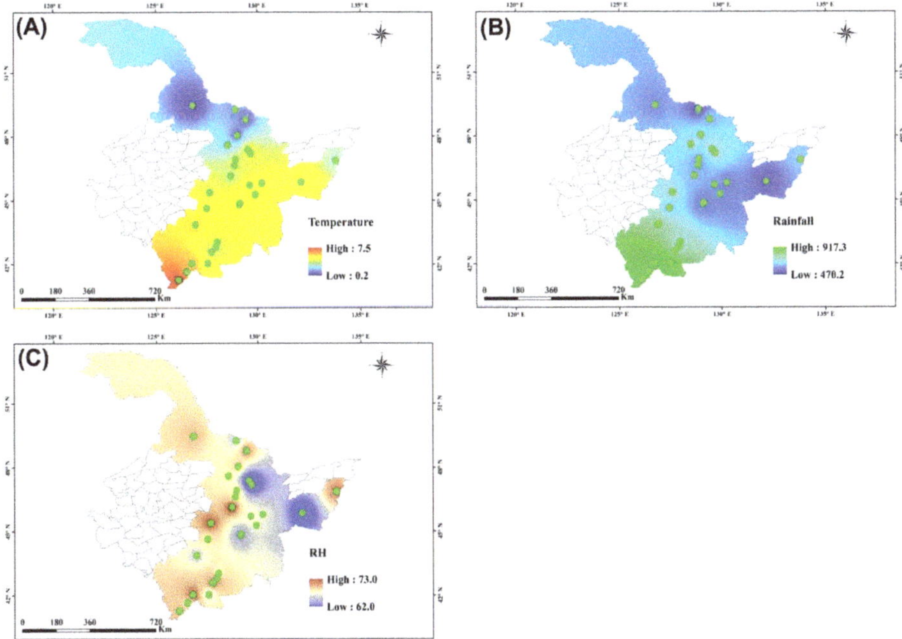

Figure 3. Interpolated mapping of geographical distributions of climatic factors in forests harboring *Eleutherococcus senticosus* populations in Northeast China. (**A**) Geographical distribution of temperature (°C); (**B**) geographical distribution of annual rainfall; (**C**) geographical distribution of relative humidity (RH) (%). Green dots indicate distribution of plots.

3.2. Spatial Distribution of Bioactive Compounds

Eleutheroside B concentration in *E. senticosus* shoot was higher in surrounding areas than in the central part of the study area (Figure 4A). Eleutheroside E concentration was higher in the southern and eastern regions of the study area than in the northern regions (Figure 4B). Shoot isofraxidin concentration was found to be higher in the surrounding areas than in central regions of the studied areas (Figure 4C).

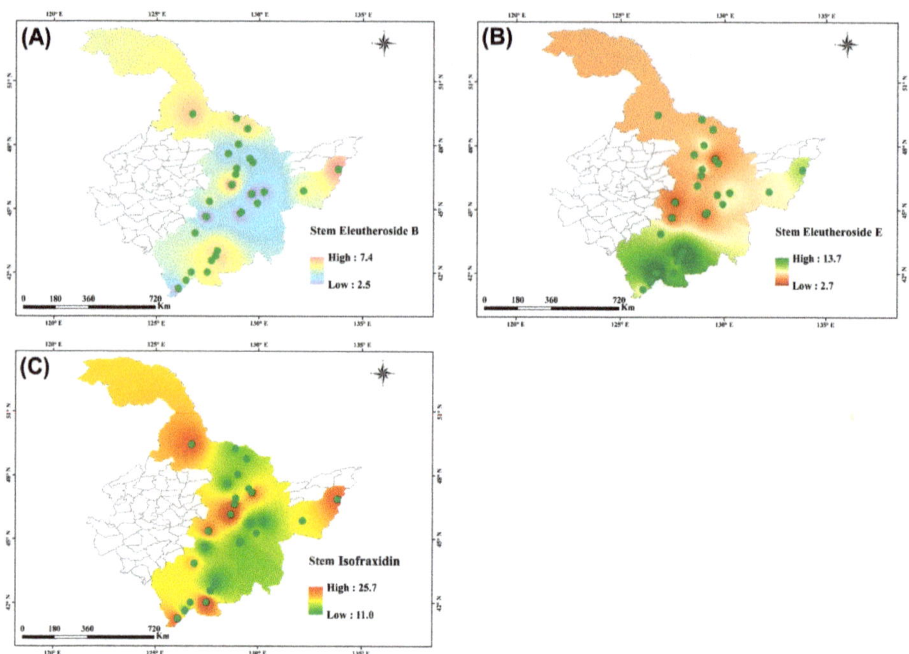

Figure 4. Interpolated mapping of geographical distributions of concentrations of bioactive compounds in aerial organs of *Eleutherococcus senticosus* individuals from populations in Northeast China. (**A**) Geographical distribution of eleutheroside B (mg g^{-1}); (**B**) geographical distribution of eleutheroside E (mg g^{-1}); (**C**) geographical distribution of Isofraxidin (mg g^{-1}). Green dots indicate distribution of plots.

Eleutheroside B and eleutheroside E concentrations in *E. senticosus* roots had similar spatial distribution patterns, which showed alternatively low and high concentrations from the northern part of the study area to the south (Figure 5A,B). These two compounds also showed higher concentrations in the eastern regions. However, root isofraxidin showed another different distribution pattern where the concentration was higher in the two regions near the western and eastern edges than in the central part (Figure 5C).

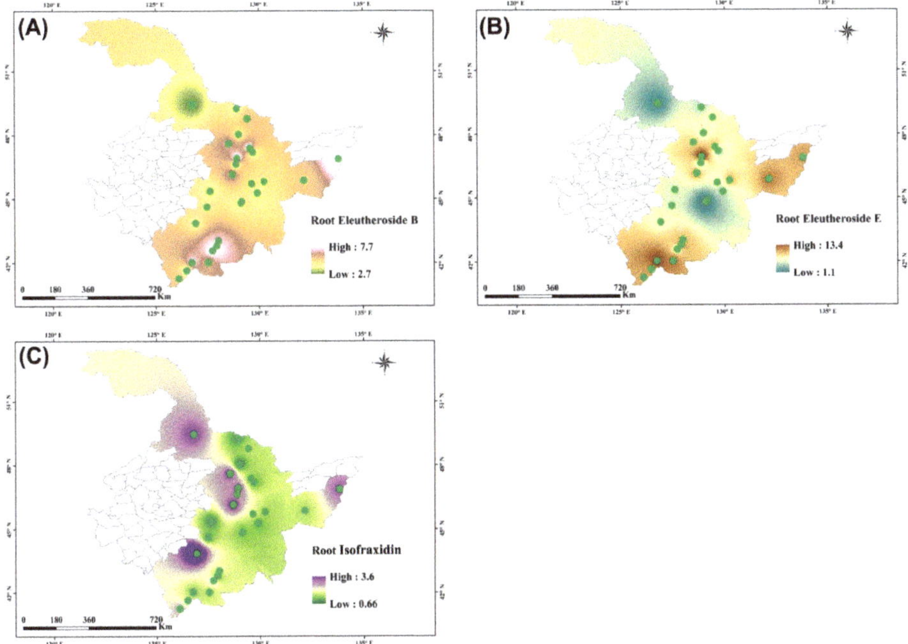

Figure 5. Interpolated mapping of geographical distributions of concentrations of bioactive compounds in underground organs of *Eleutherococcus senticosus* individuals from populations in Northeast China. (**A**) Geographical distribution of eleutheroside B (mg g^{-1}); (**B**) geographical distribution of eleutheroside E (mg g^{-1}); (**C**) geographical distribution of Isofraxidin (mg g^{-1}). Green dots indicate distribution of plots.

3.3. Relationship Between Parameters About Climate and Topography

Both longitude and latitude had a negative relationship with temperature, suggesting temperature tended to decline to the northern or to the eastern orientations in the study area (Table 3). Elevation had a positive relationship with RH, suggesting the high air humidity at high altitude. Longitude also had a positive relationship with elevation, but latitude had a negative relationship with elevation (Table 3). These results suggested that mountains tended to be higher in regions to the southeastern orientation.

3.4. Relationship Between Abiotic Factors and Bioactive Compounds

The Pearson regression indicated that RH was positively correlated to eleutheroside B concentration in shoot (Pearson correlation coefficients: $R = 0.5003$; $P = 0.0079$). Therefore, RH was further found to have a linear relationship with eleutheroside B concentration in shoot (Figure 6A). In contrast, temperature was negatively correlated to eleutheroside B concentration in shoot (Pearson correlation coefficients: $R = -0.4394$; $P = 0.0218$). Another negatively linear regression was found between temperature and shoot eleutheroside B concentration (Figure 6B). RH was positively correlated to eleutheroside B concentration in roots (Pearson correlation coefficients: $R = 0.4155$; $P = 0.0311$) (Figure 6C), but latitude was negatively correlated to eleutheroside E concentration in shoot (Pearson correlation coefficients: $R = -0.5083$; $P = 0.0068$) (Figure 6D).

Table 3. Pearson correlation between parameters about climate and topography.

	Regression Coefficient	Temperature	RH [1]	Rainfall [2]	Longitude	Latitude	Elevation [3]	Slope [4]
Temperature	R	1						
	P							
RH	R	0.22984	1					
	P	0.2488						
Rainfall	R	−0.14057	−0.22209	1				
	P	0.4843	0.2655					
Longitude	R	−0.41167	0.14211	−0.11174	1			
	P	0.0329	0.4795	0.579				
Latitude	R	−0.78313	−0.25705	0.08392	0.0885	1		
	P	<0.0001	0.1955	0.6773	0.6607			
Elevation	R	0.1373	0.52385	−0.05712	0.43327	−0.50833	1	
	P	0.4947	0.0050	0.7772	0.0240	0.0068		
Slope	R	0.07157	−0.01393	−0.27039	0.36195	−0.0009	−0.12076	1
	P	0.7228	0.945	0.1726	0.0636	0.9964	0.5485	

[1] RH, relative humidity, transformed by the square divided by 100; [2] transformed by the reciprocal times 100; [3] transformed by the logarithm; [4] transformed by the logarithm of raw data plus one.

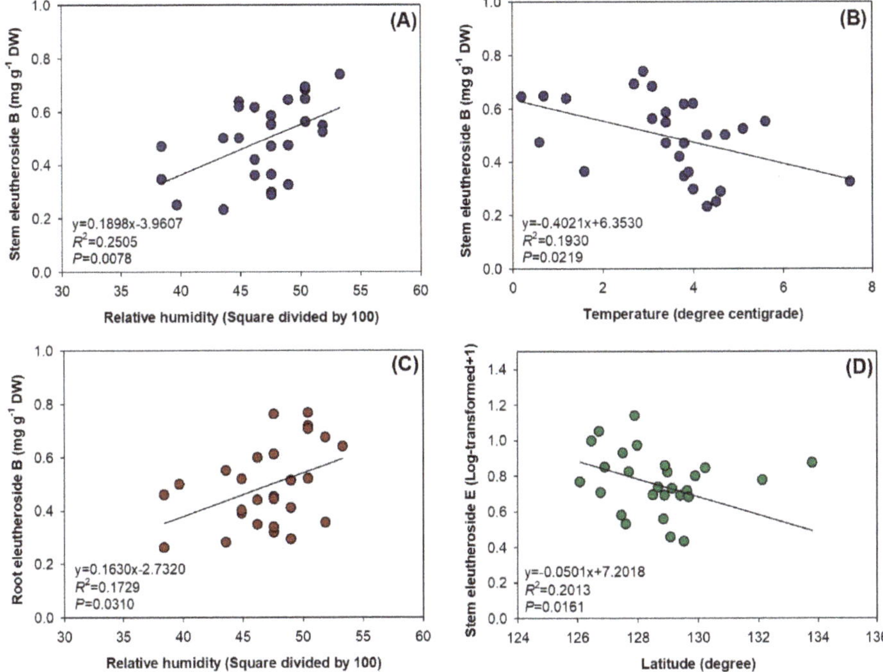

Figure 6. Linear correlation between abiotic factors and concentrations of bioactive compounds in *Eleutherococcus senticosus* individuals from populations in Northeast China. (**A**) Regression between relative humidity and shoot eleutheroside B concentration; (**B**) regression between temperature and shoot eleutheroside B concentration; (**C**) regression between relative humidity and root eleutheroside B concentration; (**D**) regression between latitude and shoot eleutheroside E concentration. R^2 and P values are coefficients from linear regression models.

3.5. Regression of Multiple Varialbles of Abiotic Factors with Bioactive Compounds

Stepwise regression indicated that temperature and latitude had negative contributions to eleutheroside B concentrations in *E. senticosus* shoot (Table 4). Although latitude also had a negative contribution to eleutheroside E concentration in shoot, the contribution of temperature to shoot eleutheroside E concentration was positive.

Table 4. Stepwise regression of temperature and latitude factors on eleutheroside B and E concentrations in *Eleutherococcus senticosus* stems.

Variable	Estimate	Standard Error	Type II SS [1]	*F* Value	*Pr > F*
		Stem eleutheroside B			
Intercept	28.06726	7.82155	18.16931	12.88	0.0015
Temperature	−0.91374	0.23469	21.38854	15.16	0.0007
Latitude	−0.43905	0.15775	10.93055	7.75	0.0103
		Stem eleutheroside E [2]			
Intercept	4.39383	0.87811	0.44527	25.04	<0.0001
Temperature	0.07197	0.02635	0.13269	7.46	0.0116
Latitude	−0.07435	0.01771	0.31342	17.62	0.0003

[1] SS, sum of squares; [2] Data about stem eleutheroside E were log-transformed before use in the regression model.

The maximum-likelihood regression indicated that both RH and longitude had positive contributions to eleutheroside B concentration in roots, but precipitation had a strongly negative contribution to the eleutheroside E concentration in roots (Table 5). Temperature and elevation had positive contributions to isofraxidin concentration in shoot, while the contributions from RH and slope were negative. Temperature, RH, and longitude all had positive contributions to isofraxidin concentration in roots, but precipitation had a negative contribution.

Table 5. Maximum-likelihood regression of variables about climate and topography on eleutheroside B and E concentrations in roots and isofraxidin content in stem and root in *Eleutherococcus senticosus* plants.

Variable	DF	Estimate	Wald Chi-square	$Pr >$ Chi-square	Estimate	Wald Chi-square	$Pr >$ Chi-square
		Root eleutheroside B			Root eleutheroside E		
Intercept	1	**−56.9266**	**4.72**	**0.0298**	−53.51	1.11	0.2929
Temperature	1	−0.208	0.48	0.4897	−0.0061	0	0.9916
RH	1	**0.1708**[1]	**5.4**	**0.0201**	0.1321	0.86	0.3546
Rainfall	1	−8.0786	0.4	0.5247	**−53.2192**	**4.66**	**0.0309**
Longitude	1	**0.5154**	**9.05**	**0.0026**	0.6154	3.42	0.0643
Latitude	1	−0.2855	1.46	0.2271	−0.2281	0.25	0.6191
Elevation	1	1.0146	0.44	0.5092	−2.4571	0.68	0.4103
Slope	1	−0.1232	0.03	0.8630	0.622	0.2	0.6537
Scale	1	1.1264			2.1869		
		Stem isofraxidin			Root isofraxidin		
Intercept	1	16.0687	2.45	0.1173	1.6175	1.74	0.1866
Temperature	1	**2.1062**	**9.67**	**0.0019**	**0.3094**	**14.64**	**0.0001**
RH	1	**−0.6087**	**4.47**	**0.0346**	**0.1628**	**22.43**	**<0.0001**
Rainfall	1	−1.2718	0.9	0.3438	**−0.5127**	**10.22**	**0.0014**
Longitude	1	0.0076	0	0.9706	**0.0485**	**3.85**	**0.0496**
Latitude	1	0.0009	0	0.9546	−0.0027	2.07	0.1499
Elevation	1	**0.7587**	**4.74**	**0.0295**	0.0266	0.41	0.5219
Slope	1	**−2.2329**	**3.89**	**0.0486**	−0.0963	0.51	0.4762
Scale	1	5.9059			0.705		

[1] Values in bold font indicate estimates that have significant contribution to dependent variables.

4. Discussion

Our results were agreeable to our hypothesis that eleutheroside B concentration in *E. senticosus* shoot tended to be higher in the northern part of the study area and lower in the southeast part. Pearson correlation clearly indicated a negative relationship between temperature and shoot eleutheroside B concentration. Our results were different from those of Shohael et al. [31], where eleutheroside B concentration increased from 18 to 24 °C with an undetectable level at 12 °C. It may be roughly concluded that eleutheroside B concentration increased with temperature in Shohael et al., which appeared to disagree to our results [31]. However, two aspects of methodology existed in the former study to generate the difference. Firstly, materials in Shohael et al. were used as somatic embryos, which were fragile and sensitive to exogenous environment [31], but ours were employed as mature individuals from the natural population. Furthermore, the temperature in Shohael et al. varied within the range of indoor conditions from 12 to 30 °C [31], but the real temperature for natural populations was as low as 0–7 °C with annual chilling periods. It was indicated that the temperature at 20 °C was suitable for the development of embryos [14], but this temperature may occur only in summer for some natural populations. However, other botanic studies revealed that low temperature can induce the synthesis of anthocyanin in several plant species [15,18]. In addition, Schmidt et al. studied the climatic influence on kale (*Brassica oleracea*) and found that the concentration of flavonoids grew in cooler temperatures ranging 0.3–9.6 °C [32].

It was surprising that the geographical change of shoot eleutheroside B concentration had no relationship with the single factor of latitude unless with the involvement of temperature as another independent variable in the stepwise regression. Thus, both geography and Pearson correlation clearly revealed that temperature declined along the latitude gradient from the south to the north. These results together suggest that the special distribution of eleutheroside B concentration in *E. senticosus* stem was shaped by the geographical distribution of temperature along the latitude gradient. On the other hand, we also found a positive relationship between RH and shoot eleutheroside B concentration. This part of results contradicted those in Guo et al., where RH was negatively correlated with betulin and lupeol concentrations in *B. platyphylla* trees [26]. However, our results concur with those findings about forest crops in the understory [23,33]. Two possible explanations may be responsible for our results. The increase of RH can induce the synthesis of secondary metabolites through depressing vapor pressure deficit [23]. Otherwise, higher RH may induce the synthesis of eleutheroside B concentration as a response to the stimulated disease explosion [33].

In contrast to the relationship between temperature and shoot eleutheroside B concentration, temperature was found to be positively correlated with shoot eleutheroside E concentration with decreasing latitudes. This coincided with the higher eleutheroside E concentration in *E. senticosus* individuals in the southwestern part of the study area and the lower one in the northern part. This special distribution matched that of regional temperature. Again, our results about shoot eleutheroside E concentration disagreed with those in Shohael et al., where *E. senticosus* embryos showed decreasing eleutheroside E concentration with the increase of temperature from 12 to 24 °C [31]. This part of the results concurs with those in Guo et al., where the concentration of secondary metabolites in *Scutellaria baicalensis* plants was positively correlated with temperature across mainland China [29]. However, our results about shoot eleutheroside E concentration failed to support the hypothesis because there was a negative relationship with latitude. Because the single factor temperature failed to have any relationship with shoot eleutheroside E concentration, we surmise other abiotic factors, such as light quality and ultraviolet (UV) light [34], may have contributed to the spatial pattern of eleutheroside E distribution.

Results about eleutheroside B concentration in *E. senticosus* roots support the other part of our hypothesis. The maximum-likelihood regression indicated that both RH and longitude contributed to the positive effect on root eleutheroside B concentration. In addition, RH was positively correlated with elevation, which was further positively correlated with longitude. Therefore, these results together suggest that *E. senticosus* individuals tended to have high accumulation of eleutheroside B concentration in roots in populations in moist forests distributed in altitudes of mountains in the eastern regions. The eastern alp in our study was mainly accounted for by Changbai mountains, where it was reported that the decline of temperature with the increase of elevation controlled the evapotranspiration, which resulted in the rising RH along the elevation gradient [35]. The high RH in the eastern mountains may have promoted the accumulation of eleutheroside B in roots.

Although we found the negative relationship between rainfall and eleutheroside E concentration in roots, no relationship was further detected between rain and other climate or topographical parameters. We surmise that abundant rainfall favored the condition for *E. senticosus* growth and controlled the synthesis of eleutheroside E in roots. This part of the results concurred with those about medical herbs in South Africa [36] and Brazil [37]. A study on *Euclea undulata* Thunb. var. *myrtina* reported that concentrations of epicatechin and 7-methyl-juglone in roots were depressed in the rainy season for the population in the winter rainfall area. The study on *Tithonia diversifolia* revealed that change of inorganic-element concentrations in soils determined the concentration of secondary metabolites in roots.

Factors of slope and temperature had strong negative and positive contributions (estimates of about ± 2.0) to the isofraxidin concentration in stem, respectively. These two factors had no relationship with each other; therefore, the two contributions were separated. The positive effect of temperature on isofraxidin concentration coincided with that on eleutheroside E, which may share the same mechanism that had been discussed in the preceding paragraphs. The slope in our study ranged between 0° and 33°. In another study, Wei et al. studied the response of *Aralia elata*, the species from another genus in Araliaceae, to different slopes and found that nearly all plant parameters decreased since the slope increased up to 9° [13]. The positive correlation between elevation and shoot isofraxidin concentration resulted from strong UV-light-induced compound synthesis and accumulation [34]. The strongly negative correlation between rainfall and isofraxidin concentration in roots concurred with the findings about root eleutheroside E concentration. Positive contributions of temperature, RH, and longitude to root isofraxidin concentration indicated that *E. senticosus* individuals in natural populations in regions at high elevation with an annual temperature of about 5 °C had the highest levels of root isofraxidin.

5. Conclusions

In this study, 27 plots were investigated to detect the special distribution and environmental correlates of eleutheroside B, eleutheroside E, and isofraxidin in individuals from natural *E. senticosus* populations in forests of Northeast China. Populations in northern regions around the Greater Khingan Mountains tended to have a higher concentration of eleutheroside B in aerial organs, which was driven by the lower temperature therein. Another region of forests that harbored *E. senticosus* individuals with higher concentration of eleutheroside B concentration in shoots was located at the eastern mountains with high elevation and high humidity. Concentrations of eleutheroside B and isofraxidin were higher in the southern and eastern edges of the study area, where higher temperature drove the concentrations in aerial organs and lower rainfall drove those in roots. Therefore, the spatial distribution of bioactive compounds in medical plants in the understory can be driven by regional climate factors. The management of non-wood forest product plants at the regional scale should consider effects from climatic geography.

Author Contributions: Conceptualization, S.G., H.W., and Z.W.; methodology, S.G. and H.W.; software, H.W. and X.C.; field investigation, S.G., J.L., R.F., and M.X.; chemical analysis, J.L. and X.C.; data curation, S.G., H.W., and X.C.; writing—original draft preparation, S.G.; writing—review and editing, H.W.; project administration, Z.W.; funding acquisition, Z.W.

Funding: This research was funded by the National Key Research and Development Program of China (grant number 2016YFC0500300), the Heilongjiang Province Foundation for The National Key Research and Development Program of China (grant number GX17C006), the Post-Doctoral Foundation of Heilongjiang Province of China (grant number LBH-Z17208), the National Natural Science Foundation of China (grant number 41971122; 41861017), and the Funding for Jilin Environmental Science (grant number 2017-16).

Conflicts of Interest: The authors declare no conflict of interest.

References

1. Walsh, E.S.; Vierling, K.T.; Strand, E.; Bartowitz, K.; Hudiburg, T.W. Climate change, woodpeckers, and forests: Current trends and future modeling needs. *Ecol. Evol.* **2019**, *9*, 2305–2319. [CrossRef] [PubMed]

2. Lemordant, L.; Gentine, P. Vegetation response to rising co2 impacts extreme temperatures. *Geophys. Res. Lett.* **2019**, *46*, 1383–1392. [CrossRef]

3. Park, H.; Jeong, S.J.; Ho, C.H.; Kim, J.; Brown, M.E.; Schaepman, M.E. Nonlinear response of vegetation green-up to local temperature variations in temperate and boreal forests in the northern hemisphere. *Remote Sens. Environ.* **2015**, *165*, 100–108. [CrossRef]

4. Wang, G.; Wang, P.; Wang, T.Y.; Zhang, Y.C.; Yu, J.J.; Ma, N.; Frolova, N.L.; Liu, C.M. Contrasting changes in vegetation growth due to different climate forcings over the last three decades in the selenga-baikal basin. *Remote Sens.* **2019**, *11*, 17. [CrossRef]

5. Ahlstrand, N.I.; Reghev, N.H.; Markussen, B.; Hansen, H.C.B.; Eiriksson, F.F.; Thorsteinsdottir, M.; Ronsted, N.; Barnes, C.J. Untargeted metabolic profiling reveals geography as the strongest predictor of metabolic phenotypes of a cosmopolitan weed. *Ecol. Evol.* **2018**, *8*, 6812–6826. [CrossRef]

6. Jochum, G.M.; Mudge, K.W.; Thomas, R.B. Elevated temperatures increase leaf senescence and root secondary metabolite concentrations in the understory herb panax quinquefolius (araliaceae). *Am. J. Bot.* **2007**, *94*, 819–826. [CrossRef]

7. Souther, S.; McGraw, J.B. Evidence of local adaptation in the demographic response of american ginseng to interannual temperature variation. *Conserv. Biol.* **2011**, *25*, 922–931. [CrossRef]
8. Vezza, M.; Nepi, M.; Guarnieri, M.; Artese, D.; Rascio, N.; Pacini, E. Ivy (hedera helix l.) flower nectar and nectary ecophysiology. *Int. J. Plant Sci.* **2006**, *167*, 519–527. [CrossRef]
9. Abraham, E.M.; Theodoropoulos, K.; Eleftheriadou, E.; Ragkos, A.; Kyriazopoulos, A.P.; Parissi, Z.M.; Arabatzisa, G.; Soutsas, K. Non-wood forest products from the understory and implications for rural development: The case of a broadleaf deciduous oak forest (quercus frainetto t e n.) in chalkidiki, greece. *J. Environ. Prot. Ecol.* **2015**, *16*, 1024–1032.
10. FAO Towards a harmonized definition of non-wood forest products. *Unasylva* **1999**, *198*, 63–66.
11. Huber, P.; Hujala, T.; Kurttila, M.; Wolfslehner, B.; Vacik, H. Application of multi criteria analysis methods for a participatory assessment of non-wood forest products in two european case studies. *For. Policy Econ.* **2019**, *103*, 103–111. [CrossRef]
12. Huber, F.K.; Ineichen, R.; Yang, Y.P.; Weckerle, C.S. Livelihood and conservation aspects of non-wood forest product collection in the shaxi valley, southwest china(1). *Econ. Bot.* **2010**, *64*, 189–204. [CrossRef]
13. Wei, H.X.; Zhao, H.T.; Chen, X. Foliar n:P stoichiometry in aralia elata distributed on different slope degrees. *Not. Bot. Horti. Agrobo.* **2018**, *47*, 887–895. [CrossRef]
14. Gao, Z.; Khalid, M.; Jan, F.; Saeed ur, R.; Jiang, X.; Yu, X. Effects of light-regulation and intensity on the growth, physiological and biochemical properties of aralia elata (miq.) seedlings. *South Afr. J. Bot.* **2019**, *121*, 456–462. [CrossRef]
15. Nadeau, I.; Olivier, A. The biology and forest cultivation of american ginseng (panax quinquefolius l.) in canada. *Can. J. Plant Sci.* **2003**, *83*, 877–891.
16. Yu, X.H.; Gao, Z.L.; Juan, J.X.; Cheng, Y.; Zhao, M.L.; Ma, D.N.; Xu, X.B.; Jiang, X.M. Digital analysis on cone-thorn density and morphology in mutant aralia elata seedlings induced by ethyl methanesulfonate. *Int. J. Agric. Biol.* **2019**.
17. Yang, X.; Skidmore, A.K.; Melick, D.R.; Zhou, Z.; Xu, J. Mapping non-wood forest product (matsutake mushrooms) using logistic regression and a gis expert system. *Ecol. Model.* **2006**, *198*, 208–218. [CrossRef]
18. Choi, S.; Kwon, Y.R.; Hossain, M.A.; Hong, S.W.; Lee, B.H.; Lee, H. A mutation in ela1, an age-dependent negative regulator of pap1/myb75, causes uv- and cold stress-tolerance in arabidopsis thaliana seedlings. *Plant Sci.* **2009**, *176*, 678–686. [CrossRef]
19. Jung, C.H.; Ahn, J.; Heo, S.H.; Ha, T.-Y. Eleutheroside e, an active compound from eleutherococcus senticosus, regulates adipogenesis in 3t3-l1 cells. *Food Sci. Biotechnol.* **2014**, *23*, 889–893. [CrossRef]
20. Wu, K.X.; Liu, J.; Liu, Y.; Guo, X.R.; Mu, L.Q.; Hu, X.H.; Tang, Z.H. A comparative metabolomics analysis reveals the tissue-specific phenolic profiling in two acanthopanax species. *Molecules* **2018**, *23*, 14. [CrossRef]
21. Sun, H.; Liu, J.H.; Zhang, A.H.; Zhang, Y.; Meng, X.C.; Han, Y.; Zhang, Y.Z.; Wang, X.J. Characterization of the multiple components of acanthopanax senticosus stem by ultra high performance liquid chromatography with quadrupole time-of-flight tandem mass spectrometry. *J. Sep. Sci.* **2016**, *39*, 496–502. [CrossRef] [PubMed]
22. Lu, F.; Sun, Q.; Bai, Y.; Bao, S.; Li, X.; Yan, G.; Liu, S. Characterization of eleutheroside b metabolites derived from an extract of acanthopanax senticosus harms by high-resolution liquid chromatography/quadrupole time-of-flight mass spectrometry and automated data analysis. *Biomed. Chromatogr.* **2012**, *26*, 1269–1275. [CrossRef] [PubMed]
23. Lihavainen, J.; Keinanen, M.; Keski-Saari, S.; Kontunen-Soppela, S.; Sober, A.; Oksanen, E. Artificially decreased vapour pressure deficit in field conditions modifies foliar metabolite profiles in birch and aspen. *J. Exp. Bot.* **2016**, *67*, 4367–4378. [CrossRef] [PubMed]
24. Lee, S.Y.; Son, D.; Ryu, J.; Lee, Y.S.; Jung, S.H.; Kang, J.I.; Lee, S.Y.; Kim, H.S.; Shin, K.H. Anti-oxidant activities of acanthopanax senticosus stems and their lignan components. *Arch. Pharm. Res.* **2004**, *27*, 106–110. [CrossRef] [PubMed]
25. Niu, X.F.; Xing, W.; Li, W.F.; Fan, T.; Hu, H.; Li, Y.M. Isofraxidin exhibited anti-inflammatory effects in vivo and inhibited tnf-alpha production in lps-induced mouse peritoneal macrophages in vitro via the mapk pathway. *Int. Immunopharmacol.* **2012**, *14*, 164–171. [CrossRef] [PubMed]
26. Guo, S.L.; Zhang, D.H.; Wei, H.Y.; Zhao, Y.N.; Cao, Y.B.; Yu, T.; Wang, Y.; Yan, X.F. Climatic factors shape the spatial distribution of concentrations of triterpenoids in barks of white birch (betula platyphylla suk.) trees in northeast china. *Forests* **2017**, *8*, 12. [CrossRef]

27. Shen, X.J.; Liu, B.H.; Xue, Z.S.; Jiang, M.; Lu, X.G.; Zhang, Q. Spatiotemporal variation in vegetation spring phenology and its response to climate change in freshwater marshes of northeast china. *Sci. Total Environ.* **2019**, *666*, 1169–1177. [CrossRef] [PubMed]

28. Dai, L.M.; Shao, G.F.; Xiao, B.Y. Ecological classification for mountain forest sustainability in northeast china. *For. Chron.* **2003**, *79*, 233–236. [CrossRef]

29. Guo, L.; Wang, S.; Zhang, J.; Yang, G.; Zhao, M.; Ma, W.; Zhang, X.; Li, X.; Han, B.; Chen, N.; et al. Effects of ecological factors on secondary metabolites and inorganic elements of scutellaria baicalensis and analysis of geoherblism. *Sci. China Life Sci.* **2013**, *56*, 1047–1056. [CrossRef] [PubMed]

30. Wan, J.Z.; Wang, C.J.; Yu, J.H.; Nie, S.M.; Han, S.J.; Liu, J.Z.; Zu, Y.G.; Wang, Q.G. Developing conservation strategies for pinus koraiensis and eleutherococcus senticosus by using model-based geographic distributions. *J. For. Res.* **2016**, *27*, 389–400. [CrossRef]

31. Shohael, A.M.; Ali, M.B.; Yu, K.W.; Hahn, E.J.; Paek, K.Y. Effect of temperature on secondary metabolites production and antioxidant enzyme activities in eleutherococcus senticosus somatic embryos. *Plant Cell Tissue Organ. Cult.* **2006**, *85*, 219–228. [CrossRef]

32. Schmidt, S.; Zietz, M.; Schreiner, M.; Rohn, S.; Kroh, L.W.; Krumbein, A. Genotypic and climatic influences on the concentration and composition of flavonoids in kale (brassica oleracea var. Sabellica). *Food Chem.* **2010**, *119*, 1293–1299. [CrossRef]

33. Shivanna, M.B.; Achar, K.G.S.; Vasanthakumari, M.M.; Mahishi, P. Phoma leaf spot disease of tinospora cordifolia and its effect on secondary metabolite production. *J. Phytopathol.* **2014**, *162*, 302–312. [CrossRef]

34. Jaakola, L.; Hohtola, A. Effect of latitude on flavonoid biosynthesis in plants. *Plant Cell Environ.* **2010**, *33*, 1239–1247. [CrossRef] [PubMed]

35. Yan, C.; Han, S.; Zhou, Y.; Zheng, X.; Yu, D.; Zheng, J.; Dai, G.; Li, M.-H. Needle δ13c and mobile carbohydrates in pinus koraiensis in relation to decreased temperature and increased moisture along an elevational gradient in ne china. *Trees* **2012**, *27*, 389–399. [CrossRef]

36. Botha, L.E.; Prinsloo, G.; Deutschländer, M.S. Variations in the accumulation of three secondary metabolites in euclea undulata thunb. Var. Myrtina as a function of seasonal changes. *South Afr. J. Bot.* **2018**, *117*, 34–40. [CrossRef]

37. Sampaio, B.L.; Edrada-Ebel, R.; Da Costa, F.B. Effect of the environment on the secondary metabolic profile of tithonia diversifolia: A model for environmental metabolomics of plants. *Sci. Rep.* **2016**, *6*, 29265. [CrossRef]

 *forests*

Article

Transcriptome Analysis of Elm (*Ulmus pumila*) Fruit to Identify Phytonutrients Associated Genes and Pathways

Luoyan Zhang, Xuejie Zhang, Mengfei Li, Ning Wang, Xiaojian Qu and Shoujin Fan *

Key Lab of Plant Stress Research, College of Life Science, Shandong Normal University, Jinan 250014, China
* Correspondence: fansj@sdnu.edu.cn; Tel.: +86-0531-8618-0178

Received: 9 August 2019; Accepted: 26 August 2019; Published: 27 August 2019

Abstract: Plant fruit is an important source of natural active phytonutrients that are profitable for human health. Elm (*Ulmus pumila*) fruit is considered as natural plant food in China that is rich in nutrients. In the present study, high-throughput RNA sequencing was performed in *U. pumila* edible fruits and leaves and 11,386 unigenes were filtered as dysregulated genes in fruit samples, including 5231 up- and 6155 downregulated genes. Hundreds of pathways were predicted to participate in seed development and phytonutrient biosynthesis in *U. pumila* by GO, MapMan, and KEGG enrichment analysis, including "seed maturation", "glycine, serine, and threonine metabolism" and "phenylpropanoid biosynthesis". ABA-mediated glucose response-related ethylene-activated signaling pathway (e.g., ABI4) were supposed to associate with elm fruit development; unsaturated fatty acids pathway (e.g., ACX2 and SAD) were predicted to participate in determination of fatty acid composition in elm fruit; flavonoid and coumarins biosynthesis (e.g., CYP98A3 and CCoAOMT1) were demonstrated to correlate with the bioactivity of elm fruits in human cancer and inflammation resistance. To provide more information about fruit developmental status, the qRT-PCR analysis for key genes of "phenylpropanoid biosynthesis" and "alpha-Linolenic acid metabolism" were conducted in samples of young fruits, ripe fruit, old fruit, and leaves. Two biosynthetic pathways for unsaturated fatty acid and Jasmonic acid (JA) were deduced to be involved in fruit development in *U. pumila* and the phenylpropanoid glycoside, syringin, was speculated to accumulate in the early development stages of elm fruit. Our transcriptome data supports molecular clues for seed development and biologically active substances in elm fruits.

Keywords: *Ulmus pumila*; transcriptome analysis; phytonutrients; seed development; phenylpropanoid biosynthesis

---

1. Introduction

Human diets are recommended to be rich in vegetables and fruits, which are beneficial to human health. The relationship between plant food intake and health has been the focus of many scientific studies to determine specific plant components that are active in conveying health benefits [1]. It is commonly accepted that plant fruit is an important source of natural active phytonutrients for human health, including amino acids, vitamins, terpenoids, flavonoids, alkaloids, phenolic compounds, and other metabolites, which lessen the risk of chronic diseases, such as metabolic syndrome, cardiovascular disease, obesity, and cancer [2–4]. Studies of dietary intervention have concluded that eating fruit helps maintain a healthy weight and reduces the risk of a wide range of cancers and cardiovascular disease [5–7].

The current work examines the fruit and leaves of elm (*Ulmus pumila*) using Illumina sequencing. This species belongs to the botanical classification of Ulmaceae and is a deciduous tree native to central Asia and is widely located in Asia, America, and southern Europe [8]. The species is a natural herb and of great economical value for use in traditional medicine in Asia. Humans use the stem and root parts of *U. pumila* for the treatment of various ailments such as edema, mastitis, gastric cancer, and inflammation in the traditional system of medicine [9,10]. Not surprisingly, a number of bioactive natural products have been identified from extracts of *U. pumila* [10–14].

The elm fruit is considered a plant food by the Chinese. The fruit contains about 3.3–4.5 g of proteins, 8.0–10.0 g of carbohydrate, and 1.0–1.5 g of dietary fiber per 100 g of fresh weight (FW), respectively [15–17]. It is also rich in vitamins (vitamin B1, vitamin B6, nicotinic acid, and ascorbic acid) and some minerals (calcium, potassium, magnesium, copper, iron) [15–17]. The elm fruit is considered useful for curing abscesses, infection, edema, cancer, and is of benefit for tonifying the spleen, treating insomnia, reducing blood glucose concentration, decreasing cholesterol and improving children's growth and development [15–17].

Although chemical, medical, and physiological evidence has been published for elm fruit, the molecular mechanisms of its phytonutrients remain partly unknown. Transcriptome sequencing is an effective method for genome-wide detection of potential participants of complex traits [18–27]. Dozens of studies in the field of fruit development and nutrition have implemented sequencing technologies [28–30]. In view of these facts, the work presented here was performed to uncover the key phytonutrient-associated genes and their regulators in elm fruit.

2. Materials and Methods

2.1. Plant Materials and RNA Extraction

Fruit and leaves of *U. pumila* were collected from the same tree in the specimen garden of the Shandong Normal University, Jinan, China, in March 2019. The elm fruits were classified by days after flowering (DAF) with different length and colors: young fruit, 20 DAF, 8–10 mm in length, and dark green color (fruit stage 1); young fruit, 30 DAF, 10–13 mm in length, and green color (fruit stage 2); ripe fruit, 40 DAF, 14–15 mm in length, and green color (fruit stage 3); and old fruit, 50 DAF, 14–15 mm in length, and yellow color (fruit stage 4). The fruits with 14–15 mm in length and green color (fruit stage 3) were identified as edible and were chosen for transcriptome analysis (Figure 1A,B). The selected fruits and leaves were collected and immediately frozen in liquid nitrogen and then stored at −80 °C until use. Three replicates of RNA extraction and transcriptome sequencing were performed for fruits and leaves. For each replicate, a total of 1 g of plant material was grinded in liquid nitrogen and TRIzol Reagent (Invitrogen, Carlsbad, CA, USA) was used for total RNA extraction according to the manufacturer's procedures. RNA Nano 6000 Assay Kit of the Agilent Bioanalyzer 2100 system (Agilent Technologies, Santa Clara, CA, USA) and a NanoDrop 2000 spectrophotometer (Thermo Scientific, Wilmington, NC, USA) were used for RNA quality assessment.

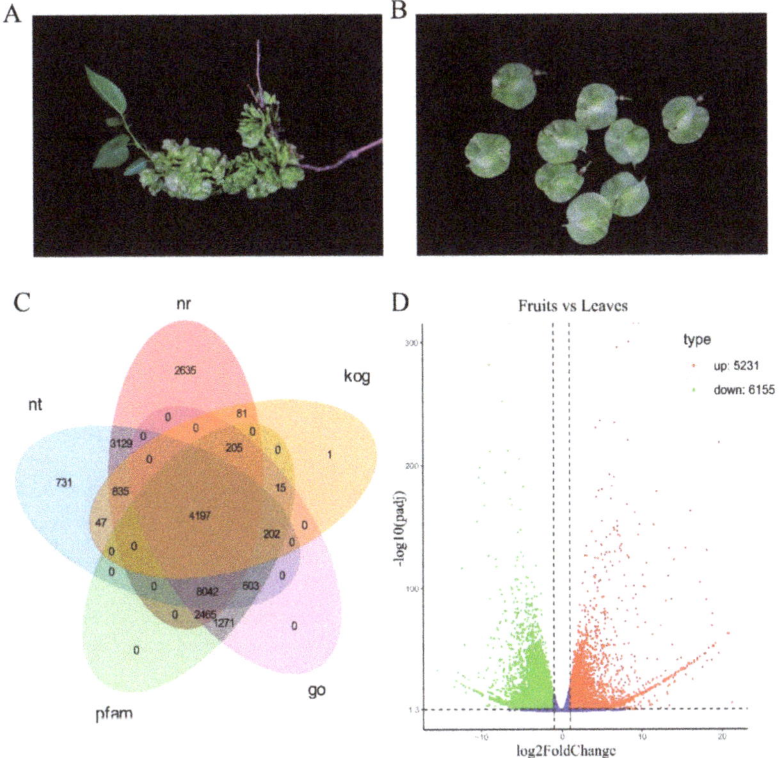

Figure 1. (**A**) The morphology of elm (*Ulmus pumila*) fruits and leaves. (**B**) The morphology of *U. pumila* fruits. (**C**) Venn diagram of functional annotations of unigenes in nt (NCBI nonredundant protein sequences), nr (NCBI nonredundant protein sequences), kog (Clusters of Orthologous Groups of proteins), go (Gene Ontology), and pfam (Protein family) databases. (**D**) Expression patterns of differentially expressed genes (DEGs) identified between fruits and leaves. Red and green dots represent DEGs, grey dots indicate genes that were not differentially expressed. In total, 11,386 unigenes were revealed to be differentially expressed (padj < 0.05) in fruit compared with control samples, with 5231 being upregulated and 6155 downregulated.

2.2. Library Construction and De Novo Assembly

NEBNext® UltraTM RNA Library Prep Kit for Illumina® (NEB, Beverly, MA, USA) were used to produce the RNA sequencing libraries. The mRNA purified by the poly-T oligo-attached magnetic beads and the synthesized cDNA (150–200 bp) were filtered by the AMPure XP system (Beckman Coulter, Beverly, MA, USA). After enriching and screening by PCR amplification, the products were sequenced by Illumina HiSeq × platform (Illumina, San Diego, CA, USA). Novogene Co., LTD were selected to perform cDNA library generation and PE150 sequencing. The clean reads were filtered by three steps: removing adapter, deleting poly-N reads and omitting low-quality reads. The Trinity software were used for de novo assembly [31]. The assembled sequences were annotated to public databases (NR, Pfam, KOG, KEGG, and Gene Ontology) by BLAST searches with an *E*-value cutoff of 1×10^{-5}. All sequencing data generated by this study have been submitted to the NCBI Sequence Read Archive (SRA) database [32], with accession: PRJNA545392.

2.3. Calculation of Genes' Expression and Enrichment Analyses in U. pumila

Six independent transcripts libraries were obtained for *U. pumila* by a PE150 sequencing. RSEM [33] were used to evaluate the gene expression levels. After the aligning the clean reads to the de novo assembled transcriptome, the fragment per kilobase of exon model per million mapped reads (FPKM) method [34] was selected to calculate genes' expression. The FPKM values between ripe fruit and leaf samples of all unigenes were compared by a threshold of |log2(foldchange)| > 1 and adjusted *p*-value < 0.05.

The unigenes of *U. pumila* were assigned to *A. thaliana* gene IDs by BLASTing the genome of *A. thaliana* with a cutoff of 1×10^{-5}. The topGO package of R and MapMan (version 3.5.1 R2) [35] were chosen for GO enrichment analysis and metabolic pathways enrichment for the differentially expressed genes (DEGs) in *U. pumila* fruit samples. The KEGG pathway enrichment analysis was performed by the software KOBAS [36].

2.4. qRT-PCR Analysis

The gene expression of four upregulated unigenes of "phenylpropanoid biosynthesis" pathway and four downregulated genes of "photosynthesis" pathway calculated by RNA-seq in ripe fruit and leave samples was validated by the qRT-PCR verification. Expression of eight key genes of "phenylpropanoid biosynthesis" (ko00940) and thirteen genes of "alpha-Linolenic acid metabolism" pathways (ko00592) were tested in samples of young fruits, ripe fruit, old fruit, and leaves by qRT-PCR analysis. The purified RNA samples were treated with DNaseI and converted to cDNA using the PrimeScript RT Reagent Kit with gDNA Eraser (Takara, Dalian, China). The ortholog of the *A. thaliana* member of Actin gene family ACT7 in *U. pumila* was used to normalize the amount of template cDNA. Premier 5.0 software was used to design gene-specific qRT-PCR primers (20–23 bp) (Table S5). qPCR analyses were performed on the ABI7500 Real-Time PCR System (ABI, USA) by using SYBR Green qPCR Master Mix (DBI, Germany). Each analysis was performed in triplicate and relative gene expression was quantified using the $2^{-(\Delta\Delta Ct)}$ method.

3. Results

3.1. Transcriptome Profiling of U. pumila

By sequencing with the Illumina HiSeq × platform, a total of 51,869,958, 54,374,336, 63,237,860, 66,499,546, 64,015,920, and 53,666,908 pair-end reads were acquired from three leaf and three fruit samples of *U. pumila*, respectively (Table 1). De novo transcriptome assembly yielded 35,416 unigenes, the average length of which was 1460 nt and N50 of 2251. Generally, 80.26% of the reads were mapped to the genome referenced for six samples of leaf and fruit (Table 1).

Table 1. Summary of mapping transcriptome reads to reference sequence in elm (*Ulmus pumila*).

Sample Name	Sample Description	Total Reads	Total Mapped	Ratio of Mapped Reads
Up_F_1	Fruits replication 1	51,869,958	41,402,822	79.82%
Up_F_2	Fruits replication 2	54,374,336	44,276,232	81.43%
Up_F_3	Fruits replication 3	63,237,860	51,316,902	81.15%
Up_L_1	Leaves replication 1	66,499,546	53,135,314	79.90%
Up_L_2	Leaves replication 2	64,015,920	50,704,484	79.21%
Up_L_3	Leaves replication 3	53,666,908	42,942,126	80.02%

3.2. Functional Annotations of Unigenes in U. pumila

BLASTX was used to perform similarity searches to annotate unigenes against various databases. All 35,416 (100%) unigenes were annotated in at least one database. A total of 21,589 (60.95%), 17,786 (50.22%), and 17,000 (48%) unigenes resembled the sequences in databases of NR, NT, and PFAM with an *E*-value threshold of 1×10^{-5} (Figure 1C). A total of 17,000 (48%) unigenes were annotated by

Blast2GO v2.5 in GO database with an *E*-value cutoff of 1×10^{-6}. A total of 21,255 unigenes of *U. pumila* were assigned to *A. thaliana* gene IDs for GO annotation mapping by BLASTX with an *E*-value cutoff of 1×10^{-5} and were used for MapMan analysis.

3.3. Differentially Expressed Genes (DEGs) Calculation in U. pumila

The relative expressional level of related genes in *U. pumila* fruits or leaves was evaluated by the FPKM values, which was computed following the uniquely mapped reads. The FPKM value for different genes ranged from 0.31 to 27,173.72, with a mean value of 30.28 detected in six samples. 5231 unigenes were filtered as upregulated and 6155 calculated as downregulated genes in fruit samples with the cutoff of |log2(foldchange)| > 1 and padj < 0.05 by comparative analysis (Figure 1D, Table S1). The 30 up- and downregulated genes are shown in Table 2. The lipids and proteins accumulation related protein oleosin 2 (Cluster-6074.11735, L_2fc = 9.880) and seed storage protein CRUCIFERINA (Cluster-6074.12126, L_2fc = 13.102) were identified as upregulated; the lipid-transfer protein (Cluster-6074.1319, L_2fc = −8.832) and pathogenesis-related thaumatin superfamily protein (Cluster-6074.23644, L_2fc = −7.223) were verified as the top downregulated genes (Table 2).

Table 2. The top 30 up- and downregulated genes.

Gene ID	L_2fc	Padj	Arabidopsis ID	Gene Description
		Upregulated		
Cluster-6074.11735	9.880	0.00	AT5G40420	oleosin 2
Cluster-6074.11841	9.147	0.00		
Cluster-6074.12126	13.102	0.00	AT2G25890	
Cluster-6074.12191	18.879	0.00	AT5G44120	CRUCIFERINA
Cluster-6074.12375	13.383	0.00		
Cluster-6074.13108	18.27	0.00	AT1G03890	
Cluster-6074.12362	8.535	3.45×10^{-301}	AT1G62710	beta vacuolar processing enzyme
Cluster-6074.13152	7.047	9.73×10^{-297}	AT5G12380	annexin 8
Cluster-6074.12791	14.590	6.74×10^{-295}	AT4G25140	oleosin 1
Cluster-6074.12340	4.909	1.58×10^{-237}	AT5G49360	beta-xylosidase 1
Cluster-6074.11747	6.724	1.65×10^{-235}	AT5G12380	annexin 8
Cluster-6074.10421	4.358	2.17×10^{-231}	AT1G21410	
Cluster-6074.12488	8.434	2.01×10^{-221}	AT4G37370	cytochrome P450, family 81, subfamily D, polypeptide 8
Cluster-6074.12147	19.796	1.96×10^{-219}	AT1G03890	
Cluster-6074.13683	6.109	4.23×10^{-193}	AT1G04560	
		Downregulated		
Cluster-6074.9536	−6.306	0.00		
Cluster-6074.1319	−8.832	8.48×10^{-283}	AT2G45180	
Cluster-6074.23644	−7.223	2.72×10^{-253}	AT1G20030	
Cluster-6074.18869	−6.508	6.11×10^{-213}	AT5G20740	
Cluster-6074.1201	−8.801	2.92×10^{-212}	AT4G11650	osmotin 34
Cluster-6074.20011	−5.099	3.94×10^{-206}	AT2G22540	short vegetative phase
Cluster-6074.25984	−9.979	6.56×10^{-199}		
Cluster-6074.18265	−6.868	2.59×10^{-195}	AT5G35630	glutamine synthetase 2
Cluster-6074.1293	−10.101	4.36×10^{-191}		
Cluster-6074.22654	−9.467	9.42×10^{-187}	AT5G59190	
Cluster-6074.1730	−4.625	1.19×10^{-183}	AT5G22430	
Cluster-6074.24974	−6.239	1.35×10^{-173}	AT5G67150	
Cluster-6074.16553	−6.034	9.25×10^{-169}	AT3G54420	homolog of carrot EP3-3 chitinase
Cluster-6074.1245	−9.437	2.26×10^{-168}		
Cluster-6074.21282	−4.960	8.82×10^{-165}	AT4G15440	hydroperoxide lyase 1

Note: L_2fc = Log_2 fold change.

3.4. GO, MapMan, and KEGG Enrichment Result of DEGs in U. pumila

To uncover the nutrient component-associated pathways in *U. pumila* fruit, the DEGs were described with GO databases. In total, 167 biological process (BP) terms were enriched with the cutoff of *p*-value < 0.05 by the 5231 upregulated unigenes, including "translation" (GO:0006412), "seed maturation" (GO:0010431), and "response to cadmium ion" (GO:0046686) (Table S2). A total of 173 BP terms were calculated to be enriched for the 6155 downregulated genes, such as "defense response" (GO:0006952), "photosynthesis" (GO:0015979), and "signal transduction" (GO:0007165) (Table S2).

As a result of the large numbers and the complex branch structure of GO categories, REVIGO was selected to uncover typical subgroups of the terms by employing a simple clustering algorithm, which is based on semantic similarity measures. The biological processes enriched by upregulated genes were pooled into eight groups (Figure 2A), 43 terms were summarized to the "translation" subset, including the BP terms "DNA demethylation" (GO:0080111), "biosynthetic process" (GO:0009058), and "unsaturated fatty acid biosynthetic process" (GO:0006636); 17 terms were classified to the "lipid storage" group; and 13 terms were assigned to the "seed maturation" group. In total, 3055 and 2960 homologs in *Arabidopsis* were assigned to up- and downregulated unigenes, respectively. A total of 1029 pathways were mapped by MapMan for these genes, of which, 117 pathways were filtered to be enriched by the dysregulated genes with the cutoff *p*-value < 0.05 (Table S3). The metabolism result of MapMan analysis is shown in Figure 2B.

The KEGG pathways enriched by upregulated unigenes are shown in Table S4 and the top 20 are represented in Figure S1A. The KEGG pathway "Ribosome" (ko03010) was enriched by 127 upregulated unigenes with rich_factor = 0.51, "Glycine, serine and threonine metabolism" (ko00052) was annotated for 29 overexpressed genes, and 49 upregulated genes were annotated in the KEGG pathway "Phenylpropanoid biosynthesis" (ko00940) (Figure S1A). The nutrition-related KEGG pathways enriched by upregulated genes are classified in Table 3. The "Amino acid" group included "Cysteine and methionine metabolism" (enriched with 29 upregulated genes), "Alanine, aspartate and glutamate metabolism" (enriched with 17 upregulated genes) and "Arginine biosynthesis" (enriched with 13 upregulated genes). A total of six fatty acid-related pathways are represented in Table 3, including "Fatty acid biosynthesis" (including 25 upregulated genes), "Biosynthesis of unsaturated fatty acids" (including nine upregulated genes) and "Glycerophospholipid metabolism" (including 24 upregulated genes). A total of 11 and seven unigenes were related to "Diterpenoid biosynthesis" and "Zeatin biosynthesis" pathways, respectively, and five unigenes related to "Vitamin B6 metabolism" pathway were upregulated in the fruit samples (Table 3, Table S4).

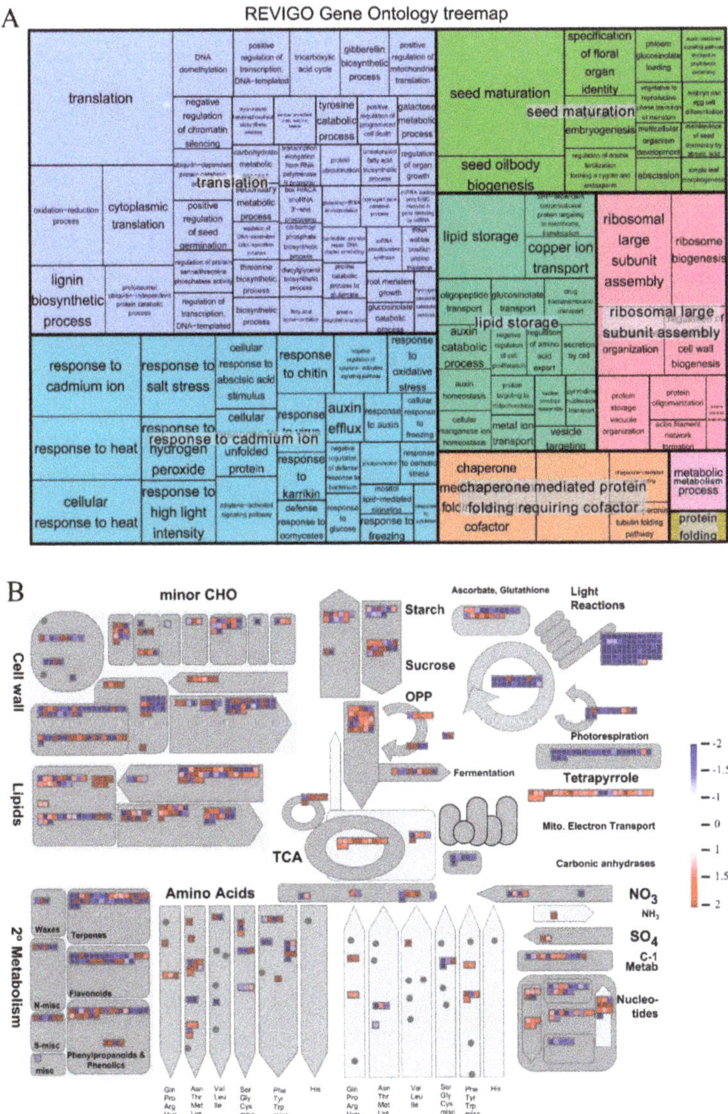

Figure 2. (**A**) REVIGO analysis for genes upregulated in *U. pumila* fruits. The representatives are classified into "superclusters" of loosely related terms, presented by different colors. The *p* value of the GO term is reflected by size of the rectangles which was calculated by TopGO. In this study, the biological processes enriched by upregulated genes were integrated into eight groups, 43 terms were summarized to the "translation" subset, 17 terms were classified to the "lipid storage" group, and 13 terms were assigned to the "seed maturation" group. (**B**) Differently expressed genes (DEGs) viewed globally, which are involved in diverse metabolic pathways. DEGs were chosen for the metabolic pathway analysis using the MapMan software (3.5.1 R2). Different colors of boxes indicate the Log$_2$ of the expression ratio of DEGs genes. In total, 3055 and 2960 homologs were assigned in *Arabidopsis* for the dysregulated unigenes, respectively. In total, 1029 pathways were mapped for these genes by MapMan, of which, 117 pathways were filtered to be enriched by the dysregulated genes with the cutoff *p*-value < 0.05 in ripe fruit.

Table 3. The nutrition-related KEGG pathways enriched by upregulated genes.

Item	Pathway	Annotated Gene Number	Enriched Gene Number
Amino acid	Glycine, serine and threonine metabolism	66	29
	Alanine, aspartate and glutamate metabolism	46	17
	Arginine biosynthesis	36	13
	Tyrosine metabolism	52	17
	Cysteine and methionine metabolism	99	29
	Cyanoamino acid metabolism	56	16
	Beta-Alanine metabolism	44	12
	Valine, leucine and isoleucine degradation	48	13
Fatty acid	Glycosphingolipid biosynthesis-globo series	10	4
	Fatty acid biosynthesis	65	25
	Selenocompound metabolism	22	7
	Biosynthesis of unsaturated fatty acids	31	9
	Glycerophospholipid metabolism	86	24
	Glycerolipid metabolism	72	19
Natural compounds	Diterpenoid biosynthesis	23	11
	Zeatin biosynthesis	18	7
	Phenylpropanoid biosynthesis	147	49
	Isoquinoline alkaloid biosynthesis	31	8
Vitamin	Biotin metabolism	19	7
	Vitamin B6 metabolism	14	5
	Ascorbate and aldarate metabolism	60	17

3.5. Real-Time Quantitative PCR Validation

To validate the RNA-Seq results in fruits of *U. pumila*, another strategy was chosen for the dysregulated unigenes. Four over- and four underregulated unigenes were selected for verification by real-time quantitative PCR (qRT-PCR) with the identical RNA samples that have been used for RNA-Seq. Primers were designed to span exon–exon junctions (Table S5). The expression trends of most test genes were similar between the RNA-Seq and qRT-PCR methods (Figure 3). For example, the homolog of SAD SSI2, Cluster-6074.5104, which was detected by RNA-Seq as an overexpressed unigene in the ripe fruit samples (L_2fc = 2.660), was also detected as significantly upregulated by qRT-PCR (Figure 3).

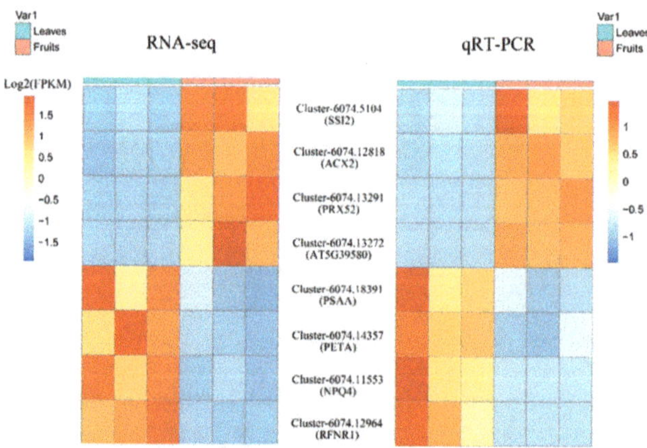

Figure 3. Validation of RNA-Seq results by qRT-PCR for eight *U. pumila* dysregulated genes. The expression of selected DEGs in fruits and leaves are shown. A red color indicates high expression for the gene in the fruit samples. Log_2(FPKM) means Log_2() value of FPKM for unigenes.

3.6. qRT-PCR Analysis for Phytonutrient-Associated Genes in Different Stages of Fruit Development

Information on the major gene expression variations that occur during fruit development has been predicted for edible ripe fruit (fruit stage 3); however, transcriptome analysis is limited to only the matured stage. To provide more information about other fruit developmental stages, the qRT-PCR analysis for key genes in two phytonutrient-associated pathways were conducted in samples of young fruits (fruit stage 1–2), ripe fruit (fruit stage 3), old fruit (fruit stage 4), and leaves. In total, eight unigenes of lignin/Coniferin/Syringin metabolism in the "phenylpropanoid biosynthesis" (ko00940) pathway and 13 genes of the "alpha-Linolenic acid metabolism" (ko00592) pathway were selected for qRT-PCR analysis (Table S5).

Most of the unigenes upregulated in the "phenylpropanoid biosynthesis" pathway screened by RNA_seq were discovered to be overexpressed in the ripe fruit stage and all other fruit development stages by qRT-PCR (Figure 4A,B), including O-methyltransferase 1 (COMT) Cluster-6074.22341 (L_2fc = 2.515 in fruit stage 3/L_2fc = 1.837 in four fruit stages), cinnamoyl coa reductase 1 (CCR) Cluster-6074.9370 (L_2fc = 2.412/L_2fc = 1.308), GroES-like zinc-binding alcohol dehydrogenase family protein (CAD) Cluster-6074.16801 (L_2fc = 2.182/L_2fc = 1.013) and peroxidase superfamily protein (peroxidase) Cluster-6074.12541 (L_2fc = 15.679/L_2fc = 19.146). In total, nine of 13 unigenes in "alpha-Linolenic acid metabolism" pathway were uncovered as overexpressed in the ripe fruit and all other fruit stages compared with leaf (Figure S2A,B), including secretory phospholipase A2 (TGL4) Cluster-6074.14136 (L_2fc = 1.705/L_2fc = 1.481), lipoxygenase (LOX2S) Cluster-6074.3481 (L_2fc = 3.304/L_2fc = 3.319), acetyl-CoA acyltransferase (ACAA1) Cluster-6074.12598 (L_2fc = 7.907/L_2fc = 9.560), and enoyl-CoA hydratase/3-hydroxyacyl-CoA dehydrogenase (MFP2) Cluster-6074.13989 (L_2fc = 4.702/L_2fc = 4.534).

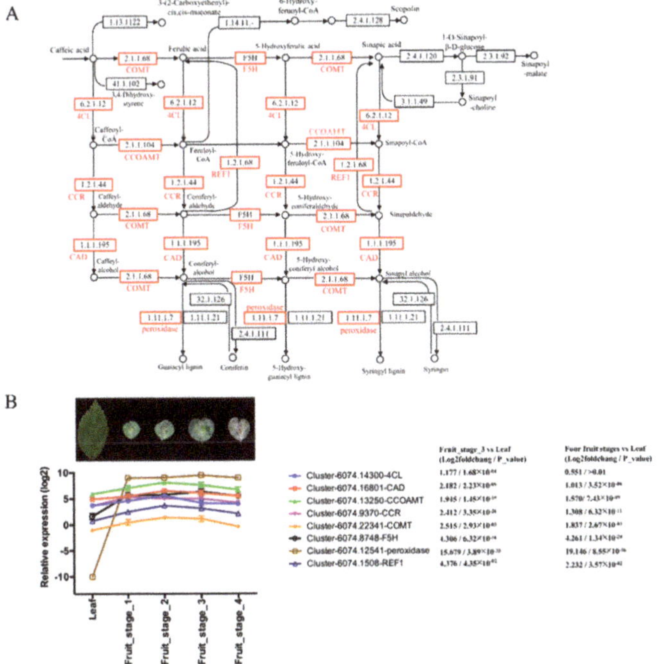

Figure 4. (**A**) The upregulated unigenes were mapped to lignin/Coniferin/Syringin metabolism in "phenylpropanoid biosynthesis" (ko00940) based on the KEGG database. The significantly upregulated genes in ko00940 are labeled with a red border. (**B**) Gene expression patterns of eight key genes were obtained by qRT-PCR analysis in young fruits (fruit stage 1–2), ripe fruit (fruit stage 3), old fruit (fruit stage 4), and leaves. The Log$_2$() fold change and *p*-value were listed for dysregulated unigenes in fruit stage 3 and four fruit stages compared with that of leaf.

4. Discussion

Plant fruits are an important source of naturally active phytonutrients which reduce the risk of cardiovascular disease, metabolic syndrome, and cancer [3,37–40]. *U. pumila* is used as traditional medicine and its fruit is considered a natural plant food in China rich in nutritional substances (proteins, dietary fiber, vitamins, and minerals) [15,41–43]. Although chemical, medical, and physiological evidence has been published for elm roots, leaves, and fruits, the molecular and genetic mechanism of its phytonutrient-associated metabolic processes remain unknown. Therefore, we attempted to uncover the key genes and pathways of elm fruit. RNA-Seq has been used to accurately monitor gene expression in hundreds of fruiting plants, such as sweet orange (*Citrus sinensis*) [44], mango (*Mangifera indica*) [28], *Cucumis melo* [45], and *Idesia polycarpa* [46], to track gene expression trends during fruit development. The synchronous dysregulation of key regulators in pathways verified by RNA-Seq and/or qRT-PCR analyses is valuable for understanding transcriptomic dynamics during elm fruit development. In this study, 11,386 unigenes were discovered to be differentially expressed in *U. pumila* fruits and leaves by RNA-seq, including 5231 up- and 6155 downregulated genes. Our transcriptome analysis contributes to understanding the molecular mechanisms for seed development and biologically active substances in *U. pumila* fruits.

Plant fruits develop primarily from the ovary following fertilization and seed development [47,48]. In this study, the processes "seed maturation" (GO:0010431), "embryo development ending in seed dormancy" (GO:0009793), and "seed oilbody biogenesis" (GO:0010344) were enriched by 50, 184, and six upregulated genes, respectively.

Several plant hormones are reported to control the regulation of fruit development and ripening, such as the gaseous hormone ethylene, abscisic acid (ABA), and gibberellin [49–51]. In this study, 59 upregulated genes were annotated in the "abscisic acid-activated signaling pathway" (GO:0009738), including *Arabidopsis* seed storage protein CRA1 (Cluster-6074.12833, $L_2fc = 19.335$) and seed maturation-associated ABA-responsive element binding protein AREB3 (Cluster-6074.15351, $L_2fc = 1.998$); 43 upregulated genes were annotated in "ethylene-activated signaling pathway" (GO:0009873), such as fatty acid accumulation during seed maturation related transcription factor WRI1 (Cluster-6074.11856, $L_2fc = 10.875$). ABA was found to promote the sensitivity to ethylene, enhancing the initiation and progression of ethylene-mediated fruit ripening processes [52]. In this study, ABA-mediated glucose response-related ethylene-activated signaling pathway member ABI4 (Cluster-6074.4471, $L_2fc = 12.728$), which participates in seed development in *Arabidopsis*, was significantly upregulated in elm fruit samples. Our transcriptome data provides evidence for the molecular mechanisms of elm fruit development.

Elm fruit contains significant amounts of vitamin B-complex, mainly B_1 (0.08–0.10 mg), B_2 (0.06–0.12 mg), and B_6 (1.20–1.50 mg) in 100 g FW [15–17]. In this study, five vitamin B_6 pathway-related unigenes were significantly upregulated in elm fruit compared with leaves, including pyridoxal phosphate synthase PDX1.2 (Cluster-6074.9816, $L_2fc = 2.399$), threonine synthase MTO2 (Cluster-6074.10873, $L_2fc = 2.246$), and pyridoxal reductase PLR1 (Cluster-6074.12916, $L_2fc = 1.348$). Vitamin B_6 with coenzymes performs various functions in the body. It is involved in amino acid/carbohydrate/lipid metabolism, cognitive development, gluconeogenesis and glycogenolysis, immune function, and hemoglobin formation [53–55]. In this study, 17 unigenes of the "Ascorbate and aldarate metabolism" (ko00053) pathway were upregulated in the fruit samples, such as GDP-l-galactose phosphorylase VTC2 (Cluster-6074.14130, $L_2fc = 1.392$) and L-Galactono-1,4-lactone dehydrogenase GLDH (Cluster-6074.25435, $L_2fc = 5.210$), which catalyzes the final step of ascorbate biosynthesis.

The lipid content of the elm fruit is restricted to the seeds and the defined fatty acids include 40–45% saturated fatty acids and 55–60% unsaturated fatty acids. In this study, nine unsaturated fatty acids pathway-associated unigenes were upregulated in elm fruit, including long chain fatty acid biosynthesis-related acyl-CoA oxidase ACX2 (Cluster-6074.12818, $L_2fc = 2.499$) and ACX4 (Cluster-6074.16191, $L_2fc = 1.727$), glyoxysomal fatty acid beta-oxidation pathway member stearoyl-acyl-carrier-protein desaturase (Cluster-6074.11259, $L_2fc = 3.019$), and peroxisomal

3-ketoacyl-CoA thiolase 3 (Cluster-6074.12598, L_2fc = 2.285). In plants, soluble stearoyl-acyl carrier protein (ACP) desaturase (SAD) was suggested to catalyze the conversion of stearoyl-ACP to oleoyl-ACP [56]. The ratio of saturated and unsaturated fatty acids is significantly influenced by the activity of SAD, and SAD is treated as a major determinant of fatty acid composition. The function of SAD in desirable fatty acid compositions has been identified in cacao (*Theobroma cacao*) [56], *Arabidopsis* [57], and peanut (*Arachis hypogaea*) [58]. In this study, the homolog of SAD SSI2 (Cluster-6074.5104, L_2fc = 2.660) was significantly upregulated in elm fruit, which indicated its function in determination of fatty acid composition.

Alpha-linolenic acid is an essential omega-3 fatty acid which is necessary for regular human growth and development. Alpha-linolenic acid protects against heart attacks, lowers blood pressure, reduces cholesterol, and reverses atherosclerosis. It is also used to treat multiple sclerosis, diabetes, ulcerative colitis, renal disease, and Crohn's disease [59]. In this study, proteins (secretory phospholipase A2, lipoxygenase) participating in metabolism of compounds alpha-Linolenic acid and (9Z,11E,15Z)-(13S)-Hydroperoxyoctadeca-9,11,15-trienoate and their coding genes were upregulated in the young and ripe elm fruits (Figure S2A,B). Downstream of the "alpha-Linolenic acid metabolism" pathway, genes for key enzymes (12-oxophytodienoic acid reductase, OPC-8:0 CoA ligase 1, acyl-CoA oxidase, enoyl-CoA hydratase and acetyl-CoA acyltransferase) which initiate jasmonic acid biosynthesis by shortening the beta-oxidative chain of its precursors were found to be synchronously overexpressed in all fruit stages (Figure S2A,B). Jasmonic acid (JA) and its derivatives are the best-studied signaling molecules derived from fatty acids. JA biosynthesis utilizes alpha-linolenic acid as a fatty acid substrate, which is released from the galactolipids of the chloroplast [60]. On the basis of gene regulation of alpha-linolenic acid and JA biosynthesis-related genes, unsaturated fatty acid biosynthesis and JA biosynthesis pathways were deduced to be involved in fruit development in *U. pumila*.

Phenylpropanoids (PPs) make up the largest portion of secondary metabolites produced by plants. As naturally occurring antioxidants, they have function in human tumors, inflammation, and cellular damage [61,62]. In this study, the "Phenylpropanoid biosynthesis" pathway was significantly enriched with 49 upregulated unigenes including many enzymes of this pathway, such as 4-coumarate-CoA ligase (Cluster-6074.14300, L_2fc = 1.077) and cinnamyl-alcohol dehydrogenase (Cluster-6074.20308, L_2fc = 4.604) (Figure S1B). Many phenolic compounds (flavonoids, coumarines, isoflavonoids, and lignans) derived from plants are secondary products of PP metabolism. Flavonoids have a favorable effect on cardiovascular disease, including anti-inflammatory functions [63,64], and coumarins derivatives have shown anticancer activity in various cancer cell lines. In this study, flavonoid and coumarins biosynthesis-associated oumarate 3-hydroxylase CYTOCHROME P450 (Cluster-6074.16505, L_2fc = 1.952), caffeoyl coenzyme A O-methyltransferase 1 (Cluster-6074.13250, L_2fc = 2.343), and lignin biosynthesis involved peroxidase 52 (Cluster-6074.13291, L_2fc = 9.155) were upregulated in the elm fruits. Our transcriptome data indicated the PPs might be associated with the bioactivity of elm fruits in human cancer and inflammation resistance.

Eleutheroside B (syringin) is a phenylpropanoid glycoside first isolated from *Acanthopanax senticosus* and has neuroprotective, tonic, adaptogenic, and immune-modulating properties. Syringin has a potent protection against LPS/D-GalN-induced fulminant hepatic failure, and it activates Nrf2 and inhibits the NF-κB signaling pathway to attenuate LPS-induced acute lung injury [62,63,65]. In this study, expression of genes encoding key enzymes (4-coumarate-CoA ligase, cinnamoyl-CoA reductase, and cinnamyl-alcohol dehydrogenase) which transform sinapic acid to intermediates of syringin (sinapoyl-CoA, sinapoyl aldehyde, and sinapyl alcohol) showed an increase from fruit stage 1 (young fruit) to fruit stage 3 (ripe fruit) and decreased in the old fruit stage, indicating syringin accumulation in the early fruit development stages.

5. Conclusions

Transcriptome analysis was performed for ripe *U. pumila* fruit. In total, 5231 unigenes were filtered as upregulated in edible fruit samples and 6155 were downregulated in *U. pumila*. Hundreds of pathways were predicted to participate in seed development and phytonutrients biosynthesis in *U. pumila* by GO, MapMan, and KEGG enrichment analysis. The pathways "seed maturation", "glycine, serine, and threonine metabolism", and "phenylpropanoid biosynthesis" were found to be highly expressed in fruits. The ABA-mediated glucose response-related ethylene-activated signaling pathway (e.g., ABI4) was associated with elm fruit development; the unsaturated fatty acids pathway (e.g., ACX2 and SAD) was predicted to participate in determinant of fatty acid composition in elm fruit; flavonoid and coumarins biosynthesis (e.g., CYP98A3 and CCoAOMT1) were found to be related to the bioactivity of elm fruit in human cancer and inflammation resistance. To provide more information about other fruit developmental stages, qRT-PCR analysis for key genes of "phenylpropanoid biosynthesis" and "alpha-Linolenic acid metabolism" were conducted in samples of young fruits, ripe fruit, old fruit, and leaves. Two biosynthetic pathways for unsaturated fatty acid and JA were involved in fruit development in *U. pumila* and syringin is speculated to accumulate in the early development stages of elm fruit.

Data Archiving Statement

All genetic data have been submitted to the NCBI Sequence Read Archive (SRA) database (https://submit.ncbi.nlm.nih.gov/subs/sra), PRJNA545392 for *U. pumila*.

Supplementary Materials: The following are available online at http://www.mdpi.com/1999-4907/10/9/738/s1, Table S1. Information of unigenes dysregulated in *elm (Ulmus pumila)*. Table S2. GO enrichment analysis of genes differentially expressed in *U. pumila* fruits. Table S3. MapMan enrichment analysis of genes differentially expressed in *U. pumila* fruits. Table S4. KEGG enrichment analysis of genes differentially expressed in *U. pumila* fruits. Table S5. qRT -PCR primers. Figure S1. (A) Scatterplot of top 20 KEGG pathways enriched by upregulated unigenes. Rich factor represents the ratio of the number of differentially expressed genes (DEG)s and the number of all genes in the pathway. The KEGG pathway "Ribosome" (ko03010) was enriched by 127 upregulated unigenes, "Glycine, serine and threonine metabolism" (ko00052) were annotated by 29 over-expressed genes and 49 upregulated genes were annotated in the KEGG pathway "Phenylpropanoid biosynthesis" (ko00940). (B) "Phenylpropanoid biosynthesis" (ko00940) pathway enriched by the upregulated unigenes based on the KEGG database. The significantly upregulated genes in ko00940 were labeled with red bracket. Figure S2. (A) The upregulated unigenes were mapped to "alpha-Linolenic acid metabolism" pathways (ko00592) based on the KEGG database. The significantly upregulated genes in ko00592 were labeled with red bracket. (B) Gene expression patterns of nine upregulated key genes obtained by qRT-PCR analysis in young fruits (fruit stage 1–2), ripe fruit (fruit stage 3), old fruit (fruit stage 4), and leaves. The $\mathrm{Log_2}()$ fold change and p-value were listed for dysregulated unigenes in fruit stage 3 and four fruit stages compared with that of leaf.

Author Contributions: Conceptualization, L.Z., X.Z., M.L. and S.F.; Data curation: L.Z., X.Z., M.L., N.W. and X.Q.; Funding acquisition, L.Z. and S.F.; Investigation, L.Z., M.L., N.W. and X.Q.; Methodology, L.Z.; Software, X.Z.; Supervision, S.F.; Writing—Original draft, L.Z., X.Z. and S.F.; Writing—Review & editing, S.F.

Funding: This work was supported by Natural Science Foundation of China (31800185, 31470298) and the Science and Technology Development Foundation of Shandong Province (2018LZGC038). A Project of Shandong Province Higher Educational Science and Technology Program (J18KA147).

Acknowledgments: This work was supported by Natural Science Foundation of China (31800185, 31470298) and the Science and Technology Development Foundation of Shandong Province (2018LZGC038). A Project of Shandong Province Higher Educational Science and Technology Program (J18KA147).

Conflicts of Interest: The authors declare no conflict of interest.

References

1. Kris-Etherton, P.M.; Hecker, K.D.; Bonanome, A.; Coval, S.M.; Binkoski, A.E.; Hilpert, K.F.; Griel, A.E.; Etherton, T.D. Bioactive compounds in foods: Their role in the prevention of cardiovascular disease and cancer. *Am. J. Med.* **2002**, *113*, 71–88. [CrossRef]

2. Lampe, J.W. Health effects of vegetables and fruit: Assessing mechanisms of action in human experimental studies. *Am. J. Clin. Nutr.* **1999**, *70*, 475s–490s. [CrossRef]

3. Neto, C.C. Cranberry and blueberry: Evidence for protective effects against cancer and vascular diseases. *Mol. Nutr. Food Res.* **2010**, *51*, 652–664. [CrossRef]

4. Fernie, A.R.; Schauer, N. Metabolomics-assisted breeding: A viable option for crop improvement? *Trends Genet.* **2009**, *25*, 39–48. [CrossRef] [PubMed]

5. Wang, Q.; Chen, Y.; Wang, X.; Gong, G.; Li, G.; Li, C. Consumption of fruit, but not vegetables, may reduce risk of gastric cancer: Results from a meta-analysis of cohort studies. *Eur. J. Cancer* **2014**, *50*, 1498–1509. [CrossRef]

6. Everitt, A.V.; Hilmer, S.N.; Brandmiller, J.C.; Jamieson, H.A.; Truswell, A.S.; Sharma, A.P.; Mason, R.S.; Morris, B.J.; Le, C.D. Dietary approaches that delay age-related diseases. *Clin. Interv. Aging* **2006**, *1*, 11–31. [CrossRef] [PubMed]

7. Bendinelli, B.; Masala, G.; Saieva, C.; Salvini, S.; Calonico, C.; Sacerdote, C.; Agnoli, C.; Grioni, S.; Frasca, G.; Mattiello, A. Fruit, vegetables, and olive oil and risk of coronary heart disease in Italian women: The EPICOR Study. *Am. J. Clin. Nutr.* **2011**, *94*, 287–288. [CrossRef]

8. Wiegrefe, S.J.; Sytsma, K.J.; Guries, R.P. Phylogeny of elms (Ulmus, Ulmaceae): Molecular evidence for a sectional classification. *Syst. Bot.* **1994**, *19*, 590–612. [CrossRef]

9. Ghosh, C.; Yang, S.H.; Hwang, S.G. Methanol extract of *Ulmus pumila*. L exerts potent anti-inflammatory effects in murine macrophages and mouse skin. *FASEB J.* **2013**, *27*, 1093.

10. Wang, D.; Xia, M.Y.; Cui, Z. New triterpenoids isolated from the root bark of *Ulmus pumila* L. *Chem. Pharm. Bull.* **2006**, *54*, 775–778. [CrossRef]

11. Feng, Z.T.; Deng, Y.Q.; Fan, H.; Sun, Q.J.; Sui, N.; Wang, B.S. Effects of NaCl stress on the growth and photosynthetic characteristics of *Ulmus pumila* L. seedlings in sand culture. *Photosynthetica* **2014**, *52*, 313–320. [CrossRef]

12. Mu, D.Y.; Zwiazek, J.J.; Li, Z.Q.; Zhang, W.Q. Genotypic variation in salt tolerance of *Ulmus pumila* plants obtained by shoot micropropagation. *Acta Physiol. Plant* **2016**, *38*, 188. [CrossRef]

13. Zhu, J.F.; Yang, X.Y.; Liu, Z.X.; Zhang, H.X. Identification and Target Prediction of MicroRNAs in *Ulmus pumila* L. Seedling Roots under Salt Stress by High-Throughput Sequencing. *Forests* **2016**, *7*, 318. [CrossRef]

14. Zhou, Z.H.; Shao, H.J.; Han, X.; Wang, K.J.; Gong, C.P.; Yang, X.B. The extraction efficiency enhancement of polyphenols from *Ulmus pumila* L. barks by trienzyme-assisted extraction. *Ind. Crop Prod.* **2017**, *97*, 401–408. [CrossRef]

15. Yu, S.L. Nutrition and health effects of elm fruits. *Food Nutr. China* **2009**, *9*, 60–62.

16. Li, Y.; Wang, Y.; Xue, H.; Pritchard, H.W.; Wang, X.F. Changes in the mitochondrial protein profile due to ROS eruption during ageing of elm (*Ulmus pumila* L.) seeds. *Plant Physiol. Biochem.* **2017**, *114*, 72–87. [CrossRef] [PubMed]

17. Wang, Y.; Li, Y.; Xue, H.; Pritchard, H.W.; Wang, X.F. Reactive oxygen species-provoked mitochondria-dependent cell death during ageing of elm (*Ulmus pumila* L.) seeds. *Plant J.* **2015**, *81*, 438–452. [CrossRef] [PubMed]

18. Yuan, F.; Lyu, M.J.A.; Leng, B.Y.; Zhu, X.G.; Wang, B.S. The transcriptome of NaCl-treated *Limonium bicolor* leaves reveals the genes controlling salt secretion of salt gland. *Plant Mol. Biol.* **2016**, *91*, 241–256. [CrossRef]

19. Zhang, H.; Zhang, Q.; Zhai, H.; Li, Y.; Wang, X.; Liu, Q.; He, S. Transcript profile analysis reveals important roles of jasmonic acid signalling pathway in the response of sweet potato to salt stress. *Sci. Rep. UK* **2017**, *7*, 40819. [CrossRef]

20. Yuan, F.; Lyu, M.J.A.; Leng, B.Y.; Zheng, G.Y.; Feng, Z.T.; Li, P.H.; Zhu, X.G.; Wang, B.S. Comparative transcriptome analysis of developmental stages of the *Limonium bicolor* leaf generates insights into salt gland differentiation. *Plant Cell Environ.* **2015**, *38*, 1637–1657. [CrossRef]

21. Xu, J.J.; Li, Y.Y.; Ma, X.L.; Ding, J.F.; Wang, K.; Wang, S.S.; Tian, Y.; Zhang, H.; Zhu, X.G. Whole transcriptome analysis using next-generation sequencing of model species *Setaria viridis* to support C-4 photosynthesis research. *Plant Mol. Biol.* **2013**, *83*, 77–87. [CrossRef] [PubMed]

22. Lin, J.; Li, J.P.; Yuan, F.; Yang, Z.; Wang, B.S.; Chen, M. Transcriptome profiling of genes involved in photosynthesis in *Elaeagnus angustifolia* L. under salt stress. *Photosynthetica* **2018**, *56*, 998–1009. [CrossRef]

23. Yang, S.; Li, L.; Zhang, J.L.; Geng, Y.; Guo, F.; Wang, J.G.; Meng, J.J.; Sui, N.; Wan, S.B.; Li, X.G. Transcriptome and Differential Expression Profiling Analysis of the Mechanism of Ca^{2+} Regulation in Peanut (*Arachis hypogaea*) Pod Development. *Front. Plant Sci.* **2017**, *8*, 1609. [CrossRef] [PubMed]

24. Yang, Z.; Wang, Y.; Wei, X.C.; Zhao, X.; Wang, B.S.; Sui, N. Transcription Profiles of Genes Related to Hormonal Regulations Under Salt Stress in Sweet Sorghum. *Plant Mol. Biol. Rep.* **2017**, *35*, 586–599. [CrossRef]

25. Du, M.F.; Ding, G.J.; Cai, Q.O. The Transcriptomic Responses of *Pinus massoniana* to Drought Stress. *Forests* **2018**, *9*, 326. [CrossRef]

26. Cai, Q.F.; Li, B.; Lin, F.R.; Huang, P.; Guo, W.Y.; Zheng, Y.Q. De Novo Sequencing and Assembly Analysis of Transcriptome in *Pinus bungeana* Zucc. ex Endl. *Forests* **2018**, *9*, 156. [CrossRef]

27. Zhao, D.Q.; Zhang, X.Y.; Fang, Z.W.; Wu, Y.Q.; Tao, J. Physiological and Transcriptomic Analysis of Tree Peony (Paeonia section Moutan DC.) in Response to Drought Stress. *Forests* **2019**, *10*, 135. [CrossRef]

28. Wu, H.X.; Jia, H.M.; Ma, X.W.; Wang, S.B.; Yao, Q.S.; Xu, W.T.; Zhou, Y.G.; Gao, Z.S.; Zhan, R.L. Transcriptome and proteomic analysis of mango (*Mangifera indica* Linn) fruits. *J. Proteom.* **2014**, *105*, 19–30. [CrossRef]

29. Munoz-Espinoza, C.; Di Genova, A.; Correa, J.; Silva, R.; Maass, A.; Gonzalez-Aguero, M.; Orellana, A.; Hinrichsen, P. Transcriptome profiling of grapevine seedless segregants during berry development reveals candidate genes associated with berry weight. *BMC Plant Biol.* **2016**, *16*, 104. [CrossRef]

30. Sweetman, C.; Wong, D.C.J.; Ford, C.M.; Drew, D.P. Transcriptome analysis at four developmental stages of grape berry (*Vitis vinifera* cv. Shiraz) provides insights into regulated and coordinated gene expression. *BMC Genom.* **2012**, *13*, 691. [CrossRef]

31. Grabherr, M.G.; Haas, B.J.; Moran, Y.; Levin, J.Z.; Thompson, D.A.; Ido, A.; Xian, A.; Lin, F.; Raktima, R.; Qiandong, Z. Full-length transcriptome assembly from RNA-Seq data without a reference genome. *Nat. Biotechnol.* **2011**, *29*, 644. [CrossRef] [PubMed]

32. NCBI Sequence Read Archive (SRA) Database. Available online: https://www.ncbi.nlm.nih.gov/sra (accessed on 27 August 2019).

33. Bo, L.; Dewey, C.N. RSEM: Accurate transcript quantification from RNA-Seq data with or without a reference genome. *BMC Bioinform.* **2011**, *12*, 323.

34. Haas, B.J.; Alexie, P.; Moran, Y.; Manfred, G.; Blood, P.D.; Joshua, B.; Matthew Brian, C.; David, E.; Bo, L.; Matthias, L. De novo transcript sequence reconstruction from RNA-seq using the Trinity platform for reference generation and analysis. *Nat. Protoc.* **2013**, *8*, 1494–1512. [CrossRef] [PubMed]

35. Thimm, O.; Blasing, O.; Gibon, Y.; Nagel, A.; Meyer, S.; Kruger, P.; Selbig, J.; Muller, L.A.; Rhee, S.Y.; Stitt, M. MAPMAN: A user-driven tool to display genomics data sets onto diagrams of metabolic pathways and other biological processes. *Plant J.* **2010**, *37*, 914–939. [CrossRef]

36. Chen, X.; Xizeng, M.; Jiaju, H.; Yang, D.; Jianmin, W.; Shan, D.; Lei, K.; Ge, G.; Chuan-Yun, L.; Liping, W. KOBAS 2.0: A web server for annotation and identification of enriched pathways and diseases. *Nucleic Acids Res.* **2011**, *39*, 316–322.

37. Latocha, P. The Nutritional and Health Benefits of Kiwiberry (*Actinidia arguta*)—A Review. *Plant Food Hum. Nutr.* **2017**, *72*, 325–334. [CrossRef]

38. Aune, D.; Giovannucci, E.; Boffetta, P.; Fadnes, L.T.; Keum, N.; Norat, T.; Greenwood, D.C.; Riboli, E.; Vatten, L.J.; Tonstad, S. Fruit and vegetable intake and the risk of cardiovascular disease, total cancer and all-cause mortality-a systematic review and dose-response meta-analysis of prospective studies. *Int. J. Epidemiol.* **2017**, *46*, 1029–1056. [CrossRef]

39. Block, G.; Patterson, B.; Subar, A. Fruit, vegetables, and cancer prevention: A review of the epidemiological evidence. *Nutr. Cancer* **1992**, *18*, 1–29. [CrossRef]

40. Steinmetz, K.A.; Potter, J.D. Vegetables, fruit, and cancer prevention: A review. *J. Am. Diet Assoc.* **1996**, *96*, 1027–1039. [CrossRef]

41. Dukic, M.; Dunisijevic-Bojovic, D.; Samuilov, S. The Influence of Cadmium and Lead on *Ulmus Pumila* L. Seed Germination and Early Seedling Growth. *Arch. Biol. Sci.* **2014**, *66*, 253–259. [CrossRef]

42. Ghosh, C.; Chung, H.Y.; Nandre, R.M.; Lee, J.H.; Jeond, T.I.; Kim, I.S.; Yang, S.H.; Hwang, S.G. An active extract of *Ulmus pumila* inhibits adipogenesis through regulation of cell cycle progression in 3T3-L1 cells. *Food Chem. Toxicol.* **2012**, *50*, 2009–2015. [CrossRef] [PubMed]

43. Qin, J.; Xi, W.M.; Rahmlow, A.; Kong, H.Y.; Zhang, Z.; Shangguan, Z.P. Effects of forest plantation types on leaf traits of *Ulmus pumila* and *Robinia pseudoacacia* on the Loess Plateau, China. *Ecol. Eng.* **2016**, *97*, 416–425. [CrossRef]

44. Yu, K.; Xu, Q.; Da, X.; Guo, F.; Ding, Y.; Deng, X. Transcriptome changes during fruit development and ripening of sweet orange (*Citrus sinensis*). *BMC Genom.* **2012**, *13*, 10. [CrossRef] [PubMed]

45. Zhang, H.; Wang, H.; Yi, H.; Zhai, W.; Wang, G.; Fu, Q. Transcriptome profiling of *Cucumis melo* fruit development and ripening. *Hortic. Res.* **2016**, *3*, 16014. [CrossRef] [PubMed]

46. Li, R.J.; Gao, X.; Li, L.M.; Liu, X.L.; Wang, Z.Y.; Lü, S.Y. *De novo* assembly and characterization of the fruit transcriptome of *Idesia polycarpa* reveals candidate genes for lipid biosynthesis. *Front. Plant Sci.* **2016**, *7*, 801. [CrossRef] [PubMed]

47. Barbosa, J.; Teixeira, P. Development of probiotic fruit juice powders by spray-drying: A review. *Food Rev. Int.* **2017**, *33*, 335–358. [CrossRef]

48. Seymour, G.B.; Ostergaard, L.; Chapman, N.H.; Knapp, S.; Martin, C. Fruit development and ripening. *Annu. Rev. Plant Biol.* **2013**, *64*, 219–241. [CrossRef]

49. Kumar, R.; Khurana, A.; Sharma, A.K. Role of plant hormones and their interplay in development and ripening of fleshy fruits. *J. Exp. Bot.* **2014**, *65*, 4561–4575. [CrossRef]

50. Zhu, Y.M.; Zheng, P.; Varanasi, V.; Shin, S.B.; Main, D.; Curry, E.; Mattheis, J.P. Multiple plant hormones and cell wall metabolism regulate apple fruit maturation patterns and texture attributes. *Tree Genet. Genomes* **2012**, *8*, 1389–1406. [CrossRef]

51. Xue, S.; Dong, M.; Liu, X.; Xu, S.; Pang, J.; Zhang, W.; Weng, Y.; Ren, H. Classification of fruit trichomes in cucumber and effects of plant hormones on type II fruit trichome development. *Planta* **2019**, *249*, 407–416. [CrossRef]

52. Jiang, Y.; Joyce, D.C.; Macnish, A.J. Effect of Abscisic Acid on Banana Fruit Ripening in Relation to the Role of Ethylene. *J. Plant Growth Regul.* **2000**, *19*, 106–111. [CrossRef] [PubMed]

53. Balk, E.M.; Raman, G.; Tatsioni, A.; Chung, M.; Lau, J.; Rosenberg, I.H. Vitamin B6, B12, and folic acid supplementation and cognitive function: A systematic review of randomized trials. *Arch. Intern. Med.* **2007**, *167*, 21–30. [CrossRef] [PubMed]

54. Mathers, J.C. Plant foods for human health: Research challenges. *Proc. Nutr. Soc.* **2006**, *65*, 198–203. [CrossRef] [PubMed]

55. Luthje, S.; Deswal, R.; Agrawal, G.K. Plant-based Foods: Seed, Nutrition and Human Health. *Proteomics* **2015**, *15*, 1638. [CrossRef] [PubMed]

56. Zhang, Y.; Maximova, S.N.; Guiltinan, M.J. Characterization of a stearoyl-acyl carrier protein desaturase gene family from chocolate tree, *Theobroma cacao* L. *Front. Plant Sci.* **2015**, *6*, 239. [CrossRef] [PubMed]

57. Bryant, F.M.; Munoz-Azcarate, O.; Kelly, A.A.; Beaudoin, F.; Kurup, S.; Eastmond, P.J. ACYL-ACYL CARRIER PROTEIN DESATURASE2 and 3 Are Responsible for Making Omega-7 Fatty Acids in the Arabidopsis Aleurone. *Plant Physiol.* **2016**, *172*, 154–162. [CrossRef]

58. Chi, X.; Yang, Q.; Pan, L.; Chen, M.; He, Y.; Yang, Z.; Yu, S. Isolation and characterization of fatty acid desaturase genes from peanut (*Arachis hypogaea* L.). *Plant Cell Rep.* **2011**, *30*, 1393–1404. [CrossRef]

59. Gerber, M. Omega-3 fatty acids and cancers: A systematic update review of epidemiological studies. *Br. J. Nutr.* **2012**, *107*, S228–S239. [CrossRef]

60. Weber, H. Fatty acid-derived signals in plants. *Trends Plant Sci.* **2002**, *7*, 217–224. [CrossRef]

61. Hemaiswarya, S.; Doble, M. Combination of phenylpropanoids with 5-fluorouracil as anti-cancer agents against human cervical cancer (HeLa) cell line. *Phytomedicine* **2013**, *20*, 151–158. [CrossRef]

62. Rochfort, S.; Parker, A.J.; Dunshea, F.R. Plant bioactives for ruminant health and productivity. *Phytochemistry* **2008**, *69*, 299–322. [CrossRef] [PubMed]

63. Le Marchand, L. Cancer preventive effects of flavonoids—A review. *Biomed. Pharmacother.* **2002**, *56*, 296–301. [CrossRef]

64. Romagnolo, D.F.; Selmin, O.I. Flavonoids and cancer prevention: A review of the evidence. *J. Nutr. Gerontol. Geriatr.* **2012**, *31*, 206–238. [CrossRef] [PubMed]

65. Krishnan, S.S.C.; Subramanian, I.P.; Subramanian, S.P. Isolation, characterization of syringin, phenylpropanoid glycoside from, *Musa paradisiaca*, tepal extract and evaluation of its antidiabetic effect in streptozotocin-induced diabetic rats. *Biomed. Prev. Nutr.* **2014**, *4*, 105–111. [CrossRef]

Article

A Comprehensive Assessment of Bioactive Metabolites, Antioxidant and Antiproliferative Activities of *Cyclocarya paliurus* (Batal.) Iljinskaja Leaves

Mingming Zhou [1], Pei Chen [1], Yuan Lin [1], Shengzuo Fang [1,2,*] and Xulan Shang [1,2]

[1] College of Forestry, Nanjing Forestry University, Nanjing 210037, China
[2] Co-Innovation Center for Sustainable Forestry in Southern China, Nanjing Forestry University, Nanjing 210037, China
*   Correspondence: fangsz@njfu.edu.cn or fangsz@njfu.com.cn; Tel./Fax: +86-258-542-7797

Received: 1 July 2019; Accepted: 22 July 2019; Published: 26 July 2019

Abstract: *Cyclocarya paliurus* (Batal.) Iljinskaja is an indigenous and multifunction tree species in China, but it is mainly used in pharmaceutical and nutraceutical ingredients. To make a comprehensive evaluation on its bioactive metabolites, antioxidant and antitumor potentials of *C. paliurus* leaves, the leaf samples were collected from 15 geographic locations (natural populations) throughout its distribution areas. High-performance liquid chromatography (HPLC) and colorimetric methods were used to detect the contents of bioactive metabolites. The antioxidant activity was evaluated by 2,2′-diphenyl-1-picrylhydrazyl (DPPH), 2,2-azino-bis (3-ethylbenzothiazoline-6-sulfonic acid) (ABTS) and reducing power assays. The antiproliferative activity on different cancer cell types was evaluated by 3-(4,5-dimethylthiazol-2-yl)-2,5-diphenyltetrazolium bromide (MTT) assay. Contents of bioactive metabolites, and antioxidant and antiproliferative activities in the extracts were significantly affected by solvent and population. In most cases, the contents of flavonoids and triterpenoids, and the antioxidant and antiproliferative activities in the ethanol extracts were higher than the water extracts. The best scavenging capacity of DPPH (IC_{50} = 0.34 mg/mL) and ABTS (IC_{50} = 0.50 mg/mL) radical occurred in the ethanol extracts of S15 and S7 population respectively, while the strongest reducing power (EC_{50} = 0.71 mg/mL) was achieved in the ethanol extracts of S14 population. The antiproliferation effects of *C. paliurus* extracts on cancer cells varied with different cell types. The HeLa cell was the most sensitive to *C. paliurus* extracts, and their IC_{50} values of the ethanol extracts varied from 0.13 to 0.42 mg/mL among *C. paliurus* populations. Redundancy analysis showed that total polyphenol had the greatest contribution to the antioxidant activity, but total flavonoid was mostly responsible for the antiproliferation effects. These results would provide important scientific evidences not only for developing *C. paliurus* as a potent antioxidant and antitumor reagent, but also for obtaining the higher yield of bioactive compounds in the *C. paliurus* plantation.

Keywords: *Cyclocarya paliurus*; flavonoid; phenolics; triterpenoid; solvent; natural population

---

1. Introduction

Cancer, a disease characterized by abnormal cells that divide and invade other tissues and organs uncontrollably, is one of the leading causes of unnatural death in the world [1]. To date, there are more than 100 types of cancer, but the common cancers are lung, breast, colorectal and stomach [2]. Surgery, chemotherapy and radiation are commonly used for cancer treatment along with side effects [3]. Over the past decades, natural products have attracted a great deal of attention due to their utilization of cancer prevention and treatment. Over 60% of anticancer drugs are derived from natural products [4]. Most anticancer drugs, such as taxol, camptothecin, vinblastine, vincristine and podophyllotoxin, originate from plant resources [4,5]. The bioactive metabolites of plant extracts, such as phenolics, alkaloids and terpenoids, are considered as potential anticancer reagents [4,6,7].

Oxidative damage plays an important role in the occurrence of various diseases including cancer. Reactive oxygen species (ROS) could result in oxidative damage of cell components such as lipids, protein and DNA, and thereby induce cancer cell formation, transformation and cancer-promoting signaling molecules [8]. Antioxidants can protect cells from oxidative damage though neutralizing ROS. In recent years, many studies have demonstrated that the intake of plant products that are rich in antioxidant phytochemicals can reduce the incidence of many diseases including cancer [9,10]. Therefore, it is important to seek natural antioxidants with anticancer potentials.

Cyclocarya paliurus (Batal.) Iljinskaja, an indigenous and multifunction tree species in China, is mainly distributed in highlands of subtropical areas. Its leaves have been made into nutraceutical tea beverage for a long time in folk tradition and served as a formula of traditional Chinese medicine [11]. Moreover, the leaves of *C. paliurus* have been listed as new food raw material by National Health and Family Planning Commission of China since 2013 [12]. Pharmacological studies on *C. paliurus* indicated that its extracts had a variety of biological activities such as antidiabetic, antioxidant and antibacterial activities, while less attention was paid to the antiproliferation effects of *C. paliurus* on different cancer cells. Previous studies had revealed that the abundant phytochemicals in the extracts of *C. paliurus* leaves are responsible for these bioactivities. For instance, hypolipidemic and hypoglycemic effects of *C. paliurus* were ascribed to the flavonoids and triterpenoids in its extracts respectively [13], while polysaccharides isolated from the water extracts of *C. paliurus* exhibited hypolipidemic effects and antiproliferation effects on HeLa cell through cell-cycle arrest in the S phase [14,15]. Additionally, *C. paliurus* polysaccharides showed significant scavenging capacity on 2,2′-diphenyl-1-picrylhydrazyl (DPPH) radicals at IC_{50} value of 52.3 µg/mL [16]. Total polyphenol among phenolics and polysaccharide exhibited the greatest contribution to antioxidant activity in aqueous extracts of *C. paliurus* leaves [17]. Quercetin and kaempferol glycosides of phenolic compounds were key antioxidant components in ethanol extracts of *C. paliurus* leaves [18]. However, there was less information available on the relationship between antioxidant activity and triterpenoid in *C. paliurus* extracts. Recently, there have been numerous studies on plants products in order to seek key components with antioxidant and antitumor properties. Although polysaccharides, phenolics and triterpenoids have antiproliferation effects on cancer cells and antioxidant activity in some plants [15,19], the main bioactive constituents with antioxidant and antiproliferative activities remain unclear in leaf extracts of *C. paliurus*. Polysaccharide content and antioxidant activity of water extracts of *C. paliurus* leaves collected from different populations were investigated in previous studies [14,17]. Moreover, no polysaccharide was found in the ethanol extracts [13]. In this context, it is crucial to study the bioactive components belonging to phenolics or triterpenoids for antioxidant and antitumor potentials in *C. paliurus* leaves.

Geographical origin has a pivotal impact on phytochemicals and biological activities. The contents of polysaccharide, phenolics, triterpenoid and their antioxidant, hypolipidemic and hypoglycemic effects in *C. paliurus* leaves have been reported to vary with different geographical locations [13,17,18,20]. Phytochemical contents, antioxidant and anticancer activities in plant extracts also vary with extraction solvents such as water, aqueous mixtures of ethanol, ethanol and methanol [21,22]. For example, antiproliferative activity on cancer cells was found in both ethanol and aqueous extracts of *Ficus beecheyana* and *Saururus chinensis* root, however the ethanol extracts showed better antioxidant activity [23,24]. The phenolic compounds, anticancer and antioxidant activities of indigo plants also varied with the extraction solvents [25]. In *C. paliurus* leaves, the chemical composition and antidiabetic effects in vivo have been evaluated in both aqueous and ethanol extracts from five locations [13]. However, to our best knowledge, there was no comprehensive assessment on antioxidant and antiproliferative activities of *C. paliurus* leaves considering both solvents and geographical origins.

In this framework, the contents of phenolics and triterpenoids, antioxidant and antiproliferation effects of both ethanol and water extracts were investigated in *C. paliurus* leaves collected from different geographical locations. Our aims were to detect which solvent was more efficient for antioxidant and antitumor potentials, seek for the components that were responsible for antioxidant and antiproliferative activities in *C. paliurus* leaves, and screen out the superior natural populations with targeted bioactivity. The results will not only offer a theoretical basis for obtaining the higher yield of bioactive compounds from *C. paliurus* leaves, but also provide a scientific basis for developing *C. paliurus* as a potential antioxidant and antitumor reagent.

2. Materials and Methods

2.1. Plant Material

The leaf samples of *C. paliurus* were collected from 15 geographical locations of its natural distribution areas in current study (Figure 1). At each location, the mature leaves from 6 to 30 dominant or co-dominant trees were sampled. And then the leaves collected from trees per location were mixed as a population sample. The sample collection and pre-treatment followed the method of Liu et al. [20]. Finally, the *C. paliurus* leaves were dried to constant weight at 70 °C and then grinded into powder. All samples were stored at room temperature prior to analysis.

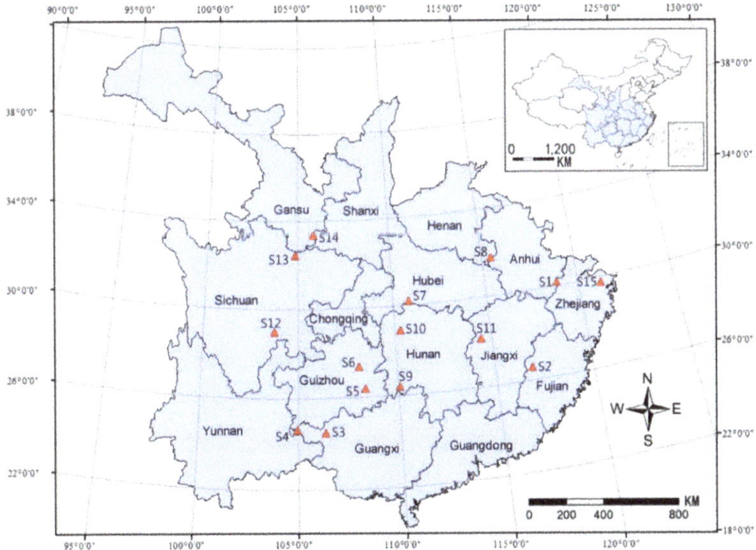

Figure 1. A map showing natural distribution across 15 province or municipalities in China and geographical locations of *C. paliurus* populations sampled in this study (red triangle). The populations codes were as follows: S1 (Jixi) from Anhui province; S2 (Mingxi) from Fujian province; S3 (Baise) and S4 (Jinzhongshan) from Guangxi province; S5 (Jianhe) and S6 (Shiqian) from Guizhou province; S7 (Wufeng) from Hubei province; S8 (Shangcheng) from Henan province; S9 (Suining) and S10 (Yongshun) from Hunan province; S11 (Xiushui) from Jiangxi province; S12 (Muchuan) and S13 (Qingchuan) from Sichuan province; S14 (Lueyang) from Shanxi province; S15 (Fenghua) from Zhejiang province.

2.2. Preparation of Extracts

Each 3.0 g leaf sample was extracted with 30 mL of 70% ethanol and water for 1 h at 90 °C, respectively. And then they were subjected to ultrasound for 45 min. The mixtures were centrifuged at 10,000 rpm for 10 min and the supernatants were retained. The obtained extracts were separated on C18 solid phase extraction column for determination. For high-performance liquid chromatography (HPLC) analysis, the extracts were filtered through a 0.22 μm syringe filter. Each sample was analyzed in triplicate.

2.3. Content Determination of Bioactive Metabolites

Total polyphenol content was determined by the Folin-Ciocalteu (FC) colorimetric method as described by Xie et al. [26], while total flavonoid content was determined using the aluminum trichloride colorimetric method as described by Li et al. [27], and total triterpenoid content was measured by the colorimetric method as described by Fan and He [28]. HPLC analysis was conducted to detect the phenolic acid, flavonoid and triterpenoid content by following the method of Cao et al. [11]. The chromatograms of the representative sample and the 16 mixed standards were shown in Supplementary Figure S1.

2.4. Antioxidant Assay

Six concentration gradients of ethanol extracts (1 mL, 0.2–1.2 mg/mL) and water extracts (1 mL, 0.4–2.4 mg/mL) were designed by equal interval to detect the antioxidant capacities. Sample solvent (70% ethanol or distilled water) was used as the control. The scavenging capacity of DPPH and 2,2-azino-bis (3-ethylbenzothiazoline-6-sulfonic acid) (ABTS)radicals, and the reducing power were determined and analyzed by the methods described by Zhou et al. [17]. The percent inhibition of radical was calculated as $((A_{control} - A_{sample})/A_{control}) \times 100\%$, where $A_{control}$ and A_{sample} are the absorbances of the control and sample, respectively.

2.5. Anticancer Assay

2.5.1. Cell Culture

All cell lines used in this study originated from human bodies. Renal cell line (HEK293) was used as human normal cells, whereas the colon cancer cell line (HCT-116), cervical cancer cell line (HeLa), liver cancer cell line (HepG2), breast cancer cell line (MCF-7), lung cancer cell line (A549), and pancreatic cancer cell line (PANC-1) were used as the human cancer cell lines in this study. Different mediums were used for the incubation of cell lines. HEK-293 and PANC-1 were incubated by DMEM (Dulbeco's Modified Eagle's Medium) after adding 10% FBS (fetal bovine serum). MCF-7 and A549 were incubated by RPMI1640 (Roswell Park Memorial Institute medium) after adding 10% FBS. HeLa and HepG2 were incubated by MEM (Minimum Essential Medium) after adding 10% FBS. HCT-116 was incubated by McCoy's 5A Medium after adding 10% FBS. Cell lines were maintained in mediums at 37 °C and 5% CO_2 incubator.

2.5.2. MTT Assay

Inhibition effects on the proliferation of cancer cells were determined using MTT assay. The cells were adjusted at a concentration of 4×10^4 cells/mL, and then 100 μL cell suspension was added to each well in 96 well microplates and incubated at 37 °C and 5% CO_2 for 24 h. Each of 100 μL extracts of different concentrations (100, 200, 400, 800, and 1600 μg/mL) were added to 96 well microplates. Then, cells were incubated for 72 at 37 °C and 5% CO_2. After adding 20 μL MTT (5 mg/mL) to per wells, the cells were incubated for 4 at 37 °C and 5% CO_2. Then culture media was removed and 150 μL dimethyl sulfoxide (DMSO) was added to per well, and then stirred for 10 min. Absorbance was measured at 490 nm with microplate reader. Sample solvent (70% ethanol or distilled water) was used as the control. The inhibition percentage of cell proliferation was calculated by the following equation: $((A_{control} - A_{sample})/A_{control}) \times 100\%$, where $A_{control}$ and A_{sample} are the absorbance of the control and sample, respectively.

2.6. Statistical Analysis

All values were expressed as means ± standard deviations (SD). One-way analysis of variance (ANOVA) was used for test the variation of phytochemicals, antioxidant and anticancer potentials among *C. paliurus* populations, followed by Turkey range test with $p = 0.05$. General linear model (GLM) analysis was performed to detect the effects of solvents and geographical origin on phytochemical contents, antioxidant and anticancer properties. The above analyses were performed by SPSS 20.0 software (SPSS Inc., Chicago, IL, USA). Redundancy analysis (RDA) was carried out to detect the correlation between phytochemicals and bioactivity by Canoco 4.5 software (Wageningen UR, Wageningen, The Netherlands).

3. Results and Discussion

3.1. Effects of Solvent and Geographical Origin on Bioactive Metabolites

The contents of bioactive metabolites in *C. paliurus* leaves were significantly affected by solvent and population. Based on the F-values, solvent had larger effects on bioactive metabolites except for 3-*O*-caffeoyluinic acid and quercetin-3-*O*-glucuronide (Supplementary Table S1). Additionally, there were significant interaction effects between solvent and population on bioactive metabolites (Supplementary Table S1). The amount of total polyphenol, total flavonoid and total triterpenoid varied from 4.85 to 52.67 mg/g, 2.81 to 20.72 mg/g and 2.61 to 50.67 mg/g, respectively. The highest contents were found in ethanol extracts of S7 population, while the lowest contents were recorded in water extracts of S9 population (Figure 2). Furthermore, the HPLC analysis showed that in most cases, the water extracts exhibited higher 3-*O*-caffeoyluinic acid (from 0.10 to 2.29 mg/g) and 4-*O*-caffeoyluinic acid (from 0.03 to 0.48 mg/g), and the lowest values appeared in S9 population (Table 1). However, the ethanol solvent was more effective for the extraction of flavonoids and triterpenoids in most *C. paliurus* populations. Quercetin-3-*O*-glucuronide (from 0.14 to 3.71 mg/g) was found to be the dominated flavonoid, followed by kaempferol-3-*O*-rhamnoside (from 0.16 to 2.73 mg/g) and kaempferol-3-*O*-glucuronide (from 0.14 to 1.77 mg/g) (Table 2). Arjunolic acid ranged from 0.23 to 4.60 mg/g was the leading component of triterpenoids in the ethanol extracts. In addition, S11 population was superior to all other populations with respect to the individual triterpenoid (Table 3).

Table 1. The content (mg/g) of phenolic acid in ethanol and water extracts of *C. paliurus* leaves.

Populations	Ethanol			Water		
	P1	P2	P3	P1	P2	P3
S1	1.23 ± 0.01^c	0.24 ± 0.00^b	0.33 ± 0.02^b	1.43 ± 0.05^{bc}	0.42 ± 0.03^a	0.03 ± 0.00^c
S2	0.14 ± 0.02^i	$0.05 + 0.01^{fg}$	nd	0.22 ± 0.02^{hi}	0.09 ± 0.01^{de}	nd
S3	1.65 ± 0.07^a	0.11 ± 0.00^e	0.19 ± 0.01^d	2.29 ± 0.06^a	0.29 ± 0.01^b	nd
S4	0.15 ± 0.01^i	0.04 ± 0.01^g	nd	0.32 ± 0.04^{hi}	0.06 ± 0.01^{de}	nd
S5	0.84 ± 0.03^e	0.16 ± 0.01^d	0.19 ± 0.00^d	0.92 ± 0.01^{ef}	0.25 ± 0.00^b	nd
S6	0.68 ± 0.01^f	0.15 ± 0.00^d	0.18 ± 0.00^d	1.12 ± 0.01^{de}	0.27 ± 0.01^b	0.01 ± 0.00^d
S7	1.07 ± 0.02^d	0.31 ± 0.01^a	0.26 ± 0.01^c	1.24 ± 0.18^{cd}	0.47 ± 0.07^a	0.04 ± 0.00^a
S8	0.49 ± 0.03^g	0.11 ± 0.01^e	0.05 ± 0.01^e	0.66 ± 0.01^{fg}	0.26 ± 0.01^b	nd
S9	0.26 ± 0.03^h	0.06 ± 0.00^f	nd	0.10 ± 0.01^i	0.03 ± 0.00^e	nd
S10	0.23 ± 0.02^{hi}	0.05 ± 0.00^{fg}	nd	0.76 ± 0.01^f	0.20 ± 0.00^{bc}	nd
S11	0.67 ± 0.06^f	0.19 ± 0.02^c	0.20 ± 0.03^d	0.41 ± 0.01^{gh}	0.10 ± 0.01^{de}	nd
S12	1.44 ± 0.01^b	0.21 ± 0.01^{bc}	0.22 ± 0.01^{cd}	2.18 ± 0.04^a	0.43 ± 0.02^a	nd
S13	0.59 ± 0.00^f	0.09 ± 0.00^e	0.19 ± 0.01^d	0.43 ± 0.01^{gh}	0.13 ± 0.00^{cd}	nd
S14	1.61 ± 0.02^a	0.33 ± 0.01^a	0.47 ± 0.04^a	1.73 ± 0.33^b	0.48 ± 0.08^a	0.04 ± 0.00^b
S15	0.15 ± 0.03^i	0.06 ± 0.01^{fg}	nd	0.43 ± 0.05^{gh}	0.20 ± 0.02^{bc}	nd

P1: 3-*O*-caffeoyluinic acid; P2: 4-*O*-caffeoyluinic acid; P3: 4,5-di-*O*-caffeoyluinic acid; nd: not detected. Values within different superscripts are different in the same column at 0.05 levels.

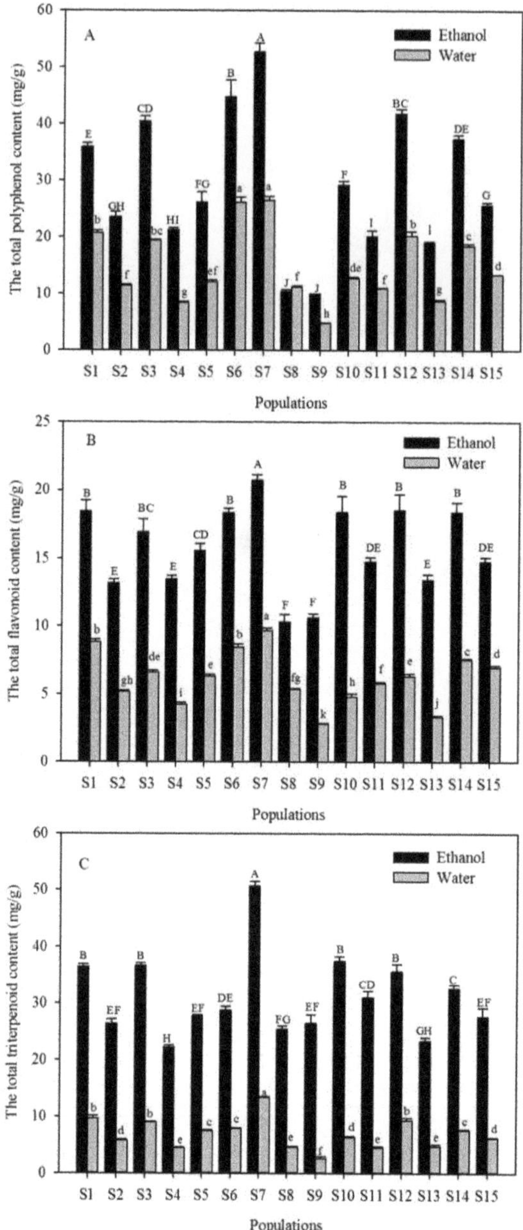

Figure 2. The contents of total polyphenol (**A**), total flavonoid (**B**) and total triterpenoid (**C**) in ethanol and water extracts of *C. paliurus* leaves. Different capital and small letters indicated significant differences in the ethanol and water extracts at 0.05 levels, respectively.

Table 2. The content (mg/g) of individual flavonoid in ethanol and water extracts of *C. paliurus* leaves.

Populations	Ethanol							Water						
	F1	F2	F3	F4	F5	F6	F7	F1	F2	F3	F4	F5	F6	F7
S1	3.71 ± 0.02[a]	0.76 ± 0.00[a]	0.23 ± 0.00[fg]	1.55 ± 0.01[abc]	0.11 ± 0.00[gh]	0.33 ± 0.00[a]	2.73 ± 0.02[a]	2.30 ± 0.03[a]	0.24 ± 0.02[b]	0.08 ± 0.00[b]	0.95 ± 0.08[bc]	0.07 ± 0.01[bc]	0.13 ± 0.01[a]	0.66 ± 0.27[a]
S2	0.20 ± 0.04[g]	0.05 ± 0.00[fg]	0.02 ± 0.00[h]	0.15 ± 0.03[h]	0.06 ± 0.00[h]	0.02 ± 0.00[f]	0.83 ± 0.06[f]	0.33 ± 0.05[h]	0.05 ± 0.00[gh]	0.01 ± 0.00[e]	0.25 ± 0.03[fg]	nd	0.02 ± 0.00[gh]	0.07 ± 0.01[hi]
S3	2.25 ± 0.06[b]	0.81 ± 0.03[a]	0.68 ± 0.02[b]	1.06 ± 0.04[e]	0.32 ± 0.01[d]	0.23 ± 0.01[b]	1.69 ± 0.05[b]	1.40 ± 0.04[c]	0.24 ± 0.01[b]	0.06 ± 0.00[bc]	0.64 ± 0.02[e]	0.06 ± 0.00[bc]	0.08 ± 0.00[bc]	0.44 ± 0.01[c]
S4	0.13 ± 0.01[g]	0.04 ± 0.00[g]	0.01 ± 0.00[h]	0.20 ± 0.08[gh]	0.06 ± 0.01[h]	0.03 ± 0.00[f]	0.16 ± 0.01[h]	0.35 ± 0.06[h]	0.04 ± 0.00[hi]	0.01 ± 0.00[e]	0.24 ± 0.03[fgh]	nd	0.03 ± 0.01[e-h]	0.11 ± 0.02[gh]
S5	1.22 ± 0.02[e]	0.55 ± 0.01[c]	0.28 ± 0.01[f]	1.30 ± 0.02[cde]	0.20 ± 0.00[ef]	0.14 ± 0.00[c]	1.07 ± 0.02[e]	0.65 ± 0.00[ef]	0.12 ± 0.00[d]	0.02 ± 0.00[e]	0.70 ± 0.00[de]	nd	0.04 ± 0.00[def]	0.21 ± 0.00[e]
S6	2.15 ± 0.06[b]	0.63 ± 0.01[b]	0.36 ± 0.01[e]	1.51 ± 0.03[bc]	0.15 ± 0.00[fg]	0.07 ± 0.00[de]	0.45 ± 0.01[g]	1.69 ± 0.02[b]	0.28 ± 0.00[a]	0.15 ± 0.00[a]	1.08 ± 0.01[ab]	0.07 ± 0.00[bc]	0.04 ± 0.00[de]	0.15 ± 0.00[fg]
S7	2.0 ± 0.07[bc]	0.34 ± 0.01[e]	0.16 ± 0.00[g]	1.37 ± 0.06[cd]	0.14 ± 0.02[fgh]	0.17 ± 0.01[c]	1.47 ± 0.06[c]	1.65 ± 0.04[b]	0.18 ± 0.02[c]	0.02 ± 0.00[de]	1.17 ± 0.02[a]	0.06 ± 0.00[bc]	0.09 ± 0.01[b]	0.56 ± 0.00[b]
S8	0.55 ± 0.04[f]	0.13 ± 0.01[f]	0.05 ± 0.00[h]	0.78 ± 0.06[f]	0.09 ± 0.00[gh]	0.09 ± 0.01[d]	0.80 ± 0.06[f]	0.42 ± 0.01[gh]	0.09 ± 0.00[ef]	0.03 ± 0.00[de]	0.62 ± 0.01[e]	0.06 ± 0.00[bc]	0.04 ± 0.00[d-g]	0.12 ± 0.00[gh]
S9	0.21 ± 0.01[g]	0.06 ± 0.00[fg]	0.02 ± 0.00[h]	0.30 ± 0.01[gh]	0.06 ± 0.01[h]	0.04 ± 0.00[ef]	0.24 ± 0.03[h]	0.07 ± 0.00[i]	0.02 ± 0.00[i]	0.01 ± 0.00[e]	0.08 ± 0.01[h]	nd	0.02 ± 0.00[h]	0.03 ± 0.00[i]
S10	0.20 ± 0.05[g]	0.05 ± 0.01[g]	0.03 ± 0.01[h]	0.14 ± 0.02[h]	0.07 ± 0.00[gh]	0.03 ± 0.00[f]	0.16 ± 0.03[h]	0.55 ± 0.01[fg]	0.05 ± 0.00[gh]	0.01 ± 0.00[e]	0.32 ± 0.02[f]	0.09 ± 0.00[b]	0.02 ± 0.00[gh]	0.18 ± 0.00[ef]
S11	1.4 ± 0.16[de]	0.63 ± 0.07[bc]	0.40 ± 0.05[de]	1.74 ± 0.19[ab]	0.25 ± 0.02[de]	0.22 ± 0.02[b]	1.26 ± 0.14[d]	0.46 ± 0.07[gh]	0.08 ± 0.01[ef]	0.01 ± 0.00[e]	0.55 ± 0.09[e]	0.05 ± 0.00[c]	0.04 ± 0.01[de]	0.13 ± 0.02
S12	1.84 ± 0.35[c]	0.30 ± 0.05[e]	0.46 ± 0.03[d]	1.19 ± 0.20[de]	0.41 ± 0.06[c]	0.16 ± 0.03[c]	1.82 ± 0.32[b]	1.53 ± 0.04[bc]	0.11 ± 0.00[de]	0.07 ± 0.00[b]	0.94 ± 0.02[bc]	0.09 ± 0.00[b]	0.07 ± 0.00[c]	0.55 ± 0.00[b]
S13	0.11 ± 0.00[g]	0.62 ± 0.02[bc]	0.58 ± 0.02[c]	0.43 ± 0.01[g]	1.43 ± 0.04[a]	0.22 ± 0.01[b]	1.05 ± 0.03[e]	0.07 ± 0.00[i]	0.08 ± 0.00[ef]	0.05 ± 0.00[cd]	0.11 ± 0.00[gh]	0.05 ± 0.00[c]	0.04 ± 0.00[de]	0.14 ± 0.00[fg]
S14	1.72 ± 0.07[cd]	0.44 ± 0.03[d]	1.02 ± 0.07[a]	1.77 ± 0.12[a]	1.34 ± 0.07[b]	0.16 ± 0.02[c]	1.72 ± 0.09[b]	0.87 ± 0.15[d]	0.12 ± 0.02[d]	0.15 ± 0.03[a]	0.86 ± 0.14[cd]	0.19 ± 0.02[a]	0.05 ± 0.01[d]	0.35 ± 0.06[d]
S15	0.20 ± 0.01[g]	0.06 ± 0.01[fg]	0.02 ± 0.00[h]	0.17 ± 0.00[gh]	0.07 ± 0.01[h]	0.02 ± 0.00[f]	1.02 ± 0.06[e]	0.73 ± 0.09[de]	0.08 ± 0.01[fg]	0.02 ± 0.00[e]	0.64 ± 0.08[e]	0.05 ± 0.00[c]	0.03 ± 0.00[fgh]	0.17 ± 0.02[efg]

F1: quercetin-3-*O*-glucuronide; F2: quercetin-3-*O*-galactoside; F3: isoquercitrin; F4: kaempferol-3-*O*-glucuronide; F5: kaempferol-3-*O*-glucoside; F6: quercetin-3-*O*-rhamnoside; F7: kaempferol-3-*O*-rhamnoside; nd: not detected. Values within different superscripts are different in the same column at 0.05 levels.

Table 3. The content (mg/g) of individual triterpenoid in ethanol and water extracts of *C. paliurus* leaves.

Populations	Ethanol						Water			
	T1	T2	T3	T4	T5	T6	T1	T2	T5	T6
S1	3.35 ± 0.03[b]	2.00 ± 0.00[a]	0.73 ± 0.01[ef]	2.38 ± 0.01[c]	0.86 ± 0.01[b]	0.47 ± 0.00[c]	0.16 ± 0.01[b]	0.17 ± 0.03[cd]	nd	0.07 ± 0.01[a]
S2	0.93 ± 0.06[ef]	0.21 ± 0.03[f]	0.76 ± 0.02[ef]	0.32 ± 0.01[de]	0.20 ± 0.03[ef]	0.14 ± 0.00[fg]	0.09 ± 0.01[cd]	0.13 ± 0.00[e]	nd	0.05 ± 0.00[cd]
S3	2.87 ± 0.10[b]	1.48 ± 0.04[b]	1.91 ± 0.05[d]	1.88 ± 0.07[c]	0.72 ± 0.02[bc]	0.36 ± 0.02[d]	0.16 ± 0.01[b]	0.16 ± 0.00[cde]	nd	0.05 ± 0.00[cd]
S4	0.26 ± 0.04[g]	0.27 ± 0.01[f]	Nd	0.07 ± 0.00[e]	0.04 ± 0.01[f]	0.05 ± 0.00[gh]	0.002 ± 0.0003[f]	0.13 ± 0.01[e]	nd	0.03 ± 0.00[fg]
S5	1.87 ± 0.05[c]	1.35 ± 0.04[b]	3.11 ± 0.08[c]	2.11 ± 0.10[c]	0.85 ± 0.07[bc]	0.42 ± 0.01[cd]	0.24 ± 0.00[a]	0.18 ± 0.00[abc]	0.04 ± 0.00[c]	0.04 ± 0.00[efg]
S6	1.51 ± 0.03[cd]	0.90 ± 0.02[cd]	0.31 ± 0.01[fg]	0.89 ± 0.02[d]	0.77 ± 0.02[bc]	0.71 ± 0.03[ab]	0.13 ± 0.00[bc]	0.17 ± 0.00[e]	nd	0.04 ± 0.00[de]
S7	1.50 ± 0.04[cd]	0.63 ± 0.02[de]	0.77 ± 0.04[ef]	0.79 ± 0.02[d]	0.31 ± 0.02[de]	0.22 ± 0.01[ef]	0.17 ± 0.03[b]	0.14 ± 0.01[de]	nd	0.04 ± 0.00[efg]
S8	1.87 ± 0.15[c]	0.91 ± 0.06[c]	5.33 ± 0.39[b]	5.00 ± 0.39[b]	0.69 ± 0.04[c]	0.37 ± 0.03[d]	0.08 ± 0.00[d]	0.21 ± 0.00[ab]	0.05 ± 0.00[b]	0.07 ± 0.00[ab]
S9	0.60 ± 0.00[fg]	0.14 ± 0.02[f]	0.22 ± 0.02[fg]	0.23 ± 0.04[de]	0.16 ± 0.00[ef]	0.02 ± 0.00[h]	nd	nd	nd	0.05 ± 0.00[cd]
S10	0.23 ± 0.04[g]	0.28 ± 0.03[f]	0.14 ± 0.02[fg]	0.19 ± 0.02[de]	0.04 ± 0.01[f]	0.05 ± 0.01[gh]	0.08 ± 0.01[de]	0.18 ± 0.02[abc]	0.03 ± 0.00[d]	0.04 ± 0.01[def]
S11	4.60 ± 0.51[a]	2.22 ± 0.23[a]	6.09 ± 0.68[a]	6.36 ± 0.72[a]	1.26 ± 0.14[a]	0.80 ± 0.09[a]	nd	0.13 ± 0.00[e]	nd	0.06 ± 0.00[bc]
S12	1.81 ± 0.32[c]	1.61 ± 0.25[b]	Nd	1.77 ± 0.33[c]	0.75 ± 0.13[bc]	0.39 ± 0.07[cd]	0.16 ± 0.01[b]	0.21 ± 0.01[a]	0.03 ± 0.00[e]	0.03 ± 0.00[g]
S13	0.71 ± 0.02[efg]	0.34 ± 0.01[f]	0.57 ± 0.02[fg]	0.46 ± 0.01[de]	0.83 ± 0.03[bc]	0.66 ± 0.02[b]	0.005 ± 0.0008[f]	nd	nd	0.03 ± 0.00[g]
S14	1.13 ± 0.09[de]	0.41 ± 0.02[ef]	1.23 ± 0.18[e]	0.61 ± 0.05[de]	0.24 ± 0.02[de]	0.19 ± 0.01[ef]	0.04 ± 0.00[e]	nd	nd	0.06 ± 0.00[bc]
S15	0.79 ± 0.02[ef]	0.26 ± 0.05[f]	0.46 ± 0.03[fg]	0.21 ± 0.00[de]	0.40 ± 0.04[d]	0.25 ± 0.01[e]	0.21 ± 0.04[a]	0.17 ± 0.01[bc]	0.07 ± 0.00[a]	0.04 ± 0.00[def]

T1: arjunolic acid; T2: cyclocaric acid B; T3: pterocaryoside B; T4: pterocaryoside A; T5: hederagenin; T6: oleanolic acid; nd: not detected. Values within different superscripts are different in the same column at 0.05 levels.

Bioactive metabolites of *C. paliurus* leaves varied with solvent and population, which corresponded with previous results [13]. Solvent extraction is a commonly used method for obtaining plant extracts. However, the extraction efficiency has a correlation to the polarity of solvents and extracted compounds [29]. The solubility of the compounds can be changed by the interactions between solvents and extracted compounds [30]. Our results were not consistent with the results from Heo et al. [25], where the water extracts of indigo leaves had higher total polyphenol and total flavonoid content than that of the ethanol extracts. Results from this study showed that ethanol solvent was more effective for the extraction of flavonoids and triterpenoids of *C. paliurus* leaves. In addition, no pterocaryoside A and pterocaryoside B were detected in the water extracts, and only minor hederagenin and 4,5-di-*O*-caffeoyluinic acid were observed in fewer populations, which basically agreed with a previous study [13]. These results indicated that the extraction solvents not only affected the contents of bioactive metabolites but also its composition in *C. paliurus* extracts. Moreover, contents of total polyphenol and total flavonoid in both ethanol and water extracts of *C. paliurus* leaves were considerably higher than those of Tossa jute leaves [31]. However, in the water extracts, they were lower than three *Centaurea* species and *Ganoderma adspersum* [32,33]. In the ethanol extracts, they are comparable with pyrola leaves from some locations in Northeast China, but lower than *Ruta chalepensis* and olive leaves [26,34,35].

3.2. Effects of Solvent and Geographical Origin on Antioxidant Activity

Solvent and natural population had significant effects on the scavenging capacity of DPPH and ABTS radical and reducing power of the *C. paliurus* extracts, and the significant interaction effects between solvent and population were also observed on antioxidant property (Supplementary Table S2). Ethanol extracts showed stronger antioxidant capacity when compared with water extracts in most *C. paliurus* populations. In the ethanol extracts, the strongest capacity of scavenging DPPH radical was observed in S15 population (IC_{50} = 0.34 mg/mL), while S3 (IC_{50} = 0.64 mg/mL), S6 (IC_{50} = 0.59 mg/mL) and S7 (IC_{50} = 0.50 mg/mL) populations were the best to eliminate ABTS radical. Moreover, the strongest reducing power occurred in the ethanol extracts of S14 population (EC_{50} = 0.71 mg/mL) (Table 4). Simultaneously, in the water extracts, S6 and S7 populations showed better performance of scavenging DPPH and ABTS radical and the strongest reducing power was achieved in S6 and S12 populations (Table 4). Moreover, S9 population had the lowest antioxidant capacity in both ethanol and water extracts of *C. paliurus* leaves (Table 4).

Our results corresponded with previous reports of Liu et al. [20] and Zhou et al. [17] who indicated that the antioxidant activity of *C. paliurus* varied with the geographical locations. The ethanol extracts of *C. paliurus* leaves showed stronger antioxidant capacity than the water extracts, which were in line with the extracts of *F. beecheyana* and *S. chinensis* [23,24]. Thus, the higher yields of bioactive compounds were likely to contribute to the stronger antioxidant property in the ethanol extracts. However, our results were not consistent with the results from Heo et al. [25], where higher contents of total flavonoid and total polyphenol were detected in the water extracts of indigo leaves when compared to the ethanol extracts. Compared with other plant species, the scavenging capacity of DPPH radical and reducing power in the ethanol extracts of *C. paliurus* leaves were weaker than the ethanol extracts of pyrola leaves, but the scavenging capacity of ABTS was similar between the two plants from specific locations [34]. Most populations were higher than goji berry and *Aspalathus Linearis* and some populations were comparable to *R. tomentosa* in terms of the scavenging capacity of DPPH radical [36–38]. The scavenging capacity of ABTS of some populations was higher than specific goji berry and Solanaceae species [36,39]. However, the antioxidant activity of the water extracts in the present study was weaker in comparison with a previous study [17]. It was likely that the extraction method, especially the solid-liquid ratio, affected the antioxidant activity, as the solid-liquid ratio was positively correlated with the yield of bioactive compounds and they might contribute to higher antioxidant activity [30].

Table 4. The antioxidant activity of ethanol and water extracts of *C. paliurus* leaves.

Populations	Ethanol			Water		
	DPPH	ABTS	Reducing Power	DPPH	ABTS	Reducing Power
S1	0.52 ± 0.01^f	0.75 ± 0.02^{cde}	1.10 ± 0.01^{fg}	1.24 ± 0.01^{ef}	1.71 ± 0.02^{ef}	2.48 ± 0.05^f
S2	$0.98 \pm 0.02^{c-f}$	$1.42 \pm 0.04^{b-e}$	1.03 ± 0.01^g	2.78 ± 0.05^d	3.68 ± 0.20^d	2.57 ± 0.03^f
S3	0.47 ± 0.00^f	0.64 ± 0.01^e	1.00 ± 0.01^g	1.41 ± 0.01^{ef}	1.81 ± 0.02^{ef}	1.95 ± 0.02^{hi}
S4	1.35 ± 0.01^{cde}	1.88 ± 0.03^{bcd}	1.82 ± 0.02^{bc}	4.44 ± 0.11^b	4.72 ± 0.12^c	3.86 ± 0.00^c
S5	0.81 ± 0.08^{ef}	$1.19 \pm 0.01^{b-e}$	1.28 ± 0.02^{ef}	1.17 ± 0.04^{ef}	1.55 ± 0.06^{ef}	1.82 ± 0.06^i
S6	0.40 ± 0.02^f	0.59 ± 0.02^e	0.99 ± 0.01^g	0.88 ± 0.00^f	1.18 ± 0.01^f	1.59 ± 0.01^j
S7	0.92 ± 0.01^{def}	0.50 ± 0.01^e	1.15 ± 0.01^{fg}	0.90 ± 0.01^f	1.26 ± 0.01^f	2.09 ± 0.10^{gh}
S8	5.87 ± 0.64^b	6.68 ± 0.92^a	1.91 ± 0.02^{bc}	3.65 ± 0.08^c	5.48 ± 0.13^b	4.26 ± 0.04^b
S9	7.23 ± 0.67^a	7.51 ± 1.15^a	3.48 ± 0.16^a	10.65 ± 0.69^a	11.05 ± 0.67^a	5.97 ± 0.11^a
S10	0.67 ± 0.00^{ef}	$0.98 \pm 0.02^{b-e}$	1.34 ± 0.01^e	1.39 ± 0.02^{ef}	2.04 ± 0.15^e	2.18 ± 0.04^g
S11	1.65 ± 0.07^c	2.03 ± 0.09^b	1.77 ± 0.01^c	3.38 ± 0.11^c	5.11 ± 0.45^{bc}	3.31 ± 0.04^d
S12	0.47 ± 0.01^f	0.74 ± 0.04^{de}	1.03 ± 0.09^g	1.35 ± 0.02^{ef}	1.69 ± 0.06^{ef}	1.45 ± 0.02^j
S13	1.53 ± 0.05^{cd}	1.90 ± 0.06^{bc}	1.99 ± 0.15^b	3.93 ± 0.02^{bc}	3.50 ± 0.09^d	3.27 ± 0.12^d
S14	0.50 ± 0.03^f	0.78 ± 0.02^{cde}	0.71 ± 0.02^h	1.47 ± 0.02^e	2.05 ± 0.02^e	3.01 ± 0.06^e
S15	0.34 ± 0.00^f	$1.44 \pm 0.09^{b-e}$	1.54 ± 0.02^d	2.25 ± 0.03^d	1.22 ± 0.01^f	2.91 ± 0.01^e

DPPH: the IC_{50} of DPPH; ABTS: the IC_{50} of ABTS; Reducing power: the EC_{50} of reducing power; IC_{50} (mg/mL): the sample concentration at which the radical was scavenged by 50%; EC_{50} (mg/mL): the absorbance was 0.5 for reducing power. Values within different superscripts are different in the same column at 0.05 levels.

3.3. Effects of Solvent and Geographical Origin on Anticancer Activity

The antiproliferative effects of *C. paliurus* leaves were investigated against A549, HCT-116, HeLa, HepG2, MCF-7, PANC-1 human cancer cell lines and one normal cell line, HEK-293 by MTT. As displayed in Supplementary Table S2, the solvent and population significantly affected the antiproliferative activity of *C. paliurus* leaves. The *C. paliurus* extracts were able to inhibit the growth of cancer cell lines in a dose-dependent manner. In general, the ethanol extracts exhibited considerable antiproliferation effects when compared with the water extracts (Table 5). However, the antiproliferation effects of *C. paliurus* extracts varied with cancer cell types. The HeLa cell was the most sensitive to the extracts of *C. paliurus* leaves, and the IC_{50} values of the ethanol and water extracts among *C. paliurus* populations varied from 0.13 to 0.42 mg/mL and from 0.15 to 2.24 mg/mL, respectively. The ethanol and water extracts of S6 population showed the best antiproliferative activity on A549 and MCF-7 cells. In the ethanol extracts, S7, S8 and S9 populations exhibited the strongest antiproliferative effects on HepG2 cell (IC_{50} = 0.25 mg/mL), while S7 and S8 populations had the best inhibition effects on the proliferation of PANC-1 cell (IC_{50} = 0.52 mg/mL).

MTT assay as calorimetric method is commonly used for detecting cell viability. To the best of our knowledge, this is the first report to demonstrate the antiproliferation effects of *C. paliurus* leaves on different cancer cell types. The *C. paliurus* leaves showed the property of noncytotoxic activity as the IC_{50} values of all cells were more than 30 µg/mL [40]. Our results were in harmony of the results from Dahham et al. [41], where the ethanol extracts of *Pandanus tectorius* fruits had stronger antiproliferation effects on human cancer cells than the water extracts. Moreover, *C. paliurus* extracts showed stronger antiproliferation effects on HCT-116 (IC_{50} = 0.84 mg/mL), MCF-7 (IC_{50} = 1.5 mg/mL), HeLa (IC_{50} = 1.4 mg/mL) and HepG2 (IC_{50} = 4.1 mg/mL) cancer cells than chlorogenic acid complex isolated from green coffee beans [42]. The antiproliferation effects of the ethanol extracts from most *C. paliurus* populations on A-549, HCT-116, HepG2, MCF-7 cancer cells was weaker than that from the three algal species, but five *C. paliurus* populations showed better antiproliferation effects on HeLa cancer cell than *Laurencia majuscule* species [43]. In terms of antiproliferation effects on HeLa cells, *C. paliurus* can be compared with of Thai medicine plants [44]. The proliferation inhibition of *C. paliurus* leaves on HeLa deserves further research due to the greatest antiproliferation effects among the cancer cells studied in this study. The *C. paliurus* extracts also inhibited the proliferation of normal kidney cell HEK-293, which was in accordance with the results from Liu et al. [45] and Heo et al. [25] who indicated that *Macleaya cordata* and indigo plants had antiproliferation effects on normal human cells (fetal lung fibroblast cell MRC5 and kidney cell HEK-293).

Table 5. The antiproliferation effects (IC_{50} values) of ethanol and water extracts of *C. paliurus* leaves on different cell lines.

Populations	Ethanol							Water						
	A549	HCT-116	HEK-293	HeLa	HepG2	MCF-7	PANC-1	A549	HCT-116	HEK-293	HeLa	HepG2	MCF-7	PANC-1
S1	0.72 ± 0.03[de]	0.66 ± 0.02[b]	0.51 ± 0.02[b]	0.15 ± 0.01[g]	0.41 ± 0.00[e]	0.32 ± 0.00[def]	0.63 ± 0.00[cd]	1.34 ± 0.05[cd]	1.04 ± 0.04[efg]	8.73 ± 1.17[a]	0.68 ± 0.03[c]	1.60 ± 0.05[ef]	0.72 ± 0.02[efg]	1.89 ± 0.09[cde]
S2	0.87 ± 0.02[b]	0.79 ± 0.00[a]	0.40 ± 0.00[d]	0.42 ± 0.01[a]	0.41 ± 0.00[e]	0.33 ± 0.01[de]	0.85 ± 0.01[a]	3.43 ± 0.52[b]	1.71 ± 0.27[d]	1.33 ± 0.08[bcd]	0.80 ± 0.00[bc]	1.52 ± 0.07[f]	1.26 ± 0.03[bc]	4.82 ± 0.48[b]
S3	0.53 ± 0.01[j]	0.51 ± 0.00[de]	0.33 ± 0.01[fg]	0.28 ± 0.03[bcd]	0.34 ± 0.01[g]	0.38 ± 0.02[bc]	0.54 ± 0.02[fg]	2.24 ± 0.46[bc]	1.01 ± 0.03[efg]	1.05 ± 0.07[bcd]	0.22 ± 0.01[fg]	2.08 ± 0.03[def]	0.75 ± 0.02[ef]	2.79 ± 0.29[c]
S4	0.96 ± 0.03[a]	0.80 ± 0.00[a]	0.35 ± 0.02[ef]	0.22 ± 0.02[def]	0.57 ± 0.01[a]	0.41 ± 0.01[ab]	0.86 ± 0.03[a]	-	-	1.04 ± 0.05[bcd]	0.38 ± 0.01[ef]	2.72 ± 0.14[cd]	1.14 ± 0.08[cd]	-
S5	0.60 ± 0.01[gh]	0.51 ± 0.01[def]	0.68 ± 0.02[a]	0.13 ± 0.00[g]	0.44 ± 0.00[d]	0.35 ± 0.01[cd]	0.57 ± 0.01[efg]	-	1.85 ± 0.14[cd]	-	2.10 ± 0.17[a]	3.65 ± 0.57[c]	1.45 ± 0.05[a]	2.70 ± 0.27[cd]
S6	0.35 ± 0.01[k]	0.48 ± 0.00[efg]	0.11 ± 0.01[i]	0.22 ± 0.02[ef]	0.38 ± 0.01[f]	0.22 ± 0.04[g]	0.65 ± 0.01[bc]	0.54 ± 0.01[d]	0.51 ± 0.02[g]	-	0.30 ± 0.002[efg]	1.03 ± 0.01[f]	0.34 ± 0.01[h]	1.08 ± 0.02[e]
S7	0.64 ± 0.02[fg]	0.47 ± 0.01[g]	0.20 ± 0.02[h]	0.28 ± 0.00[b-e]	0.25 ± 0.00[k]	0.38 ± 0.01[bc]	0.52 ± 0.02[g]	0.87 ± 0.04[cd]	0.83 ± 0.03[efg]	8.22 ± 0.53[a]	0.46 ± 0.02[de]	1.24 ± 0.00[f]	0.60 ± 0.02[g]	1.24 ± 0.03[e]
S8	0.77 ± 0.01[c]	0.52 ± 0.01[d]	0.38 ± 0.01[de]	0.34 ± 0.03[b]	0.25 ± 0.00[k]	0.30 ± 0.01[ef]	0.52 ± 0.02[g]	1.33 ± 0.07[cd]	2.56 ± 0.48[b]	2.08 ± 0.24[b]	0.39 ± 0.03[def]	5.21 ± 0.55[b]	1.40 ± 0.11[ab]	4.86 ± 0.77[b]
S9	0.76 ± 0.00[cd]	0.49 ± 0.01[efg]	0.43 ± 0.01[c]	0.31 ± 0.02[bc]	0.25 ± 0.00[k]	0.38 ± 0.02[bc]	0.54 ± 0.01[fg]	-	-	1.16 ± 0.06[bcd]	0.96 ± 0.02[b]	5.21 ± 0.55[b]	1.28 ± 0.07[bc]	-
S10	0.62 ± 0.01[g]	0.53 ± 0.02[d]	0.22 ± 0.00[h]	0.29 ± 0.02[bc]	0.30 ± 0.00[i]	0.30 ± 0.01[ef]	0.53 ± 0.01[fg]	6.84 ± 1.28[a]	1.36 ± 0.17[de]	1.25 ± 0.09[bcd]	0.20 ± 0.00[fg]	5.16 ± 0.25[b]	1.22 ± 0.05[cd]	2.82 ± 0.19[c]
S11	0.71 ± 0.01[e]	0.48 ± 0.01[fg]	0.49 ± 0.02[b]	0.27 ± 0.03[cde]	0.32 ± 0.01[h]	0.44 ± 0.03[a]	0.54 ± 0.00[fg]	-	2.34 ± 0.15[bc]	1.74 ± 0.15[bc]	0.68 ± 0.06[c]	9.41 ± 0.37[a]	1.16 ± 0.03[cd]	9.50 ± 0.46[a]
S12	0.56 ± 0.00[hij]	0.53 ± 0.01[d]	0.21 ± 0.01[h]	0.17 ± 0.01[fg]	0.47 ± 0.01[c]	0.40 ± 0.01[b]	0.58 ± 0.01[def]	0.90 ± 0.02[cd]	0.87 ± 0.01[efg]	0.47 ± 0.01[d]	0.15 ± 0.01[g]	1.20 ± 0.01[f]	0.66 ± 0.02[fg]	1.44 ± 0.21[de]
S13	0.66 ± 0.01[f]	0.44 ± 0.00[h]	0.36 ± 0.01[ef]	0.30 ± 0.05[bc]	0.28 ± 0.00[j]	0.28 ± 0.01[f]	0.55 ± 0.02[efg]	-	3.48 ± 0.30[a]	0.76 ± 0.01[cd]	2.24 ± 0.20[a]	5.80 ± 0.95[b]	-	9.57 ± 1.04[a]
S14	0.55 ± 0.04[ij]	0.56 ± 0.01[c]	0.31 ± 0.01[g]	0.13 ± 0.01[g]	0.53 ± 0.01[b]	0.29 ± 0.02[ef]	0.71 ± 0.03[b]	0.97 ± 0.01[cd]	0.78 ± 0.01[fg]	0.74 ± 0.02[cd]	0.60 ± 0.02[cd]	1.12 ± 0.02[f]	1.08 ± 0.04[d]	1.43 ± 0.08[e]
S15	0.57 ± 0.01[hi]	0.58 ± 0.01[c]	0.36 ± 0.01[ef]	0.13 ± 0.00[g]	0.31 ± 0.00[hi]	0.35 ± 0.01[cd]	0.60 ± 0.01[de]	3.60 ± 0.70[b]	1.15 ± 0.04[ef]	1.46 ± 0.12[bcd]	0.23 ± 0.01[fg]	2.68 ± 0.021[cde]	0.84 ± 0.03[e]	1.10 ± 0.02[e]

A549, HCT-116, HEK-293, HeLa, HepG2, MCF-7, PANC-1: the IC_{50} of cells; IC_{50} (mg/mL): the sample concentration at which the growth of cell was inhibited by 50%. Values within different superscripts are different in the same column at 0.05 levels.

3.4. Correlation between Phytochemicals and Bioactivity

In order to understand which compounds could be responsible for the antioxidant and antiproliferative activities in *C. paliurus* extracts, the RDA models were performed in the present study. The result indicated that the canonical axes was significant ($p = 0.004$) and explained 90.0% of the total variation in antioxidant activity (Figure 3A). In the RDA model, the first axes was significant ($p = 0.006$), and 83.6% of the variation was explained by RDA1 and 5.4% was explained by RDA2 (Figure 3A). Based on the correlation coefficients, total polyphenol (−0.83) showed the greatest correlation with the first axes, followed by total flavonoid (−0.75) and total triterpenoid (−0.62). Apart from 4-*O*-caffeoyluinic acid, kaempferol-3-*O*-glucoside, pterocaryoside A and pterocaryoside B, the explanatory variables were significantly ($p < 0.05$) associated with antioxidant activity by Monte Carlo permutation test (Table 6). Total polyphenol had the highest explanation of 64.1% and total flavonoid had the second highest explanation of 52.3% (Table 6). Moreover, among the significant variables, phenolic contributed more to antioxidant activity in comparison with triterpenoid with respect to individual compounds (Table 6). Kaempferol-3-*O*-rhamnoside was predominant compounds of antioxidant potential among the individual compounds studied. All the compounds studied were negatively correlated with antioxidant properties except for pterocaryoside A and pterocaryoside B (Figure 3A).

On the other hand, the canonical axes explained 79.1% of the total variation in antiproliferative activity, but the first two axes explained 63.2% and 10.2% of the total variation, respectively (Figure 3B). Based on the correlation coefficients in the RDA model, the first axis was predominantly correlated with total flavonoid (−0.81) and total triterpenoid (−0.77). As indicated in Table 6, the explanatory variables showed significant ($p < 0.05$) correlation with antiproliferative activity except for 3-*O*-caffeoyluinic acid, 4-*O*-caffeoyluinic acid, kaempferol-3-*O*-glucoside, pterocaryoside A and pterocaryoside B, and total flavonoid and total triterpenoid explained 45.2% and 41.6% of the total variation, respectively. Hederagenin showed the greatest explanation to antiproliferative activity among individual compounds, followed by arjunolic acid, kaempferol-3-*O*-rhamnoside and oleanolic acid. The explanatory variables were positively correlated with the antiproliferation effects on the cancer cells studied except for 4-*O*-caffeoyluinic acid (Figure 3B).

As hydrogen donor, reactive oxygen quenchers, reducing agent and free radical scavenging agent in redox reactions, phenolic compounds are very important for antioxidant activity [38]. Our results showed that phenolics had a greater contribution to antioxidant activity than triterpenoid in the extracts of *C. paliurus* leaves, which confirmed the above viewpoint. Our results were also corresponded with the information given by Zhou et al. [14], where total polyphenol gave the highest explanation for antioxidant property in the extracts of *C. paliurus* leaves. Meanwhile, the total triterpenoid, arjunolic acid, cyclocaric acid B, hederagenin and oleanolic acid also contributed to antioxidant activity, supporting the point that terpenoid compounds were active principles of antioxidant property [46]. In previous studies, Valente et al. [47] and Shaikh et al. [46] reported that the terpenoids of plant extracts could prevent metabolic pathways contributing to cancer, and may be responsible for antiproliferation effects. Gao et al. [48] indicated the antiproliferative activity of oats was closely associated with phenolic compounds. Our results confirmed that terpenoids and phenolic compounds played a vital role in antiproliferation effects in the extracts of *C. paliurus* leaves. Plants possessing abundant phenolics and higher antioxidant capacity often showed higher antiproliferation effects on cancer cells [49]. However, *Pandanus tectorius* showed no cytotoxicity property against HeLa and MCF-7 cell lines even if abundant phenolics and stronger antioxidant properties were observed in its fruits [40]. In the present study, total polyphenol, total flavonoid and total triterpenoid were the top three components contributing to antioxidant and antiproliferative activities. However, no outstanding antiproliferation effects were detected in the ethanol extracts of S7 population, even if the highest total polyphenol, total flavonoid and total triterpenoid and higher scavenging capacity of ABTS radical and reducing power were observed in S7 population. Accordingly, there were other compounds not examined in *C. paliurus* leaves, which might contribute to the antioxidant and antiproliferative effects [50]. For instance, the *C. paliurus* water-soluble polysaccharide exhibited antioxidant and antiproliferation effects on HeLa cancer cells [15,16]. Besides, the possible synergistic or antagonistic effects between the bioactive metabolites should also be taken into consideration. *C. paliurus* extracts are a complex mixture; their biological activity is attributed to the comprehensive effects of bioactive metabolites. Indeed, the individual compounds would deserve future investigation, as they are likely to show higher bioactivity than the *C. paliurus* extracts.

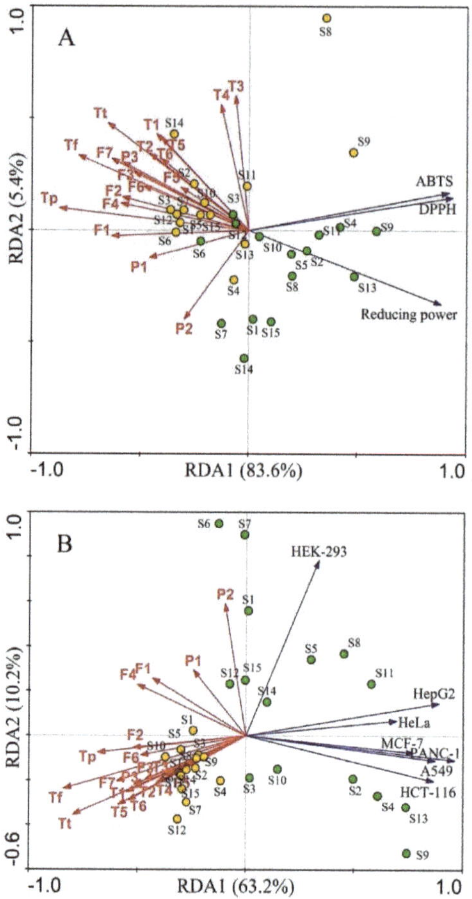

Figure 3. Redundancy Analysis (RDA) ordination diagram of bioactive metabolites and antioxidant activity (**A**), bioactive metabolites and antiproliferation effects on cells (**B**). The yellow and green circles represent the ethanol and water extracts of *C. paliurus* population samples, respectively. Abbreviations: Tp: total polyphenol; Tf: total flavonoid; Tt: total triterpenoid; P1: 3-*O*-caffeoyluinic acid; P2: 4-*O*-caffeoyluinic acid; P3: 4,5-di-*O*-caffeoyluinic acid; F1: quercetin-3-*O*-glucuronide; F2: quercetin-3-*O*-galactoside; F3: isoquercitrin; F4: kaempferol-3-*O*-glucuronide; F5: kaempferol-3-*O*-glucoside; F6: quercetin-3-*O*-rhamnoside; F7: kaempferol-3-*O*-rhamnoside; T1: arjunolic acid; T2: cyclocaric acid B; T3: pterocaryoside B; T4: pterocaryoside A; T5: hederagenin; T6: oleanolic acid;DPPH: the IC_{50} of DPPH; ABTS: the IC_{50} of ABTS; Reducing power: the EC_{50} of reducing power; A549, HCT-116, HEK-293, HeLa, HepG2, MCF-7, PANC-1: the IC_{50} of cells.

Table 6. The explanation and significance of variables in RDA models.

Explanatory Variable	Antioxidant			Antiproliferation		
	Variance Explained%	F Value	*p*-Value	Variance Explained%	F	*p*-Value
Tp	64.1%	49.974	0.002 **	29.7%	11.846	0.002 **
Tf	52.3%	30.705	0.002 **	45.2%	23.132	0.002 **
Tt	36.9%	16.354	0.002 **	41.6%	19.908	0.002 **
P1	17.9%	6.087	0.014 *	4.9%	1.445	0.216
P2	8.3%	2.544	0.096	4.4%	1.303	0.248
P3	29.1%	11.503	0.004 **	17.4%	5.879	0.008 **
F1	33.7%	14.236	0.006 **	12.6%	4.043	0.028 *
F2	28.8%	11.352	0.006 **	17.2%	5.828	0.014 *
F3	24.1%	8.887	0.006 **	14.1%	4.601	0.022 *
F4	29.4%	11.664	0.002 **	16.5%	5.535	0.012 *
F5	8.6%	2.627	0.080	8.4%	2.582	0.086
F6	20.1%	7.050	0.016 *	15.1%	4.972	0.018 *
F7	34.2%	14.554	0.002 **	22.9%	8.313	0.004 **
T1	16.4%	5.487	0.034 *	19.9%	6.976	0.010 **
T2	18.3%	6.259	0.008 **	17.3%	5.847	0.014 *
T3	2.3%	0.651	0.458	9.5%	2.955	0.058
T4	3.1%	0.905	0.364	11.3%	3.571	0.052
T5	14.7%	4.844	0.014 *	23.0%	8.360	0.004 **
T6	14.4%	4.700	0.030 **	19.4%	6.749	0.006 **

Tp: total polyphenol; Tf: total flavonoid; Tt: total triterpenoid; P1: 3-*O*-caffeoyluinic acid; P2: 4-*O*-caffeoyluinic acid; P3: 4,5-di-*O*-caffeoyluinic acid; F1: quercetin-3-*O*-glucuronide; F2: quercetin-3-*O*-galactoside; F3: isoquercitrin; F4: kaempferol-3-*O*-glucuronide; F5: kaempferol-3-*O*-glucoside; F6: quercetin-3-*O*-rhamnoside; F7: kaempferol-3-*O*-rhamnoside; T1: arjunolic acid; T2: cyclocaric acid B; T3: pterocaryoside B; T4: pterocaryoside A; T5: hederagenin; T6: oleanolic acid. * and ** indicated significant at 0.05 and 0.01 levels, respectively.

4. Conclusions

In summary, both extraction solvent and geographic origin had significant effects on bioactive metabolites, antioxidant and antiproliferative activities in *C. paliurus* leaves. In most cases, the ethanol solvent was more effective for the extraction of flavonoids and triterpenoids, and higher antioxidant and antiproliferative activities were observed in ethanol extracts. Total polyphenol showed the greatest contribution to the antioxidant activity, while total flavonoid was most responsible for the antiproliferation effects. Moreover, the HeLa cell was the most sensitive to *C. paliurus* extracts among the cancer cell types studied. Further researches should be performed to detect the antioxidant and antitumor effects of individual compounds on HeLa cancer cell, as well as the bioactive mechanisms in *C. paliurus* leaves.

Supplementary Materials: The following are available online at http://www.mdpi.com/1999-4907/10/8/625/s1, Figure S1: HPLC chromatograms of the representative sample solution (top) and the responding standard solution containing the 16 quantitative compounds (bottom). 1: 3-*O*-caffeoyluinic acid; 2: 4-*O*-caffeoyluinic acid; 3: quercetin-3-*O*-glucuronide; 4: quercetin-3-*O*-galactoside; 5: isoquercitrin; 6: kaempferol-3-*O*-glucuronide; 7: kaempferol-3-*O*-glucoside; 8: quercetin-3-*O*-rhamnoside; 9: 4,5-di-*O*-caffeoyluinic acid; 10: kaempferol-3-*O*-rhamnoside; 11: arjunolic acid; 12: cyclocaric acid B; 13: pterocaryoside B; 14: pterocaryoside A; 15: hederagenin; 16: oleanolic acid, Table S1: F-values and probability levels from the general linear model (GLM) analysis of bioactive metabolite contents in *C. paliurus* leaves, Table S2: F-values and probability levels from the general linear model (GLM) analysis of antioxidant and anticancer activities in *C. paliurus* leaves.

Author Contributions: Conceived and designed the experiment: M.Z. and S.F. Performed the experiment: M.Z., P.C. and Y.L. Analyzed the data: M.Z., P.C. and Y.L. Conceived the paper, wrote the first draft and edited the manuscript: M.Z. and S.F. Supervised the manuscript: X.S.

Funding: This work was financially supported by the Jiangsu Province Science Foundation for Youths (No. BK20160926), National Natural Science Foundation of China (No. 31470637), the Priority Academic Program Development of Jiangsu Higher Education Institutions (PAPD) and Doctorate Fellowship Foundation of Nanjing Forestry University. The funders had no role in study design, data collection and analysis, decision to publish, or preparation of the manuscript.

Acknowledgments: We acknowledge Xiangxiang Fu, Wanxia Yang, Yang Liu, Bo Deng and Yanni Cao, and Qingliang Liu from Nanjing Forestry University for field and laboratory assistance. We would like to thank Zhiqi Yin from Department of Natural Medicinal Chemistry & State Key Laboratory of Natural Medicines, China Pharmaceutical University for her thoughtful comments to the manuscript.

Conflicts of Interest: The authors declare that there are no conflict of interest.

References

1. Lee, S.R.; Roh, H.S.; Lee, S.; Park, H.B.; Jang, T.S.; Ko, Y.J.; Baek, K.H.; Kim, K.H. Bioactivity-guided isolation and chemical characterization of antiproliferative constituents from morel mushroom (*Morchella esculenta*) in human lung adenocarcinoma cells. *J. Funct. Foods* **2018**, *40*, 249–260. [CrossRef]

2. Roleira, F.M.F.; Tavares-da-Silva, E.J.; Varela, C.L.; Costa, S.C.; Silva, T.; Garrido, J.; Borges, F. Plant derived and dietary phenolic antioxidants: Anticancer properties. *Food Chem.* **2015**, *183*, 235–258. [CrossRef]

3. Wang, D.D.; Wang, S.; Feng, Y.; Zhang, L.; Li, Z.; Ma, J.; Luo, Y.Q.; Xiao, W. Antitumor effects of Bulbus *Fritillariae cirrhosae* on Lewis lung carcinoma cells in vitro and in vivo. *Ind. Crop. Prod.* **2014**, *54*, 92–101. [CrossRef]

4. Motta, L.B.; Furlan, C.M.; Santos, D.Y.A.C.; Salatino, M.L.F.; Negri, G.; Carvalho, J.E.D.; Monteiro, P.A.; Ruiz, A.L.T.G.; Caruzo, M.B.; Salatino, A. Antiproliferative activity and constituents of leaf extracts of *Croton sphaerogynus* Baill. (Euphorbiaceae). *Ind. Crop. Prod.* **2013**, *50*, 661–665. [CrossRef]

5. Elkady, W.M.; Ayoub, I.M. Chemical profiling and antiproliferative effect of essential oils of two Araucaria species cultivated in Egypt. *Ind. Crop. Prod.* **2018**, *118*, 188–195. [CrossRef]

6. Carocho, M.; Calhelha, R.C.; Queiroz, M.R.P.; Bento, A.; Morales, P.; Sokovic, M.; Ferreira, I.C.F.R. Infusions and decoctions of *Castanea sativa* flowers as effective antitumor and antimicrobial matrices. *Ind. Crop. Prod.* **2014**, *62*, 42–46. [CrossRef]

7. Ashraf, A.; Sarfraz, R.A.; Mahmood, A.; Din, M.U. Chemical composition and in vitro antioxidant and antitumor activities of *Eucalyptus camaldulensis* Dehn. Leaves. *Ind. Crop. Prod.* **2015**, *74*, 241–248. [CrossRef]

8. Olejnik, A.; Kaczmarek, M.; Olkowicz, M.; Kowalska, K.; Juzwa, W.; Dembczyński, R. ROS-modulating anticancer effects of gastrointestinally digested *Ribesnigrum* L. fruit extract in human colon cancer cells. *J. Funct. Foods* **2018**, *42*, 224–236. [CrossRef]

9. Li, W.; Liu, J.; Guan, R.; Chen, J.P.; Yang, D.P.; Zhao, Z.M.; Wang, D.M. Chemical characterization of procyanidins from *Spatholobus suberectus* and their antioxidative and anticancer activities. *J. Funct. Foods* **2015**, *12*, 468–477. [CrossRef]

10. Dienaitė, L.; Pukalskienė, M.; Matias, A.A.; Pereira, C.V.; Pukalskas, A.; Venskutoni, P.R. Valorization of six *Nepeta* species by assessing the antioxidant potential, phytochemical composition and bioactivity of their extracts in cell cultures. *J. Funct. Foods* **2018**, *45*, 512–522. [CrossRef]

11. Cao, Y.N.; Fang, S.Z.; Yin, Z.Q.; Fu, X.X.; Shang, X.L.; Yang, W.X.; Yang, H.M. Chemical fingerprint and multicomponent quantitative analysis for the quality evaluation of *Cyclocarya paliurus* Leaves by HPLC-Q-TOF-MS. *Molecules* **2017**, *22*, 1927. [CrossRef]

12. Xie, J.H.; Wang, Z.J.; Shen, M.Y.; Nie, S.P.; Xie, M.Y. Sulfated modification, characterization and antioxidant activities of polysaccharide from *Cyclocarya paliurus*. *Food Hydrocoll.* **2016**, *53*, 7–15. [CrossRef]

13. Liu, Y.; Cao, Y.N.; Fang, S.Z.; Wang, T.L.; Yin, Z.Q.; Shang, X.L.; Yang, W.X.; Fu, X.X. Antidiabetic effects of *Cyclocarya paliurus* leaves depends on the contents of antihyperglycemic flavonoids and antihyperlipidemic triterpenoids. *Molecules* **2018**, *23*, 1042. [CrossRef]

14. Yang, Z.W.; Ouyang, K.H.; Zhao, J.; Chen, H.; Xiong, L.; Wang, W.J. Structural characterization and hypolipidemic effect of *Cyclocarya paliurus* polysaccharide in rat. *Int. J. Biol. Macromol.* **2016**, *91*, 1073–1080. [CrossRef]

15. Xie, J.H.; Liu, X.; Shen, M.Y.; Nie, S.P.; Zhang, H.; Li, C.; Gong, D.M.; Xie, M.Y. Purification, physicochemical characterisation and anticancer activity of a polysaccharide from *Cyclocarya paliurus* leaves. *Food Chem.* **2013**, *136*, 1453–1460. [CrossRef]

16. Xie, J.H.; Xie, M.Y.; Nie, S.P.; Shen, M.Y.; Wang, Y.X.; Li, C. Isolation, chemical composition and antioxidant activities of a water-soluble polysaccharide from *Cyclocarya paliurus* (Batal.) Iljinskaja. *Food Chem.* **2010**, *119*, 1626–1632. [CrossRef]

17. Zhou, M.M.; Lin, Y.; Fang, S.Z.; Liu, Y.; Shang, X.L. Phytochemical content and antioxidant activity in aqueous extracts of *Cyclocarya paliurus* leaves collected from different populations. *PeerJ* **2019**, *7*, e6492. [CrossRef]

18. Liu, Y.; Chen, P.; Zhou, M.M.; Wang, T.L.; Fang, S.Z.; Shang, X.L.; Fu, X.X. Geographic variation in the chemical composition and antioxidant properties of phenolic compounds from *Cyclocarya paliurus* (Batal.) Iljinskaja Leaves. *Molecules* **2018**, *23*, 2440. [CrossRef]

19. Oueslati, S.; Ksouri, R.; Falleh, H.; Pichette, A.; Abdelly, C.; Legault, J. Phenolic content, antioxidant, anti-inflammatory and anticancer activities of the edible halophyte *Suaedafruticosa* Forssk. *Food Chem.* **2012**, *132*, 943–947. [CrossRef]

20. Liu, Y.; Fang, S.Z.; Zhou, M.M.; Shang, X.L.; Yang, W.X.; Fu, X.X. Geographic variation in water-soluble polysaccharide content and antioxidant activities of *Cyclocarya paliurus* leaves. *Ind. Crop. Prod.* **2018**, *121*, 180–186. [CrossRef]

21. Farràs, A.; Cásedas, G.; Les, F.; Terrado, E.M.; Mitjans, M.; López, V. Evaluation of anti-tyrosinase and antioxidant properties of four fern species for potential cosmetic applications. *Forests* **2019**, *10*, 179. [CrossRef]

22. Graça, V.C.; Barros, L.; Calhelha, R.C.; Dias, M.I.; Carvalho, A.M.; Santos-Buelga, C.; Santos, P.F.; Ferreira, I.C.F.R. Chemical characterization and bioactive properties of aqueous and organic extracts of *Geranium robertianum* L. *Food Funct.* **2016**, *17*, 3807–3814. [CrossRef]

23. Yen, G.C.; Chen, C.S.; Chang, W.T.; Wu, M.F.; Cheng, F.T.; Shiau, D.K.; Hsu, C.L. Antioxidant activity and anticancer effect of ethanolic and aqueous extracts of the roots of *Ficus beecheyana* and their phenolic component. *J. Food Drug Anal.* **2018**, *26*, 182–192. [CrossRef] [PubMed]

24. Alaklabi, A.; Arif, I.A.; Ahamed, A.; Kumar, R.S.; Idhayadhulla, A. Evaluation of antioxidant and anticancer activities of chemical constituents of the *Saururus chinensis* root extracts. *Saudi J. Biol. Sci.* **2018**, *25*, 1387–1392. [CrossRef] [PubMed]

25. Heo, B.G.; Park, Y.J.; Park, Y.S.; Bae, J.H.; Cho, J.Y.; Park, K.; Jastrzebski, Z.; Gorinstein, S. Anticancer and antioxidant effects of extracts from different parts of indigo plant. *Ind. Crop. Prod.* **2014**, *56*, 9–16. [CrossRef]

26. Xie, P.J.; Huang, L.X.; Zhang, C.H.; Zhang, Y.L. Phenolic compositions, and antioxidant performance of olive leaf and fruit (*Olea europaea* L.) extracts and their structure-activity relationships. *J. Funct. Foods* **2015**, *16*, 460–471. [CrossRef]

27. Li, F.M.; Tan, J.; Nie, S.P.; Dong, C.J.; Li, C. The study on determination methods of total flavonoids in *Cyclocarya paliurus*. *Food Sci. Technol.* **2006**, *4*, 34–37. [CrossRef]

28. Fan, J.P.; He, C.H. Simultaneous quantification of three major bioactive triterpene acids in the leaves of *Diospyros kaki* by high-performance liquid chromatography method. *J. Pharm. Biomed. Anal.* **2006**, *41*, 950–956. [CrossRef]

29. Lu, Y.; Luthria, D. Influence of postharvest storage, processing, and extraction methods on the analysis of phenolic phytochemicals. *Instrum. Method. Anal. Ident. Bioact. Mol.* **2014**, *1185*, 3–31. [CrossRef]

30. Castro-López, C.; Ventura-Sobrevilla, J.M.; González-Hernández, M.D.; Rojas, R.; Ascacio-Valdés, J.A.; Aguilar, C.N.; Martínez-Ávila, G.C. Impact of extraction techniques on antioxidant capacities and phytochemical composition of polyphenol-rich extracts. *Food Chem.* **2017**, *237*, 1139–1148. [CrossRef]

31. Yakoub, A.R.B.; Abdehedi, O.; Jridi, M.; Elfalleh, W.; Nasri, M.; Ferchichi, A. Flavonoids, phenols, antioxidant, and antimicrobial activities in various extracts from Tossa jute leave (*Corchorusolitorus* L.). *Ind. Crop. Prod.* **2018**, *118*, 206–213. [CrossRef]

32. Aktumsek, A.; Zengin, G.; Guler, G.O.; Cakmak, Y.S.; Duran, A. Antioxidant potentials and anticholinesterase activities of methanolic and aqueous extracts of three endemic *Centaurea* L. species. *Food Chem. Toxicol.* **2013**, *55*, 290–296. [CrossRef] [PubMed]

33. Tel-Çayan, G.; Öztürk, M.; Duru, M.E.; Rehman, M.U.; Adhikari, A.; Türkoglu, A.; Choudhary, M.I. Phytochemical investigation, antioxidant and anticholinesterase activities of *Ganoderma adspersum*. *Ind. Crop. Prod.* **2015**, *76*, 749–754. [CrossRef]

34. Zhang, D.Y.; Luo, M.; Wang, W.; Zhao, C.J.; Gu, C.B.; Zu, Y.G.; Fu, Y.J.; Yao, X.H.; Duan, M.H. Variation of active constituents and antioxidant activity in pyrola (*P. incarnata* Fisch.) from different sites in Northeast China. *Food Chem.* **2013**, *141*, 2213–2219. [CrossRef] [PubMed]

35. Gali, L.; Bedjou, F. Antioxidant and anticholinesterase effects of the ethanol extract, ethanol extract fractions and total alkaloids from the cultivated *Ruta chalepensis*. *S. Afr. J. Bot.* **2019**, *120*, 163–169. [CrossRef]

36. Skenderidis, P.; Kerasioti, E.; Karkanta, E.; Stagos, D.; Kouretas, D.; Petrotos, K.; Hadjichristodoulou, C.; Tsakalof, A. Assessment of the antioxidant and antimutagenic activity of extracts from goji berry of Greek cultivation. *Toxicol. Rep.* **2018**, *5*, 251–257. [CrossRef] [PubMed]

37. Bhebhe, M.; Chipurura, B.; Muchuweti, M. Determination and comparison of phenolic compound content and antioxidant activity of selected local Zimbabwean herbal teas with exotic *Aspalathus Linearis*. *S. Afr. J. Bot.* **2015**, *100*, 213–218. [CrossRef]

38. Hamid, H.A.; Mutazah, R.; Yusoff, M.M.; Karim, N.A.A.; Razis, A.F.A. Comparative analysis of antioxidant and antiproliferative activities of *Rhodomyrtus tomentosa* extracts prepared with various solvents. *Food Chem. Toxicol.* **2017**, *108*, 451–457. [CrossRef]

39. Almoulah, N.F.; Voynikov, Y.; Gevrenova, R.; Schohn, H.; Tzanova, T.; Yagi, S.; Thomas, J.; Mignard, B.; Ahmed, A.A.A.; Siddig, M.A.E.; et al. Antibacterial, antiproliferative and antioxidant activity of leaf extracts of selected Solanaceae species. *S. Afr. J. Bot.* **2017**, *112*, 368–374. [CrossRef]

40. Andriani, Y.; Ramli, N.M.; Syamsumir, D.F.; Kassim, M.N.I.; Jaafar, J.; Aziz, N.A.; Marlina, L.; Musa, N.S.; Mohamad, H. Phytochemical analysis, antioxidant, antibacterial and cytotoxicity properties of keys and cores part of *Pandanus tectorius* fruits. *Arab. J. Chem.* **2015**, *50*, 519–521. [CrossRef]

41. Dahham, S.S.; Al-Rawi, S.S.; Ibrahim, A.H.; Majid, A.S.A.; Majid, A.M.S.A. Antioxidant, anticancer, apoptosis properties and chemical composition of black truffle *Terfezia claveryi*. *Saudi J. Biol. Sci.* **2018**, *25*, 1524–1534. [CrossRef] [PubMed]

42. Gouthamchandra, K.; Sudeep, H.V.; Venkatesh, B.J.; Prasad, K.S. Chlorogenic acid complex (CGA7), standardized extract from green coffee beans exerts anticancer effects against cultured human colon cancer HCT-116 cells. *Food Sci. Hum. Well.* **2017**, *6*, 147–153. [CrossRef]

43. Al-Enazi, N.M.; Awaad, A.S.; Zain, M.E.; Alqasoumi, S.I. Antimicrobial, antioxidant and anticancer activities of *Laurencia catarinensis*, *Laurencia majuscule* and *Padina pavonica* extracts. *Saudi Pharm. J.* **2018**, *26*, 44–52. [CrossRef] [PubMed]

44. Siriwatanametanon, N.; Fiebich, B.L.; Efferth, T.; Prieto, J.M.; Heinrich, M. Traditionally used Thai medicinal plants: In vitro anti-inflammatory, anticancer and antioxidant activities. *J. Ethnopharmacol.* **2010**, *130*, 196–207. [CrossRef] [PubMed]

45. Liu, M.; Lin, Y.L.; Chen, X.R.; Liao, C.C.; Poo, W.K. In vitro assessment of *Macleaya cordata* crude extract bioactivity and anticancer properties in normal and cancerous human lung cells. *Exp. Toxicol. Pathol.* **2013**, *65*, 775–787. [CrossRef] [PubMed]

46. Shaikh, R.; Pund, M.; Dawane, A.; Iliyas, S. Evaluation of anticancer, antioxidant, and possible anti-inflammatory properties of selected medicinal plants used in Indian traditional medication. *J. Tradit. Complement. Med.* **2014**, *4*, 253–257. [CrossRef] [PubMed]

47. Valente, M.J.; Pinho, P.G.D.; Henrique, R.; Pereira, J.A.; Carvalho, M. Further insights into chemical characterization through GC-MS and evaluation for anticancer potential of *Dracaena draco* leaf and fruit extracts. *Food Chem. Toxicol.* **2012**, *50*, 3847–3852. [CrossRef] [PubMed]

48. Gao, Y.; Guo, X.B.; Liu, Y.; Zhang, M.W.; Zhang, R.F.; Abbasi, A.M.; You, L.J.; Li, T.; Liu, R.H. Comparative assessment of phytochemical profile, antioxidant capacity and anti-proliferative activity in different varieties of brown rice (*Oryza sativa* L.). *LWT-Food Sci. Technol.* **2018**, *96*, 19–25. [CrossRef]

49. Li, K.; Yang, X.; Hu, X.S.; Han, C.; Lei, Z.F.; Zhang, Z.Y. In vitro antioxidant, immunomodulatory and anticancer activities of two fractions of aqueous extract from *Helicteres angustifolia* L. root. *J. Taiwan Inst. Chem. Eng.* **2016**, *61*, 75–82. [CrossRef]

50. Yang, H.M.; Yin, Z.Q.; Zhao, M.G.; Jiang, C.H.; Zhang, J.; Pan, K. Pentacyclic triterpenoids from *Cyclocarya paliurus* and their antioxidant activities in FFA-induced HepG2 steatosis cells. *Phytochemistry* **2018**, *151*, 119–127. [CrossRef]

 *forests*

Article

Seasonal Variation in Phenolic Compounds and Antioxidant Activity in Leaves of *Cyclocarya paliurus* (Batal.) Iljinskaja

Yanni Cao [1,2], **Shengzuo Fang** [1,3,*], **Xiangxiang Fu** [1,3], **Xulan Shang** [1,3] and **Wanxia Yang** [1,3]

[1]   College of Forestry, Nanjing Forestry University, Nanjing 210037, China
[2]   Research Center of Forestry and Fruit, Lianyungang Academy of Agricultural Science, Lianyungang 222000, China
[3]   Co-Innovation Center for Sustainable Forestry in Southern China, Nanjing Forestry University, Nanjing 210037, China
*   Correspondence: fangsz@njfu.edu.cn; Tel.: +86-025-8542-7797

Received: 6 June 2019; Accepted: 22 July 2019; Published: 26 July 2019

Abstract: *Cyclocarya paliurus* (Batal.)   Iljinskaja is a plant with nutraceutical importance since its leaves have been used historically as folk medicines for hundreds of years. The content of 10 phenolic compounds was determined throughout the growing season by high-performance liquid chromatography (HPLC) with UV detector, while the antioxidant activities of *C. paliurus* leaf extracts were evaluated by 1,1-diphenyl-2-picrylhydrazyl (DPPH), 2,2′-azino-bis (3-ethylbenzothiazoline-6-sulfonic acid) diammonium salt radical cation (ABTS), and ferric reducing antioxidant power (FRAP) methods. Seasonal variations in phenolic concentration and antioxidant activity as well as linkage between the phenolic composition and antioxidant activity were assessed. A significant seasonal variation of phenolic compounds was observed in the leaves and the highest content appeared in May, July, and November. Seventy percent ethanol extract of *C. paliurus* leaves possessed a good radical scavenging potency. Meanwhile, a significant correlation between antioxidant activities and contents of phenolics was detected. Results of the relationship between molecular structures and their antioxidant activities showed that both the number and configuration of H-donating hydroxyl groups are the main structural features influencing the antioxidant capacity of phenolics, while glycosylation may reduce the antioxidant capacity. The information provided by this study not only revealed the accumulative dynamics of phenolic compounds, but also established a basis for determining the optimal time for harvesting to improve the content of beneficial compounds in the leaves of *C. paliurus* in the future.

Keywords: *Cyclocarya paliurus*; seasonal dynamic; phenolic acids; flavonoids; antioxidant activity; structure-activity relationship

---

1. Introduction

Phenolic compounds are known to be food constituents of health-beneficial nature. As a large group of plant secondary metabolites, they are present in most plants. So far, more than 8000 dietary phenolic substances have been identified [1,2]. Phenolic compounds have shown diverse pharmaceutical and health-promoting effects, including antibacterial, anticarcinogenic, antioxidant, antimutagenic, anti-inflammatory, antiallergic, anti-obesity, and antidiabetic activities [3–5]. Among them, flavonoids are the largest and predominant group with important health value. There is a lot of research interest in the antioxidant activity of flavonoids and other plant phenolic compounds owing to their tremendous potential in health promotion and disease prevention [6–9]. Antioxidant substances are able to scavenge free radicals through a variety of mechanisms, thereby helping to protect biologically important cellular components, such as DNA, proteins, and membrane lipids, from free radical attacks leading to cell damage, which has been linked to aging, inflammation, atherosclerosis, ischemic injury, and cancer [7,10,11]. Consequently, phenolic compounds from plants, which were not only used as functional food ingredients but also for other preparations of health-promoting products, have become a hot topic for research and development [12]. As a kind of major secondary metabolites, the biosynthesis and accumulation of phenolic compounds are not only dependent on intrinsic factors and developmental stage, but they can be greatly influenced also by external factors such as light, temperature, and wounding, as well as a cultivation technique and the soil permeability and its depth [13–16].

Cyclocarya paliurus (Batal.) Iljinskaja, a member of Juglandaceae, is naturally scattered in the highland areas in southern China. Traditionally, *C. paliurus* leaves have been used as a nutraceutical tea for a long time in China due to its health promotion effects and special flavor and taste [17,18]. Moreover, the leaves of *C. paliurus* have been widely used for the treatment of obesity, hypertensive, lipid peroxidation, and diabetes in Chinese folk medicine [12,19–21]. Additionally, their immunity enhancement, anti-aging, anti-bacterial, and anti-cancer activities have also been reported [22–24]. It is believed that the physiologically active substances in *C. paliurus* leaves are responsible for its therapeutic effects. Recently, 10 phenolic compounds have been isolated and identities from the 70% ethanol extract of *C. paliurus* leaves by HPLC-Q-TOF-MS [25]. Previous phytochemical investigation also reported the presence of some other bioactive nature compounds, such as triterpenoids, steroids, alkaloids, and polysaccharides, from *Cyclocarya* species [26–28]. Due to its various benefits to health, a huge production of *C. paliurus* leaves is required [29]. Thus, recently, attempts have been made to develop plantations of *C. paliurus* as a functional food or an important raw material for pharmaceutical industry [30]. The accumulation of physiologically active substances in *C. paliurus* are related to many factors, but determination of the best harvesting time is one of most important silvicultural practices in the plantation management.

Temporal variations are quite common for natural compounds. For example, the camptothecin content in leaves of *Camptotheca acuminata* showed a reduction trend during the growth season [31], while the contents of saponins, dencichine, flavonoid, and polysaccharide in root of *Panax notoginseng* showed a seasonal variation [32]. The contents of vitamin C and flavonoids in lemon tree were significantly affected by harvest times [33]. Apparently, temporal variations are related to the plant species and can contribute to variations of the finished botanical products. Our previous studies also showed the seasonal variations of water soluble polysaccharides, selected flavonoids, and microelement contents in *C. paliurus* leaves [30,34,35]. However, little information about variation in quantities of phenolic constituents in leaves of *C. paliurus* during the whole year is available. The objective of the present study was to investigate temporal variation of phenolic acids and flavonoids contents as well as their antioxidative effectiveness of *C. paliurus* leaves collected over the whole growth period and to illustrate the structure–activity relationship of phenolic compounds in *C. paliurus* leaves. The results of the present study may shed light on the accumulative dynamics of phenolic compounds in the leaves of *C. paliurus* and provide a valuable reference for determining the appropriate harvesting time.

2. Materials and Methods

2.1. Plant Materials

Seeds of 10 *C. paliurus* families were collected from natural forests of Lushan, Jiangxi province (29°33′ N, 116°30′ E) in late October 2006. The collected seeds were prepared according to the method proposed by Fang et al. [17]. After stratification treatment, the germinated seeds were first sown in plastic containers (5 cm in diameter and 15 cm in height) before being transplanted to the *C. paliurus* germplasm nursery when the seedlings were about 6 cm in height. Following one year of growth in the nursery, seedlings of the 10 families were transplanted to Nanjing Forestry University Base (31°66′ N, 119°01′ E, 368 m about sea level), with an average annual temperature of 15.5 °C, average annual precipitation of 1037 mm, annual sunshine hour of 2146 h, and an annual frost-free period of 237 days. Briefly, the plantation was established with a planting spacing of 3 × 4 m in 2008 and each family consisted of 10 to 30 seedlings.

2.2. Sample Collection

To investigate the accumulative dynamic of phenolic in the leaves of *C. paliurus* during the growing period, approximately 200 g fresh fully developed leaves of the 10 families were sampled at 4 weeks intervals from May to November (24 May, 21 June, 19 July, 16 August, 13 September, 11 October, and 9 November in 2017). For each family, samples were randomly collected from 3 individual trees with similar canopy, which were mixed to form a pool representing the family.

2.3. Chemical Reagents and Reference

Acetonitrile was of HPLC grade from Tedia (Fairfied, OH, USA); Water was deionized using a Milli-Q water purification system (Millipore, Millford, MA, USA); formic acid was purchased from Aladdin Co., Ltd. (Shanghai, China), and other reagents were all of analytical reagent grade. The following standards were used for quantification of phenolic compounds including 3-*O*-caffeoylquinic acid (3-CQA), 4-*O*-caffeoylquinic acid (4-CQA), 4,5-di-*O*-caffeoylquinic acid (4,5-CQA), quercetin-3-*O*-glucuronide (Q-3-Glu), quercetin-3-*O*-galactoside (Q-3-Gal), quercetin-3-*O*-glucoside (Q-3-Glc) quercetin-3-*O*-rhamnoside (Q-3-Rha), kaempferol-3-*O*-glucuronide (K-3-Glu), kaempferol-3-*O*-glucoside (K-3-Glc), kaempferol-3-*O*-rhamnoside (K-3-Rha), quercetin (Quer), and kaempferol (Kaempf), which were purchased from Shanghai Yuanye Biotechnology Co., Ltd. (Shanghai, China) with purities of over 98%. 1,1-diphenyl-2-picrylhydrazyl (DPPH) was obtained from Sigma-Aldrich (St. Louis, MO, USA), while the ferric reducing antioxidant power (FRAP) assay kit and 2,2′-azino-bis (3-ethylbenzothiazoline-6-sulfonic acid) diammonium salt radical cation (ABTS) assay kit were purchased from the Beyotime Institute of Biotechnology (Nantong, China).

All solvents and samples were filtered through 0.22 μm filter before injecting into HPLC.

2.4. Sample and Standard Solutions Preparation

All samples were oven-dried to constant weight at 60 °C. Subsequently, the dried leaves were ground to a fine powder with a tissue grinder. Then the powder samples were stored at room temperature prior to analysis. Approximately 1 g of leaf powder sample was extracted with 100 mL of petroleum ether in a soxhlet extractor and refluxed for 4 h in a water bath at 80 °C to remove the fat soluble impurities such as pigment contained in the sample. The extract was discarded, while the residues were retained and dried at the room temperature. Phenolic compounds in the residue were extracted using an ultrasonic-assisted method [25]. Briefly, 15 mL of 70% ethanol was added to each sample and the samples were sonicated in an ultrasonic cleaner (ultrasonic instruments, Kunshan, China) at 70 °C for 45 min and then centrifuged for 10 min in a high-speed centrifuge (10,000 rpm).

Standard solutions were prepared by weighing the 10 reference compounds accurately and dissolving them in methanol. Then the stock solutions of the 10 reference compounds were further diluted to appropriate concentrations for establishment of calibration curves. The external standard calibrations were constructed at six data points covering the concentration range of each compound according to the levels of these compounds reported by our previous survey for this species. The calibration regression was plotted after linear regression of the peak areas versus concentrations. All the solutions were stored in a refrigerator at 4 °C and brought to room temperature before use, and all solvents were filtered with a 0.22 μm organic phase filter into an HPLC vial and subjected to HPLC analysis.

2.5. HPLC Determination of Phenolic Compounds

The quantification of the individual phenolic compounds of *C. paliurus* was performed on a Waters e2695 Alliance High Performance Liquid Chromatography (HPLC) system (Waters Crop., Milford, MA,

USA), consisting of a Waters 2695 separation unit (a quaternary pump solvent management system, an autosampler, an online degasser, a column heater and a gasket cleaning system), an ultraviolet detector (Waters 2489) and an Empower 3 data processing system. The quantification of the individual phenolic compounds was achieved on a reversed-phase X-Bridge C18 column (250 × 4.6 mm internal diameter, 5 μm particle size) with flow rate of 1.0 mL/min and the column was operated at 45 °C. The detection wavelength was kept at 360 nm and the injection volume was 10.0 μL.

The mobile phases were (A) acetonitrile with 0.01% formic acid and (B) water with 0.01% formic acid. For the analysis of phenolic compounds, a gradient elution protocol as follows was used: 8%–19% A at 0–13 min, 19%–21% A at 13–28 min, 21%–50% A at 28–40 min, and re-equilibration over 10 min to the initial composition. Contents of individual phenolics were quantified from their external standards. The representative chromatograms of a sample of *C. paliurus* and mixed standards of phenolic compounds are presented in Figure 1.

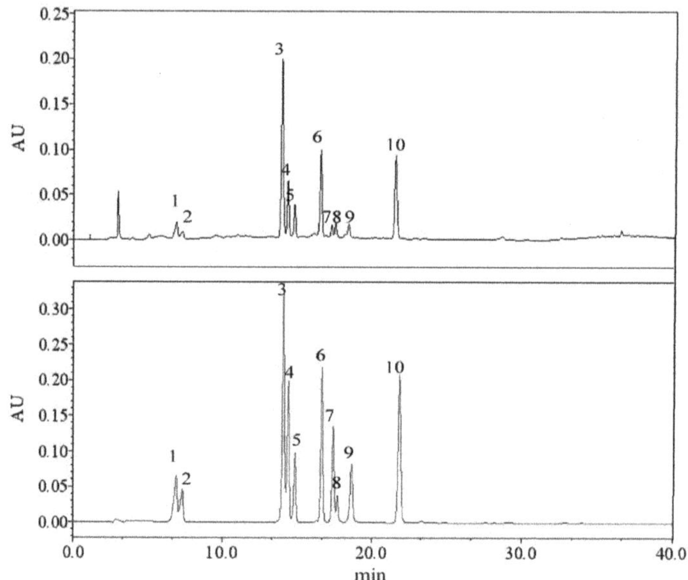

Figure 1. The representative HPLC chromatograms of *C. paliurus* leaf extract (**top**) and a mixed standards (**bottom**). 1. 3-*O*-caffeoylquinic acid; 2. 4-*O*-caffeoylquinic acid; 3. quercetin-3-*O*-glucuronide; 4. quercetin-3-*O*-galactoside; 5. quercetin-3-*O*-glucoside; 6. kaempferol-3-*O*-glucuronide; 7. kaempferol-3-*O*-glucoside; 8. quercetin-3-*O*-rhamnoside; 9. 4,5-di-*O*-caffeoylquinic acid; 10. kaempferol-3-*O*-rhamnoside.

2.6. Measurement of Antioxidant Activities

Leaves collected from family 4# monthly were used as a representative sample to determine the seasonal variation of antioxidant activity of *C. paliurus* leaf extracts. Additionally, in order to assess to what extent these detected phenolics contribute to the antioxidative effectiveness of the leaf extracts; antioxidant activities were performed on each individual phenolic compound. For this purpose, pure standard molecules were used. For antioxidant activity assay, all determinations were carried out in triplicate.

2.6.1. DPPH Radical Scavenging Activity

The DPPH radical scavenging capacities of *C. paliurus* leaf extracts and phenolic individuals were determined by using a colorimetric method according to a previous study with slight modification [24].

Briefly, 100 μL of the extracts 70% ethanol solution and phenolic individuals with gradient concentrations were mixed with ethanol (1.4 mL), respectively. Then all the mixtures were added to 0.004% DPPH (1 mL, Sigma-Aldrich) in ethanol. The mixtures were shaken vigorously and left to stand in the dark at room temperature for 30 min, and then the reduction of DPPH radical was evaluated spectrophotometrically by monitoring the decrease in absorbance at 517 nm against a blank of pure ethanol. The DPPH radical scavenging activities of each sample were calculated as the percent inhibition according to the following equation: DPPH radical scavenging (%) = $((A_0 - A_1)/A_0) \times 100$, where A_0 was the absorbance of the DPPH solution without the sample and A1 was the absorbance of the tested samples.

Finally, the data obtained from above experiments were used to establish calibration curves and to calculate the IC_{50} values (μg/mL), which were defined as the concentration of the test material producing 50% reduction of the DPPH free radical. IC_{50} was used as an index to compare the antioxidant activity of individuals.

2.6.2. Ferric Reducing Antioxidant Power (FRAP)

The assay was performed according to the instruction by Beyotime Institute of Biotechnology. Stock solutions included detective buffer, TPTZ (2,4,6-tripyridyl-s-triazine) solution, TPTZ dilution, 0.5 mL 10 mM $FeSO_4$ solution and 0.1 mL 10 mM Trolox solution. A working solution was prepared freshly by mixing TPTZ dilution, detective buffer, and TPTZ solution in a ratio of 10:1:1 (*v/v*), respectively. The working solution was warmed to 37 °C before use. A sample (5 μL) was mixed with 180 μL of FRAP working solution and kept for 5 min at 37 °C. The absorbance of the reaction mixture was then monitored at 593 nm. Trolox, a water-soluble analogue of vitamin E, was used as the reference compound to prepare a calibration curve for a concentration range of 0.15–1.5 mM. Results were expressed as trolox equivalent antioxidant capacity (TEAC), which was defined as the mmol of trolox whose antioxidant activity is equivalent to the activity of 1 g of extracts. Higher TEAC values demonstrate higher antioxidant activity.

2.6.3. ABTS Radical Cation Scavenging Activity

The total antioxidant capacity of each sample was determined by a total antioxidant capacity assay kit with the ABTS method (Beyotime Institute of Biotechnology, Shanghai, China). ABTS assay is based on the inhibition by antioxidants of the absorbance of $ABTS^+$ and this inhibition depends on the antioxidant capacity of the tested sample. The stock solutions included ABTS solution and oxidant solution. The working solution was prepared by mixing the two stock solutions at a ratio of 1:1 (*v/v*), and the mixture was incubated in the dark at room temperature for 12 to 16 h before use. The resulting $ABTS^+$ solution was then diluted with 80% ethanol to obtain an absorbance value of 0.70 ± 0.05 at 734 nm. A fresh $ABTS^+$ solution was prepared for each assay. Plant extracts or phenolic individuals (10 μL) were mixed with 200 μL of the diluted $ABTS^+$ solution for 6 min in the dark at room temperature. Then the absorbance of the mixture was recorded at 734 nm. The ABTS radical scavenging activity of the sample was calculated as follows: ABTS radical scavenging (%) = $((A_0 - A_1)/A_0) \times 100$, where A_0 is the absorbance of the control (ABTS solution without test sample) and A_1 is the absorbance in the presence of the test sample. Trolox was used as the reference compound to prepare a calibration curve for a concentration range of 0.15 to 1.5 mM. Results were expressed as trolox equivalent antioxidant capacity (TEAC), too. Higher TEAC values demonstrate higher antioxidant activity.

2.7. Data Analysis

Data are expressed as the mean + standard deviation (SD) of samples of the ten families. All statistical was performed using the SPSS 19.0 statistical software program (SPSS Inc., Chicago, IL, USA). A one-way analysis of variance (ANOVA) was conducted to compare the contents of phenolics in *C. paliurus* leaves collected monthly, followed by Duncan's multiple-range test. Different superscripts within columns represent significant differences at $p < 0.05$. Pearson correlation coefficient was used to reflect relationship between contents of phenolics and antioxidant activities.

Contents of total phenolic acids (TPA), total quercetin glycoside (TQ), total kaempferol glycosides (TK), and total flavonoids (TF) were calculated as the sums of their corresponding individuals.

3. Results

3.1. Seasonal Variation of Phenolic Acids Contents

The seasonal fluctuation patterns of phenolic acids content were almost synchronous among the 10 families we investigated. Therefore, the seasonal dynamics of leaf phenolic acids contents in *C. paliurus* are presented as the means of ten families and are given in Figure 2. The results revealed that the accumulation of phenolic acids in leaves of *C. paliurus* was significantly affected by the sampling time. 3-*O*-caffeoylquinic acid (3-CQA) was always the major phenolic acid during the whole growth period, and the content of 4-*O*-caffeoylquinic acid (4-CQA) was always higher than the content of 4,5-di-*O*-caffeoylquinic acid (4,5-CQA), except that of November.

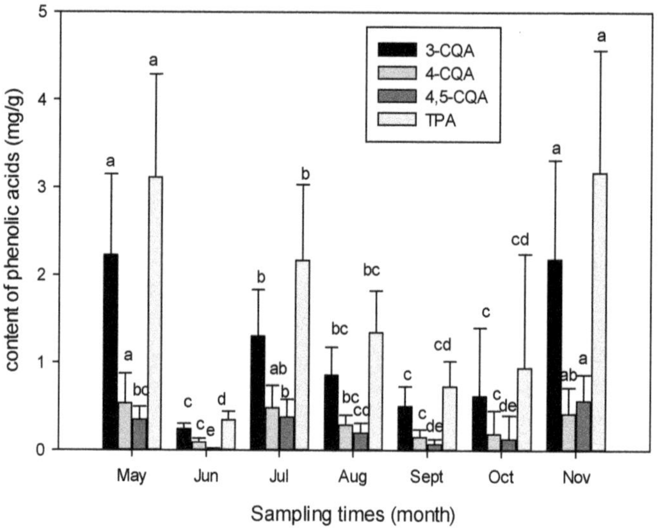

Figure 2. Seasonal variation of phenolic acids in leaves of *C. paliurus* (mean ± SD). 3-CQA, 4-CQA, 4,5-CQA, and TPA represent 3-*O*-caffeoylquinic acid, 4-*O*-caffeoylquinic acid, 4,5-di-*O*-caffeoylquinic acid, and total phenolic aids, respectively. Different letters in this figure indicate significant differences between the sampling times for the same category according to Duncan's test ($p < 0.05$). Each value presented is the mean of the 10 sampled families.

The 3-CQA contents showed a clear seasonal variation, ranging from 0.24 to 2.22 mg/g. The highest content of 3-CQA was measured in November, followed by July, while the lowest one was monitored in June and September. 4-CQA exhibited generally similar seasonal variation patterns as that of 3-CQA. The highest content of 4-CQA was found in May, up to 0.54 mg/g, followed by July and November, while the lowest one was measured in June, which was less than 0.10 mg/g. Among the three tested phenolic acids, 4,5-CQA levels varied most noticeably, approximately 64-fold. The highest content of 4,5-CQA was recorded in November, reaching 0.57 mg/g, whereas the lowest one was detected in June with 0.009 mg/g. The contents of 4,5-CQA in May and July were in the moderate level. Moreover, the total phenolic acid (TPA) content was calculated by summing the individual phenolic acid compounds based on HPLC determination of individuals. TPA also changed considerably during the year. The highest content of TPA was found in May and November, more than 3.10 mg/g, followed by July, while the lowest one was measured in June, less than 0.35 mg/g.

3.2. Seasonal Variation of Flavonoids Content

In general, flavonoids are present as glycosides in plants, usually conjugated with glucose. In our experiment, seven individual flavonoid compounds, namely quercetin-3-*O*-glucuronide (Q-3-Glu), quercetin-3-*O*-galactoside (Q-3-Gal), quercetin-3-*O*-glucoside (Q-3-Glc), quercetin-3-*O*-rhamnoside (Q-3-Rha), kaempferol-3-*O*-glucuronide (K-3-Glu), kaempferol-3-*O*-glucoside (K-3-Glc), and kaempferol-3-*O*-rhamnoside (K-3-Rha), were detected in *C. paliurus* leaves. The seasonal variation patterns of the investigated flavonoids were almost synchronous among the 10 sampled families. Thus, the seasonal fluctuations of leaf flavonoids content are shown in Figure 3 based on the means of the 10 families. The results of ANOVA indicated that the contents of flavonoids in *C. paliurus* leaves were statistically significantly influenced by seasonal progression. Q-3-Glu was considered as the major flavonoid compound and the crucial component in *C. paliurus* leaves, ranging from 1.61 (June) to 4.36 (July) mg/g. However, the seasonal variation of the other three quercetin glycosides in *C. paliurus* leaves was found to a slight difference (Figure 3). The contents of Q-3-Gal, Q-3-Glc and Q-3-Rha were at a low level until the last sampling time. Their highest contents were all achieved in November, and the lowest in June. Among the three quercetin glycosides, content of Q-3-Glc varied to a relatively large extent during the year, and its content in November was 83-fold larger compared to the content in June. For kaempferol glycosides, the contents of K-3-Glu and K-3-Rha were notably higher than that of K-3-Glc. The contents of K-3-Glu and K-3-Rha were at a high level in July, August and November. Compared to K-3-Glu and K-3-Rha contents, the content of K-3-Glc showed a relatively large extent variation throughout the growing period, ranging from 0.05 (June) to 0.5 (November) mg/g.

The total content of the four quercetin glycosides determined in this study was defined as TQ. Similarly, TK was defined as the total content of the three kaempferol glycosides investigated. Therefore, the total content of all the flavonoid individuals measured in this study was defined as TF. As shown in Figure 3C, contents of TQ, TK, and TF in November were the highest, and the lowest contents were detected in June and September. Interestingly, the content of TQ was always higher than TK during the seasonal progression except for June, and the difference was statistically significant in May, July, August, and November.

3.3. Seasonal Variation of Antioxidant Activities

Significant seasonal fluctuations of antioxidant activities were observed in *C. paliurus* leaf extracts (Figure 4). Samples collected in May, July, August, and November showed good performance in the three antioxidant activities assays (DPPH, FRAP, and ABTS). Generally, these three assays showed consistent results with the seasonal variations of phenolic acids and flavonoids, as shown in Figures 2 and 3. The higher antioxidant activities of the samples collected in May, July, August, and November might be attributed to higher phenolic contents at those periods.

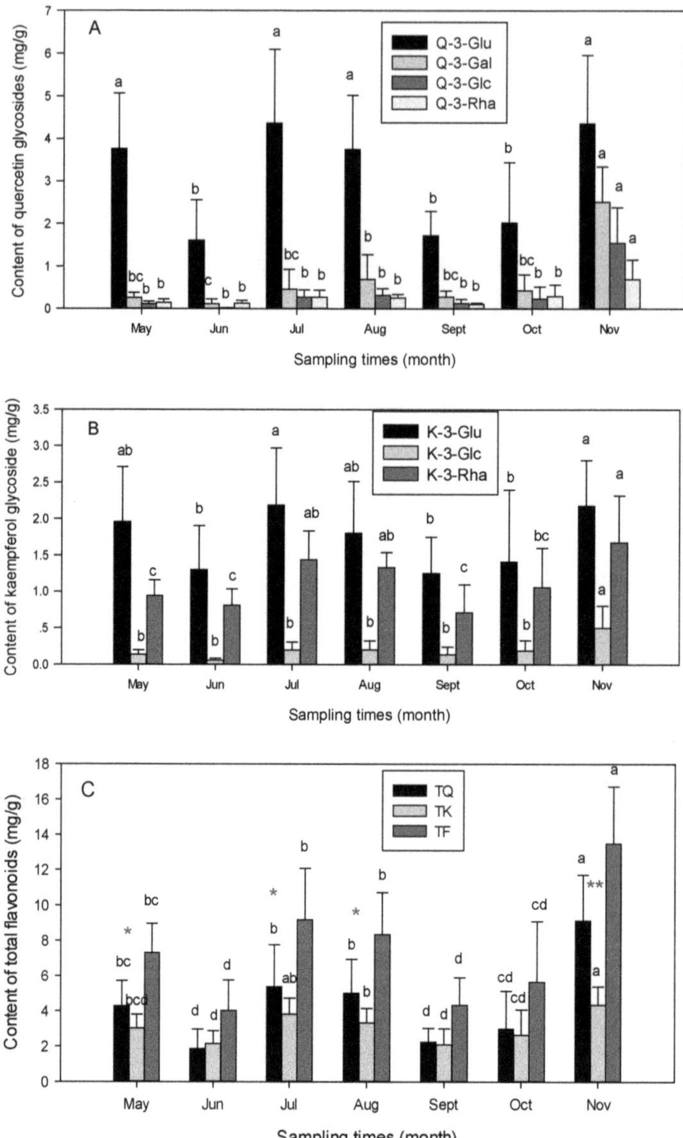

Figure 3. Seasonal variation of quercetin glycosides (**A**), kaempferol glycosides (**B**) and total flavonoids (**C**) in leaves of *C. paliurus* (mean ± SD). Q-3-Glu, Q-3-Gal, Q-3-Glc, Q-3-Rha, K-3-Glu, K-3-Glc, K-3-Rha, TK, TQ, and TF represent quercetin-3-*O*-glucuronide, quercetin-3-*O*-galactoside, quercetin-3-*O*-glucoside, quercetin-3-*O*-rhamnoside, kaempferol-3-*O*-glucuronide, kaempferol-3-*O*-glucoside, kaempferol-3-*O*-rhamnoside, total quercetin glycosides, total kaempferol glycosides, and total flavonoids, respectively. Different letters in this figure indicate significant differences between the sampling times for the same category according to Duncan's test ($p < 0.05$). Each value presented is the mean of 10 families. * and ** indicate variation between total content of quercetin glycosides and total content of kaempferol glycosides in the same month is significant at the 0.05 and 0.01 level, respectively.

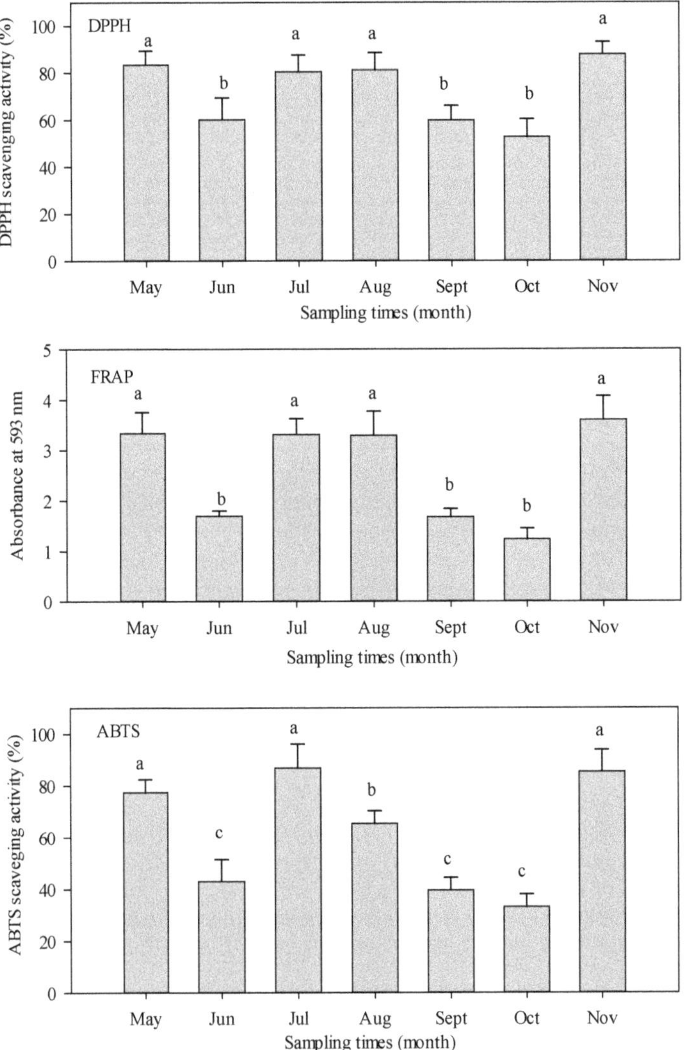

Figure 4. Seasonal variation of antioxidant activities of *C. paliurus* leaf extracts (mean ± SD). Different letters in this figure indicate significant differences between the sampling times for the same category according to Duncan's test ($p < 0.05$). Each value presented is the mean of triplicate of family of 4#.

3.4. Correlation between Contents of Phenolic Compounds and Antioxidant Activity

In order to determine the possible association between phenolic contents and antioxidant capacities of the leaf extracts, Pearson's correlation analysis was conducted based on the results of phenolic contents and antioxidant activities we investigated at seven sampling times (Table 1 and Supplementary Materials Table S1). Strong correlations between antioxidant activities and contents of phenolic acids as well as flavonoids were observed in DPPH, FRAP, and ABTS assays. Moreover, Pearson's correlation coefficients between antioxidant activities and total phenolic acid contents or total quercetin glycosides contents were higher than that between antioxidant activities and total kaempferol glycosides, indicating that the contribution of quercetin glycosides and phenolic acids to antioxidant activity was greater than that of kaempferol glycosides.

Table 1. Pearson correlation coefficients between phenolic contents in *C. paliurus* leaves and antioxidant activities of *C. paliurus* leaf extracts ($n = 21$).

Compounds	DPPH	FRAP	ABTS
Total phenolic aids	0.71 *	0.80 *	0.85 **
Total quercetin glycosides	0.73 *	0.86 **	0.91 **
Total kaempferol glycosides	0.6	0.76 *	0.74 *
Total flavonoids	0.7	0.85 **	0.90 **

* and ** indicate correlation is significant at the 0.05 and 0.01 level, respectively. DPPH: 1,1-diphenyl-2-picrylhydrazyl. ABTS: 2,2′-azino-bis (3-ethylbenzothiazoline-6-sulfonic acid) diammonium salt radical cation. FRAP: ferric reducing antioxidant power.

3.5. Effects of Individual Phenolic Compounds on Antioxidant Capacity

In order to evaluate to what extent these identified phenolic substances contribute to the antioxidative effectiveness of *C. paliurus* extracts, antioxidant capacity assays were performed on each individual phenolic compound by DPPH, ABTS, and FRAP methods (Table 2). It was observed that quercetin exerted the strongest antioxidant activity in DPPH radical scavenging assay with the lowest IC_{50} value of 0.008 mg/mL, followed by quercetin glycosides. Kaempferol and three phenolic acids also showed good DPPH radical scavenging capacity, while a striking difference between DPPH assay results of kaempferol and kaempferol-glycosides was observed, and kaempferol-glycosides were practically invisible to DPPH. Consistent results were observed in the FRAP and ABTS assays. These results indicated that quercetin and its glycosides as well as phenolic acids might be the main contributors to the antioxidant capacity for *C. paliurus* extracts.

Table 2. The antioxidant performances of different phenolic compounds in *C. paliurus* leaves.

Compounds	DPPH (IC_{50} mg/mL)	FRAP (mmol TEAC/g)	ABTS (mmol TEAC/g)
3-*O*-caffeoylquinic acid	0.24 ± 0.003	3.51 ± 0.02	4.58 ± 0.03
4-*O*-caffeoylquinic acid	0.17 ± 0.001	3.38 ± 0.06	4.91 ± 0.07
4,5-di-*O*-caffeoylquinic acid	0.24 ± 0.002	4.70 ± 0.04	5.13 ± 0.03
Quercetin-3-*O*-glucuronide	0. 12 ± 0.000	3.77 ± 0.02	4.44 ± 0.04
Quercetin-3-*O*-galactoside	0.13 ± 0.002	3.33 ± 0.01	4.41 ± 0.01
Quercetin-3-*O*-glucoside	0.13 ± 0.002	2.80 ± 0.03	4.27 ± 0.06
Quercetin-3-*O*-rhamnoside	0.13 ± 0.000	2.65 ± 0.02	4.39 ± 0.06
Kaempferol-3-*O*-glucuronide	inactive	0.22 ± 0.00	0.55 ± 0.01
Kaempferol-3-*O*-glucoside	inactive	0.30 ± 0.0	0.57 ± 0.00
Kaempferol-3-*O*-rhamnoside	inactive	0.30 ± 0.03	0.68 ± 0.04
Quercetin	0,008 ± 0,000	11.25 ± 0.24	6.37 ± 0.21
Kaempferol	0. 26 ± 0.002	3.36 ± 0.05	3.67 ± 0.09

4. Discussion

4.1. Effects of Environment and Development Phase on Leaf Phenolic Accumulation

Phenolic compounds distribute ubiquitously in the plants kingdom [1,2,36,37]. Many studies have reported phenolic compounds possess various biological activities [3–5,23,24]. For the plant itself, one of the well-known and important functions of phenolic compounds is their role in plant defense mechanisms [14,38,39]. As in the case of other plant secondary metabolites, the biosynthesis and accumulation of phenolic compounds can be greatly influenced by many endogenous and exogenous factors. The intrinsic factors contain genotype and physiological condition [15,30,33]. The exogenous factors reflect the biotic and abiotic environmental stimuli that occur during the plant growing period, including feeding of phytophagous insects or herbivorous animals, the availability of light and water, soil composition, temperature, and interaction with pathogens and parasites [13–16,39,40]. In our study, the phenolic profiles were similar in terms of chemical composition throughout the growing season in *C. paliurus* leaves, but they were significantly different in terms of quantity. All families we investigated presented higher values of phenolic substances in May and July. The temporal variation pattern in phenolic compounds presented here is partly in tune with the results of Amaral et al., who attributed the increase in phenolics in walnut in July to the higher solar radiation level [15]. In our previous study, the accumulation of selected flavonoids in *C. paliurus* was found to be stimulated under higher radiation [41]. Tsormpatsisidis et al. (2008) found the contents of total flavonoids in Lollo Rosso lettuce "Revolution" in June was lower compared to that of July, due to the higher radiation levels in July [42]. Phenolics are responsible for protecting the plants from damage caused by radiation, in particular UV-B. Accordingly, exposure to global radiation stimulated the synthesis and accumulation of these shielding compounds, such as flavonoids, especially in the epidermis of fully developed leaves [43]. It is reported that *Ligustrum vulgare* leaves grown under full sunlight exposure contained three-fold more polyphenols than those from the shade side of a bush [44]. Wang et al. also found that the contents of anthocyanin and flavanol in *Vaccinium uliginosum* berries show a growing trend with increase in altitude [45]. In leaves of bilberry (*Vaccinium myrtillus*) exposed to sunlight, an increasing expression of genes encoding phenylalanine ammonium lyase, chalcone synthase, and flavanone 3-β-hydroxylase was found, owing to higher level of reactive oxygen species (ROS) [46].

Reyes et al. reported that the phenolic content of potato tubers can be affected both by longer days and cooler temperature [47]. An enhancement of the total flavonoids in winter wheat was also determined at lower temperatures during cultivation [48], as we found for flavonoids in *C. paliurus* leaves gathered in November. Lower temperatures are related to higher concentrations of flavonoids due to the higher level of ROS [48]. Flavonoids act directly as antioxidants, as they are strong scavengers of ROS. It's reported the mRNAs of phenylalanine ammonia lyase and chalcone synthase are enhanced or accumulated in maize seedlings and *Arabidopsis thaliana* at lower temperatures [49,50]. Herein, it could be concluded that cooler climate might promote the formation of flavonoids in the leaves of *C. paliurus*.

It is well known that plants produce ROS during normal metabolism and under various environmental stress conditions. Additionally, with ongoing plant development an increasing ROS is accompanied by physiological aging [51]. There is increasing evidence that phenolic compounds may act as antioxidants under certain physiological conditions and, thereby, protect plants against oxidative stress [52]. Camptothecin contents in *Camptotheca acuminata* leaves exhibited a decreasing trend as leaves age as well as seasonal progression [53]. Both concentration and composition of secondary metabolites are usually different during tree ontogeny, and can interfere with the quality of medicinal plants [30]. In our study, the higher phenolics content in early November infers that increased phenolic accumulation is likely in senescent leaves of *C. paliurus*.

Previous studies have demonstrated that primary and secondary metabolisms share the common precursors and intermediates, which results in competition for common substrates in the processes of phenolic biosynthesis and growth [54–56]. Ma et al. found the quercetin contents decreased significantly in the leaves of both *Apocynum venetum* and *Poacynum hendersonii* after flowering, because a considerable amount of photosynthates flows into reproductive organs when plants come into reproductive growth [57]. Amaral et al. also indicated the association between the decline of phenolic content in the walnut leaves in June and the rapid development of the fruits at the time, because most of the nutrients and photosynthetic products are employed in fruit growth [40]. In *C. paliurus*, reduced accumulation of phenolics in June implies that higher quantities of photoassimilates are allocated for vegetative growth and reproductive growth. Taken the growth of *C. paliurus* into consideration, the collection of *C. paliurus* leaves should be performed in early November when the weather began to turn cold.

The contents of Q-3-Glu, Q-3-Glc, Q-3-Rha, K-3-Glu, K-3-Glc, and K-3-Rha have already been described in *C. paliurus* leaves, but no information is published about their ratios. In the current study, we investigated seasonal changes of ratios of Q-3-Glu/K-3-Glu, Q-3-Glc/K-3-Glc, and Q-3-Rha/K-3-Rha. As shown in Figure 5, the ratios of Q-3-Glu/K-3-Glu, Q-3-Glc/K-3-Glc, and Q-3-Rha/K-3-Rha ranged from 1.43 to 2.46, 0.22 to 3.34, 0.15 and 0.40, respectively, demonstrating a wider range of seasonal variation with exception of Q-3-Rha/K-3-Rha. These results revealed that the flavonoid composition ratios in *C. paliurus* leaves were markedly affected by environmental factors, even if the detailed mechanisms affecting this are not very clear. The value of Q-3-Glu/K-3-Glu was always more than 1.0 during the whole growing season, indicating that Q-3-Glu became the predominant flavonoid accompanied with changed environment. It is reported that a more pronounced increase of quercetin glycosides than that of kaempferol glycosides was induced by ultraviolet light [58]. Schmidt et al. also found the responding of quercetin and kaempferol to radiation was different, and quercetin may be more sensitive to the variation of environmental factors, as shown in the present studies, but the detailed mechanisms are not very clear [59].

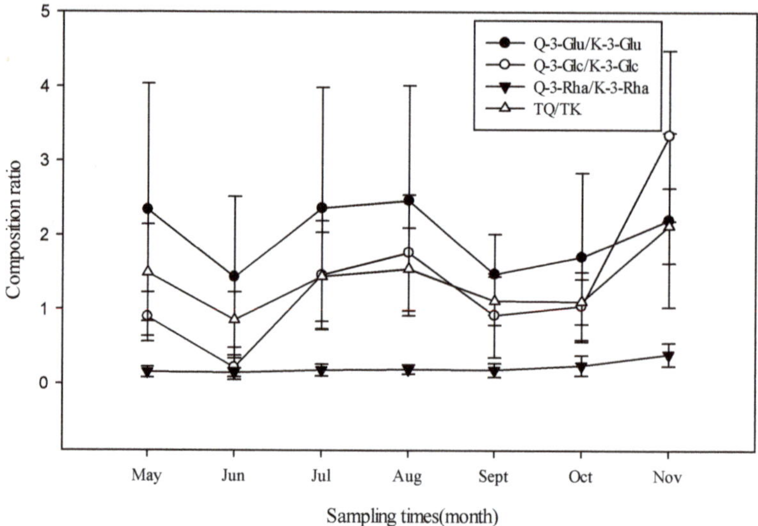

Figure 5. Seasonal variation in ratios of quercetin glycosides to kaempferol glycosides for the leaves of *C. paliurus*. The meaning of Q-3-Glu, Q-3-Glc, Q-3-Rha, K-3-Glu, K-3-Glc, K-3-Rha, TQ, and TK is the same as Figure 3. Each value presented is the mean of 10 families.

4.2. Relationship between Phenolic Composition and Antioxidant Capacity

As an ingredient of dietary or traditional medicine, both flavonoids and phenolic acids possess high antioxidant capacities and hence protect humans against free radical-induced diseases [59,60]. In *C. paliurus*, flavonoids are generally present as quercetin glycosides and kaempferol glycosides, while the aglycones of quercetin and kaempferol are found in quantitatively non-detectable traces [25]. Our results found the antioxidant activity of flavonoids glycosides was dramatically reduced compared to its aglycone (Table 2). Furthermore, a positive correlation between antioxidant activity and contents of phenolic acids as well as flavonoids was observed in DPPH, FRAP, and ABTS assays (Table 1). Our result was in accordance with the recent publications, which suggested the higher the phenolic content, the stronger the antioxidant activity [61,62]. Therefore, the *C. paliurus* leaves with the high content of phenolic compounds showed a great potential as a natural source of antioxidant for health promotion.

The different antioxidant capacity presented by phenolic compounds is closely related to their chemical structures in regard to the number and location of phenolic hydroxyl groups [63]. The flavonoids are a series of compounds consisting of a backbone with two benzene rings linked by a pyran chain (C_6-C_3-C_6) (Figure 6). It is well-known that hydroxyl substitution is essential for the antioxidant activity of a flavonoid [64]. Generally, compounds with more hydroxyl substitutions on the B ring might show stronger antioxidant activity [65]. Accordingly, quercetin with a 3′,4′-catechol substitution in the B ring revealed more effective radical scavenging capacity than kaempferol that has a single 4′-hydroxy in the B ring (Table 2). However, flavonoids are generally present as glycosides in plants, whereas glycosylated flavonoids were found have lower antioxidant capacity than their corresponding aglycones, which was conformable to our result [64,66]. In addition, there were likely synergistic interactions among the phenolics in *C. paliurus* leaves that contribute to its excellent antioxidant capacity, but further research is needed to illuminate the synergistic effects of phenolics in *C. paliurus* leaf extracts.

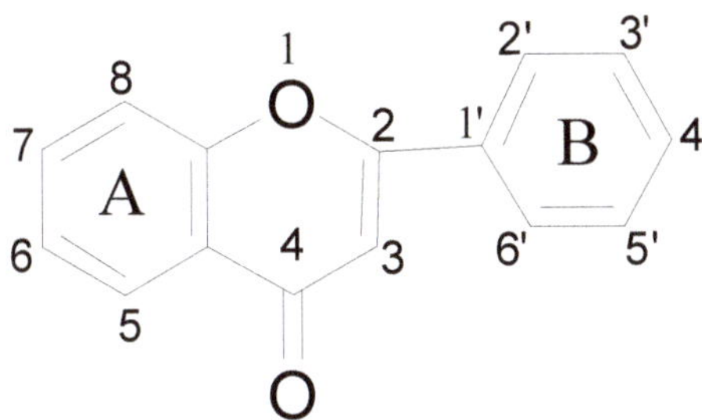

Figure 6. Chemical structure of flavonoids used in the present study.

5. Conclusions

In conclusion, our results clearly demonstrated that there existed higher content of phenolics in leaves of *C. paliurus*, and relatively higher antioxidant activities were observed in *C. paliurus* leaf extract. Both phenolic concentrations and antioxidant activities in the *C. paliurus* extracts showed a significant seasonal variation, and variation patterns were similar. The contents of phenolic compounds in *C. paliurus* leaf extract were strongly correlated with antioxidant activities, indicating that extracts with higher phenolic contents had higher antioxidant activities. Moreover, the antioxidant capacity of phenolics is dependent upon the arrangement of functional group on the nuclear structure. Overall, harvest season significantly impacts the contents of phenolics in the leaves of *C. paliurus*, and consequently affects the antioxidant activity. Based on the results, we suggest early November is the optimal time to harvest *C. paliurus* leaves. As a potential source of natural antioxidants, *C. paliurus* could be very useful supplements for pharmaceutical products and functional food ingredients in both nutraceutical and food industries.

Supplementary Materials: The following are available online at http://www.mdpi.com/1999-4907/10/8/624/s1, Table S1: Pearson correlation coefficients between contents of individual phenolics in *C. paliurus* leaves and antioxidant activities of *C. paliurus* leaf extracts (n = 21).

Author Contributions: S.F. and X.F. conceived and designed the experiments; X.S. and W.Y. collected the leaf samples; Y.C. performed the experiments, analyzed the data, and wrote the manuscript. S.F. revised the manuscript.

Funding: This work was funded by the National Natural Science Foundation of China (No. 31470637) and the Priority Academic Program Development of Jiangsu Higher Education Institutions (PAPD).

Conflicts of Interest: The authors declare no conflict of interest.

References

1. Del Rio, D.; Rodriguez-Mateos, A.; Spencer, J.P.E.; Tognolini, M.; Borges, G.; Crozier, A. Dietary (poly) phenolics in human health: Structures, bioavailability, and evidence of protective effects against chronic diseases. *Antioxid. Redox Signal.* **2013**, *18*, 1818–1892. [CrossRef] [PubMed]

2. Crozier, A.; Jaganath, I.B.; Clifford, M.N. Dietary phenolics: Chemistry, bioavailability and effects on health. *Nat. Prod. Rep.* **2009**, *26*, 1001–1043. [CrossRef] [PubMed]

3. Cicerale, S.; Lucas, L.J.; Keast, R.S.J. Antimicrobial, antioxidant and anti-inflammatory phenolic activities in extra virgin olive oil. *Curr. Opin. Biotechnol.* **2012**, *23*, 129–135. [CrossRef] [PubMed]

4. Noratto, G.D.; Bertoldi, M.C.; Krenek, K.; Talcott, S.T.; Stringheta, P.C.; Mertens-Talcott, S.U. Anticarcinogenic effects of polyphenolics from mango (*Mangifera indica*) varieties. *J. Agric. Food Chem.* **2010**, *58*, 4104–4112. [CrossRef] [PubMed]

5. Yang, D.J.; Chang, Y.Y.; Hsu, C.L.; Liu, C.W.; Lin, Y.L.; Lin, Y.H.; Liu, K.C.; Chen, Y.C. Antiobesity and hypolipidemic effects of polyphenol-rich longan (*Dimocarpus longans* Lour.) flower water extract in hypercaloric-dietary rats. *J. Agric. Food Chem.* **2010**, *58*, 2020–2027. [CrossRef] [PubMed]

6. Dillard, C.J.; German, J.B. Phytochemicals: nutraceuticals and human health. *J. Sci. Food Agric.* **2000**, *80*, 1744–1756. [CrossRef]

7. Su, L.; Yin, J.J.; Charles, D.; Zhou, K.; Moore, J.; Yu, L. Total phenolic contents, chelating capacities, and radical-scavenging properties of black peppercorn, nutmeg, rosehip, cinnamon and oregano leaf. *Food Chem.* **2007**, *100*, 990–997. [CrossRef]

8. Pereira, D.M.; Valentão, P.; Pereira, J.A.; Andrade, P.B. Phenolics: From chemistry to biology. *Molecules* **2009**, *14*, 2202–2211. [CrossRef]

9. Parr, A.J.; Bolwell, G.P. Phenols in the plant and in man. The potential for possible nutritional enhancement of the diet by modifying the phenols content or profile. *J. Sci. Food Agric.* **2000**, *80*, 985–1012. [CrossRef]

10. Ghosh, S.; Padilla-González, G.F.; Rangan, L. *Alpinia nigra* seeds: A potential source of free radical scavenger and antibacterial agent. *Ind. Crops Prod.* **2013**, *49*, 348–356. [CrossRef]

11. Stanojević, L.; Stanković, M.; Nikolić, V.; Nikolić, L.; Ristić, D.; Čanadanovic-Brunet, J.; Tumbas, V. Antioxidant activity and total phenolic and flavonoid contents of *Hieracium pilosella* L. extracts. *Sensors* **2009**, *9*, 5702–5714. [CrossRef] [PubMed]

12. Zhang, J.; Shen, Q.; Lu, J.C.; Li, J.Y.; Liu, W.Y.; Yang, J.J.; Li, J.; Xiao, K. Phenolic compounds from the leaves of *Cyclocarya paliurus* (Batal.) Ijinskaja and their inhibitory activity against PTP1B. *Food Chem.* **2010**, *119*, 1491–1496. [CrossRef]

13. Treutter, D. Biosynthesis of phenolic compounds and its regulation in apple. *Plant Growth Regul.* **2001**, *34*, 71–89. [CrossRef]

14. Bennett, R.N.; Wallsgrove, R.M. Secondary metabolites in plant defense mechanisms. *New Phytol.* **1994**, *127*, 617–633. [CrossRef]

15. Amaral, J.S.; Seabra, R.M.; Andrade, P.B.; Valentão, P.; Pereira, J.A.; Ferreres, F. Phenolic profile in the quality control of walnut (*Juglans regia* L.) leaves. *Food Chem.* **2004**, *88*, 373–379. [CrossRef]

16. Radix, P.; Bastien, C.; Jay-Allemand, C.; Charlot, G.; Seigle-Murandi, F. The influence of soil nature on polyphenols in walnut tissues. A possible explanation of differences in the expression of walnut blight. *Agronomie* **1998**, *18*, 627–637. [CrossRef]

17. Fang, S.Z.; Wang, J.Y.; Wei, Z.Y.; Zhu, Z.X. Methods to break seed dormancy in *Cyclocarya paliurus* (Batal.) Iljinskaja. *Sci. Hortic.* **2006**, *110*, 305–309. [CrossRef]

18. Birari, R.B.; Bhutani, K.K. Pancreatic lipase inhibitors from natural source: unexplored potential. *Drug Discov. Today* **2007**, *12*, 879–889. [CrossRef]

19. Hou, X.L.; Liu, X.X.; Wang, S.; Zhou, X.L.; Wei, F.; Miao, J.H. Effect of the flavonoids from *Cyclocarya paliurus* on spontaneous hypertension rats. *Pharm. Clin. Chin. Mater. Med.* **2014**, *30*, 62–69.

20. Kurihara, H.; Fukami, H.; Kusumoto, A.; Toyoda, Y.; Shibata, H.; Matsui, Y. Hypoglycemic action of *Cyclocarya paliurus* (Batal.) Iljinskaja in normal and diabetic mice. *Biosci. Biotechnol. Biochem.* **2003**, *67*, 877–880. [CrossRef]

21. Wu, Z.F.; Gao, T.H.; Zhong, R.L.; Lin, Z.; Jiang, C.H.; Ouyang, S.; Zhao, M.; Che, C.T.; Zhang, J.; Yin, Z.Q. Antihyperlipidaemic effect of triterpenic acid-enriched fraction from *Cyclocarya paliurus* leaves in hyperlipidaemic rats. *Pharm. Biol.* **2017**, *55*, 712–721. [CrossRef] [PubMed]

22. Liu, X.; Xie, J.H.; Jia, S.; Huang, L.X.; Wang, Z.J.; Li, C.; Xie, M.Y. Immunomodulatory effects of an acetylated *Cyclocarya paliurus* polysaccharide on murine macrophages raw264.7. *Int. J. Biol. Macromol.* **2017**, *98*, 576–581. [CrossRef] [PubMed]

23. Xie, J.H.; Liu, X.; Shen, M.Y.; Nie, S.P.; Zhang, H.; Li, C.; Gong, D.M.; Xie, M.Y. Purification, physicochemical characterization and anticancer activity of a polysaccharide from *Cyclocarya paliurus* leaves. *Food Chem.* **2013**, *136*, 1453–1460. [CrossRef] [PubMed]

24. Xie, J.H.; Shen, M.Y.; Xie, M.Y.; Nie, S.P.; Chen, Y.; Li, C.; Huang, D.F.; Wang, Y.X. Ultrasonic-assisted extraction, antimicrobial and antioxidant activities of *Cyclocarya paliurus* (Batal.) Iljinskaja polysaccharides. *Carbohydr. Polym.* **2012**, *89*, 177–184. [CrossRef] [PubMed]

25. Cao, Y.N.; Fang, S.Z.; Yin, Z.Q.; Fu, X.X.; Shang, X.L.; Yang, W.X.; Yang, H.M. Chemical fingerprint and multicomponent quantitative analysis for the quality evaluation of *Cyclocarya paliurus* leaves by HPLC–Q–TOF–MS. *Molecules* **2017**, *22*, 1927. [CrossRef] [PubMed]

26. Shu, R.G.; Xu, C.R.; Li, L.N.; Yu, Z.L. Cyclocariosides II and III: two secodammarane triterpenoid saponins from *Cyclocarya paliurus*. *Planta Med.* **1995**, *61*, 551–555. [CrossRef] [PubMed]

27. Kennelly, E.J.; Cai, L.; Long, L.; Shamon, L.; Zaw, K.; Zhou, B.N.; Pezzuto, J.M.; Kinghorn, A.D. Novel highly sweet secodammarane glycosides from *Pterocarya paliurus*. *J. Agric. Food Chem.* **1995**, *43*, 2602–2607. [CrossRef]

28. Xie, J.H.; Xie, M.Y.; Nie, S.P.; Shen, M.Y.; Wang, Y.X.; Li, C. Isolation, chemical composition and antioxidant activities of a watersoluble polysaccharide from *Cyclocarya paliurus* (Batal.) Iljinskaja. *Food Chem.* **2010**, *119*, 1626–1632. [CrossRef]

29. Deng, B.; Fang, S.Z.; Shang, X.L.; Fu, X.X.; Li, Y. Influence of provenance and shade on biomass production and triterpenoid accumulation in *Cyclocarya paliurus*. *Agrofor. Syst.* **2019**, *93*, 483–492. [CrossRef]

30. Cao, Y.N.; Deng, B.; Fang, S.Z.; Shang, X.L.; Fu, X.X.; Yang, W.X. Genotypic variation in tree growth and selected flavonoids in leaves of *Cyclocarya paliurus*. *South. For.* **2018**, *80*, 67–74. [CrossRef]

31. Liu, Z.; Carpenter, S.B.; Bourgeois, W.J.; Yu, Y.; Constantin, R.J.; Falcon, M.J.; Adams, J.C. Variations in the secondary metabolite camptothecin in relation to tissue age and season in *Camptotheca acuminata*. *Tree Physiol.* **1998**, *18*, 265–270. [CrossRef] [PubMed]

32. Dong, T.T.; Cui, X.M.; Song, Z.H.; Zhao, K.J.; Ji, Z.N.; Lo, C.K.; Tsim, K.W. Chemical assessment of roots of *Panax notoginseng* in China: regional and seasonal variations in its active constituents. *J. Agric. Food Chem.* **2003**, *51*, 4617–4623. [CrossRef] [PubMed]

33. González-Molina, E.; Moreno, D.A.; García-Viguera, C. Genotype and harvest time influence the phytochemical quality of Fino lemon juice (*Citrus limon* (L.) Burm. F.) for industrial use. *J. Agric. Food Chem.* **2008**, *56*, 1669–1675. [CrossRef] [PubMed]

34. Fu, X.X.; Zhou, X.D.; Deng, B.; Shang, X.L.; Fang, S.Z. Seasonal and genotypic variation of water-soluble polysaccharide content in leaves of *Cyclocarya paliurus*. *South. For.* **2015**, *77*, 231–236. [CrossRef]

35. Fang, S.Z.; Yang, W.X.; Chu, X.L.; Shang, X.L.; She, C.Q.; Fu, X.X. Provenance and temporal variations in selected flavonoids in leaves of *Cyclocarya paliurus*. *Food Chem.* **2011**, *124*, 1382–1386. [CrossRef]

36. Moyer, R.A.; Hummer, K.E.; Finn, C.E.; Frei, B.; Wrolstad, R.E. Anthocyanins, phenolics, and antioxidant capacity in diverse small fruits: vaccinium, rubus, and ribes. *J. Agric. Food Chem.* **2002**, *50*, 519–525. [CrossRef] [PubMed]

37. Kalt, W.; Ryan, D.A.; Duy, J.C.; Prior, R.L.; Ehlenfeldt, M.K.; Kloet, S.P. Interspecific variation in anthocyanins, phenolics, and antioxidant capacity among genotypes of highbush and lowbush blueberries (Vaccinium section cyanococcus spp.). *J. Agric. Food Chem.* **2001**, *49*, 4761–4767. [CrossRef] [PubMed]

38. Rühmann, S.; Leser, C.; Bannert, M.; Treutter, D. Relationship between growth, secondary metabolism, and resistance of apple. *Plant Biol.* **2002**, *4*, 137–143. [CrossRef]

39. Dixon, R.A.; Paiva, N.L. Stress-induced phenylpropanoid metabolism. *Plant Cell* **1995**, *7*, 1085–1097. [CrossRef]

40. Lindroth, R.L.; Peterson, S.S. Effects of plant phenols of performance of southern armyworm larvae. *Oecologia* **1988**, *75*, 185–189. [CrossRef]

41. Deng, B.; Shang, X.L.; Fang, S.Z.; Li, Q.Q.; Fu, X.X.; Su, J. Integrated effects of light intensity and fertilization on growth and flavonoid accumulation in *Cyclocarya paliurus*. *J. Agric. Food Chem.* **2012**, *60*, 6286–6292. [CrossRef] [PubMed]

42. Tsormpatsidis, E.; Henbest, R.G.C.; Davis, F.J.; Battey, N.H.; Hadley, P.; Wagstaffe, A. UV irradiance as a major influence on growth, development and secondary products of commercial importance in Lollo Rosso lettuce 'Revolution' grown under polyethylene films. *Environ. Exp. Bot.* **2008**, *63*, 232–239. [CrossRef]

43. Edreva, A. The importance of non-photosynthetic pigments and cinnamic acid derivatives in the photoprotection. *Agric. Ecosyst. Environ.* **2005**, *106*, 135–146. [CrossRef]

44. Tattini, M.; Galardi, C.; Pinelli, P.; Massai, R.; Remorini, D.; Aggati, G. Differential accumulation of flavonoids and hydroxycinnamates in leaves of *Ligustrum vulgare* under excess light and drought stress. *New Phytol.* **2004**, *163*, 547–561. [CrossRef]

45. Wang, L.J.; Su, S.; Wu, J.; Du, H.; Li, S.S.; Huo, J.W.; Zhang, Y.; Wang, L.S. Variation of anthocyanins and flavonols in *Vaccinium uliginosum* berry in lesser khingan mountains and its antioxidant activity. *Food Chem.* **2014**, *160*, 357–364. [CrossRef] [PubMed]

46. Jaakola, L.; Määttä-Riihinen, K.; Kärenlampi, S.; Hohtola, A. Activation of flavonoid biosynthesis by solar radiation in bilberry (*Vaccinium myrtillus* L.) leaves. *Planta* **2004**, *218*, 721–728. [PubMed]

47. Reyes, L.F.; Miller, J.C.; Cisneros-Zevallos, L. Environmental conditions influence the content and yield of anthocyanins and total phenolics in purple- and red-flesh potatoes during tuber development. *Am. J. Potato Res.* **2004**, *81*, 187–193. [CrossRef]

48. Klimov, S.V.; Burakhanova, E.A.; Dubinina, I.M.; Alieva, G.P.; Sal'nikova, E.B.; Olenichenko, N.A.; Zagoskina, N.V.; Trunova, T.I. Suppression of the source activity affects carbon distribution and frost hardiness of vegetating winter wheat plants. *Russ. J. Plant Physiol.* **2008**, *55*, 308–314. [CrossRef]

49. Christie, P.J.; Alfenito, M.R.; Walbot, V. Impact of low-temperature stress on general phenylpropanoid and anthocyanin pathways: Enhancement of transcript abundance and anthocyanin pigmentation in maize seedlings. *Planta* **1994**, *194*, 541–549. [CrossRef]

50. Leyva, A.; Jarillo, J.A.; Salinas, J.; Martinez-Zapater, J.M. Low-temperature induces the accumulation of *Phenylalanine Ammonia-Lyase* and *Chalcone Synthase* m-RNAs of *Arabidopsis thaliana* in a light-dependent manner. *Plant Physiol.* **1995**, *108*, 39–46. [CrossRef]

51. Vogt, T.; Gul, P.G. Accumulation of flavonoids during leaf development in *Citrus laurifolius*. *Phytochemistry* **1994**, *36*, 591–597. [CrossRef]

52. Crespo-Sempere, A.; Selma-Lázaro, C.; Palumbo, J.D.; González-Candelas, L.; Martínez-Culebras, P.V. Effect of oxidant stressors and phenolic antioxidants on the ochratoxigenic fungus *Aspergillus carbonarius*. *J. Sci. Food Agric.* **2016**, *96*, 169–177. [CrossRef] [PubMed]

53. Liu, Z.J.; Zhou, G.M.; Xu, S.Y.; Wu, J.S.; Yin, E.Q. Provenance variation in camptothecin concentrations of *Camptotheca acuminata* grown in China. *New For.* **2002**, *24*, 215–224. [CrossRef]

54. Margna, U. Control at the level of substrate supply-an alternative in the regulation of phenylpropanoid accumulation in plant cells. *Phytochemistry* **1977**, *16*, 419–426. [CrossRef]

55. Margna, U.; Margna, E.; VainjÄrv, T. Influence of nitrogen nutrition on the utilization of L-phenylalanine for building flavonoids in buckwheat seedling tissues. *J. Plant Physiol.* **1989**, *134*, 697–702. [CrossRef]

56. Phillips, R.; Henshaw, G.G. The regulation of synthesis of phenolics in stationary phase cell cultures of *Acer pseudoplatanus* L. *J. Exp. Bot.* **1977**, *28*, 785–794. [CrossRef]

57. Ma, M.; Hong, C.L.; An, S.Q.; Li, B. Seasonal, spatial, and interspecific variation in quercetin in *Apocynum venetum* and *Poacynum hendersonii*, Chinese traditional herbal teas. *J. Agric. Food Chem.* **2003**, *51*, 2390–2393. [CrossRef]

58. Hofmann, R.W.; Swinny, E.E.; Bloor, S.J.; Markham, K.R.; Ryan, K.G.; Campbell, B.D.; Jordan, B.R.; Fountain, D.W. Responses of nine *Trifolium repens* L. populations to ultraviolet-B radiation: Differential flavonol glycoside accumulation and biomass production. *Ann. Bot.* **2000**, *86*, 527–537. [CrossRef]

59. Schmidt, S.; Zietz, M.; Schreiner, M.; Rohn, S.; Kroh, L.W.; Krumbein, A. Genotypic and climatic influence on the concentration and composition of flavonoids in kales (*Brassica oleracea* var. sabellica). *Food Chem.* **2010**, *119*, 1293–1299. [CrossRef]

60. André, C.M.; Oufir, M.; Hoffmann, L.; Hausman, J.F.; Rogez, H.; Larondelle, Y.; Evers, D. Influence of environment and genotype on polyphenol compounds and vitro antioxidant capacity of native Andean potatoes (*Solanum tuberosum* L.). *J. Food Compos. Anal.* **2009**, *22*, 517–524. [CrossRef]

61. Zhang, W.N.; Zhao, X.Y.; Sun, C.D.; Li, X.; Chen, K.S. Phenolic composition from different loquat, (*Eriobotrya japonica* Lindl.) cultivars grown in China, and their antioxidant properties. *Molecules* **2015**, *20*, 542–555. [CrossRef] [PubMed]

62. Kim, Y.; Goodner, K.L.; Park, J.D.; Choi, J.; Talcott, S.T. Changes in antioxidant phytochemicals and volatile composition of *Camellia sinensis* by oxidation during tea fermentation. *Food Chem.* **2011**, *129*, 1331–1342. [CrossRef]

63. Csepregi, K.; Neugart, S.; Schreiner, M.; Hideg, É. Comparative evaluation of total antioxidant capacities of plant polyphenols. *Molecules* **2016**, *21*, 208–224. [CrossRef] [PubMed]

64. Rice-Evans, C.A.; Miller, N.J.; Paganga, G. Structure antioxidant activity relationships of flavonoids and phenolic acids. *Free Radic. Biol. Med.* **1996**, *20*, 933–956. [CrossRef]

65. Cao, G.H.; Sofic, E.; Prior, R.L. Antioxidant and prooxidant behavior of flavonoids: Structure–activity relationships. *Free Radic. Biol. Med.* **1997**, *22*, 749–760. [CrossRef]
66. Burda, S.; Oleszek, W. Antioxidant and antiradical activities of flavonoids. *J. Agric. Food Chem.* **2001**, *49*, 2774–2779. [CrossRef]

Article

Analysis of the Essential Oils of *Chamaemelum fuscatum* (Brot.) Vasc. from Spain as a Contribution to Reinforce Its Ethnobotanical Use

Marcos Fernández-Cervantes [1], María José Pérez-Alonso [1], José Blanco-Salas [2,*],
Ana Cristina Soria [3] and Trinidad Ruiz-Téllez [2]

[1] Department of Biodiversity, Ecology and Evolution. Universidad Complutense, 28071 Madrid, Spain
[2] Department of Vegetal Biology, Ecology and Earth Science, Faculty of Sciences, University of Extremadura, 06071 Badajoz, Spain
[3] Instituto de Química Orgánica General (CSIC), Juan de la Cierva, 3, 28006 Madrid, Spain
* Correspondence: blanco_salas@unex.es; Tel.: +34-9242-89300

Received: 24 May 2019; Accepted: 24 June 2019; Published: 27 June 2019

Abstract: *Chamaemelum fuscatum* (Brot.) Vasc. is a south west Iberian chamomile that has been traditionally used as folk medicine in its natural distribution area but currently it is underestimated regarding its biological activities. For this reason, it is proposed in this paper to get insight into the scientific validation of the traditional knowledge of this plant with the aim of taking advantage of its anti-inflammatory, gastroprotective and antinociceptive activities, among others. To this aim, the chemical composition of the essential oil from the whole plant, the flowers and the green parts of this plant has been evaluated by gas chromatography–mass spectrometry (GC–MS). Plant materials were collected in Badajoz (Spain). A total of 61 components including monoterpenoids, sesquiterpenoids and aliphatic esters were identified. (E)-2-Methyl-2-butenyl methacrylate (27.57%–18.53%) and 2-methylallyl isobutyrate (9.79%–7.51%) were the most abundant compounds in the essential oils of flowers and of the whole plant, whereas α-curcumene, trans-pinocarveol, α-bergamotene and pinocarvone were the major terpenoids irrespective of the plant part considered. Certain compounds showing a relative high abundance as isobutyl methacrylate, isoamyl butyrate, α-bergamotene and pinocarvone were identified for the first time in this species. Finally, we have reviewed the bioactivity of several compounds to relate the ethnobotanical use of this plant in Spain with its volatile profile. This work is a preliminary contribution to reinforce the use to this Mediterranean endemic plant as a natural source of bioactives.

Keywords: *Chamaemelum fuscatum*; chamomile; essential oil; aliphatic esters; methacrylate; Compositae; Mediterranean

---

1. Introduction

Ethnobotanical studies often bring to light interesting uses to be further validated. This is the case of a little-known west Mediterranean chamomile, *Chamaemelum fuscatum* (Brot.) Vasc., Asteraceae, frequent in south west of the Iberian Peninsula, forming part of terophytic pastures on wet or temporarily flooded substrates, from 100 to 900 m [1,2]. It is an annual and aromatic little daisy (2–7 cm) easy to identify by its involucral bracts and interfloral scales. It flowers from October to May but mainly in the late winter [1,2].

C. *fuscatum* has traditionally been used as a medicinal plant in Iberian rural areas. The flowers of the species are the most frequently used part [2]. Most of the verified uses of this species have been described for its natural distribution area that roughly coincides with the Luso-Extremadurean region (Table 1).

Table 1. Traditional uses for *Chamaemelum fuscatum* in Spain.

Uses Upon IECTB [1]	Reference	Location	Part Used	Formulation	Preparation	Popular Uses
HUMAN CONSUMPTION						
Non-alcoholic beverage	[3]	Cáceres	Inflorescence	Infusion		Drink
ANIMAL FEEDING						
Fodder	[4]	Jaén	Inflorescence			Birdseed
MEDICINAL USE						
Digestive system	[5]	Badajoz	Inflorescence	Decoction	Mouthwash	Swollen gums
Digestive system	[3,5]	Cáceres Badajoz	Inflorescence	Infusion	Drink (sweetened with honey or sugar)	Digestive problems
Digestive system	[3]	Cáceres	Inflorescence	Infusion	Drink (mixed with olive oil)	Indigestion
Digestive system	[3]	Cáceres	Inflorescence	Infusion	Drink (mixed with olive oil)	Colic pains
Digestive system	[3,5]	Cáceres Badajoz	Inflorescence	Infusion	Drink	Laxative
Genito-urinary system	[5]	Badajoz	Inflorescence	Decoction	Washing of affected area	Relieve vaginal itching
Respiratory system	[3]	Cáceres	Inflorescence	Decoction	Syrup (mixed with lemon and honey)	
Musculature and skeleton	[6]	Badajoz	Inflorescence	Decoction	Drink	Antirheumatic
Nervous system and mental illness	[3,6]	Cáceres Badajoz	Inflorescence	Infusion	Drink (in cáceres sweetened with honey or sugar)	Sedative
Sense organs	[3,5]	Cáceres Badajoz	Inflorescence	Infusion (Cáceres); decoction (Badajoz)	Eye drops	Eye irritation, conjunctivitis
Other infectious and parasitic diseases	[5]	Badajoz	Inflorescence	Infusion	Drenching of affected area	Herpes ("feve")
VETERINARY USE						
Digestive system	[7]	Cáceres	Inflorescence	Infusion	Drink (mixed with olive oil, brandy or bicarbonate)	Digestive problems, lack of rumination, accumulation of gases
Musculature and skeleton	[7]	Badajoz	Green parts and leaves	Decoction	Rubbing	Lameness and inflammation
Skin and subcutaneous tissue	[7]	Badajoz	Green parts and leaves	Decoction	Rubbing	Wounds

Table 1. *Cont.*

Uses Upon IECTB [1]	Reference	Location	Part Used	Formulation	Preparation	Popular Uses
HUMAN CONSUMPTION						
Non-alcoholic beverage	[3]	Cáceres	Inflorescence	Infusion		Drink
ANIMAL FEEDING						
Fodder	[4]	Jaén	Inflorescence			Birdseed
INDUSTRY AND CRAFTS						
Cosmetics, perfumes and cleaning products	[3]	Cáceres	Whole plant	Decoction		Dyed blond hair
Clothing and personal adornment	[3]	Cáceres	Inflorescence			Personal adornment

[1] IECTB: Spanish Inventory of Traditional Knowledge Concerning Biodiversity.

The medicinal applications of *C. fuscatum* for digestive pathologies, as antiseptic or anti-inflammatory remedy for external use were those traditionally in force [2]. However, and despite its multiple bioactivities, the current use of this species is less than that of other chamomiles. Its harvesting for marketing today is insignificant and it remains someway underestimated [2].

Regarding the chemical composition of *C. fuscatum*, it started to be studied in the eighties at Salamanca University (Spain). De Pascual Teresa et al. [8] analyzed the hexane extract of the aerial parts of the plant, and first identified the methacrylic esters of 2-methyl-2-*(E)*-butenol, 2-hydroxy-2-methyl-3-butenol and 2-hydroxy-2-methyl-3-oxobutanol by spectral measurements and analysis of the corresponding standards. After that, these authors used spectroscopic methods and chemical transformations to elucidate the structures of different eudesmanolides in the chloroform extract of the aerial parts of this plant [9], and two years later, they reported four new eudesmanolides in the same extract and a new aliphatic ester (2-methylene-3-oxobutyl methacrylate) in the essential oil isolated from flowers [10]. Since then, and as far as we know, no information has been published on the chemical profile of *C. fuscatum*. Therefore, it is proposed as the main objective of this work to undertake a detailed characterization by gas chromatography–mass spectrometry (GC–MS) of *C. fuscatum* essential oils, whose composition is still deeply unknown. Moreover, a comprehensive bibliographic review of the bioactivity of the compounds here determined has been done to relate the ethnobotanical use of this plant in Spain with its volatile profile. It is expected that getting insight into the scientific validation of the traditional use of this plant is a contribution to confer more value to *C. fuscatum* as a new source of bioactives and, therefore, it also contributes to guarantee its preservation.

2. Materials and Methods

2.1. Bibliographic Prospection

After an exhaustive review of the Spanish ethnobotanical literature following the methodology of the Spanish Inventory of Traditional Knowledge Concerning Biodiversity (IECTB) [11], the traditional uses of this species in Spain was summarized in this paper. This information was complemented with a bibliographic review aimed to get insight into the chemical composition and pharmacological activities of this species. The following resources were accessed: Academic Search Complete, Agricola, Agris, Biosis, CAB Abstracts, Cochrane, Cybertesis, Dialnet, Directory of Open Access Journals, Embase, Espacenet, Google Academics, Google Patents, Medline, PubMed, Science Direct, Scopus, Teseo and Web of Science by the Institute for Scientific Information (ISI).

As for the systematic review of pharmacologically active components of *C. fuscatum*, the general procedures of Prisma 2009 Flow Diagram [12] were followed. Keywords selected as search terms were the compounds listed in Table 2 with a concentration higher than 1% and "activity" or "pharmacol*" or "biological activity".

2.2. Plant Material

C. fuscatum samples were collected in March 2017 in Cerro del Viento (Badajoz, Spain) at the following coordinates: 29SPD70, 38°51′46.2″ N, 6°58′35.15″ W. A voucher specimen with reference number HSS 68,118 was lodged at the Herbarium of the Research Center Finca La Orden, CicyTex, Junta de Extremadura, Badajoz (Spain). Samples collected at their flowering stage were air-dried and stored in the dark at 10–15 °C until distillation.

2.3. Isolation of C. fuscatum Essential Oil

The essential oils from the different parts (flowers (F), whole plant (WP) and green parts (GP), which includes a mixture of leaves and stems), of *C. fuscatum* were isolated by hydrodistillation with cohobation for 8 h, using a Clevenger modified apparatus, according to the method recommended in the Spanish Pharmacopeia. The oils were dried over anhydrous magnesium sulphate and stored at 4

°C in the dark until analysis. The extraction yield (%) was calculated as the amount of essential oil (in g) extracted by hydrodistillation from 100 g of dry plant.

2.4. Gas Chromatography–Mass Spectrometry (GC–MS) Analysis

GC–MS analyses were carried out on a 7890A gas chromatograph coupled to a 5975C quadrupole mass detector (both from Agilent Technologies, Palo Alto, CA, USA), using He at ~1 mL min^{-1} as carrier gas. Injections were carried out in split mode (1:20) at 250 °C. Separations were performed using a Zebron 5% phenylmethyl silicone column (30 m × 0.25 mm, 0.25 µm film thickness) from Phenomenex (Madrid, Spain). Oven temperature program was raised from 70 °C (0.5 min) to 290 °C (30 min) at 6 °C min^{-1}. Mass spectra were recorded in electron impact (EI) mode at 70 eV, scanning the 35–450 *m/z* range. Interface and source temperature were set at 280 °C and 230 °C, respectively. Acquisition was done using HP ChemStation software (Agilent Technologies).

Qualitative analysis was based on the comparison of experimental mass spectra with data from the Wiley mass spectral library [13] and was confirmed, when possible, by using linear retention indices (I^T) [14,15]. Semiquantitative data (percentage of total volatiles) were directly calculated from peak areas of total ion current (TIC) profiles.

Table 2. Essential oil composition of the different parts of *C. fuscatum* (F: Flowers, S: Green parts, WP: Whole plant). Class of identified compound is marked with a superscript ([a]: Aliphatic esters, [b]: Monoterpenes, [c]: Sesquiterpenes, [d]: Others). [e]: Retention indices on the DB5 column were taken from [14] except those marked with '*', which were taken from [15]. -: Not found; n.i.: Non identified. Unidentified components less than 0.5% were not included in this table.

				Percent Composition (%)		
Peak	Compound	I^T (exp.)	I^T (lit.) [e]	F	WP	GP
1	2-methylpropyl isobutyrate [a]	895	892	1.54	0.54	-
2	2-methylallyl isobutyrate [a]	927	-	9.79	7.51	-
3	α-pinene [b]	939	932	0.71	1.11	-
4	isobutyl methacrylate [a]	960	-	2.38	2.14	-
5	isobutyl 2-methylbutyrate [a]	1004	1002 *	0.20	0.16	-
6	2-methylbutyl isobutyrate [a]	1015	1014 *	2.03	1.66	-
7	limonene [b]	1031	1031	0.34	0.10	-
8	1,8-cineol [b]	1033	1032	0.50	0.30	-
9	γ-terpinene [b]	1060	1054	0.10	0.40	-
10	isoamyl butyrate [a]	1060	1060	3.60	1.64	-
11	(E)-2-methyl-2-butenyl methacrylate [a]	1087	-	27.57	18.53	0.73
12	3-methylbutyl-2-methyl-butyrate [a]	1100	1100	0.27	0.15	-
13	2-methylbutyl-2-methyl-butyrate [a]	1106	1103	0.30	0.17	-
14	3-methyl-3-butenyl isovalerate [a]	1118	1116 *	1.22	1.10	-
15	α-canfolenal [b]	1119	1122	0.33	0.15	-
16	*trans*-pinocarveol [b]	1139	1135	5.14	2.90	0.20
17	camphor [b]	1143	1141	0.20	0.25	-
18	pinocarvone [b]	1162	1160	4.39	2.62	0.45
19	3-pinanone [b]	1173	1172	0.43	0.28	-
20	terpinen-4-ol [b]	1179	1174	0.20	0.20	-
21	myrtenol [b]	1194	1194	1.17	0.89	-
22	amyl tiglate [a]	1229	1126 *	0.82	0.44	-
23	(-)-carvone [b]	1242	1239	0.53	0.35	-
24	nonanoic acid [d]	1280	1267	0.16	0.10	-
25	*cis*-myrtenal [b]	1289	1295	0.19	0.15	-

Table 2. *Cont.*

Peak	Compound	I^T (*exp.*)	I^T (*lit.*) [e]	F	WP	GP
				Percent Composition (%)		
26	geranyl formate [b]	1300	1298	0.35	-	-
27	(E,E)-2,4-decadienal[d]	1318	1316	0.38	-	-
28	myrtenyl acetate [b]	1326	1326	0.20	0.20	-
29	benzyl methacrylate [a]	1357	-	0.39	0.10	-
30	decanoic acid [d]	1369	1364	0.56	0.76	-
31	(E)-β-damascenone [b]	1380	1384	0.19	0.49	-
32	phenylethyl isobutyrate [a]	1396	1396 *	0.35	0.19	-
33	*trans*-caryophyllene [c]	1404	1408	0.33	0.32	-
34	α-bergamotene [c]	1438	1412	5.08	4.94	2.18
35	α-curcumene [c]	1483	1480	9.21	8.06	4.69
36	α-muurolene [c]	1499	1500	0.23	0.27	0.62
37	(E,E)-α-farnesene [c]	1509	1505	0.15	-	-
38	γ-cadinene [c]	1513	1513	0.50	0.86	0.31
39	(Z)-γ-bisabolene [c]	1515	1515	0.28	0.36	-
40	δ-cadinene [c]	1524	1523	1.39	1.73	1.30
41	cadina-1,4-diene [c]	1532	1534	0.39	0.20	-
42	(Z)-nerolidol [c]	1534	1532	0.60	0.43	0.30
43	α-cadinene [c]	1538	1538	0.38	0.20	-
44	neryl isovalerate [c]	1576	1583	1.32	2.98	6.80
45	spathulenol [c]	1578	1578	0.84	0.64	0.83
46	caryophyllene oxide [c]	1581	1578	0.10	0.27	-
47	guaiol [c]	1595	1600	0.77	0.76	0.89
48	isoamyl nerolate [c]	1601	1602	0.69	0.23	1.11
49	1,10-di-*epi*-cubenol[c]	1614	1619	0.33	0.28	0.72
50	γ-eudesmol [c]	1630	1632	0.10	0.44	1.77
51	himachalol [c]	1647	1653	0.10	0.38	0.96
52	α-cadinol [c]	1652	1654	0.10	0.28	0.26
53	β-eudesmol [c]	1659	1650	0.10	0.10	-
54	α-bisabolol [c]	1682	1685	0.39	0.71	0.72
55	xanthorrhizol [c]	1751	1753	-	-	0.21
56	pentadecanol [c]	1778	1774	-	-	0.75
57	β-bisabolenol [c]	1786	1789	0.39	0.28	0.75
58	1-octadecene [d]	1793	1790	0.52	0.41	1.06
59	hexadecanol [d]	1879	1875	-	0.20	1.24
60	hexadecanoic acid [d]	1972	1960	2.49	10.74	23.89
61	octadecanoic acid [d]	2180	2180	0.75	8.65	18.68
	Aliphatic esters ([a])			50.46	34.33	0.73
	Monoterpenoids ([b])			14.97	10.39	0.65
	Sesquiterpenoids ([c])			23.77	24.72	24.42
	Others ([d])			4.86	20.86	45.62
	% identified			94.06	90.30	71.42

3. Results

3.1. Yield

C. fuscatum gave yellow essential oils with yields of 0.40%, 0.11% and 0.25% for flowers (F), green parts (GP) and whole plant (WP), respectively. The volatiles in the essential oils here analyzed (94.06% F, 71.42% S, 90.30% WP) were identified based on their mass spectra and chromatographic retention data. Table 2 lists these compounds according to their linear retention indices (I^T).

3.2. Chemical Composition of C. fuscatum Essential Oils

Aliphatic esters were the predominant class of compounds identified in the essential oils of flowers and the whole plant (50.46% and 34.33%, respectively), followed by sesquiterpenoids (23.77%–24.72%) and monoterpenoids (10.39%–14.97%). There was a noticeable absence of aliphatic esters in the green parts of this species, with (E)-2-methyl-2-butenyl methacrylate representing only 0.73% of the essential oil of green parts.

We have identified for the first time in this species new aliphatic esters with relative abundances higher than 1% and showing very similar structures to those of compounds previously reported in [8–10]. It is also worth noting that several new aliphatic esters with relative abundances higher than 1% and showing very similar structures to compounds previously reported in [8–10] were also identified for the first time in this paper (see Figure 1).

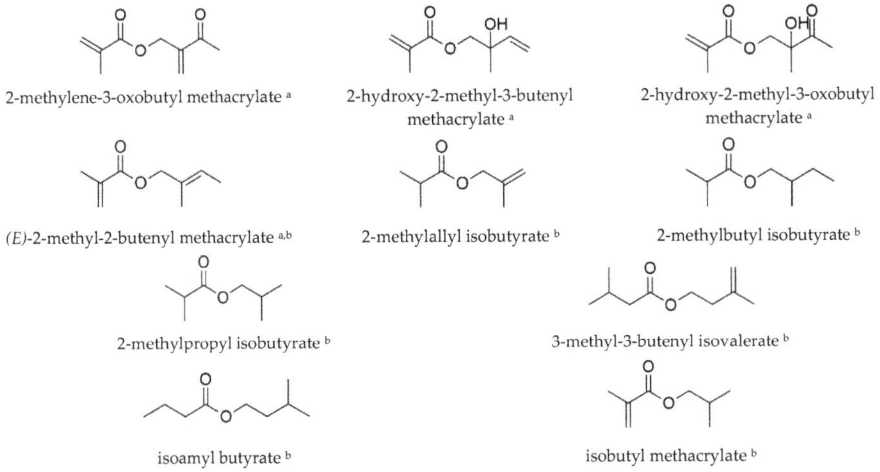

2-methylene-3-oxobutyl methacrylate [a]

2-hydroxy-2-methyl-3-butenyl methacrylate [a]

2-hydroxy-2-methyl-3-oxobutyl methacrylate [a]

(E)-2-methyl-2-butenyl methacrylate [a,b]

2-methylallyl isobutyrate [b]

2-methylbutyl isobutyrate [b]

2-methylpropyl isobutyrate [b]

3-methyl-3-butenyl isovalerate [b]

isoamyl butyrate [b]

isobutyl methacrylate [b]

Figure 1. Chemical structures of selected aliphatic esters identified in *C. fuscatum*. [a]: Aliphatic esters previously reported by De Pascual et al. [8–10], [b]: Aliphatic esters with relative abundances higher than 1% identified in this work.

A total of 42 terpenoids have also been identified in the essential oils of the aerial parts of *C. fuscatum*. α-Curcumene (9.21%), trans-pinocarveol (5.14%), α-bergamotene (5.08%) and pinocarvone (4.39%) represent a high percentage of the terpenoid fraction of the flower essential oil. Neryl isovalerate was detected in green parts in percentages higher than those found for F and WP essential oils (6.80% vs. 1.32% and 2.98%, respectively). Sesquiterpenoids were the major terpenoid class (24.42%) in the essential oil of green parts, with a noticeable contribution (45.62%) of other compounds mainly long chain fatty acids as hexadecanoic and octadecanoic acids.

3.3. Review of the Pharmacological Activities of C. fuscatum Essential Oil Compounds

Although most of essential oils isolated from *C. fuscatum* showed high relative contents of methacrylic and butyric esters, the medicinal value of this kind of compounds is still deeply unknown. (E)-2-methyl-2-butenyl methacrylate has been reported to have an intense odor [10]; however, as far as we know, no information is available on its effects over animal or human physiology.

On the other hand, the main terpenoids with reported pharmacological activities present in the essential oils of *C. fuscatum* are summarized in Table 3. Antinociceptive (α-curcumene, myrtenol), anti-inflammatory (α-curcumene, myrtenol, spathulenol, α-bisabolol, α-bergamotene), anxiolytic ((Z)-nerolidol), antimicrobial (*trans*-pinocarveol, α-pinene, γ-terpinene, δ-cadinene) and even anticarcinogen (spathulenol) compounds have been detected.

Table 3. Review of the main terpenoids with reported pharmacological activities present in *C. fuscatum* essential oils.

Compound	Reported Pharmacological Activities	References
α-Pinene	Antibacterial, antifungal	[16]
	Hypotensive	[17]
	Gastroprotective	[18]
Limonene	Anti-inflammatory, antioxidant	[19]
	Antimicrobial	[20]
1,8-Cineole	Gastroprotective	[21]
	Anti-inflammatory	[22]
γ-Terpinene	Antioxidant	[23]
	Antibacterial	[24]
Trans-pinocarveol	Antibacterial	[25]
Camphor	Analgesic, anti-inflammatory	[26]
Myrtenol	Anti-inflammatory, antinociceptive	[27]
(E)-β-Damascenone	Antispasmodic	[28]
α-Bergamotene	Anti-inflammatory	[29]
α-Curcumene	Stomachic, antinociceptive, anti-ulcer	[30]
	Anti-inflammatory	[31]
δ-Cadinene	Anti-inflammatory	[29]
	Antibacterial, antioxidant	[32]
(Z)-Nerolidol	Antinociceptive	[33]
	Sedative	[34]
Spathulenol	Anti-inflammatory, antinociceptive, anticarcinogen	[35]
α-Bisabolol	Anti-inflammatory	[36]
	Digestive	[37]

4. Discussion

The medicinal properties of chamomiles are usually attributed to their essential oil components. *C. fuscatum* mainly concentrates its essential oil in the flowers (0.40%), and in smaller proportions in green parts (0.11%). For this reason, flowers are the part of the plant generally used as folk medicine. As compared to other chamomile species, the yield of *C. fuscatum* flower essential oil (0.40%) was similar to that of other relative species. Roman chamomile (*Chamaemelum nobile*), its closest relative, has been described to usually yield around 0.70% of essential oil and a yield range between 1.90%–0.24% has been reported for German chamomile (*Matricaria recutita*) [38].

As mentioned, aliphatic esters had been previously reported in *C. fuscatum* [8–10] and also in related chamomile species as *Chamaemelum nobile* and in several species of the *Anthemis* genus [39]. (*E*)-2-methyl-2-butenyl methacrylate had been described in the hexane extract of the aerial parts by De Pascual et al. [8], together with other methacrylates (2-hydroxy-2-methyl-3-butenyl methacrylate and 2-hydroxy-2-methyl-3-oxobutyl methacrylate) that have not been detected here. 2-Methylene-3-oxobutyl methacrylate, a methacrylic ester previously isolated from the essential oil of *C. fuscatum* [10] was not found either in the essential oils here analyzed. However, we have identified in a relative high abundance (>1%) the esters isobutyl methacrylate and isoamyl butyrate, which had not been cited in previous literature of this species [8–10]. These differences could be attributed to the different extraction procedures with respect to [8–10], but also to the different environments since the population here analyzed was collected in Badajoz (Spain) and the samples analyzed in the cited studies were from Cáceres (Spain).

The pharmacological activities of the methacrylic and butyric esters are still unknown. As these compounds are present in high proportions in the volatile fraction of common medicinal species such as chamomiles, investigation on this topic would be worthy of consideration. In addition, it would be interesting to investigate if the structural similitudes among these classes of compounds could support that they act over the same biological paths.

From the 42 terpenoids identified in the aerial parts of *C. fuscatum*, only α-curcumene, neryl isovalerate and trans-pinocarveol had been previously reported [8–10]. Moreover, several terpenoids with a relative high abundance such as pinocarvone and *α*-bergamotene have been identified for the first time in this species. Together with aliphatic esters, *trans*-pinocarveol and pinocarvone have been reported as major compounds of *C. nobile* essential oils [39].

Despite its hydrophobic nature, terpenoids have also been detected in chamomile teas [40] and, therefore, they could play a role in the medicinal value of chamomile tea prepared from *C. fuscatum* flowers. In this work, we have summarized the reported pharmacological activities of several terpenoids that occur in the essential oils of this species. These activities could be related with the traditional uses of this species, which is also interesting in the context of validating its medicinal value. As an example, the presence of sesquiterpenoids with stomachic and gastroprotective activity such as *α*-curcumene, *α*-bisabolol and 1,8-cineole could support the fact that *C. fuscatum* has been traditionally used as digestive. In fact, *α*-bisabolol has been previously related to the digestive action of German chamomile [37].

As previously mentioned, the composition of *C. fuscatum* essential oils showed similitudes with that of essential oils isolated from *C. nobile*. However, there are chemical differences between both species that could result in different physiological responses. Therefore, comparative studies regarding bioactivity must be carried out before *C. fuscatum* can be raised as an interesting alternative to this plant. In contrast, the volatile profile of essential oils here analyzed differs drastically from that of German chamomile, since this plant is especially rich in bisaboloids (> 50% according to [38]), which are found in low relative contents in the essential oil obtained from *C. fuscatum* flowers. Nevertheless, the presence of *α*-bisabolol, one of the major constituents of German chamomile, in *C. fuscatum* essential oils here analyzed make worthy of investigation the role of this compound in the digestive action of both species.

5. Conclusions

In this work, a comprehensive evaluation regarding chemical composition and ethnobotanical use of a scarcely studied species, *C. fuscatum*, was carried out. In addition to a number of different monoterpenoids, sesquiterpenoids, etc., methacrylates and butyrates were determined as major components of the essential oils of this plant, being (*E*)-2-methyl-2-butenyl methacrylate the compound showing the highest percent content in both flower and whole plant essential oils. Moreover, several compounds with high relative abundances such as isobutyl methacrylate, isoamyl butyrate, *α*-bergamotene and pinocarvone were also identified for the first time in this plant. The published literature on the bioactivity of these compounds has been correlated with GC–MS data to support the popular knowledge and ethnobotanical use of this Mediterranean chamomile. Despite studies regarding the evaluation of the effect of different factors such as cultivation region, climate, extraction method, etc. will be further addressed, results of this preliminary study highlighted the potential of this plant as a natural source of bioactives of application in the pharmacological, cosmetic and food industries, among others.

Author Contributions: Conceptualization, T.R.-T.; Methodology, M.J.P.-A. and J.B.-S.; Validation, M.J.P.-A.; Formal Analysis, M.F.-C. and M.J.P.-A.; Investigation, M.F.-C., A.C.S., M.J.P.-A.; Data Curation, J.B.-S. and M.F.-C.; Writing-Original Draft Preparation, T.R.-T.; Writing-Review and Editing, J.B.-S. and T.R.-T.; Visualization, J.B.-S.; Supervision, M.J.P.-A.; Project Administration, T.R.-T.; Funding Acquisition, M.J.P.A and T.R.-T.

Funding: This research was partially funded by IB16003 project financed by the Junta of Extremadura (Spain) and the European Regional Development Fund.

Acknowledgments: To Manuel Pardo de Santayana and his team (Inventario Español de Conocimientos Tradicionales Relativos a la Biodiversidad, IECTB) for helping with the bibliographic ethnobotanical review.

References

1. Blanca, G. Chamaemelum Mill. In *Flora Vascular de Andalucía Oriental*; Blanca, G., Cabezudo, B., Cueto, M., Fernández López, C., Morales Torres, C., Eds.; Consejería de Medio Ambiente, Junta de Andalucía: Sevilla, Spain, 2011; pp. 1655–1656. ISBN 978-84-92807-12-3.

2. Blanco-Salas, J.; Ruiz-Téllez, T.; Vázquez-Pardo, F.M. Chamaemelum fuscatum (Brot.) Vasc. In *Inventario Español de los Conocimientos Tradicionales Relativos a la Biodiversidad. Fase II (Tomo 1)*; Pardo de Santayama, M., Morales, R., Tardío, J., Molina, M., Eds.; Ministerio de Agricultura y Pesca, Alimentación y Medio Ambiente: Madrid, Spain, 2018; pp. 332–334. ISBN 978-84-491-1472-4.

3. Tejerina, Á. *Usos y Saberes Sobre las Plantas de Monfragüe*; Itomonfragüe: Cáceres, Spain, 2010; ISBN 978-84-15820-10-9.

4. Casado Ponce, D. Revisión de la Flora y Etnobotánica de la Campiña de Jaén (del Guadalbullón a la Cuenca del Salado de Porcuna). Doctoral Thesis, Universidad de Jaén, Jaén, Spain, 2003.

5. Gregori, P. Medicina Popular en Valencia de Mombuey. Doctoral Thesis, Universidad de Extremadura, Badajoz, Spain, 2007.

6. Vázquez, F.M.; Suarez, M.A.; Pérez, A. Medicinal plants used in the Barros Area, Badajoz Province (Spain). *J. Ethnopharmacol.* **1997**, *55*, 81–85. [CrossRef]

7. Penco, A.D. Medicina Popular Veterinaria en la Comarca de Zafra. Doctoral Thesis, Universidad de Extremadura, Extremadura, Spain, 2005.

8. De Pascual Teresa, J.; Caballero, E.; Caballero, C.; Anaya, J.; González, M.S. Four aliphatic esters of Chamaemelum fuscatum essential oil. *Phytochemistry* **1983**, *22*, 1757–1759. [CrossRef]

9. De Pascual Teresa, J.; Caballero, E.; Anaya, J.; Caballero, C.; Gonzalez, M.S. Eudesmanolides from Chamaemelum fuscatum. *Phytochemistry* **1986**, *25*, 1365–1369. [CrossRef]

10. De Pascual Teresa, J.; Anaya, T.J.; Caballero, E.; Caballero, M.C. Sesquiterpene lactones and aliphatic esters from Chamaemelum fuscatum. *Phytochemistry* **1988**, *27*, 855–860. [CrossRef]

11. Aceituno-Mata, L.; Acosta, R.; Alcaraz, F.; Álvarez Escobar, A.; Amich, F.; Anllo Naveiras, J.; Barroso, E.; Benítez Cruz, G.; Blanco, E.; Blanco-Salas, J.; et al. Metodología para la elaboración del Inventario Español de los Conocimientos Tradicionales Relativos a la Biodiversidad. In *Inventario Español de los Conocimientos Tradicionales Relativos a la Biodiversidad. Primera Fase: Introducción, Metodología y Fichas*; Pardo de Santayana, M., Morales, R., Aceituno-Mata, L., Molina, M., Eds.; Ministerio de Agricultura, Alimentación y Medio Ambiente: Madrid, Spain, 2014; pp. 31–49. ISBN 978-84-491-1401-4.

12. Moher, D.; Liberati, A.; Tetzlaff, J.; Altman, D.G. PRISMA Group Preferred reporting items for systematic reviews and meta-analyses: The PRISMA statement. *Int. J. Surg.* **2010**. [CrossRef] [PubMed]

13. McLafferty, F.W.; Stauffer, D.B. The Wiley/NBS Registry of Mass Spectral Data, Volumes 1–7. *J. Chem. Educ.* **1989**, *66*, A256. [CrossRef]

14. Adams, R.P. *Identification of Essential Oil Components by Gas Chromatography/Quadrupole Mass Spectroscopy*; Academic Press: New York, NY, USA, 2001; ISBN 0931710855.

15. Shen, V.K.; Siderius, D.W.; Krekelberg, W.P.; Hatch, H.W. (Eds.) *NIST Standard Reference Simulation Website, NIST Standard Reference Database Number 173*; National Institute of Standards and Technology: Gaithersburg, MD, USA, 2012.

16. Rivas da Silva, A.C.; Lopes, P.M. Barros de Azevedo, M.M.; Costa, D.C.; Alviano, C.S.; Alviano, D.S. Biological activities of α-pinene and β-pinene enantiomers. *Molecules* **2012**, *17*, 6305–6316. [CrossRef]

17. Menezes, I.A.C.; Barreto, C.M.N.; Antoniolli, Â.R.; Santos, M.R.V.; de Sousa, D.P. Hypotensive activity of terpenes found in essential oils. *Zeitschrift für Naturforschung C* **2010**, *65*, 562–566. [CrossRef]

18. de Souza, M.C.; Vieira, A.J.; Beserra, F.P.; Pellizzon, C.H.; Nóbrega, R.H.; Rozza, A.L. Gastroprotective effect of limonene in rats: Influence on oxidative stress, inflammation and gene expression. *Phytomedicine* **2019**, *53*, 37–42. [CrossRef]

19. Yu, L.; Yan, J.; Sun, Z. D-limonene exhibits anti-inflammatory and antioxidant properties in an ulcerative colitis rat model via regulation of iNOS, COX-2, PGE2 and ERK signaling pathways. *Mol. Med. Rep.* **2017**, *15*, 2339–2346. [CrossRef]

20. Rancic, A.; Soković, M.; Van Griensven, L.; Vukojevic, J.; Brkic, D.; Ristic, M. Antimicrobial action of limonene. *Lekovite Sirovine* **2003**, *23*, 83–88.

21. Caldas, G.F.R.; da Silva Oliveira, A.R.; Araújo, A.V.; Lafayette, S.S.L.; Albuquerque, G.S.; Silva-Neto, J.C.; Costa-Silva, J.H.; Ferreira, F.; da Costa, J.G.; Wanderley, A.G. Gastroprotective mechanisms of the monoterpene 1, 8-cineole (eucalyptol). *PLoS ONE* **2015**, *10*, 1–17. [CrossRef]

22. Rocha, D.M.; Caldas, A.P.; Oliveira, L.L.; Bressan, J.; Hermsdorff, H.H. Saturated fatty acids trigger TLR4-mediated inflammatory response. *Atherosclerosis* **2016**, *244*, 211–215. [CrossRef] [PubMed]

23. Foti, M.C.; Ingold, K.U. Mechanism of inhibition of lipid peroxidation by γ-terpinene, an unusual and potentially useful hydrocarbon antioxidant. *J. Agric. Food Chem.* **2003**, *51*, 2758–2765. [CrossRef] [PubMed]

24. Carson, C.F.; Riley, T.V. Antimicrobial activity of the major components of the essential oil of *Melaleuca alternifolia*. *J. Appl. Bacteriol.* **1995**, *78*, 264–269. [CrossRef] [PubMed]

25. Elaissi, A.; Rouis, Z.; Mabrouk, S.; Salah, K.B.; Aouni, M.; Khouja, M.L.; Farhat, F.; Chemli, R.; Harzallah-Skhiri, F. Correlation between chemical composition and antibacterial activity of essential oils from fifteen *Eucalyptus* species growing in the Korbous and Jbel Abderrahman arboreta (North East Tunisia). *Molecules* **2012**, *17*, 3044–3057. [CrossRef] [PubMed]

26. Ghori, S.S.; Ahmed, M.I.; Arifuddin, M.; Khateeb, M.S. Evaluation of analgesic and anti-inflammatory activities of formulation containing camphor, menthol and thymol. *Int. J. Pharm. Pharm. Sci.* **2016**, *8*, 271–274.

27. Gil Silva, R.O.; Salvadori, M.S.; Sousa, F.B.M.; Santos, M.S.; Carvalho, N.S.; Sousa, D.P.; Gomes, B.S.; Oliveira, F.A.; Barbosa, A.L.R.; Freitas, R.M.; et al. Evaluation of the anti-inflammatory and antinociceptive effects of myrtenol, a plant-derived monoterpene alcohol, in mice. *Flavour Fragr. J.* **2014**, *29*, 184–192. [CrossRef]

28. Pongprayoon, U.; Baeckström, P.; Jacobsson, U.; Lindström, M.; Bohlin, L. Antispasmodic activity of β-damascenone and E-phytol isolated from *Ipomoea pes-caprae*. *Planta Med.* **1992**, *58*, 19–21. [CrossRef]

29. Veiga, V.F.; Rosas, E.C.; Carvalho, M.V.; Henriques, M.G.M.O.; Pinto, A.C. Chemical composition and anti-inflammatory activity of copaiba oils from Copaifera cearensis Huber ex Ducke, *Copaifera reticulata* Ducke and *Copaifera multijuga* Hayne-A comparative study. *J. Ethnopharmacol.* **2007**, *112*, 248–254. [CrossRef]

30. Yamahara, J.; Hatakeyama, S.; Taniguchi, K.; Kawamura, M.; Yoshikawa, M. Stomachic principles in ginger. II. Pungent and anti ulcer effects of low polar constituents isolated from ginger, the dried rhizoma of *Zingiber officinale* Roscoe cultivated in Taiwan. The absolute stereostructure of a new diarylheptanoid. *Yakugaku Zasshi* **1992**, *112*, 645–655. [CrossRef] [PubMed]

31. Pulla Reddy, A.C.; Lokesh, B.R. Effect of dietary turmeric (*Curcuma longa*) on iron-induced lipid peroxidation in the rat liver. *Food Chem. Toxicol.* **1994**, *32*, 279–283. [CrossRef]

32. González, A.M.; Tracanna, M.I.; Amani, S.M.; Schuff, C.; Poch, M.J.; Bach, H.; Catalán, C.A.N. Chemical composition, antimicrobial and antioxidant properties of the volatile oil and methanol extract of *Xenophyllum poposum*. *Nat. Prod. Commun.* **2012**, *7*, 1663–1666. [CrossRef] [PubMed]

33. Chan, W.K.; Tan, L.T.H.; Chan, K.G.; Lee, L.H.; Goh, B.H. Nerolidol: A sesquiterpene alcohol with multi-faceted pharmacological and biological activities. *Molecules* **2016**, *21*, 529. [CrossRef] [PubMed]

34. Goel, R.; Kaur, D.; Pahwa, P. Assessment of anxiolytic effect of nerolidol in mice. *Indian J. Pharmacol.* **2016**, *48*, 450–452. [CrossRef] [PubMed]

35. do Nascimento, K.F.; Moreira, F.M.F.; Alencar Santos, J.; Kassuya, C.A.L.; Croda, J.H.R.; Cardoso, C.A.L.; do, C.; Vieira, M.; Góis Ruiz, A.L.T.; Ann Foglio, M.; et al. Antioxidant, anti-inflammatory, antiproliferative and antimycobacterial activities of the essential oil of *Psidium guineense* Sw. and spathulenol. *J. Ethnopharmacol.* **2018**, *210*, 351–358. [CrossRef] [PubMed]

36. Maurya, A.; Singh, M.; Dubey, V.; Srivastava, S.; Luqman, S.; Bawankule, D. (−)-bisabolol reduces pro-inflammatory cytokine production and ameliorates skin inflammation. *Curr. Pharm. Biotechnol.* **2014**, *15*, 173–181. [CrossRef] [PubMed]

37. Moura Rocha, N.F.; Venâncio, E.T.; Moura, B.A.; Gomes Silva, M.I.; Aquino Neto, M.R.; Vasconcelos Rios, E.R.; De Sousa, D.P.; Mendes Vasconcelos, S.M.; De França Fonteles, M.M.; De Sousa, F.C.F. Gastroprotection of (−)-α-bisabolol on acute gastric mucosal lesions in mice: The possible involved pharmacological mechanisms. *Fundam. Clin. Pharmacol.* **2010**, *24*, 63–71. [CrossRef] [PubMed]

38. Mann, C.; Staba, E.J. The chemistry, pharmacology, and commercial formulation of chamomile. In *Herbs, Spices, and Medicinal Plants: Recent Advances in Botany, Horticulture, and Pharmacology*; Craker, L., Simon, J., Eds.; Food Products Press: New York, NY, USA, 1992; Volume 1, pp. 235–281. ISBN 9781560220435.

39. Radulović, N.S.; Blagojević, P.D.; Zlatković, B.K.; Palić, R.M. Chemotaxonomically important volatiles of the genus *Anthemis* L.—A detailed GC and GC/MS analysis of *Anthemis segetalis* Ten. from Montenegro. *J. Chin. Chem. Soc.* **2009**, *56*, 642–652. [CrossRef]
40. Tschiggerl, C.; Bucar, F. Guaianolides and volatile compounds in Chamomile tea. *Plant Foods Hum. Nutr.* **2012**, *67*, 129–135. [CrossRef] [PubMed]

Article

Contribution of Mangrove Forest to the Livelihood of Local Communities in Ayeyarwaddy Region, Myanmar

Wai Nyein Aye [1], Yali Wen [1,*], Kim Marin [1], Shivaraj Thapa [1] and Aung W. Tun [2]

[1]   School of Economics and Management, Beijing Forestry University, Beijing 100083, China; wainyeinaye@yahoo.com (W.N.A.); marinkim310@yahoo.com (K.M.); shivaraj_thapa@outlook.com (S.T.)

[2]   Environmental Conservation Department, Ministry of Natural Resources and Environmental Conservation, Dawei 14011, Myanmar; aungwunnnatunwunna@gmail.com

*   Correspondence: wenyali2018@bjfu.edu.cn; Tel.: +86-106-233-8455

Received: 15 March 2019; Accepted: 25 April 2019; Published: 13 May 2019

Abstract: Myanmar's forests are socially and economically significant to the country because over 70% of the country's population depends on natural resources for daily needs. We conducted this study with the aim of assessing the extent to which direct and indirect (tangible) benefits of mangrove forest contribute to local livelihoods in the Ayeyarwaddy Region, Myanmar. We used a questionnaire survey (n = 185 households), interview and group discussion for data collection. The study shows that 43% of total household income is generated through selling of forest products collected from the mangrove forest such as firewood, fishes, crabs and prawn, whereas agricultural and non-farm incomes were found to be 25% and 32% of total income, respectively. The result prevails that income from the mangrove forest products for fish, crab, prawn and firewood is specifically 36%, 28%, 9% and 27%, respectively. Hence, we confirmed that local livelihood mainly depends on the mangrove forest ecosystem.

Keywords: mangrove forest; local communities; Ayeyarwaddy region; Myanmar; economic; livelihoods

---

1. Introduction

According to the World Alas of Mangrove, Myanmar is the seventh largest mangrove area covering 3.3% of the world's landmass [1]. Mangroves cover an estimated area of 4629.64 sq km [2] making Myanmar the fourth largest mangrove coverage in Asia after Malaysia, Bangladesh and Papua New Guinea [3]. Myanmar shares common maritime boundaries in the Bay of Bengal with Bangladesh, India and Thailand. The continental shelf covers approximately 230,000 sq km with a relatively wider portion in the central and southern parts. The most extensive mangroves thrive in the Ayeyarwaddy Delta, the Thanintharyi Coastline and the Rakhine Coastline. Mangroves in Myanmar extensively grow throughout the coastal strip of the country, providing ecosystem goods and services to coastal communities as well as all other parts of the country. Mangrove forest in Myanmar is rich in biodiversity. It has 34 mangrove tree species out of the global total of around 70 mangrove tree species. Of the total Myanmar primary mangroves, the majority is located on Ayeyarwady floodplains, with the remainder in Tanintharyi and a lesser portion in the Rakhine area.

Similarly, mangroves provide shelter and nursery habitat to the aquatic animals. Win [4] stated that the importance of mangrove to fisheries is apparent especially for the white (banana) shrimp (*Penaeus merguiensis*) which is the most important shrimp species in Myanmar. It depends on mangrove forests for shelter during its juvenile stage. Some species such as tiger prawn (*Penaeus monodon*), *Penaeus indicus* and *Metapenaeus spp* also depend on mangroves at certain phases of their life cycle and the larvae, post larvae and juveniles of some penaeids species enter the estuarine mangrove areas in Myanmar [4]. Over 90% of marine species were found in the mangroves during some parts of their life cycles which shows a positive correlation between mangrove area and aquatic animals [5].

The majority of poor people in developing countries rely on forests and woodlands for their livelihood because of low income and lack of other alternative means to support their subsistence [6]. While the contribution of environmental goods and services to rural livelihoods are widely documented [7,8], their significance within forest-dependent communities remains insufficiently explored. It contributes significantly to the local economy of the people living around the mangrove forests as well as people living far from it.

The term "nutraceutical" is the combination of "nutrition" and "pharmaceutical" and was introduced by Stephen DeFelice in 1989 [9]. Mangroves are important natural resources that are able to provide a wide range of goods and services for the local community. Further, chemical compounds and extracts of mangroves can be used mainly for folk medicine [10]. Rhizophora seedlings are able to cure a sore mouth [10]. The bark extract of *Brugueria sexangula* (Lour.) Poir. is effective against two tumors of Sarcoma 180 and Lewis Lung Carcinoma [11]. Extracts from the bark of *Rhizophora mucronata* Lamk. and the leaf of *Brugueria cylindricall* (Linn.) show antiviral activity against all the viruses tested [11]. And extracts from the leaves, barks, stems and roots of *Ceriops tagal* (Perr.) C.B.Rob., *Ceriops decandra* (Griffith) Ding Hou, *Xylocarpus granatum* Koen. (Meliaceae), *Xylocarpus moluccensis* (Lam.) M. Roem., *Rhizophora mucronata* and *Rhizophora apiculata* Blume. have shown to have antistringent, antdiarrhoea and haemostatic properties [11]. Extracts from the mangroves have been applied in the treatment of health disorders for centuries. Furthermore, in coastal areas, land is a scarce resource for the local community to fulfill food demands. So, interests have been emphasized on the utilization potential of mangroves. For example, in parts of Papua New Guinea, seedlings of Bruguiera species are the staple food [12] and propagule of *Bruguiera sexangula* (Lour.) is able to be eaten after peeling, soaking and boiling [13]. Priya and Niranjana mention that *B. gymnorrhiza L.* (Lamk) and *B. cylindrica L.* (Blume) are rich in nutritional value such as calcium, iron and magnesium and should be considered famine foods in the coastal areas [12].

Mangrove forests can provide a wide range of tangible and intangible benefits such as clean, safe and healthy environments, many forest products and a wide variety of seafood. Mangrove ecosystem services are worth an estimated US\$ 33–57 thousand per hectare per year to the national economies of developing countries with mangroves [14]. Vo et al. [15] confirmed that both goods and services provided by mangrove ecosystems contribute to human well-being directly and indirectly. Similarly, Andy et al. [16] supported that knowing the economic value of ecosystem services is an important asset because a major demand is to support human well-being, sustainability and distributional fairness. If there was no mangrove forest, people who rely on the mangroves would suffer from a lack of forest products and food security, especially in fisheries, reduced crop yield and the direct impact of natural disasters. Therefore, many development factors would be negatively affected by the loss of a mangrove forest.

Concerning the economic value of mangrove forests in Myanmar, Wai [17] conducted a study on economic dependency of local communities on mangroves: a case study in Bogalay, Myanmar. Compared to other areas in the country, mangrove depletion and degradation rate is relatively greater in the Ayeyarwady region of Myanmar due to the higher population, easier access to the forest, conversion to salt-producing land and the devastating impacts of Cyclone Nargis. In Myanmar, since the past three decades, over 58% of mangroves have been undergoing over-exploitation, illegal felling, agricultural expansion and conversion to fishponds and shrimp ponds [18]. Mangrove coverage estimated in 2010 has significantly decreased in the past three decades. Major sources of livelihood activities in that area include paddy cultivation, livestock raising, small- and medium-scale agricultural and fish processing, small-scale forest activities (firewood, charcoal production and timber extraction) and salt production. Most of the livelihoods in that area are not sustainable. To give them a sustainable livelihood, alternatives should be provided for the habitat.

Particularly, the degradation of mangroves in the Ayeyarwady Region is due to extremely high demand of fuelwood for Yangon and cities in adjacent areas. Given a lower population and the fact that the coastal landscape is more sheltered in Rakhine State and the Tanintharyi Region, mangroves are in better condition [19] and there is an increasing need of fuelwood from Yangon city to meet an annual demand of 700,000 tons [20]. In addition to household consumption, fuelwood and charcoal are also supplied to cottage industries, restaurants and tea shops [18]. Cultivation of paddy fields is also another main threat of mangrove conversion to other land use though soil condition is not suitable for agriculture. Agricultural expansion into mangrove areas to meet the requirements of regional food security is also common in the other two coastal regions, especially in the Rakhine region [20].

Moreover, a worrying trend for mangroves in that area is the conversion of mangrove forests into shrimp ponds and agricultural lands as well as to other uses such as salt production. Another serious issue is the extraction of trees for fuelwood and charcoal making. As a consequence of many drivers, mangrove deforestation in the Ayeyarwaddy region is recognized as a critical environmental issue for the country. The Ayeyarwady region is the most populated state of Myanmar where 88% of people live in rural areas. Local people, particularly landless labors, generate their subsistence and income from the mangrove forest through the collection of firewood, production of charcoal, harvesting of fisheries and collection of material for shelter. Furthermore, this area was seriously affected by Cyclone Nargis in 2008 and rural communities living around the mangrove area heavily depend on mangrove products directly as well as indirectly either for subsistence use or commercial purposes. Thus, the quantification of mangrove forest contribution to rural livelihood is important for the conservation of this area.

The study was conducted to assess the extent to which direct and indirect (tangible) benefits of mangrove forest resources contribute to the livelihoods of adjacent communities in the Ayeyarwaddy Region in Myanmar. This study seeks to answer the following questions: (1) What are the mangrove forest products that the local communities receive from mangrove forest? (2) How to access these mangrove forest products for the livelihoods, directly or indirectly? (3) How much will daily income generated from the livelihood system be? (4) What are the major income sources in that area? (5) How important is mangrove forest for the livelihood of the local community?

2. Materials and Methods

2.1. Study Area

The research was carried out in the western part of Meinmahla Kyun Wildlife Sanctuary, Bogalay Township, Ayeyarwaddy Region of Myanmar. The location of Meinmahla Kyun Wildlife Sanctuary is 15° 57.822′ N and 95° 17.988′ E. This wetland reserve is on Meinmahla Kyun and is classified as a mangrove reserve. This reserve area has 136.72 sq km and was established in 1986. It is the third Ramsar site of Myanmar and was designated in 2017. It is a coastal wetland in the southern part of the Ayeyarwaddy Delta which is also an ASEAN Heritage Park. It supports one of the largest remaining mangrove areas of the Delta where mangrove ecosystems have been declining due to activities including logging, charcoal and firewood production, fishing and development of shipping lanes. At present, the mangrove species are being replaced by mangrove date palm (*Phoenix paludosa* Roxb.). It supports globally threatened species such as hawksbill turtle (*Eretmochelys imbricate*), mangrove terrapin (*Batagur basaka*), the endangered great knot (*Calidris tenuirostris*), Nordmann's greenshank (*Tringa guttifer*), green turtle (*Chelonia mydas*), dhole (*Cuon alpinus*) and vulnerable species of the Pacific ridley turtle (*Lepidochelys olivacea*), fishing cat (*Prionailurus viverrinus*), lesser adjutant (*Leptoptilos javanicus*) and the Irrawaddy dolphin (*Orcaella brevirostris*) [18]. The site is also the last estuarine habitat in Myanmar for the saltwater crocodile (*Crocodylus porosus*). It holds significant cultural and historic value for the people of Myanmar based on myths and pilgrimages which closely connect them to their environment [21].

Two villages—Padekaw village (15° 59.232′ N and 95° 15.765′ E) and Lawinekyun (A Nauk) (16° 0.586′ N and 95° 15.866′ E)—were selected as study areas to analyze the socio-economic conditions. These two villages were selected with the criteria of accessibility, near the Meinmahla Kyun Wildlife reserve, and affected by cyclone Nargis in 2008. Because most of the villages in that area can only be accessed by boat, we selected these two research areas as they are a little easier to access by boat.

Table 1 shows the general information of the two villages. Local communities living in that area survive by working cultivation of paddy fields and fisheries. Sources of employment include crop farming (mainly paddy rice cultivation), horticulture (mostly fruit trees), paid agricultural labor, fishing (fishponds, shrimp farms, inland and offshore fisheries), small- and medium-scale agricultural and fish processing and small-scale forestry activities (firewood, charcoal and timber). Some income is derived from commerce and small-scale local trade which is indirectly reliant on the environment as the target customers of those trades and businesses are natural resource-dependent. Most of the economic activities in that area are at the subsistence level. Most men in the coastal areas are fishermen while women and children are collectors of inter-tidal mollusks, fish and prawns. These products are an important source of income to both fishermen and those engaged in processing and trading. The Delta region has many challenges such as capacity development and infrastructure for hygienic drinking water and better education and healthcare. Most of the households use firewood as fuel energy for cooking and heating. Access to the villages of the studied area was difficult because of poor transport infrastructure and few all-season roads that travel between villages that are often conducted by boat. Households in the studied villages use rainwater for drinking and cooking, however they often cannot collect enough rainwater. Figure 1 shows the location of the study area.

Table 1. General information of the two villages.

Village Name	Township Name	Total Households
Padekaw	Bogalay	245
Lawinekyun (A Nauk)	Bogalay	100
Total sampling households		345

La Waing Kyun(Ah Nauk) and Pe De Kaw Villages, Ayeyarwady Region

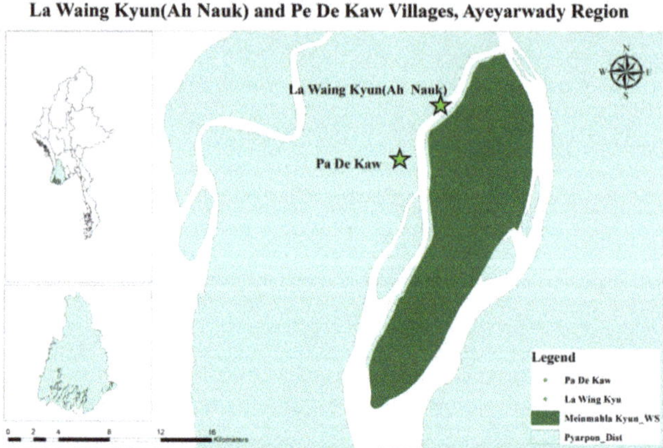

Figure 1. Map of the study area showing La Waing Kyun, Pa De Kaw Villages and the Ayeyarwady Region.

2.2. Framework of the Study

This study is based on the sustainable livelihood framework (SLF) for the analysis and assets. In SLF analysis, the relative importance of five types of assets (natural, physical financial, social and human) is evaluated. These assets (both material and social resources) constitute means that households use in the pursuit of their livelihood strategies. Factors that influence the mangrove forest income were selected based on the five that were assessed for sustainable livelihoods. In SLF, wealth, age, gender and skills are considered as financial, social and human assets.

We selected 185 sample respondents as a sample size (at 95% confidence interval and 5% marginal error) for this study by using Taro Yamane Formula [22].

$$n = \frac{N}{(1 + Ne^2)} \tag{1}$$

where n = sample size; N = total population of household; e = allowable error (5% = 0.05).

2.3. Methodology

This section contains the discussion on the method of data collection and data analysis that was used in this study.

The introduction of the research was guided to know the general information and socio-economic condition of the research area. Then, the survey was conducted during the periods of August and September 2018. For this research, a cross-sectional research design was used because this method is suitable for time and financial limitations. This design is also accurate and provides quick results. This study mainly relied on primary data and data was collected by face to face interviews. For an interview, well-prepared structured questionnaires (both open-ended and close-ended) were used. The local dwellers living and depending directly and indirectly on mangrove forest products for their daily livelihood system were selected as respondents for the interview survey. The simple random sampling method was used in selecting a sample household for the interview. A total of 185 (~185.24) households with 95% precision level were used as the sample size by calculating Taro Yamane Formula [22].

The main focus of this study was to analyze the contribution of mangrove forest products' income on the livelihood of local dwellers. To show the link between the contribution effect of mangrove forest resources and its impact on livelihoods, education, household composition, age, land and sources of family income, multiple linear regression models were used. The multiple linear regression model was estimated by using the ordinary least squares estimation technique (OLS) after the data was checked for different econometric tests. We used OLS multiple regressions to build models of the household characteristics associated with mangrove forest product earnings. The income generated from mangrove forest resources was regressed as a function of other socio-economic characters; the economic model was as follows.

$$Y = \beta_0 + \beta_1 x_i + u, \tag{2}$$

where Y = the income from the mangrove forest; β_0 = intercept, β_1 = estimated coefficients of the explanatory variable; x_i; x_i = explanatory variables (socio-economic characteristics); u = error term.

Total household income was estimated as follows:

$$\text{Total Household income} = \Sigma \text{ (income from agriculture + Non-farm income + Income from mangrove forest)} \tag{3}$$

In this study, income was calculated in the currency of Myanmar (Kyat). According to the exchange rate by the central bank of Myanmar (2018), US$ 1 is equal to about 1428.6 Kyats. Major income sources are agriculture, non-farm activities and collection of mangrove forest products. Agricultural income includes income from the cultivation of crops for purposes of both household consumption and selling. Information on crop yields was gathered from a household respondent through the questionnaire survey. Prices of crops were obtained from the local market. For non-farm income, it includes all income from wage labor, employment such as government staff and private shops. Wage labor in the study area was mostly in a mangrove forest plantation. The daily wages for men and women were not the same. The wage rate and the number of working days/hours reported by the respondents were used for the estimation. Income from private shops was obtained from the individual household respondent through the interview. And the final income source is income from the collection of mangrove forest products. Information about the collection and sale of mangrove forest products (firewood, shrimp, fish and crab) was obtained from the household questionnaire. In addition, data regarding different kinds of mangrove forest products and their price was obtained using the key informant survey and the questionnaire interviews. Monthly income was gathered from respondents through questionnaire and it was converted into annual values.

Both qualitative and quantitative techniques were used for analyzing data. Before processing the responses, the completed questionnaires were revised for completeness and consistency. Qualitative data were summarized by way of text analyses, while quantitative data were analyzed by descriptive statistics and OLS regression analysis. Descriptive statistics such as frequencies, percentages, mean value and standard deviation were computed for all the quantitative variables and information and were presented in the form of tables and graphs. Descriptive statistics were used because they enabled the research to meaningfully describe a distribution of scores or measurements using a few indices. The collected data was classified, tabulated and analyzed in Microsoft Excel and STATA version-13.

3. Results

3.1. Basic Characteristics of Sample Households and Annual Household Income

A simple random sampling survey was conducted in the study sites with prepared questionnaires during the two-month period from August to September 2018. A total of 185 sample households were interviewed from two villages of Bogalay Township, Ayeyarwaddy Region of Myanmar. Table 2 shows a descriptive analysis of household respondents. According to the results of the study, the gender distribution of the household head shows that 91.89% are male and the remaining 8.11% are female. Regarding the age distribution of respondents, only 1% of the respondents are 19 or less than 19 years old, 19% of the respondents are aged between 20 and 29 years, 27% of the respondents are aged between 30 and 39 years, 25% of the respondents are aged between 40 and 49 years and the remaining 28% of the respondents are above 49 years old.

Family size varied from 1 to 10 members with a mean value of 4.14 (standard deviation, sd = 1.65). In terms of education levels, 1% of respondents were graduates, 3% had attended high school, 22% studied at secondary school, 30% studied up to primary school and 44% of respondents had traditional Buddhist monastic education. So, this means that most of the respondents did not have a formal education. For old people, they had access to traditional monastic education. Crop farming (mainly paddy rice cultivation) is a major source of livelihood in Myanmar. The Ayeyarwaddy region is also well known as the "rice bowl of Myanmar". Among the respondents, however, 26% of households owned agricultural land and the remaining households were agriculturally landless. Minimum and maximum agricultural land holding sizes of respondents were 0 and 100 acres, respectively, with a mean value of 4.33 (standard deviation of 10.90). In the research area, 77% of households derived their income from various sources such as causal and seasonal labor in agriculture, wage labor in mangrove forest plantations, small scale trade, shop keeping, collection of firewood, fish proceeding and crafts. Then, 83% of households were non-native villagers. Further, 17% of respondent's houses were made of metal roofing and timber flooring, 8% were made of metal roofing, brick and concrete and wood, 74% of houses were made of Nypa roof, timber and bamboo flooring and the remaining 1% do not own a house and live on a boat. Some respondents harvested construction materials for their houses from the forest. For example, nipa palms (*Nypa fruticans*) along riversides are over-harvested for thatching, while Palmyra species on riverbanks and around paddy fields are utilized for both thatching and timber.

The sources of income in the study sites are farm activities, non-farm activities and collection of mangrove forest products. Among the respondents, 53% of the respondents generated their income from the collection of mangrove forest products, 23% from agricultural activities and the remaining 24% from non-farm activities. According to the results of the Table 3, mangrove income makes up 43% of the total household income. It included both subsistence and cash income. Agricultural income shares 25% of the total household income and non-farm income accounts for 32% of the total household income.

Table 2. Descriptive analysis of household respondents.

Household Characteristics		Frequency	Percentage (%)
	19 or Less than 19	2	1
Age	20–29	36	19
(years old)	30–39	50	27
	40–49	47	25
	More than 49	50	27
(Minimum = 19; Mean = 41; Maximum = 72)			
Gender	Male	15	8
	Female	170	92
	Graduate	1	1
	High School	5	3
Education	Secondary School	41	22
	Primary School	56	30
	Non formal Education	82	44
	Farmer	43	23
Occupation	Non-farm activity	98	53
	Collection of mangrove based products	44	24
Access to	Directly	121	65
mangrove products	Indirectly	64	35
Native village	Yes	154	83
	No	31	17
Agricultural land	Yes	49	26
	No	136	74
(Minimum land size = 0, Mean land size = 4.33, Maximum land size = 100)			

Table 3. Average annual household income.

Types of Income	Average Income Per Year (Kyats/year/household)	Standard Deviation	Standard Error	Income Share (%)
Mangrove Forest Income	1,119,957	1,659,339	121,997	43%
Agricultural Income	642,573	1,521,448	111,859	25%
Non-Farm Income	833,157	2,985,505	219,499	32%
Total Income	2,595,687	3,365,090	247,406	100%

3.2. The Contribution of Major Mangrove Forest Products

Income from the collection of mangrove forest products was the highest income source in the study. About 53% of sample households generated their livelihood income from the use of different mangrove forest products. The result indicates that mangrove products harvested by the local people from the study area are fuelwood, fish, crab and prawn. The main contributions in both villages are fish and crab. Their usage of timber products collected from mangrove forest is mostly for subsistence purposes such as building material and firewood because in that area they mainly use firewood as fuel for cooking. These two studied villages are dependent on the mangrove forest resources provided by a Meinmahla Kyun wetland reserve. Timbers from mangrove trees are used as poles, firewood and charcoal making for domestic purposes such as cooking, heating and ironing. Mangroves are the source of fuelwood for cooking in the rural area. Fisheries and prawn catch in particular depend on intact mangrove ecosystems. Villagers in the studied sites engage in traditional fish collection from the mangrove areas. The mangrove dwellers basically understand the daily tidal conditions by calculating the days based on the Myanmar lunar calendar [23]. Almost all of the respondents in two villages collected fish, crab and prawns at 10 to 15 days per month as they were dependent on a tidal cycle. Tidal inundation is more frequent and widespread during the period of the lunar cycle. During that period, people could harvest more fish, crab and prawn. Income share from fishing, crab, prawn and firewood were 36%, 28%, 9% and 27%, respectively. Figure 2 presents the major mangrove forest products in the study area.

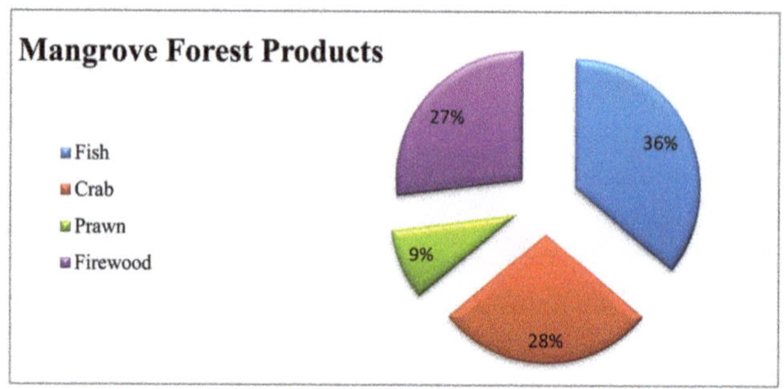

Figure 2. Major mangrove forest products in the studied area.

3.3. Mangrove Forest Income among Different Income Level

According to the results obtained as shown in Table 4, we can see clearly that the income from the mangrove product is higher at the low and middle level. To differentiate households in the form of wellbeing, we separated households based on their total income as; high, medium and low income level households. Most of the local poor people in that study area were landless and they do not have a regular income. So, in this study, it was decided that income level of less than US$ 1000 per year constitutes being poor, an income level of US$ 1000–1700 per year is medium and an income level of more than US$ 1700 per year is the high level. The income sharing from mangrove products in the high-income level was 32.8% and in the middle-income level was about 52.8%. Local communities with lower income levels generated the most mangrove products which contributes 79.4% to the total income. Most of the middle-income and low-income level households were landless and are absolutely dependent on mangrove forest resources for their livelihood activities. So, farm incomes at low- and middle-income levels were 9.4% and 10.5%, respectively. At the highest income level, households own agricultural land and have better off-farm jobs such as a private shop. Farm income and non-farm income sharing at a high level were 33.3% and 33.9% of the total income, respectively. This means that forest dependency will reduce if there are other better alternatives.

Table 4. Income sharing among different income levels.

Income Source	High Income (n = 56)		Medium Income (n = 65)		Low Income (n = 64)		Kruskal-Wallis Test
	Income	%	Income	%	Income	%	
Mangrove Product income	1,791,000	32.8	956,862	52.8	698,438	79.4	$p = 0.1930$ X2 = 3.290, df = 2
Agricultural Income	1,819,571	33.3	169,692	9.4	92,969	10.5	$p = 0.0001$ X2 = 33.753, df = 2
Non-Farm Income	1,855,179	33.9	685,569	37.8	88,782	10.1	$p = 0.0001$, X2 = 19.482, df = 2

Note: 1US$ = 1428 MM Kyats in 2018.

3.4. Mangrove Forest Income against Socio-Economic Characteristics

The relation of socioeconomic characters of the respondents and mangrove forest resources use was addressed by using a multiple linear regression model which showed mangrove forest income was regressed as a function of different socio-economic characters (factors that affect the level of forest dependency) of the respondents such as agricultural land size, household size, ways to access mangrove forest products, native village, occupation, age, gender of household head and education. The multiple linear regression model was estimated using ordinary least square estimation technique (OLS) after the data was checked for different econometric tests.

The regression result is shown in the Table 5 and many explanatory variables have expected influence on forest dependency. The model explained 35% of the variance on mangrove forest income (F = 10.68, $p < 0.000$). While coefficients on the agricultural land size, the way of accessing mangrove forest products, the occupation of farm activities were statistically significant at (1%), variable, household size was statistically significant at (10%) and variable, non-farm activities was statistically significant at (5%).

This showed that mangrove forest income was negatively correlated with native village, education and occupation activities such as farm activities and non-farm activities. On the other hand, agricultural land size, household size, the ways of getting mangrove forest product, age and household head are positively related to mangrove forest income. However, the variables such as gender of household head, age, education and native village are not statistically significant at any level of significance which shows that those variables were the least important determinants of the household dependence on forest resources.

Table 5. Ordinary least squares estimation (OLS) regression of mangrove forest income against household characteristics. Number of obs = 185; F (8,176) = 10.68; Prob > F = 0.0000; R-squared = 0.3545; Adj R-squared = 0.3213.

Variable	Estimated Coefficient	T Ratio	$p > (t)$
Agricultural land size	87,001 *** (12,945)	6.72	0.000
Household size	109,253 * (63,621)	1.72	0.088
Way of accessing mangrove products	1.1×10^6 *** (305,294)	3.60	0.000
Native village	−310,169 (281,029)	−1.10	0.271
Non-farm activities	−671,806 ** (331,192)	−2.03	0.044
Farm activities	-1.857×10^6 *** (385,256)	−4.82	0.000
Age	27,057 (101,631)	0.27	0.790
Gender of HH Head	181,432 (403,874)	0.45	0.654
Education	−141,803 (125,741)	−1.13	0.261
Constant	705,632 (692,604)	1.02	0.310

Notes: Standard errors in parentheses, *** $p < 0.01$, ** $p < 0.05$, * $p < 0.1$.

4. Discussion

4.1. Dependence on Mangrove Forest Resources

In the study sites, the majority of the people's livelihoods were at subsistence level. They heavily depended on natural resources for their livelihoods. Major livelihood activities in the study sites were agriculture, non-farm activities and mangrove forest product collection. Among them, mangrove forest resources were the major income source and most of the coastal communities have relied on them. The main provisioning services of mangroves are timber, charcoal and firewood as energy sources, shelter, fodder, medicines and a fishery which is important for subsistence, livelihood and commercial fisheries for the communities living in coastal and delta areas. The income for the local poor communities living in rural area of the developing countries was less than US$ 1 per day and they rely on the ecosystem services-ES [24]. Their income (43% to total household income) were generated by selling forest products collected from the mangrove forest such as fishes, crabs and prawn. So, half of the respondents were engaged in mangrove based occupations because they are poor and predominantly live in the delta region. The average annual household income from mangrove forest products per year was Kyats 1,119,957 (approximately US$ 784). Wai [17] stated that the economic value of the mangrove was USD$ 1497.6 (approximated Kyats 2,139,471) in her research of "Economic Dependency of Local Communities on Mangroves: A Case Study in Bogalay, Myanmar". This means that the economic value of mangrove forests is gradually decreasing due to the deforestation and degrading of mangrove forest. Levels of dependence on forest resources around the world among households with access to forests vary from 6 to 65% depending on the local circumstances [25–42]. Singh [43] in Bangladesh estimated that the contribution of non-timber forest products-NTFPs is 79% on average to the annual income of the collector's family. Clinton, U.I and Okujagu, C.M.D [44] inferred that in their study, 85% of households depended on mangrove resources for their income. In this study, agricultural income estimated about 25% of total income. Paddy field is the major cultivation in the study sites. Seaweed cultivation has rapidly emerged as another cash crop in the coastal area; women were mainly involved in seaweed cultivation. Non-farm incomes accounted 32% of total household income. Major non-farm activities were wage labor in mangrove forest plantation, causal and seasonal labor in agriculture, salary, private shop, etc. Furthermore, mangrove forest dependencies vary among different income levels. According to the result, households with middle-income and low-income levels are the most dependent on forest resources with 52.8% and 79.4% of total household income because most of the middle income and low-income level households are landless and they do not have other alternative income activities. This finding was similar to the finding of Abu Nasar Mohammad Abdullah [45] wherein lower income households were relatively more dependent on forest incomes than the better off households.

4.2. Factors Influencing Forest Dependency

The mangrove forest dependence level of rural households was calculated using the relative forest income as a share of total household income account derived from the consumption and sale of mangrove forest resources. The level of dependence (the ratio of mangrove forest income from the total household income) was 43% in the study area on average. So, local households in the studied areas are mainly dependent on the forest resources for their livelihood activities. In this research, socio-economic characteristics that influenced forest dependency were also explained. Agricultural land size is positively correlated with mangrove forest income. This result is contradictory to the general findings of other studies. Lebmeister et al. [46] observed that NTFPs dependency in the rural household was significantly decreased with increasing farmland. In Ethiopia, the relative income from the forest was negatively correlated with cropland [34]. In parts of the Ayeyarwady Delta, land degradation and declining soil fertility due to exploitative farming practices have contributed to decreasing agricultural yields. As a result, in order to maintain agricultural incomes and food production, farmers have resorted to cultivating even more land [47]. For instance, in coastal areas, converting mangrove areas to rice farms has resulted in seawater encroachment and salinization of soils, providing a source of income for only a short period of time before yields drop below economic levels [47]. Household size is directly related to forest income. As the household size increased, the dependency on mangrove forest resources of the household also increased. Ways of accessing mangrove forest products are the main determinant of being dependent on mangrove forest products. According to the result of the survey, 65% of respondents produced mangrove forest products directly and the remaining 35% produced indirectly. Education level in this study negatively impacted mangrove forest dependence because they have less access to alternative income sources. Mulatie Chanie and Tesfaye Yirsaw [48] also found in the study that education level has a negative impact on the forest dependence of an individual. This means that forest income of the non-educated household is greater than the educated one and shows that a household with educated members is less dependent on forest resources as a means of livelihood income. In this study, most respondents were extremely dependent on the forest regardless of the gender of the head of the household, a similar to the finding of Abdullah [45]. Similarly, it found a negatively significant correlation with mangrove forest income. So, if the households have other alternative livelihood sources, their dependency on mangrove forest will decrease.

5. Conclusions

Income from mangrove forest products, agricultural income and non-farm income are the sources of local people for fulfilling their subsistence needs. However, the local people living nearby mangrove forest reserve depend much more on mangrove forest as they can access the mangrove forest products easily in order to generate their income. Income from mangrove forest products is the main income sources of their livelihood income and generates 43% of the total income of the household income. So, households are significantly dependent on mangrove products. The lower level household income group had neither land for agriculture nor off-farm employment for generating their income, increasing their dependency on the forest resources for survival. People are generating their livelihood income from the use of different mangrove resources like fish, crab, prawn and firewood. Firewood is a source of energy for cooking where some households collect firewood for commercial purposes. The second largest source of income is off-farm income which accounts for 32% of the total livelihood income. And agricultural income shares 25% of household income. Lower and middle income level households are more dependent on mangrove forest products when compared to high income levels. Lower income level groups are normally landless and mostly depend on mangrove forest products for their subsistence.

Mangrove forest resources are a major income contribution in the livelihoods of local communities, although few households engage in other alternative livelihood activities, such as agriculture and non-farm employment.

Mangrove forest resources provide an important contribution to local livelihood, therefore issues on forest resource dependency and subsistence level of rural livelihood should not be ignored in policy level decisions and other interventions. In addition to the forest resource use, other income generation activities should be incorporated so that livelihood strategies can be diversified to sustain local livelihood and reduce their dependency on forest resources. To avoid deforestation and inefficient utilization of forest resources, the government needs to implement alternative income generation and forest rehabilitation activities for the protection, conservation and utilization of mangrove resources. Improvement of mangrove forest condition will assure benefit optimization and sustainable management of forest resources.

Author Contributions: For research, W.N.A. collected and organized all the data. The analyses were conducted by W.N.A. and guided by Y.W.; K.M., S.T. and A.W.T. helped with the conceptualization and partly advised in the process of writing the paper. All authors considered the outline and contributed to writing the manuscript.

Funding: This research is supported by the Natural Science Foundation of China (71861147001, 71373024).

Acknowledgments: I would like to express sincere gratitude to APFNet (Asia Pacific Network for Sustainable Forest Management and Rehabilitation) for supporting me with a scholarship to pursue a master degree in the School of Economics and Management, Beijing Forestry University, China. I wish to express my thanks to the Forest Department of Pyaphon District, Bogalay Township for their kind support during my fieldwork. Finally, thanks go to the Forest Department, Ministry of Natural Resources and Environmental Conservation for giving me a chance to study abroad.

Conflicts of Interest: The authors declare no conflicts of interest.

References

1. Spalding, M. *World Atlas of Mangroves*; Taylor & Francis: New York, NY, USA, 2010; p. 10017.
2. FD (Forest Department). *National Biodiversity Strategy and Action Plan 2015-2020*; Forest Department: Naypyitaw, Myanmar, 2015.
3. FAO (Food and Agricultural Organization of the United Nations). *Agro-Maps: Global Spatial Database of Agricultural Land-use Statistics*; Food and Agricultural Organization of the United Nations: Rome, Italy, 2013.
4. Sulit, V.T. Use of mangroves for aquaculture: Myanmar. In *Promotion of mangrove-friendly shrimp aquaculture in Southeast Asia*; Aqualture Department: lloilo, Philippines, 2004; pp. 145–150.
5. Snedaker, S.C. Mangrove: Their value and perpetuation. *Nat. Resour.* **1978**, *14*, 6–13.
6. Ngomela, A. The contribution of mangrove forests to the livelihoods of adjacent communities in Tanga and Pangani districts. Ph.D. Thesis, Sokoine University of agriculture, Morogoro, Tanzania, June 2007.
7. Chhetri, B.B.K.; Larsen, H.O.; Smith-Hall, C. Environmental resources reduce income inequality and the prevalence, depth and severity of poverty in rural Nepal. *Environ. Dev. Sustain.* **2015**, *17*, 513–530. [CrossRef]
8. Pagdee, A.; Hornchuen, S.; Sang-arome, P.; Sasaki, Y. Community forest: A local attempt in natural resource management, economic value of ecosystem services, and contribution to local livelihoods. *Warasan Wichai Mokho* **2008**, *13*, 1129–1134.
9. Kalra, E.K. Nutraceutical-Definition and Introduction. *AAPS PharmSci.* **2003**, *5*, 27–28. [CrossRef]
10. Bandaranayake, W. Traditional and medicinal uses of mangroves. *Mangroves Salt Marshes* **1998**, *2*, 133–148. [CrossRef]
11. Tirupathi, K.S.C.; Vishnuvardhan, Z.; Krishna, R.H. A Review on Chemistry of Mangrove Plants and Prospects of Mangroves as Medicinal Plants. *IJGHC* **2013**, *2*, 943–953.
12. Selvam, V. *Trees and Shrubs of the Maldives*; FAO Regional Office for Asia and the Pacific: Bangkok, Thailand, 2007.
13. Patil, P.D.; Chavan, N.S. A need of conservation of mangrove genus Bruguiera as a famine food. *Ann. Food Sci. Technol.* **2013**, *14*, 294–297.
14. Duke, N.; Nagelkerken, I.; Agardy, T.; Wells, S.; van Lavieren, H. *The Importance of Mangroves to People: A Call to Action*; Van Bochove, J., Sullivan, E., Nakamura, T., Eds.; United Nations Environment Programme World Conservation Monitoring Centre: Cambridge, UK, 2014.
15. Vo, Q.T.; Kuenzer, C.; Vo, Q.M.; Moder, F.; Oppelt, N. Review of valuation methods for mangrove ecosystem services. *Ecol. Indic.* **2012**, *23*, 431–446. [CrossRef]
16. Mojiol, A.R.; Guntabid, J.; Lintangah, W.; Ismenyah, M.; Kodoh, J.; Chiang, L.K.; Sompud, J. Contribution of Mangrove Forest and Socio-Economic Development of Local Communities in Kudat District, Sabah Malaysia. *Int. J. Agric. For. Plant.* **2016**, *2*, 122–129.
17. Wai, T.T. Economic Dependency of Local Communities on Mangroves: A Case Study in Bogalay, Myanmar. Available online: http://www.eepseapartners.org/economic-dependency-local-communities-mangroves-case-study-bogalay-myanmar/ (accessed on 26 April 2019).
18. FD (Forest Department). *Myanmar National Strategy and Action Plan*; Forest Department: Naypyitaw, Myanmar, 2016.
19. Oo, N. Present state and problems of mangrove management in Myanmar. *Trees* **2002**, *16*, 218–223. [CrossRef]
20. Zöckler, C.; Delany, S.; Barber, J. *Scoping paper: Sustainable coastal zone management in Myanmar*; ArcCona Ecological Consultants ArcCona Ecological Consultants: Cambridge, UK, 2013.
21. Meinmahla Kyun Wildlife Sanctuary. Retrieved from Ramsar Site Information Service. Available online: https://rsis.ramsar.org/ris/2280.2017 (accessed on 2 February 2017).
22. Yamane, T. *Statistics: An Introductory Analysis*, 2nd ed.; Harper and Row: New York, NY, USA, 1967.
23. Bol, E.T. Status of Mangrove Ecosystem Conservation in Myanmar. *Glob. J. Biosci. Biotechnol.* **2013**, *2*, 423–426.
24. Barbier, E.B. Natural Capital, Ecological Scarcity and Rural Poverty. World Bank: Washington, DC, USA, 2012; p. 38.
25. McSweeney, K. Who Is "Forest-Dependent"? Capturing Local Variation in Forest-Product Sale, Eastern Honduras. *Prof. Geogr.* **2002**, *54*, 158–174. [CrossRef]
26. Ambrose-Oji, B. The contribution of NTFPs to the livelihoods of the 'forest poor': evidence from the tropical forest zone of south-west Cameroon. *Int. For. Rev.* **2003**, *5*, 106–117. [CrossRef]
27. Fisher, M. Household welfare and forest dependence in Southern Malawi. *Environ. Dev. Econ.* **2004**, *9*, 135–154. [CrossRef]

28. Mamo, G.; Sjaastad, E.; Vedeld, P. Economic dependence on forest resources: A case from Dendi District, Ethiopia. *For. Policy Econ.* **2007**, *9*, 916–927. [CrossRef]

29. Shackleton, C.M.; Shackleton, S.E.; Buiten, E.; Bird, N. The importance of dry woodlands and forests in rural livelihoods and poverty alleviation in South Africa. *For. Policy Econ.* **2007**, *9*, 558–577. [CrossRef]

30. Illukpitiya, P.; Yanagida, J.F. Role of income diversification in protecting natural forests: Evidence from rural households in forest margins of Sri Lanka. *Agrofor. Syst.* **2008**, *74*, 51–62. [CrossRef]

31. Mcelwee, P.D. Forest environmental income in Vietnam: Household socioeconomic factors influencing forest use. *Environ. Conserv.* **2008**, *35*, 147–159. [CrossRef]

32. Quang, N.V.; Noriko, S. Forest Allocation Policy and Level of Forest Dependency of Economic Household Groups: A Case Study in Northern Central Vietnam. *Small-scale For.* **2008**, *7*, 49–66. [CrossRef]

33. Kamanga, P.; Vedeld, P.; Sjaastad, E. Forest incomes and rural livelihoods in Chiradzulu District, Malawi. *Ecol. Econ.* **2009**, *68*, 613–624. [CrossRef]

34. Babulo, B.; Muys, B.; Nega, F.; Tollens, E.; Nyssen, J.; Deckers, J.; Mathijs, E. The economic contribution of forest resource use to rural livelihoods in Tigray, Northern Ethiopia. *For. Policy Econo.* **2009**, *11*, 109–117. [CrossRef]

35. Yemiru, T.; Roos, A.; Campbell, B.; Bohlin, F. Forest incomes and poverty alleviation under participatory forest management in the Bale Highlands, Southern Ethiopia. *Int. For. Rev.* **2010**, *12*, 66–77. [CrossRef]

36. Heubach, K.; Wittig, R.; Nuppenau, E.-A.; Hahn, K. The economic importance of non-timber forest products (NTFPs) for livelihood maintenance of rural west African communities: A case study from northern Benin. *Ecol. Econ.* **2011**, *70*, 1991–2001. [CrossRef]

37. Bosma, R.; Sidik, A.S.; van Zwieten, P.; Aditya, A.; Visser, L. Challenges of a transition to a sustainably managed shrimp culture agro-ecosystem in the Mahakam delta, East Kalimantan, Indonesia. *Wetl. Ecol. Manag.* **2012**, *20*, 89–99. [CrossRef]

38. Kar, S.P.; Jacobson, M.G. NTFP income contribution to household economy and related socio-economic factors: Lessons from Bangladesh. *For. Policy Econo.* **2012**, *14*, 136–142. [CrossRef]

39. Tieguhong, J.C.; Nkamgnia, E.M. Household dependence on forests around lobeke National Park, Cameroon. *Int. For. Rev.* **2012**, *14*, 196–212.

40. Hogarth, N.J.; Belcher, B.; Cambell, B.; Stacey, N. The Role of Forest-Related Income in Household Economies and Rural Livelihoods in the Border-Region of Southern China. *World Dev.* **2013**, *43*, 111–123. [CrossRef]

41. Angelsen, A.; Jagger, P.; Babigumira, R.; Belcher, B.; Hogarth, N.J.; Bauch, S. Environmental Income and Rural Livelihoods: A Global-Comparative Analysis. *World Dev.* **2014**, *64*, 12–28. [CrossRef]

42. Rayamajhi, S.; Smith-Hall, C.; Helles, F. Empirical evidence of the economic importance of central Himalayan forests to rural households. *For. Policy Econo.* **2012**, *20*, 25–35. [CrossRef]

43. Singh, A.; Bhattacharya, P.; Vyas, P.; Roy, S. Contribution of NTFPs in the Livelihood of Mangrove Forest Dwellers of Sundarban. *J. Hum. Ecol.* **2010**, *29*, 191–200. [CrossRef]

44. Clinton, U.I.; Diepiriye, P.M.; Okujagu, C. Contribution of the Mangrove Forest Resources to the Livelihood of the Andoni People of Rivers State, Nigeria. *Res. J. Geogr.* **2016**, *3*, 1–11.

45. Abdullah, A.N.M.; Stacey, N.; Garnett, S.T.; Myers, B. Economic dependence on mangrove forest resources for livelihoodsin the Sundarbans, Bangladesh. *For. Policy Econo.* **2016**, *24*, 15–24. [CrossRef]

46. Leßmeister, A.; Heubach, K.; Lykke, A.M.; Thiombiano, A.; Wittig, R.; Hahn, K. The contribution of non-timber forest products (NTFPs) to rural household revenues in two villages in south-eastern Burkina Faso. *Agrofor. Syst.* **2018**, *92*, 139–155. [CrossRef]

47. UNEP (United Nations Environment Program). *Learning from Cyclone Nargis: Investing in the Environment for Livelihoods and Disaster Risk Reduction*; UNEP: Nairobi, Kenya, 2009.

48. Mulatie, C.; Tesfaye, Y. Economic Contribution of Forest Resources to Sustainable Rural Livelihoods in Bench Maji Zone, South West Ethiopia. *Inter. J. Adv. Res.* **2018**, *6*, 1–10. [CrossRef]

Article

Consuming Blackberry as a Traditional Nutraceutical Resource from an Area with High Anthropogenic Impact

Ioana Andra Vlad [1], Győző Goji [2], Florin Dinulică [3,*], Szilard Bartha [4],
Maria Magdalena Vasilescu [3] and Tania Mihăiescu [5]

[1] Department of Food Engineering, University of Oradea, Oradea, Bihor 410048, Romania;
     ioana_andravlad@yahoo.co.uk
[2] Technological High School Ștefan Manciulea, Blaj, Alba 515400, Romania; ggyozo2000@yahoo.com
[3] Transilvania University of Brașov, Faculty of Silviculture and Forest Engineering, 500123 Brașov, Romania;
     vasilescumm@unitbv.ro
[4] Department of Forestry and Forest Engineering, University of Oradea, Oradea, Bihor 410048, Romania;
     barthaszilard10@yahoo.com
[5] Plant Protection Department, University of Agricultural Sciences and Veterinary Medicine, Cluj-Napoca,
     Cluj 400372, Romania; tmihaiescu@yahoo.com
*   Correspondence: dinulica@unitbv.ro; Tel.: +40-751-137-007

Received: 31 January 2019; Accepted: 5 March 2019; Published: 11 March 2019

Abstract: The most serious quality issue of natural resources for human consumption or medicinal purposes is the contamination with pollutants harmful to consumers. Common blackberry (*Rubus fruticosus* L.) is a sought-after nutraceutical and an important component in herbal medicine in many places around the globe. The present study aims to analyze the level of heavy metal bioaccumulation in blackberry organs, as well as its spatial distribution in two consecutive years immediately after the interruption of the extended activity of the industrial source of pollution. The research was conducted in one of the most polluted areas in Romania and Eastern Europe, within a 26 km radius of the source of pollution. The Pb, Cd, Cu, and Zn concentrations in the leaves, flowers, and unwashed blackberry fruits were analyzed spectrophotometrically through flame atomic absorption spectroscopy (FAAS). The results show that blackberry is an important bioaccumulator of these heavy metals—71% of the Pb concentration values and 100% of the Cd concentration values exceeded the World Health Organization thresholds by up to 29 and 15 times, respectively. Also, the leaves are the largest reservoirs of Pb and Zn (the median values: 51.4 mg/kg dry weight and 105.2 mg/kg d.w., respectively), and the flowers contained the largest quantities of Cd and Cu (2.54 mg/kg d.w. and 11.3 mg/kg d.w., respectively). The Pb concentrations decreased by a power function in relation to the distance from the source of pollution. The implications of these results on the safety of the use of blackberry are discussed. The urgent necessity for food education of the local population which consumes contaminated nutraceutical products is emphasized.

Keywords: heavy metal contamination; herbal medicine; historically polluted area; wild food; blackberry

---

1. Introduction

In spite of all the qualitative changes which human activity has experienced over time, the attraction for products provided directly and generously by nature has not diminished [1]. The diversity of uses which every natural resource offers is the most convincing evidence of its value. For instance, blackberry is simultaneously edible, medicinal, and melliferous, thus it can be classified as a highly interesting nutraceutical (Table 1).

Table 1. The spectrum of traditional uses of blackberry.

Resource			Utilization		Information Source
Part of Plant	**Species**	**Range**	**Product**	**Uses/Disease**	
Whole plant	*Rubus fruticosus*	Romanian folk medicine	tea, decoction	leukorrhea	[2]
	Rubus sp.	Native American folk medicine	extract from fruit, root, and leaves	hair and fabric dye	[3]
Aerial parts	*Rubus fruticosus*	Europe	various	hypoglycemia	[4]
Stem	*Rubus* sp.	Native American practices	rope	transport	[3]
Young shoots	*Rubus fruticosus, R. ulmifolius* Schott	Sardinian and Sicilian traditional medicine	decoction	menstrual pain	[5]
	Rubus fruticosus	Romanian folk medicine	decoction	bronchitis, diarrhea, dysentery	[2]
Leaves	*Rubus fruticosus, R. ulmifolius*	Sardinian and Sicilian traditional medicine	fresh leaves for chewing	strengthening spongy gums	[5]
	Rubus fruticosus	Central Italy folk medicine	maceration	cicatrizant for skin, fungal infections, skin abscesses	[6]
	Rubus villosus Aiton.	around the world	leaves for chewing	bleeding gums	[7]
	Rubus fruticosus	European folk medicine	mouthwash, decoction	strengthening spongy gums, mouth ulcers, sore throats, diarrhea, hemorrhoids	[8]
Fruits	*Rubus* sp.	around the world	jam, syrup, jelly, marmalade, cake stuffing, wine, liqueur, ice cream, in yoghurt, drink and chewing gum dye		[3,9]
	Rubus fruticosus	Romanian folk medicine	decoction in lard, wine	tuberculosis, leukorrhea	[2]
	Rubus fruticosus	Ancient Greeks	fresh fruit	gout	[3]
Fruits and leaves	*Rubus fruticosus*	Pakistani traditional medicine	various	skin diseases, itching, scabies, eczema	[10]
Roots	*Rubus villosus*	around the world	dried root tea used for edema, leaves and roots used for diarrhea, enteritis, chronic appendicitis, leukorrhea, expectorant properties		[7]

Since the time of Hippocrates [11], the belief that food has therapeutic properties has gradually consolidated [12] and has engaged a rich terminology, with interchangeable notions that have created confusion and controversies [13]. A nutraceutical is a hybrid concept better located at the boundary between food and drug [14]. It was introduced by DeFelice in 1989 and brought about a revolution in nutrition [15]. In contrast to other food-derived products claimed to have benefits on human health, nutraceuticals have a proven clinic efficiency in preventing and even treating certain pathological conditions [16]. At least in Europe, the lack of nutraceuticals identity and insufficient clinical evidence from in vivo experiments has kept down the formulation of shared regulatory framework for nutraceuticals [17] that would guarantee consumers the efficacy and safety of this pharma-food.

Despite increasing research on the properties of bioactive compounds [18], nutraceuticals, as rich substance mixtures [17], require: (1) a supplementary chemometric effort to identify the dietary markers which enable the quality control of the products [19] and (2) robust clinical evidence to support their use [20] and thus the transition from potential nutraceutical to established nutraceutical [15].

Taking into account the blackberry, the most important bioactive ingredients are: (1) the ellagitannins, which, besides their usual antidiarrheal and antidysenteric astringency, inhibit the growth of cancerous cells [3,21–23] and (2) the anthocyanins and other polyphenols, with their significant antiradical, antioxidant, and chemoprotective activities [8,21–35].

In many places around the world, especially in rural and tribal areas, exploiting natural, food, and medicinal resources is a survival issue, therefore a social factor [36] or, in any case, an alternative source of income. In 2005, 14,837 t of wild blackberries were harvested, in addition to 154,578 t of cultivated blackberries [37]. In Romania, 1 ha of forest land can yield up to 12.5 t of blackberries per year [9]. In our researched area, there are over 17,400 people who have access to contaminated natural products. With the cease of pollutant activity, 80% of employees were fired in 2009 and directed towards other fields. Consequently, the interest for the exploitation of the agricultural, medicinal, and nutritional potential of the area increased.

The large number of uses of vegetal products and their composition raises the often-times vital issue of product safety. The contamination with pollutants, either local, regional, or cross-border, endangers the health of consumers of such bio-products. Heavy metals, resulting from metallurgical activities by means of the refining and burning of fossil fuels or fertilizing agricultural soils, enter the food chain via the air, water, and soil, manifesting toxicity even in very small concentrations [38,39]. For instance, lead, cadmium, and zinc poisoning attacks the nervous system, causing a decrease in intellectual performance, as well as aggressiveness, delinquency, and narcomania in youths [40–45]. Blackberry is more prone to cadmium accumulation than other fruit [46].

Impact studies on historical pollution of the chemical composition of blackberry were carried out in Sudety Mountains SW, Poland [47]; Pirdop, Bulgaria [48]; Vladivostok, Russia [49]; Lori region, Armenia [50]; Berlin, Germany [46]; Middle Spis, Slovakia [51]; and Moldova Nouă, Romania [52]. The level of heavy metal contamination in leaves, fruits, and products derived from blackberries (blackberry leaf tea, blackberry wine) was also determined [53,54].

Our research aims to determine the recent pollution level through the concentration of certain heavy metals in blackberry vegetal material (*Rubus fruticosus* L.), and to characterize its spatial distribution in relation to the distance from the source of pollution and the site geomorphology. The investigations were carried out in one of the most polluted areas in Romania and Eastern Europe. The age and seriousness of the pollution in these areas prompted a variety of impact studies on the environment and the living organisms—revised by Smejkal [55] and Micu [56]. However, blackberry was not analyzed in these studies, in spite of its wide popularity among local and national consumers [9].

2. Materials and Methods

2.1. The History of the Pollution

The source of the pollution whose effects are analyzed in this article is the industrial park in the town Copşa Mică (46°06′59.10″ N and 24°13′15.43″ E), in the center of Romania. Until now, it produced large quantities of carbon black (for 58 years: 1935–1993), metallurgical and refined zinc, electrolytic lead, bismuth, antimony, iron, cadmium powder, sulphuric acid, sulfur dioxide, sulfates, sulfurs, carbon monoxide, nitrogen oxides, volatile arsenic compounds, and ammonia (for 70 years: 1939–2009).

The location of the industrial park on the wide valley of the Târnava river, which channels the local circulation of air mass, allowed the pollutants to distribute over large distances. The hydrographic fragmentation of the territory extended the pollution transversally to the secondary valleys. At the nearest weather station, according to the climatic data provided by the National Meteorological Agency [57], the mean annual temperature is 8.4 °C, the mean annual rainfall is 625.6 mm·year^{-1}, the annual wind frequency is 65.5%, and the speed of the wind with the highest frequency is 3.1 m·s^{-1}. The low amount of rainfall leads to the persistence of pollutants in the atmosphere and the high percentage of atmospheric calm allows air mass stagnation and pollutant deposition.

The plant material was collected in two consecutive years, starting with the year when the activity on the polluting industrial platform ceased. In the two years of sampling the mean temperatures were 9.4 and 9.1 °C and the rainfall levels were 648.4 and 782.3 mm·year^{-1}, respectively [57].

2.2. Sampling Design

The distribution of pollutants in blackberry organs was examined in nine sampling plots, eight of which were grouped in the first 8 km from the source of pollution (Figure 1), and one control plot, located 26 km from the industrial park in Copşa Mică (type of site D). The target was the study of pollution in various topoclimates. Each plot was identified geographically and geomorphologically, using Global Positioning System coordinates, the side aspect, the exposure to the circulation of polluted air (Table 2), and the distance to the main flue-gas stack for emissions—which is 250 m tall.

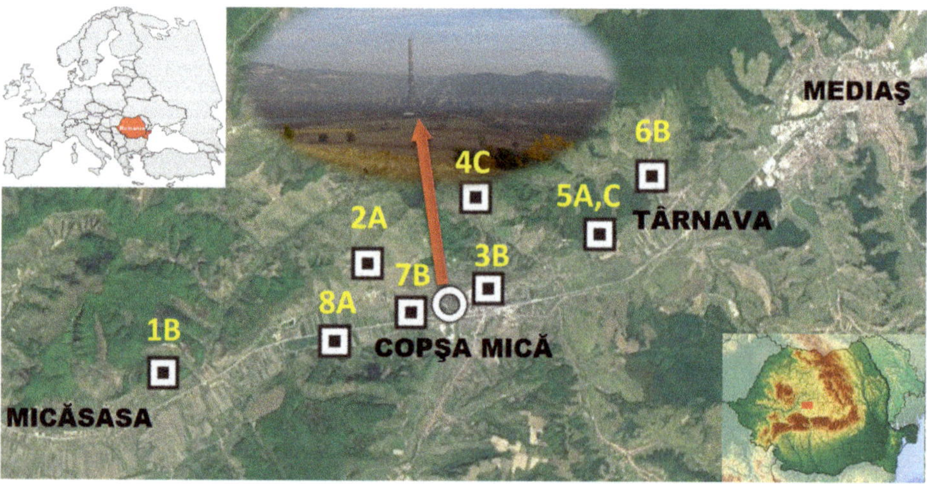

Figure 1. The sampling plots area: the circle marks the source of pollution; the squares mark the sample plots (numbered from 1 to 8 and identified with the type of site).

Table 2. Classification of sampled sites according to the location in relation to the source of pollution.

Type of Site	Site Description
A	Site located in the main valley (where the source of pollution is found) with frontal exposure to the source of pollution (slope facing the flue-gas stack).
B	Site located in the main valley (where the source of pollution is found) with tangential exposure to the source of pollution (slope not facing the flue-gas stack).
C	Site located in a secondary valley with frontal exposure to the local circulation of air mass.
D	Site located in a secondary valley, partially protected from the source of pollution.

The vegetal material samples were collected according to the regulations of the United Nations Economic Commission for Europe-International Co-operative Programme on Assessment and Monitoring of Air Pollution Effects on Forests- [58]. At least five blackberry dominant bushes were chosen from every sampling plot. The material (20–30 g leaves, flowers/sampling plot and 100–200 g fruits/plot) was collected systematically from the four cardinal sides of the bush. Only healthy samples were considered [59], and great care was taken to avoid touching or contaminating them with the tools used. The flowers were collected no later than 2–3 days after bloom or in the budding stage, to prevent loss of pollen due to insect pollination. The leaves were collected in the second half of the growing season, but before the autumnal senescence, when the heavy metal concentration peaks [60]. The ripe fruits were harvested in the firm stage.

2.3. Processing the Material

The vegetal material was not washed, so as to identify the total pollutant concentrations in the state in which the resource is used [61]. For instance, the blackberry long-lived leaves are eaten by game, particularly by cervids [9], and blackberries are not washed before consumption. To avoid pollen removal, which bioconcentrates an important fraction of heavy metals, the flowers were not washed either.

The laboratory investigations followed Kelp's [62] recommendations. The samples were oven-dried to a constant mass at 60 °C, which did not affect the sanogenetic qualities of the product [9]. Mineralization was achieved after wet digestion [58], using the Berghof MWS-2 microwave oven. The mixture of 0.3–0.5 g dried plant powder, 2 mL concentrated HNO_3 (65% concentration, Merck extra pure), and 3 mL H_2O_2 (30% concentration, Merck, Darmstadt, Germany) were introduced in the microwave system (Berghof MWS-2, Eningen, Germany). Mineralization was carried out in three steps, at temperatures of 145, 180, and 100 °C (Table 3).

Table 3. The settings for mineralization of samples.

Temperature (°C)	145	180	100
Power (%)	75	90	40
Time (min)	5	10	10

After mineralization, samples were filtered through a 0.45 mm filter and brought to a volume of 50 mL in a volumetric flask with ultrapure water with a specific resistance of 18.2 MΩ/cm obtained from a Direct Q3UV Smart (Millipore SAS, Molsheim, France). The digested samples were analyzed by flame atomic absorption spectroscopy (FAAS) with ZEEnit 700 Atomic Absorption Spectrometer (Analytik Jena AG, Jena, Germany). Calibrating standard solutions of Cd, Cu, Pb, and Zn were prepared daily by the accurate dilution of the respective stock standard solutions (1000 mg/L). Ultrapure water with a specific resistance of 18.2 MΩ/cm obtained from a Direct Q3UV Smart (Millipore SAS, Molsheim, France) was used to prepare the standard solutions. For quality control purpose, blanks and triplicates samples ($n = 3$) were analyzed during the procedure. The variation coefficient was under 5%. The operation conditions were those recommended for each metal in the instrument's method (Table 4).

Table 4. Instrumental parameters for metal determination by flame atomic absorption spectroscopy (FAAS).

Standard Conditions	Element			
	Cd	Cu	Pb	Zn
Wavelength, λ (nm)	228.8	324.8	283.3	213.9
Slit width (nm)	1.2	1.2	1.2	0.5
Hollow-cathode lamp current (mA)	3	3	3	4
Background correction	Deuterium	Deuterium	Deuterium	Deuterium
Flame	C_2H_2/air	C_2H_2/air	C_2H_2/air	C_2H_2/air
Fuel flow (N L/h)	50	50	65	50

The sensitivity of the FAAS method was estimated using the limit of detection (LOD) and the limit of quantification (LOQ). The LOD and LOQ (Table 5) were calculated based on the standard deviation of the response and the slope [63–66]. A total number of 171 spectrometric determinations were carried out.

Table 5. Limit of detection (LOD) and limit of quantification (LOQ) of the flame atomic absorption spectroscopy method.

Parameter	Element			
	Cd	Zn	Pb	Cu
Linear working range (mg/L)	0–1	0–1	0–1	0–3
Limit of detection (mg/L)	0.012	0.013	0.083	0.036
Limit of quantification (mg/L)	0.039	0.042	0.276	0.119

2.4. Data Processing

Data analysis was performed using Microsoft EXCEL 2007 and STATISTICA 8.0. The results were related to the World Health Organization [67] limits for heavy metals in products with ecosanogenetic qualities (Table 6).

Table 6. Tolerable limits for heavy metals in food supplements and herbal drugs.

Reference	Pb (mg/kg)	Cd (mg/kg)	Zn (mg/kg)	Cu (mg/kg)
[68]	10.0	0.5	-	-
[67]	10.0	0.3	-	-
[62]	5	4	-	-
[69]	-	-	-	5 (berries and small fruits)
[62]	5	0,5	-	-

3. Results and Discussions

3.1. The Level of Heavy Metal Contamination in Blackberry

The concentrations of the studied microelements were found to be strongly scattered around the mean (high coefficients of variation—Table 7). Thus, the arithmetic mean was no longer relevant and was replaced with the median to express the central tendency. Most of the lead and cadmium concentration values greatly exceeded the toxicity thresholds (Table 7). Furthermore, these thresholds were exceeded in the control plot as well, which was believed to be unaffected by the influence of the pollution caused by the industrial park in Copşa Mică. As such, 40% of the measured lead concentrations, 100% of the cadmium concentrations, and 67% of the copper concentrations in the control plot exceeded the WHO permissible limit. This result is proof of the area expansion of heavy metal pollution. The other sampling plots are located up to 8 km from the source of pollution and have pollutant concentrations which exceeded the permissible limit for lead by up to 29.1 times, the permissible limit for cadmium by up to 14.9 times, and the permissible limit for copper by up to 38.8 times. Approximately a quarter of the values of lead concentration exceeded the permissible limit by at least 5 times. More than half of the values of cadmium concentration exceeded the permissible limit by at least 5 times.

Table 7. Statistics of heavy metal content in the blackberry samples from Copsa Mică area, Romania.

Metal	The Significance of the Differences between Individual Values *		Range	Arithmetic Mean	Median	Coefficient of Variation (%)	Relative Frequency (%) of Values Which Exceed the World Health Organization Threshold	The Significance of the Differences between Blackberry Organs ** (Kruskal–Wallis Test)	
	t	p						H	p
Pb (mg/kg dry weight)	4.64	<0.001	1.67–291.39	34.72	20.27	141.59	70.5	14.27	<0.001
Cd (mg/kg d.w.)	11.49	<0.001	0.32–4.46	1.86	1.61	57.08	100.0	8.18	0.02
Zn (mg/kg d.w.)	10.01	<0.001	10.91–193.54	76.03	70.29	65.49	-	23.05	<0.001
Cu (mg/kg d.w.)	7.84	<0.001	1.23–34.08	8.51	7.18	69.89	83.3	9.34	0.01

* All values of heavy metals concentrations either from leaves, either from fruits or flowers were merged; ** The differences refer to the concentrations of Pb, Cd, Zn and Cu grouped according to the three blackberry organs (leaves, fruits, flowers).

Based on the blackberry average yield in Romania [9], this means that a hectare of blackberry shrubs from the Copşa Mică area sequesters yearly through leaves, flowers, and fruits: 2.17 kg Pb, 0.17 kg Cd, 7.52 kg Zn, and 0.77 kg Cu.

The fact that these discovered values are greater than those highlighted in pollution literature is worrisome for the local consumers. Gasser et al. [70] processed the database of the German Medicines Manufacturers' Association and indicated that the following values of Cd and Pb concentrations range in blackberry leaves: <0.07–0.32 mg/kg dry weight and <0.4–2.8 mg/kg d.w., respectively. Shikhova [49] highlighted average concentrations of 15.07 mg/kg d.w. Pb in *Rubus sachalinensis* H. Lév. from the suburban forest phytocenosis in Vladivostok. After analyzing samples of *Rubus fruticosus* harvested from different sampling plots in Berlin, von Hoffen and Säumel [46] found average cadmium concentrations of 0.0081 mg/kg d.w., and lead concentrations of 0.0595 mg/kg d.w.

Investigations of heavy metal content in blackberry were also carried out in areas with historical pollution of mining or metallurgical origin. Micu et al. [52] identified average concentrations of 12 ppm Cu, 0.03 mg/kg d.w. Cd, and 19 mg/kg d.w. Pb in the blackberry leaves on the spoil heaps of Moldova Nouă (Romania). Wisłocka et al. [47] found in washed *Rubus idaeus* L. leaves grown on uranium mine dumps in the Sudety Mountains range heavy metal concentrations of 17.6–41.0 mg/kg d.w. for Pb, 0.40–1.60 mg/kg d.w. for Cd, 1.20–10.50 mg/kg d.w. for Cu, and 27–88 mg/kg d.w. for Zn, which were consistent with their concentrations in the soil. In the Middle Spis (Slovakia), which was affected by acid and heavy metal pollution for decades, Vollmannova et al. [51] found the following range of toxic metal concentrations: 0.30–1.19 mg/kg d.w. Pb, 0.18–0.42 mg/kg d.w. Cd, 5.50–6.50 mg/kg d.w. Cu, and 16.1–30.7 mg/kg d.w. Zn in dry blackberry leaves, as well as 0.03 mg/kg d.w. Pb, 0.03–0.05 mg/kg d.w. Cd, 0.48–0.99 mg/kg d.w. Cu, and 2.08–3.13 mg/kg d.w. Zn in fresh blackberries. Compared to our results, the investigations carried out by Teofilova et al. [48] in the area of the copper foundry in Pirdop (Bulgaria) reported higher average values of copper content (62.5 mg/kg d.w.), the same values for zinc (80 mg/kg d.w.), and lower values for lead (8.5 mg/kg d.w.) and cadmium (0.275 mg/kg d.w.) in blackberry fruits.

The large discrepancy with the literature data is partly due to the way our samples were prepared, i.e., without pre-washing. We intended to quantify the total bioaccumulation of heavy metals in the blackberry organs, as they are used directly by consumers (humans, cervids, bees) and thus the input to the trophic chain through the contaminants is more widely dispersed in the ecosystem structure. The data from the literature of other species reveal that, by washing, the heavy metal concentrations were reduced by 3.09–85.79% for Pb, 4.00–86.11% for Cd, 0.78–84.85% for Zn, and 0.76–86.41% for Cu, varying by species, organ harvested, culture system, sampling period, and degree and type of pollution [71–76]. We assume that even in the case of blackberry, as a species with hirsute organs, the deposition of heavy metals at least at the surface of the leaves is considerable. It has been shown that Pb and Cd concentrations are 10 times higher in hirsute plants than in those with a smooth surface [77].

The non-parametric Kruskal–Wallis test (Table 8) led to the stratification of the values of metal concentration according to the blackberry organs.

Table 8. The significance of the differences between the heavy metal concentration values by some nonbiological factors.

	Dependent Variable			
Independent Variable	**Pb**	**Cd**	**Zn**	**Cu**
	p **from Kruskal–Wallis Test (0.05 is the Threshold Value for Statistical Significance) ***			
Distance from source of pollution	0.005	0.13	0.36	0.80
Altitude	0.01	0.31	0.27	0.07
Aspect	0.08	0.92	0.59	0.70
Exposure to air circulation	0.09	0.01	0.15	0.48
Year of sampling	0.26	0.41	0.89	-

* Kruskal-Wallis *p*-values for the effects of independent nonbiological factors on heavy metal concentration in blackberry organs sampled across air pollution gradients in Copșa Mică area, Romania.

The values in the content of lead, cadmium, zinc, and copper in fruits are noticeably different compared to those in flowers and leaves (Figure 2). Of the sampled blackberry organs, the fruits retained the smallest metal quantities, except for copper—the copper content had the highest variation in the fruits. The leaves contained 4.5 times more lead than the fruits. The flowers contained 3 times more cadmium and 3.8 times more zinc than the fruits. The flowers contained 2.2 times more copper than the leaves (Figure 2).

This means that the risk to consumers of such resources, quantified for a portion of 100 grams of fresh blackberries, with an average moisture content of 91.4% (own data), consists in the ingestion of 8.51 mg Pb, 0.74 mg Cd, 19.64 mg Zn, and 5.71 mg Cu.

Yedoyan and Yedoyan [50] found notable differences between blackberry organs which were polluted anthropogenically, especially in terms of lead and copper content. For instance, the root was found to be an important copper reservoir.

Concerning other species besides blackberry, from the Copșa Mică area, Alexa et al. [78] spectrometrically measured the heavy metal content. The comparisons emphasized the fact that trees are more important heavy metal bioaccumulators than blackberry. In June 2001, in full industrial season, up to 620 mg/kg d.w. lead and up to 8.5 mg/kg d.w. cadmium were found in locust leaves [78].

The net differentiation of blackberry organs in heavy metals storage is the consequence of their different and asynchronous lifespan, the morphological characteristics of their surface, and the exposure to the pollutant flow. Blackberry leaves—long-living, hirsute on both sides, and more exposed to atmospheric pollutants—are the largest reservoirs of heavy metals (Figure 2). The consistency and morphology of the floral tissue and the synchronization of the flowering stage with the rainiest months reduce the retention of heavy metals in flowers compared to leaves. Blackberries themselves, sheltered by leaves and slowly ripening, accumulate smaller amounts of heavy metals (Figure 2).

The unnoticeable differences of heavy metal concentrations between the two years of sampling (Table 8) suggest a long-lasting soil pollution and a strong sequestration of these pollutants in the blackberry organs, which goes beyond the rainfall increase in the second year.

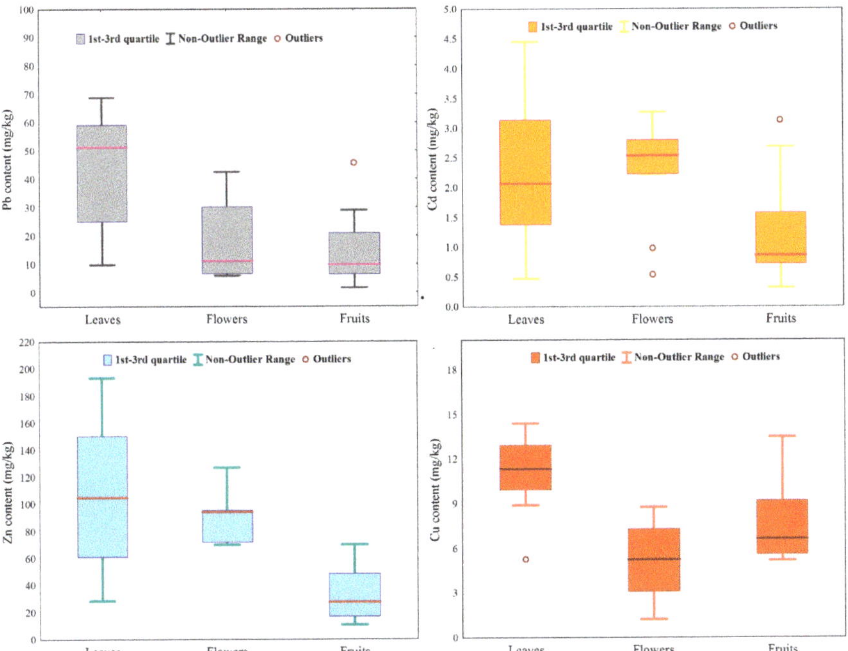

Figure 2. The stratification of heavy metal content according to the sampled blackberry organs.

3.2. The Spatial Variability of the Heavy Metal Content in Blackberry

The factors of influence on the metal content in the vegetal material were identified by using the non-parametric Kruskal–Wallis test, when the majority of dependent variables had non-Gaussian distributions. The results show that only the lead content is more sensitive to location change (Table 8). It decreases by a power function according to the distance from the source of pollution (Figure 3). In the first 3 km, the higher dispersion of lead concentrations and the more pronounced decay with the distance in the case of the leaves were noted. The matrix of the sign test (not listed) indicated that, in fact, the variation of lead content according to altitude is due only to the lowest altitude, which had the largest lead quantities. The zinc and copper concentrations were changeless from one sampling plot to another. The cadmium concentration depended on the exposure of the location to the local circulation of polluted air. The differences between the sampled years were not statistically significant (Table 8).

3.3. Safety in Herbal Medicine

In spite of the growing popularity of natural products, one must be realistic and admit that none can be completely free from various contaminants. The risk of contaminated nutraceuticals intake is much higher in the absence of specific legislation, as manufacturers are not compelled to oversee the nature, safety, and therapeutic and nutritional efficacy of these products [11]. Furthermore, the preference for nutraceuticals is fueled by consumers' false belief that the natural product is inevitably healthy and safe [19].

Food products consumed by people undoubtedly contain metals and metalloids [38,79]. Even if it comes up in a biotope where the anthropogenic pressure is low, the collected raw material may be contaminated due to certain non-hygienic harvesting techniques or poor storage and conditioning. Heavy metals can bioaccumulate in plants in concentrations which exceed the maximum limits permitted by the environmental regulations, where they can reach the human or animal organism directly or indirectly through the food chain. A lot of metals give rise to toxicity even with reduced concentrations.

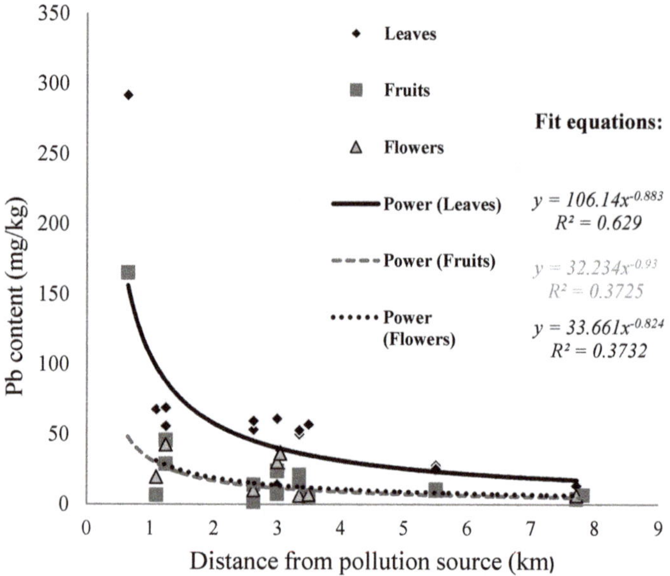

Figure 3. The variation of lead content in relation to the distance from the main source of pollution (control plot is excluded).

Important lead, cadmium, and zinc concentrations were found in consumer finished food or medicinal products, such as tea and blackberry wine [53,54]. Small quantities of these metals in *Rubus* species were found in Himalaya as well [34].

The smaller the quantities of non-essential heavy metals in traditional nutraceutical products, the lower will be the risk to the consumers health Small quantities of non-essential heavy metals in traditional nutraceutical products as their absence eliminates the risk of noxious effects on health [80]. Consequently, it is necessary to implement a qualitative assessment of wild resources consumed directly or used in ethnomedicine, before using or processing them, by determining the heavy metal content [81].

Resources, such as blackberry in Copşa Mică, are consumed by the local population in raw or processed forms. At the observed lead and cadmium concentrations (Table 7, Figure 2), the therapeutic value of the blackberry active ingredients decreases.

The International Agency for Research on Cancer classifies the anorganic compounds of Pb into group 2A—probably carcinogenic to humans. The symptoms of lead poisoning are abdominal pain, constipation, nausea, cramps, vomiting, anorexia, and weight loss [82]. Chronic exposure to high levels of Pb produces significant accumulations in the bones, as well as disorders of the central nervous system, hepatic and renal disorders, gout, and high blood pressure. Furthermore, it affects the optimal functions of the male and female reproductive system, with negative effects on pregnancy [83–88].

As a non-essential metal, Cd accumulates in the environment continuously, with one of its main sources being the atmospheric deposit. Chronic exposure to Cd causes kidney failure, increased risk of pre-diabetes and diabetes, high blood pressure, osteoporosis, and cancer [89–92]. In our researched area children represent the age range most exposed to the risk of contamination by eating blackberries. The poor education of some people maintains this risk. Hence, an acute need for food education for all social categories from the area is felt.

4. Conclusions

Blackberry is a popular nutraceutical, but unfortunately it is also an important heavy metal bioaccumulator. The extended industrial activity (which began in 1935) of metallurgical and chemical production in Copşa Mică led to the remnant contamination of blackberry with lead, cadmium, zinc, and copper. Shortly after the interruption of the pollution emission, the lead concentrations in blackberry were found to exceed the recommended threshold by up to 29 times in 71% of cases. Furthermore, all the cadmium concentrations exceeded the WHO threshold by up to 15 times, and 83% of the values of copper concentration exceeded the permissible limit by up to 39 times. The organs of blackberry store these elements differently—the flowers and leaves are the largest bioaccumulators. The lead bioaccumulation was found to have a definite spatial distribution. Conversely, the zinc and copper concentrations were changeless from one sampling plot to another. The results indicate a wide geographic expansion of pollution with these metals, within a radius of at least 26 km.

Author Contributions: Conceived and designed the experiment: G.G., F.D., and I.A.V. Performed the experiment: G.G. and T.M. Analyzed the data: F.D. and T.M. Interpreted the results: I.A.V., F.D., and G.G. Conceived the paper, wrote the first draft and edited the manuscript: F.D., M.M.V., and S.B. Supervised the manuscript: F.D.

Conflicts of Interest: The authors declare no conflict of interest.

References

1. Ekor, M. The Growing Use of Herbal Medicines: Issues Relating to Adverse Reactions and Challenges in Monitoring Safety. *Front. Pharmacol.* **2014**, *4*, 177. [CrossRef] [PubMed]
2. Butură, V. *Encyclopaedia of Romanian Ethnobotany*; Ştiinţifică şi Enciclopedică Publishing House: Bucharest, Romania, 1979; pp. 159–202.
3. Lim, T.K. *Edible Medicinal and Non-Medicinal Plants—Fruits*; Springer: Berlin, Germany, 2012; pp. 559–569.
4. Alonso, R.; Cavidad, I.; Calleja, J.M. A preliminary study of hypoglycemic activity of *Rubus fruticosus*. *Planta Med.* **1980**, *40*, 102–106. [CrossRef] [PubMed]
5. Leonti, M.; Casu, L.; Sanna, F.; Bonsignore, L. A comparison of medicinal plant use in Sardinia and Sicily-De Materia Medica revisited? *J. Ethnopharmacol.* **2009**, *121*, 55–67. [CrossRef] [PubMed]
6. Guarrera, P.M. Food medicine and minor nourishment in the folk traditions of Central Italy (Marche, Abruzzo and Latium). *Fitoterapia* **2003**, *74*, 515–544. [CrossRef]
7. Lust, J. *The Herb Book*; Dover Publications Inc.: New York, NY, USA, 2014; pp. 298–300.
8. Chevallier, A. *Encyclopedia of Herbal Medicine*, 3rd ed.; Dorling Kindersley Publishing: New York, NY, USA, 2016; pp. 264–265.
9. Beldeanu, E.C. *Forest Species of Sanogenic Interest*; Transilvania University Press: Braşov, Romania, 2004; pp. 145–147.
10. Sher, H. Ethnoecological evaluation of some medicinal and aromatic plants of Kot Malakand Agency, Pakistan. *Sci. Res. Essays* **2011**, *6*, 2164–2173.
11. Daliu, P.; Santini, A.; Novellino, E. From pharmaceuticals to nutraceuticals: Bridging disease prevention and management. *Expert Rev. Clin. Pharmacol.* **2018**, *28*, 1–7. [CrossRef] [PubMed]
12. Santini, A.; Novellino, E.; Armini, V.; Ritieni, A. State of the art of Ready-to-Use Therapeutic Food: A tool for nutraceuticals addition to foodstuff. *Food Chem.* **2013**, *140*, 843–849. [CrossRef] [PubMed]
13. Santini, A.; Novellino, E. To Nutraceuticals and Back: Rethinking a Concept. *Foods* **2017**, *6*, 74. [CrossRef] [PubMed]
14. Santini, A.; Novellino, E. Nutraceuticals: Shedding light on the grey area between pharmaceuticals and food. *Expert Rev. Clin. Pharmacol.* **2018**, *11*, 545–547. [CrossRef] [PubMed]
15. De Felice, S.L. The nutraceutical revolution: Its impact on food industry R&D. *Trends Food Sci. Technol.* **1995**, *6*, 59–61.
16. Santini, A.; Tenore, G.C.; Novellino, E. Nutraceuticals: A paradigm of proactive medicine. *Eur. J. Pharm. Sci.* **2017**, *96*, 53–61. [CrossRef] [PubMed]
17. Santini, A.; Cammarata, S.M.; Capone, G.; Ianaro, A.; Tenore, G.C.; Pani, L.; Novellino, E. Nutraceuticals: Opening the debate for a regulatory framework. *Br. J. Clin. Pharmacol.* **2018**, *84*, 659–672. [CrossRef] [PubMed]

18. Yeung, A.W.K.; Tzvetkov, N.T.; El-Tawil, O.S.; Bungău, S.G.; Abdel-Daim, M.M.; Atanasov, A.G. Antioxidants: Scientific Literature Landscape Analysis. *Oxid. Med. Cell. Longev.* **2019**, *2019*, 8278454. [CrossRef] [PubMed]

19. Durazzo, A.; D'Addezio, L.; Camilli, E.; Piccinelli, R.; Turrini, A.; Marletta, L.; Marconi, S.; Lucarini, M.; Lisciani, S.; Gabrielli, P.; et al. From Plant Compounds to Botanicals and Back: A Current Snapshot. *Molecules* **2018**, *23*, 1844. [CrossRef] [PubMed]

20. Andrew, R.; Izzo, A.A. Principles of pharmacological research of nutraceuticals. *Br. J. Pharmacol.* **2017**, *174*, 1177–1194. [CrossRef] [PubMed]

21. Seeram, N.P.; Adams, L.S.; Zhang, Y.; Lee, R.; Sand, D.; Scheuller, H.S.; Heber, D. Blackberry, black raspberry, blueberry, cranberry, red raspberry, and strawberry extracts inhibit growth and stimulate apoptosis of human cancer cells in vitro. *J. Agric. Food Chem.* **2006**, *54*, 9329–9339. [CrossRef] [PubMed]

22. Dai, J.; Patel, J.D.; Mumper, R.J. Characterization of blackberry extract and its antiproliferative and anti-inflammatory properties. *J. Med. Food* **2007**, *10*, 258–265. [CrossRef] [PubMed]

23. Hager, T.J.; Howard, L.R.; Liyanage, R.; Lay, J.O.; Prior, R.L. Ellagitannin composition of blackberry as determined by HPLC-ESI-MS and MALDI-TOF-MS. *J. Agric. Food Chem.* **2008**, *56*, 661–669. [CrossRef] [PubMed]

24. Heinonen, I.M.; Meyer, A.S.; Frankel, E.N. Antioxidant activity of berry phenolics on human low-density lipoprotein and liposome oxidation. *J. Agric. Food Chem.* **1998**, *46*, 4107–4112. [CrossRef]

25. Wang, S.Y.; Lin, H.S. Antioxidant activity in fruits and leaves of blackberry, raspberry and strawberry varies with cultivar and developmental stage. *J. Agric. Food Chem.* **2000**, *48*, 140–146. [CrossRef] [PubMed]

26. Moyer, R.A.; Hummer, K.E.; Finn, C.E.; Frei, B.; Wrolstad, R.E. Anthocyanins, phenolics, and antioxidant capacity in diverse small fruits: *Vaccinium*, *Rubus*, and *Ribes*. *J. Agric. Food Chem.* **2002**, *50*, 519–525. [CrossRef] [PubMed]

27. Ding, M.; Feng, R.; Wang, S.Y.; Bowman, L.; Lu, Y.; Qian, Y.; Castranova, V.; Jiang, B.H.; Shi, X. Cyanidin-3-glucoside, a natural product derived from blackberry, exhibits chemopreventive and chemotherapeutic activity. *J. Biol. Chem.* **2006**, *281*, 17359–17368. [CrossRef] [PubMed]

28. Pantelidis, G.E.; Vasilakakis, M.; Manganaris, G.A.; Diamantidis, G.R. Antioxidant capacity, phenol, anthocyanin and ascorbic acid contents in raspberries, blackberries, red currants, gooseberries and cornelian cherries. *Food Chem.* **2007**, *102*, 777–783. [CrossRef]

29. Elisia, I.; Kitts, D.D. Anthocyanins inhibit peroxyl radical-induced apoptosis in Caco-2 cells. *Mol. Cell. Biochem.* **2008**, *312*, 139–145. [CrossRef] [PubMed]

30. Bowen-Forbes, C.S.; Zhang, Y.; Nair, M.G. Anthocyanin content, antioxidant, anti-inflammatory and anticancer properties of blackberry and raspberry fruits. *J. Food Compos. Anal.* **2010**, *23*, 554–560. [CrossRef]

31. Jing, P.; Giusti, M. Contribution of berry anthocyanins to their chemopreventive properties. In *Berries and Cancer Prevention*; Stoner, G.D., Seeram, N.P., Eds.; Springer: New York, NY, USA, 2011; pp. 1–38.

32. Kolevski, G.; Ivic-Kolevska, S. Antioxidants in fruits and human medical research: An overview. *J. Hyg. Eng. Des.* **2012**, *1*, 271–274.

33. Tavares, L.; Figueira, I.; McDougall, G.J.; Vieira, H.L.A.; Stewart, D.; Alves, P.M.; Ferreira, R.B.; Santos, C.N. Neuroprotective effects of digested polyphenols from wild blackberry species. *Eur. J. Nutr.* **2013**, *52*, 225–236. [CrossRef] [PubMed]

34. Ahmad, M.; Masood, S.; Sutana, S.; Ben Hadda, T.; Bader, A.; Zafar, M. Antioxidant and nutraceutical value of wild medicinal *Rubus* berries. *Pak. J. Pharm. Sci.* **2015**, *28*, 241–247. [PubMed]

35. Predná, L.; Habánová, M. Antioxidant potential in selected species of small berry fruits. *Acta Fytotech. Zootech.* **2015**, *18*, 116–118. [CrossRef]

36. Bhattarai, N.; Karki, M. Medicinal and aromatic plants: Ethnobotany and conservation status. In *Encyclopedia of Forest Sciences*; Burley, J., Evans, J., Youngquist, J.A., Eds.; Elsevier Ltd.: Kidlington, UK, 2004; pp. 523–532.

37. Strik, B.C. Berry crops: Worldwide area and production systems. In *Berry Fruit: Value-Added Products for Health Promotion*, 1st ed.; Zhao, Y., Ed.; CRC Press: Boca Raton, FL, USA, 2007; pp. 3–51.

38. Farmaki, E.G.; Thomaidis, N.S. Current status of the metal pollution of the environment of Greece—A review. *Glob. NEST J.* **2008**, *10*, 366–375.

39. Mohammed, A.S.; Kapri, A.; Goel, R. Heavy metal pollution: Source, impact and remedies. In *Biomanagement of Metal-Contaminated Soils, Environmental Pollution*; Khan, M.S., Zaidi, A., Goel, R., Musarrat, J., Eds.; Springer: Dordrecht, The Netherlands, 2011; Volume 20, pp. 1–29.

40. Dietrich, K.N.; Ris, M.D.; Succop, P.A.; Berger, O.G.; Bornschein, R.L. Early exposure to lead and juvenile delinquency. *Neurotoxicol. Teratol.* **2001**, *23*, 511–518. [CrossRef]

41. Needleman, H.L.; McFarland, C.; Ness, R.B.; Fienberg, S.E.; Tobin, M.J. Bone lead levels in adjudicated delinquents - a case control study. *Neurotoxicol. Teratol.* **2002**, *24*, 711–717. [CrossRef]

42. Kim, S.; Moon, C.; Eun, S.; Ryu, P.; Jo, S. Identification of ASK1, MKK4, JNK, c-Jun, and caspase-3 as a signaling cascade involved in cadmium-induced neuronal cell apoptosis. *Biochem. Biophys. Res. Commun.* **2005**, *328*, 326–334. [CrossRef] [PubMed]

43. Monroe, R.K.; Halvorsen, S.W. Cadmium blocks receptor-mediated Jak/STAT signaling in neurons by oxidative stress. *Free Radic. Biol. Med.* **2006**, *41*, 493–502. [CrossRef] [PubMed]

44. Zhu, L.; Ji, X.J.; Wang, H.D.; Pan, H.; Chen, M.; Lu, T.J. Zinc neurotoxicity to hippocampal neurons in vitro induces ubiquitin conjugation that requires p38 activation. *Brain Res.* **2012**, *1438*, 1–7. [CrossRef] [PubMed]

45. Caito, S.; Aschner, M. Developmental neurotoxicity of lead. *Adv. Neurobiol.* **2017**, *18*, 3–12. [PubMed]

46. von Hoffen, L.P.; Säumel, I. Orchards for edible cities: Cadmium and lead content in nuts, berries, pome and stone fruits harvested within the inner city neighbourhoods in Berlin, Germany. *Ecotoxicol. Environ. Saf.* **2014**, *101*, 233–239. [CrossRef] [PubMed]

47. Wisłocka, M.; Krawczyk, J.; Klink, A.; Morrison, L. Bioaccumulation of heavy metals by selected plant species from uranium mining dumps in the Sudety Mts., Poland. *Pol. J. Environ. Stud.* **2006**, *15*, 811–818.

48. Teofilova, T.; Kodzabashev, N.; Gherasimov, S.; Markova, E. Comparative characterization of the heavy metal contents in samples from two regions in Bulgaria with different anthropogenic load. *Nat. Montenegrina* **2010**, *9*, 897–912.

49. Shikhova, N.S. Some regularities in the accumulation of lead in urban plants (by example of Vladivostok). *Contemp. Probl. Ecol.* **2012**, *5*, 215–222. [CrossRef]

50. Yedoyan, R.; Yedoyan, T.V. The study of heavy metals (Ni, Zn, Cu, Pb) in the vegetative organs, harvest and growing soil of potatoes, wheat, and wild blackberry. *Food Environ. Saf. J. Fac. Food Eng.* **2012**, *11*, 38–42.

51. Vollmanova, A.; Zupka, S.; Bajcan, D.; Medvecky, M.; Daniel, J. Dangerous heavy metals in soil and small forest fruit as a result of old environmental loads. In Proceedings of the 14th International Conference on Environmental Science and Technology, Rhodes, Greece, 3–5 September 2015; pp. 698–703.

52. Micu, L.M.; Petanec, D.I.; Iosub-Ciur, M.D.; Andrian, S.; Popovici, R.A.; Porumb, A. The heavy metals content in leave of the forest fruits (*Hippophae rhamnoides* and *Rubus fruticosus*) from the tailings dumps mining. *Rev. Chim.* **2016**, *67*, 64–68.

53. Kekedy-Nagy, L.; Ionescu, A. Characterization and classification of tea herbs based on their metal content. *Acta Univ. Sapientiae Agric. Environ.* **2009**, *1*, 11–19.

54. Amidžić, D.; Klarić, I.; Velić, D.; Vedrina Dragojević, I. Evaluation of mineral and heavy metal contents in Croatian blackberry wines. *Czech J. Food Sci.* **2011**, *29*, 260–267. [CrossRef]

55. Smejkal, G. *The Forest and the Industrial Pollution*; Ceres: Bucharest, Romania, 1982; p. 195.

56. Micu, M.O. The Influence of Pollution in the Area Copşa Mică and Its Ecological Implications. Ph.D. Thesis, Transilvania University of Braşov, Braşov, Romania, 2001.

57. National Meteorological Agency. Catalog. Available online: http://www.meteoromania.ro/catalog/?tip= 1&cod_geo=614436&cod_clasa=CLIMATOLOGICA&cod_subclasa=1\T1\textbar{}18&pas=5&tipulLor= LUNARE&pagina=2 (accessed on 22 February 2019).

58. Stefan, K.; Raitio, H.; Bartels, U.; Fürst, A.; Rautio, P. Sampling and analysis of needles and leaves—Manual part IV. In *Manual on Methods and Criteria for Harmonized Sampling, Assessment, Monitoring and Analysis of the Effects of Air Pollution on Forests*; UNECE-ICP Forests Programme Co-ordinating Centre: Hamburg, Germany, 2005; pp. 1–25.

59. Luyssaert, S.; Raitio, H.; Vervaeke, P.; Mertens, J.; Lust, N. Sampling procedure for the foliar analysis of decidous trees. *J. Environ. Monit.* **2002**, *4*, 858–864. [CrossRef] [PubMed]

60. Djingova, R.; Kuleff, I. Instrumental techniques for trace analysis. In *Trace Elements: Their Distribution and Effects in the Environment*; Markert, B., Friese, K., Eds.; Elsevier Science Ltd.: London, UK, 2000; pp. 137–185.

61. Hansen, M.D.; Nøst, T.H.; Heimstad, E.S.; Evenset, A.; Dudarev, A.A.; Rautio, A.; Myllynen, P.; Dushkina, E.V.; Jagodic, M.; Christensen, G.N.; et al. The impact of a Nickel-Copper smelter on concentrations of toxic elements in local wild food from the Norwegian, Finnish, and Russia border regions. *Int. J. Environ. Res. Public Health* **2017**, *14*, 694. [CrossRef] [PubMed]

62. Council of Europe. *Kelp, Monograph 1426*, 6th ed.; Council of Europe: Strasbourg, France, 2007; Volume 2.

63. Sun, H.; Li, L. Investigation of Distribution for Trace Lead and Cadmium in Chinese Herbal Medicines and Their Decoctions by Graphite Furnace Atomic Absorption Spectrometry. *Am. J. Anal. Chem.* **2011**, *2*, 217–222. [CrossRef]
64. Thomsen, V.; Schatzlein, D.; Mercuro, D. Limits of Detection in Spectroscopy. *Spectroscopy* **2003**, *18*, 112–114.
65. Chan, C.C. Analytical method validation: Principles and practices. In *Pharmaceutical Manufacturing Handbook: Regulations and Quality*; Gad, S.C., Ed.; John Wiley & Sons: Hoboken, NJ, USA, 2008; pp. 727–742.
66. Frank, V.; Tölgyessy, J. The chemistry of soil. In *Chemistry and Biology of Water, Air and Soil: Environmental Aspects*; Tölgyessy, J., Ed.; Elsevier: Amsterdam, The Netherlands, 1993; pp. 621–698.
67. World Health Organization. *WHO Guidelines for Assessing Quality of Herbal Medicines with Reference to Contaminants and Residues*; World Health Organization: Geneva, Switzerland, 2007.
68. Kabelitz, L. Heavy metals in herbal drugs. *Pharm. Ind.* **1998**, *60*, 444–451.
69. European Commission. *Regulation (EC) No 396/2005 of the European Parliament and of the Council of 23 February 2005 on Maximum Residue Levels in or on Food and Feed of Plant and Animal Origin and Amending Council Directive 91/414/EEC*; European Commission: Brussels, Belgium, 2005; pp. 1–16.
70. Gasser, U.; Klier, B.; Kühn, A.V.; Steinhoff, B. Current findings on the heavy metal content in herbal drugs. *Pharm. Sci. Notes* **2009**, *1*, 37–50.
71. Mohite, R.D.; Basavaiah, N.; Singare, P.U.; Reddy, A.V.R.; Singhal, R.K.; Blaha, U. Assessment of Heavy Metals Accumulation in Washed and Unwashed Leafy Vegetables Sector-26 Vashi, Navi Mumbai, Maharashtra. *J. Chem. Biol. Phy. Sci. Sec. D* **2016**, *6*, 1130–1139.
72. Ataabadi, M.; Hoodaji, M.; Najafi, P. Assessment of washing procedure for determination some of airborne metal concentrations. *Afr. J. Biotechnol.* **2012**, *11*, 4391–4395. [CrossRef]
73. Aksoy, A.; Åžahin, U. Elaeagnus angustifolia L. as a biomonitor of heavy metal pollution. *Turk. J. Bot.* **1999**, *23*, 83–87.
74. Yusuf, K.A.; Oluwole, S.O. Heavy Metal (Cu, Zn, Pb) Contamination of Vegetables in Urban City: A Case Study in Lagos. *Res. J. Environ. Sci.* **2009**, *3*, 292–298. [CrossRef]
75. Felix-Henningsen, P.; Urushadze, T.; Steffens, D.; Kalandadze, B.; Narimanidze, E. Uptake of heavy metals by food crops from highly-polluted Chernozem-like soils in an irrigation district south of Tbilisi, eastern Georgia. *Agron. Res.* **2010**, *8*, 781–795.
76. Aghaei, A.; Khademi, H.; Eslamian, S. Comparison of Three Tree Leaves as Biomonitors of Heavy Metals Contamination in Dust, A Case Study of Isfahan. *Helix Int. J.* **2017**, *7*, 1873–1887.
77. Hoffman, D.J.; Rattner, B.A.; Burton, G.A.; Cairns, J. *Handbook of Ecotoxicology*; CRC Press: Boca Raton, FL, USA, 2003; p. 1315.
78. Alexa, B.; Cotârlea, I.; Bărbătei, R. *The Pollution of Forests in Mediaş Forest District and the Ecological Restoration Done*; Constant Publishing House: Sibiu, Romania, 2004; p. 145.
79. Mantovi, P.; Bonazzi, G.; Maestri, E.; Marmiroli, N. Accumulation of copperand zinc from liquid manure in agricultural soils and crop plants. *Plant Soil* **2003**, *50*, 249–257. [CrossRef]
80. Sadhu, A.; Upadhyay, P.; Singh, P.K.; Agrawal, A.; Ilango, K.; Karmakar, D.; Singh, G.P.I.; Dubey, G.P. Quantitative analysis of heavy metals in medicinal plants collected from environmentally diverse locations in India for use in a novel phytopharmaceutical product. *Environ. Monit. Assess.* **2015**, *187*, 542. [CrossRef] [PubMed]
81. Nookabkaew, S.; Rangkadilok, N.; Satayavivad, J. Determination of trace elements in herbal tea products and their infusions consumed in Thailand. *Agric. Food Chem.* **2006**, *54*, 6939–6944. [CrossRef] [PubMed]
82. World Health Organization. *IARC Monographs on the Evaluation of the Carcinogenic Risk of Chemicals to Humans—Inorganic and Organic Lead Compounds*; World Health Organization: Lyon, France, 2006; Volume 87.
83. Goyer, R. Lead toxicity: Current concerns. *Environ. Health Perspect.* **1993**, *100*, 177–187. [CrossRef] [PubMed]
84. Silbergeld, E.K.; Sauk, J.; Somerman, M.; Todd, A.; McNeill, F.; Fowler, B.; Fontaine, A.; van Buren, J. Lead in bone: Storage site, exposure source, and target organ. *Neurotoxicology* **1993**, *14*, 225–236. [PubMed]
85. Sakai, T. Biomarkers of lead exposure. *Ind. Health* **2000**, *38*, 127–142. [CrossRef] [PubMed]
86. Kalia, K.; Flora, S.J. Strategies for safe and effective therapeutic measures for chronic arsenic and lead poisoning. *J. Occup. Health* **2005**, *47*, 1–21. [CrossRef] [PubMed]
87. Navas-Acien, A.; Guallar, E.; Silbergeld, E.K.; Rothenberg, S.J. Lead exposure and cardiovascular disease - a systematic review. *Environ. Health Perspect.* **2007**, *115*, 472–482. [CrossRef] [PubMed]

88. Flora, S.J.S.; Pachauri, V.; Saxena, G. Arsenic, cadmium and lead. In *Reproductive and Developmental Toxicology*; Gupta, R.C., Ed.; Academic: New York, NY, USA, 2011; Volume 33, pp. 415–438.
89. Schwartz, G.G.; Ilyasova, D.; Ivanova, A. Urinary cadmium, impaired fasting glucose, and diabetes in the NHANES III. *Diabetes Care* **2003**, *26*, 468–470. [CrossRef] [PubMed]
90. Eum, K.D.; Lee, M.S.; Paek, D. Cadmium in blood and hypertension. *Sci. Total Environ.* **2008**, *407*, 147–153. [CrossRef] [PubMed]
91. World Health Organization. *IARC Monographs on the Evaluation of the Carcinogenic Risk of Chemicals to Humans—Beryllium, Cadmium, Mercury, and Exposures in the Glass Manufacturing Industry*; World Health Organization: Lyon, France, 1993; Volume 58.
92. World Health Organization. *IARC Monographs on the Evaluation of the Carcinogenic Risk of Chemicals to Humans—Arsenic, Metals, Fibres and Dusts*; World Health Organization: Lyon, France, 2012; Volume 100.

Article

Evaluation of Anti-Tyrosinase and Antioxidant Properties of Four Fern Species for Potential Cosmetic Applications

Adrià Farràs [1,2], Guillermo Cásedas [1], Francisco Les [1,3], Eva María Terrado [1], Montserrat Mitjans [2] and Víctor López [1,3,*]

[1] Department of Pharmacy, Faculty of Health Sciences, Universidad San Jorge, 50830 Villanueva de Gállego (Zaragoza), Spain; afarrama7@alumnes.ub.edu (A.F.); gcasedas@usj.es (G.C.); fles@usj.es (F.L.); emterrado@usj.es (E.M.T.)

[2] Department of Biochemistry and Physiology, Faculty of Pharmacy and Food Sciences, Universitat de Barcelona, 08028 Barcelona, Spain; montsemitjans@ub.edu

[3] Instituto Agroalimentario de Aragón-IA2, CITA-Universidad de Zaragoza, 50013 Zaragoza, Spain

*   Correspondence: ilopez@usj.es; Tel.: +34-976-060-100

Received: 4 January 2019; Accepted: 11 February 2019; Published: 19 February 2019

Abstract: Ferns are poorly explored species from a pharmaceutical perspective compared to other terrestrial plants. In this work, the antioxidant and tyrosinase inhibitory activities of hydrophilic and lipophilic extracts, together with total polyphenol content, were evaluated in order to explore the potential cosmetic applications of four Spanish ferns collected in the Prades Mountains (*Polypodium vulgare* L., *Asplenium adiantum-nigrum* L., *Asplenium trichomanes* L., and *Ceterach officinarum* Willd). The antioxidant activity was evaluated using the 2,2-diphenyl-1-picrylhydrazyl (DPPH) radical, oxygen radical absorbance capacity (ORAC) and xanthine/xanthine oxidase (X/XO) assays. The potential to avoid skin hyperpigmentation was tested by inhibiting the tyrosinase enzyme, as this causes melanin synthesis in the epidermis. All ferns were confirmed as antioxidant and anti-tyrosinase agents, but interestingly hydrophilic extracts (obtained with methanol) were more potent and effective compared to lipophilic extracts (obtained with hexane). *Polypodium vulgare*, *Asplenium adiantum-nigrum*, and *Ceterach officinarum* methanolic extracts performed the best as antioxidants. *Polypodium vulgare* methanolic extract also showed the highest activity as a tyrosinase inhibitor.

Keywords: pteridophytes; ferns; antioxidant; tyrosinase inhibition; cosmetics; Polypodiopsida

1. Introduction

The incidence of cutaneous disorders and melanoma has increased worldwide [1]; in fact, non-melanoma skin cancer has become the principal skin cancer among fair-skinned people [2]. Sun radiation is known to accelerate photodamage of the skin, and ultraviolet radiation is one of the main factors that causes skin hyperpigmentation and skin aging [3].

Melanin, which is obtained by irreversible tyrosine catalyzed reactions, is an important epidermal agent that blocks ultraviolet radiation [4]. It has been noted that melanocyte cultures from black skin-types increase melanogenesis and melanosis more than in fair skin-types [5]. Consequently, lighter and thinner skin is 6–33 times more susceptible to developing minimal perceptible erythema than darker and thicker skin [6,7]. Ectopic dermal melanocytes, a result of successive erythemas, are shown to be directly dependent on increased melanin in the epidermis (hypermelanosis), which can trigger sun spots [8].

Primary photoprotection, also called non-systemic photoprotection, has traditionally been considered the main strategy against the harmful effects of sun radiation [9]. This method is based on having healthy habits towards sun exposure and the use of physical photoprotective agents; however, some disadvantages have been described [10]. Antioxidant oral supplements from secondary metabolites of plants are an adjunctive to primary photoprotection [11]. This most recent strategy on photoprotection is known as secondary photoprotection, or systemic photoprotection [12,13], in which standardized aqueous extract from the fronds of *Polypodium leucotomos* L., which is marketed under the trade name Fernblock® (Cantabria Labs, Santander, Spain), has been one of the most popular systemic and topical photoprotective oral agents in cosmetic science [14,15]. The effectiveness and safety of the use of this fern is a consequence of its multiple pathways of action described by Palomino et al. [16].

Ferns (Polypodiopsida), formerly considered pteridophytes, have been reported as one of the least understood classes of tracheophyte plants from a phylogenetic perspective [17,18]. Recent reviews of the *Polypodium* genus have been published since the commercialization of Fernblock®, as shown in Berman et al. [19]. Other authors have recently published updated reviews on the phytochemistry and ethnopharmacology of ferns, highlighting the presence of polyphenols (particularly flavonoids), terpenoids, steroids, and alkaloids [20,21]. Most of these bioactive compounds are described as natural enzyme inhibitors in biomedical research drug discovery due to anticancer, antidiabetic, and antiaging properties [22–24]. The selected ferns in this study (*Polypodium vulgare* L., *Asplenium adiantum-nigrum* L., *Asplenium trichomanes* L., and *Ceterach officinarum* Willd) are some of the most common leptosporangiate ferns reported on the Prades Mountains (Spain, 41°18′43″ N 1°05′09″ E) [25].

Considering Fernblock® as a reference in skin photoprotection [26,27], the potential anti-aging and skin-whitening properties of four Spanish ferns collected in the Prades Mountains have been studied. In vitro antioxidant activities against different free radicals (2,2-diphenyl-1-picrylhydrazyl—DPPH, oxygen radical absorbance capacity—ORAC, and xanthine/xanthine oxidase—X/XO methods) and in vitro inhibition of the tyrosinase enzyme were evaluated.

2. Material and Methods

2.1. Chemicals and Reagents

All reagents used were of analytical grade. Methanol, hexane, tyrosinase, and L-dihydroxyphenylalanine (L-DOPA) were acquired through Vidrafoc® (Barcelona, Spain). Dimethyl sulfoxide (DMSO) was obtained from Fisher Scientific® (Madrid, Spain). The provider of 5-hydroxy-2-(hydroxymethyl) pyran-4-one (kojic acid) was Alfa Aesar® (Karlsruhe, Germany). The reagents used to determine the antioxidant activity such as 2,2-diphenyl-1-picrylhydrazyl (DPPH radical), 2,6-dihydroxypurine (xanthine), xanthine oxidase, and (±)-6-hydroxy-2,5,7,8-tetramethylchromane-2-carboxylic acid (trolox) were supplied by Sigma-Aldrich® (Madrid, Spain). Sodium carbonate anhydrous (Na_2CO_3) and nitrotetrazolium blue chloride (NBT) were purchased from Laboaragon® (Cartuja Baja, Spain) and Sumalsa® (Zaragoza, Spain), respectively. All aqueous solutions were prepared with ultra-pure water.

2.2. Plant Material

It was checked that the subject species had been described in The Plant List [28] and by Banco de Datos de Biodiversidad de Cataluña [29].

The whole fresh fronds of selected fern species were identified and collected from the Prades Mountains, in the province of Tarragona (Spain), in November 2016 by Adrià Farràs and Josep Mª Farràs using botanical keys [30]. The samples were dried in the shade at room temperature.

A dried voucher specimen has been deposited at the Herbarium of Universidad San Jorge, Zaragoza, Spain (*Polypodium vulgare*: voucher no. 003-2016; *Asplenium adiantum-nigrum*: voucher no. 004-2016; *Asplenium trichomanes*: voucher no. 005-2016; *Ceterach officinarum* voucher no. 006-2016). These examples were authenticated by Dr. J.A. Vicente Orellana from Universidad CEU San Pablo (Madrid, Spain).

2.3. Extracts Preparation

Hydrophilic (= polar) and lipophilic (= non polar) extracts were prepared using methanol or hexane, respectively. Dried fronds of the four species were powdered mechanically until obtaining 40 mg of each. The powdered fronds of each species were split equally into two erlenmeyers of 500 mL. Each 20 mg of fern powder was macerated with 250 mL of solvent (hexane or methanol) at room laboratory temperature for 24 h. The extract was filtered using Whatman N°4 filter paper, and the solvent was evaporated using rotatory evaporator with a thermostatic bath at 30 °C. This process was completed two more times until exhaustion of plant material; extracts were stored at -20 °C until further experiments. Yields were calculated in percentages from the dry weight of fronds used and the quantity of dry mass obtained by extraction.

2.4. Phytochemical Screening by Thin Layer Chromatography (TLC) and Total Phenolic Content (TPC)

Silica gel TLC plates coated with fluorescent indicator F254 were used in order to detect phenolic compounds (flavonoids and phenolic acids) in the samples. 10 µL of hexane and methanolic extracts of the samples at concentrations of 10 mg/mL were run on the plates with EtOAc/MeOH/H_2O (65:15:5, *v:v:v*) as mobile phase. After eluting the samples, plates were dried, sprayed with the Natural Products polyethylene glycol (PEG) reagent, observed at 365 nm and retention factors (Rf) calculated [31].

TPC was quantified by the Folin Method as previously described using gallic acid for the standard calibration curve [32].

2.5. Determination of Antioxidant Activities

Antioxidant capacity was assessed by three complementary methods that were DPPH, ORAC, and superoxide radicals generated by X/XO.

2.5.1. DPPH Radical Scavenging Activity

The neutralization of DPPH radicals as antioxidant method was reported the first time by Blois et al. [33]. In this case, the assay was carried out according to the modifications described by Casedas et al. [34]. In 96-well microplates, each well contained 150 µL of extract and 150 µL of DPPH (0.04 mg/mL methanol solution). Antioxidant activity was determined measuring absorbance (Abs) at 515 nm after 30 min of dark incubation. Blank and control wells were also considered. The highest concentration of extracts tested was 1 mg/mL. Trolox, a water soluble derivate of vitamin E, was used as positive standard. Background interferences from solvents and samples were deducted from the activities prior to calculating radical scavenging capacity (RSC) as follows: RSC (%) = [(Abs$_{control}$ − Abs$_{sample}$)/Abs$_{control}$] × 100.

2.5.2. ORAC Assay

ORAC assay was carried out to measure the capacity of extracts to scavenge peroxyl radicals. Samples and trolox were dissolved in PBS and methanol (50:50, *v:v*). Samples were incubated with fluorescein (0.07 µM) in 96-well plates for 10 min at 37 °C. Afterwards, AAPH (0.012 M) was supplemented and fluorescence was measured for 98 min at 485 nm of excitation and 520 nm of emission, in a FLUOstar Optima fluorimeter (BMG Labtech, Ortenberg, Germany) [35]. Results were expressed as µmol trolox equivalents (TE)/mg sample.

2.5.3. Superoxide Radicals Generated by Xanthine/Xanthine Oxidase (X/XO)

Xanthine oxidase and xanthine as substrate are responsible for the production of superoxide radicals [36]. The effects of fern extracts on superoxide radicals generated by X/XO were evaluated by measuring the formation of the NBT (nitrotetrazolium blue chloride)-radical superoxide complex [37] using a described procedure [34]. The reaction mixture was prepared every day as a consequence of reduced stability. This mixture was composed of 90 µM xanthine, 16 mM Na_2CO_3, and 22.8 µM NBT in phosphate buffer (pH 7.0). 240 µL of the reaction mixture in each well with 30 µL of extract solution and XO was incubated in the dark for 2 min at 37 °C and absorbance read at 560 nm. Blank and control wells were also considered, and background interferences from solvents and samples were also deducted from the activities previous to calculating the RSC (%). The reference substance (trolox) was the same used in DPPH and ORAC assays.

2.6. Inhibition of Tyrosinase Activity

The inhibition of tyrosinase was performed following a previous method [38]. Samples were mixed with 40 µL L-DOPA and 80 µL potassium phosphate buffer (pH 6.8). Finally, 40 µL of tyrosinase (200 U/mL) was added in the wells. L-DOPA and tyrosinase were solved in buffer. The inhibition of tyrosinase was determined at 475 nm. Methanolic extracts were dissolved in methanol, and hexane extracts were dissolved in DMSO. Kojic acid was the reference inhibitor substance. Background interferences from solvents and samples were previously deducted from the activities to calculate the percentage of enzymatic inhibition (compared to control activity). Control wells had the same mix except the sample/inhibitor, which was replaced by the solvents of these.

2.7. Statistical Analysis

All samples were analyzed in triplicates ($n = 3$), at least, on different days. Statistical significance was analyzed by using GraphPad Prism version 6, San Diego, CA, USA. Data are presented as mean +/- standard error. The half maximal inhibitory concentration (IC_{50}) values were obtained by non-linear regression. Activities have been compared using a one-way analysis of variance (ANOVA). Statistical differences were considered as follows: $p \leq 0.05$ (*), $p \leq 0.01$ (**), and $p \leq 0.001$ (***). Correlations were performed between TPC and ORAC values, and TPC and tyrosinase IC_{50} values; Pearson values were also obtained using GraphPad Prism version 6.

3. Results

3.1. Plant Material and Yields

Table 1 shows the scientific names and Spanish common names of the collected samples as well as yields of extraction in each case. The hydrophilic extracts obtained with methanol have higher yields (ranging from 16.27% to 29.55%) than lipophilic extracts obtained with hexane (ranging from 1.49% to 2.34%).

3.2. Polyphenol Content by Thin Layer Chromatography (TLC) and Folin Method

TLC plates sprayed with Natural Products-PEG reagent revealed the presence of flavonoids in methanolic samples obtained from *Asplenium trichomanes* (ATM, Rf = 0.97, 0.92, 0.86) and *Ceterach officinarum* (COM, Rf = 0.66, 0.5). Spots corresponding to phenolic acids were also detected with similar retention factor (Rf) values in the methanolic samples from *Polypodium vulgare* (PVM, Rf = 0.81), *Asplenium adiantum-nigrum* (AAM, Rf = 0.81), *Ceterach officinarum* (COM, Rf = 0.81), and hexane extract of *Ceterach officinarum* (COH, Rf = 0.81). An image of the TLC plate sprayed with Natural Products-PEG reagent can be downloaded from Supplementary Materials (Figure S1). Total Phenolic Content (TPC) was quantified using the Folin–Ciocalteu reagent; as observed in Table 1, methanolic extracts contained higher amounts of polyphenols, with PVM and COM showing the highest values. As predicted, methanol seems to be a better solvent to extract polyphenols.

Table 1. Fern samples, botanical names, yields of extraction, and total polyphenol content (TPC).

Species	Spanish Common Name	Methanol Extract			Hexane Extract		
		Code	Yield (%) [a]	TPC (µGAE/mg)	Code	Yield (%) [a]	TPC (µGAE/mg)
Polypodium vulgare L. (*Polypodiaceae*)	"Polipodio"	PVM	23.53	172.8 ± 3.8	PVH	1.49	74.7 ± 5.8
Asplenium adiantum-nigrum L. (*Aspleniaceae*)	"Culantrillo negro"	AAM	16.27	113.5 ± 5.8	AAH	1.57	96.0 ± 3.8
Asplenium trichomanes L. (*Aspleniaceae*)	"Culantrillo rojo"	ATM	29.55	100.4 ± 0.7	ATH	2.01	70.3 ± 6.2
Ceterach officinarum Willd (*Aspleniaceae*)	"Doradilla"	COM	28.04	193.2 ± 3.8	COH	2.34	70.3 ± 7.6

[a] The yield is a relation between the weight of the dried extract and the weight of dried plant material expressed as percentage (%). GAE: gallic acid equivalents; PVM: *P. vulgare* methanol extract; AAM: *A. adiantum-nigum* methanol extract; ATM: *A. trichomanes* methanol extract; COM: *C. officinarum* methanol extract; PVH: *P. vulgare* hexane extract; AAH: *A. adiantum-nigum* hexane extract; ATH: *A. trichomanes* hexane extract; and COH: *C. officinarum* hexane extract.

3.3. Antioxidant Activity

3.3.1. DPPH Radical Scavenging Activity

As seen in Figure 1, methanolic extracts had a more powerful capacity for DPPH reduction than hexane extract for each fern. Methanolic extracts had a very similar profile of antiradical activity reaching 100% of radical inhibition at concentrations between 0.01 and 0.1 mg/mL. PVM and COM had the lowest IC_{50} values and can therefore be considered as the best antioxidants. Extracts obtained with hexane were also antioxidants, but the concentrations needed to scavenge 100% of DPPH radicals were superior compared to samples extracted with methanol as solvent.

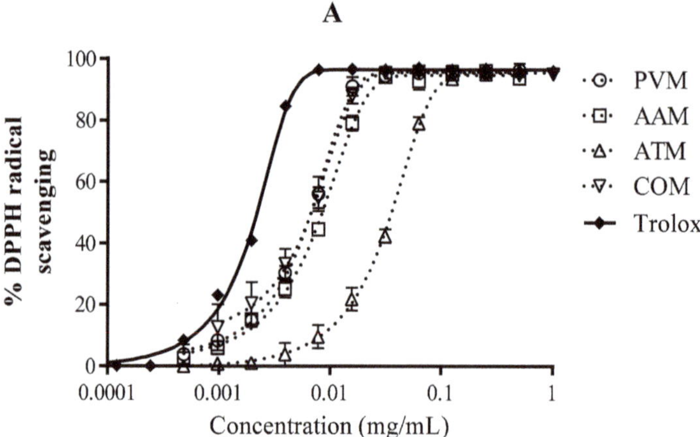

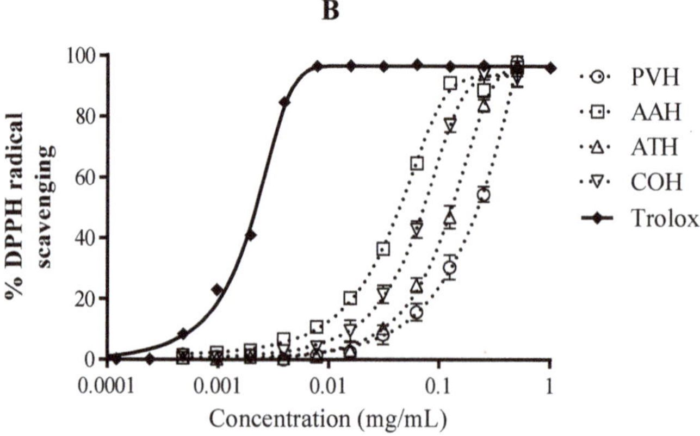

Figure 1. Antioxidant activity against DPPH radicals of methanol extracts (**A**) and hexane extracts (**B**) using trolox as a reference.

3.3.2. ORAC Assay

The ORAC assay is an internationally recognized method to measure antioxidant capacity. In Table 2, the ORAC values were also higher for methanolic extracts, particularly for PVM and COM, which is in accordance with data obtained in the DPPH assay. In Figure 2A, there is a positive correlation between ORAC values and TPC, which seems to indicate that the antioxidant activity may be mediated by polyphenols.

Table 2. ORAC values of methanolic and hexane extracts of fern species.

Ferns	ORAC (µmol Trolox Equivalents/mg Sample)	
	Methanolic Extract	**Hexane Extract**
Polypodium vulgare (PV)	2.34 ± 0.04	0.38 ± 0.02
Asplenium adiantum-nigrum (AA)	2.25 ± 0.03	0.34 ± 0.11
Asplenium trichomanes (AT)	2.25 ± 0.14	0.44 ± 0.01
Ceterach officinarum (CO)	2.93 ± 0.23 *	0.84 ± 0.06 #

* $p < 0.05$ versus PV, AA, and AT methanol extracts. # $p < 0.05$ versus PV, AA, and AT hexane extracts. Data analyzed using a one-way ANOVA and Tukey post-hoc test.

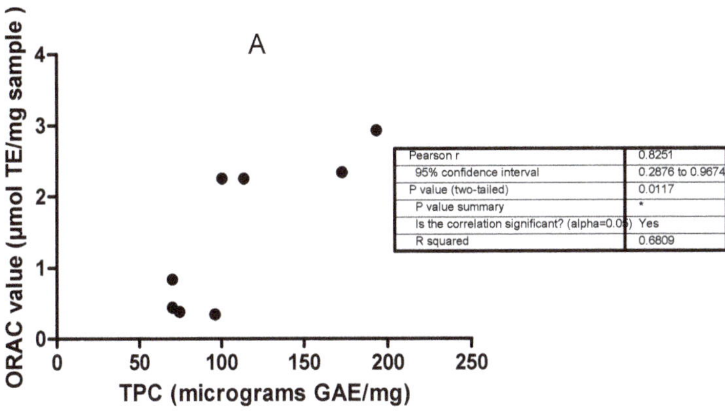

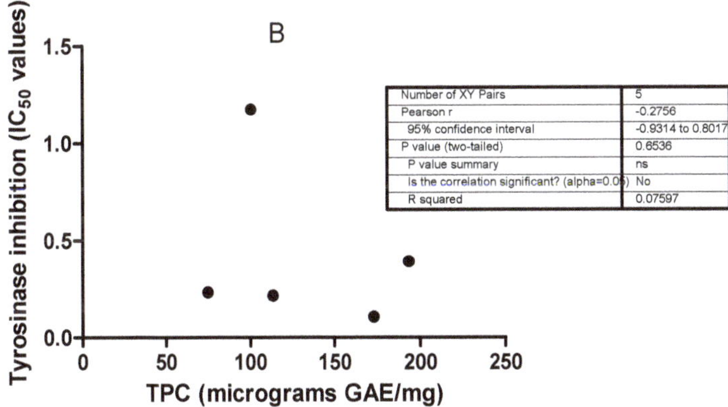

Figure 2. Correlation studies between polyphenol content and antioxidant activity (**A**) and between polyphenol content and tyrosinase inhibition (**B**). Pearson r values confirm that there is a correlation between polyphenol content and antioxidant activity measured by the ORAC method, whereas no correlation exists between polyphenol content and the inhibition of the tyrosinase enzyme.

3.3.3. Superoxide Radicals Generated by Xanthine/Xanthine Oxidase (X/XO)

In order to determine if the extracts were able to scavenge physiological radicals like superoxide anion (O_2^-) generated by X/XO, the extracts were tested using this methodology at various concentrations [15]. There were significant differences between methanolic and hexane extracts (Figure 3), but surprisingly, the activity of certain methanolic extracts was superior to the reference compound trolox (Figure 3A). Table 3 reveals that PVM, AAM, and COM showed lower IC_{50} values than trolox, which confirms their potential as antiradical agents.

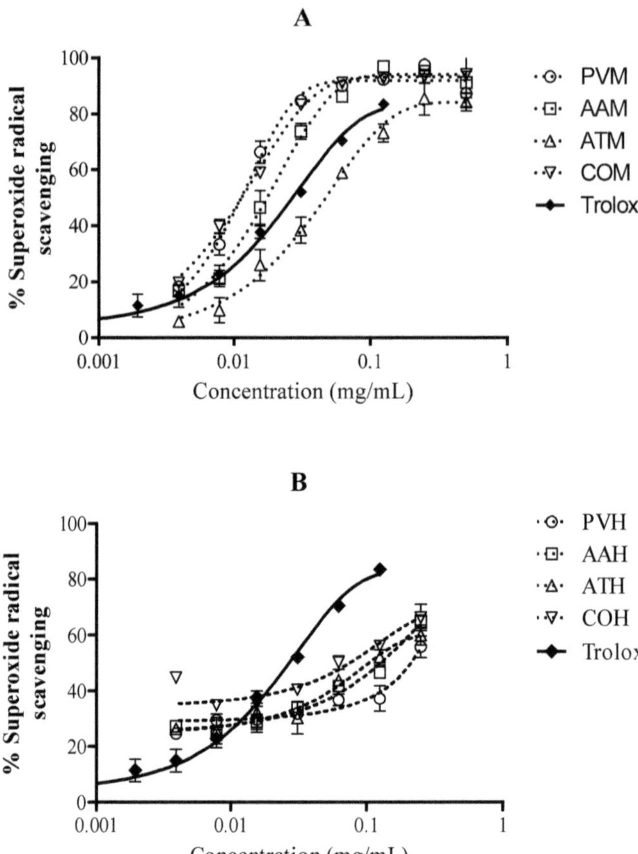

Figure 3. Antioxidant activity of fern methanolic (**A**) and hexane (**B**) extracts against superoxide radicals generated by xanthine/xanthine oxidase using trolox as a reference. *** $p < 0.001$ versus the hexane extract in the same species.

Table 3. Summary of IC$_{50}$ values of methanolic and hexane extracts in the DPPH, xanthine/xanthine oxidase, and tyrosinase assays.

Species	IC$_{50}$ (mg/mL) [a]					
	DPPH Radical		O$_2{}^-$ Radical		Tyrosinase Inhibition	
	Methanol Extract	Hexane Extract	Methanol Extract	Hexane Extract	Methanol Extract	Hexane Extract
Polypodium vulgare (PV)	0.007	0.233	0.011	0.201	0.107	0.233
Asplenium adiantum-nigrum (AA)	0.008	0.044	0.011	0.128	0.216	ND
Asplenium trichomanes (AT)	0.036	0.129	0.047	0.090	1.175	ND
Ceterach officinarum (CO)	0.007	0.072	0.012	0.073	0.392	ND
Kojic acid	-		-		0.063	
Trolox	0.002		0.026		-	

[a] Each value is expressed as the mean of at least three independent measurements. ND: Not determined at assayed concentration (consequence of low activity). IC$_{50}$ value is defined as the effective concentration of extract at which 50% DPPH radicals, 50% of superoxide radicals generated by xanthine/xanthine oxidase, or 50% of the tyrosinase enzyme are inactivated. IC$_{50}$ value was obtained by interpolation from non-linear regression analysis using GraphPad Prism version 6.

3.4. Tyrosinase Inhibition

Figure 4 shows that methanolic extracts were also better than hexane extracts as anti-tyrosinase agents. All methanolic samples presented IC_{50} values between 0.107 mg/mL and 1.175 mg/mL, whereas in the case of hexane samples, only PVH was able to reach the IC_{50} value (Table 3). According to the lowest IC_{50} values, PVM was the best sample as tyrosinase inhibitor, followed by AAM. In this case, tyrosinase inhibition was not correlated with total phenolic content (Figure 2B).

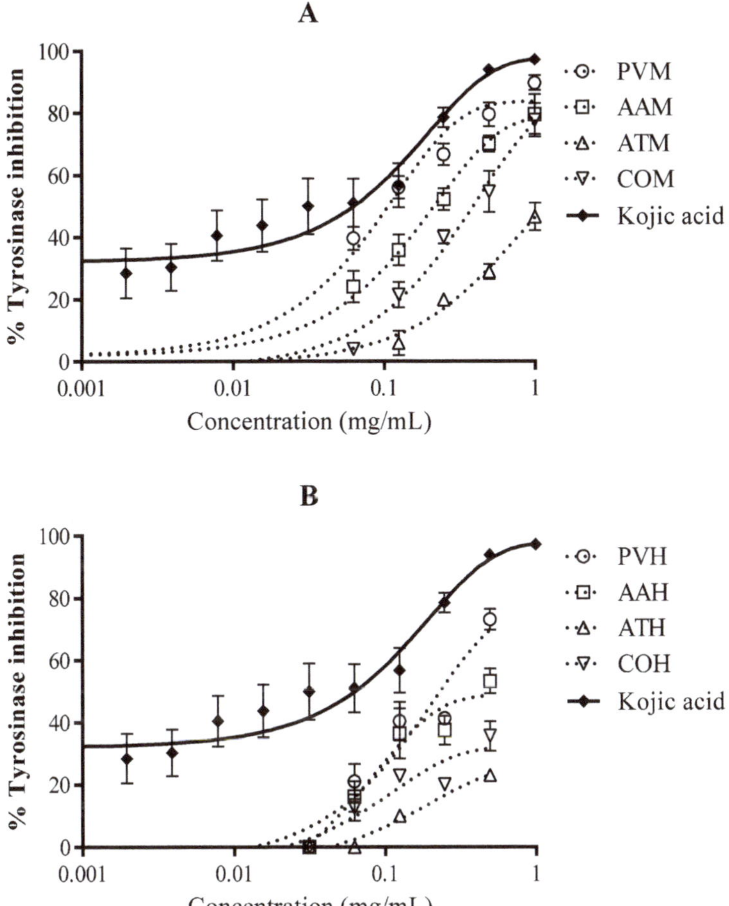

Figure 4. Inhibition of tyrosinase by methanolic (**A**) and hexane (**B**) extracts. Kojic acid was used as reference.

4. Discussion

Ferns have been used in traditional medicine in Central and South America. In fact, two of the selected species (*P. vulgare* and *A. adiantum-nigrum*) are known to possess anti-inflammatory and expectorant properties and are used in traditional medicine for colds [39]. In 2008, the European Medicines Agency (EMA) also approved the monograph of *Polypodium vulgare* rhizome for the treatment of cough and colds [40].

The antioxidant properties of ferns are not new; in fact, extracts of some ferns, such as *Pityrogramma calomelanos* and *Polypodium leucotomos*, were recently reported as antioxidants due to the presence of polyphenols and flavonoids [21,41]. Different types of flavonoids have been described on selected Aspleniaceae ferns [42] and in ferns of the *Dryopteris* genus [43]. It has also been described that the antioxidant mechanisms of flavonoids can be based on hydrogen atom transfer (HAT), single electron transfer (SET), and transition metals chelation (TMC) [44]; however, the antioxidant mechanisms for phenolic acids are predominantly HAT rather than SET [45]. The differing nature of the antioxidant methods tested in this article allows us to determine and characterize the antiradical activity [46,47].

All methanolic samples gave strong positive results in the three tested antioxidant methods. This fact suggests the presence of phenolic acids and flavonoids with hydroxyls groups in the B-ring in the samples [48]. However, the corresponding hexane extracts displayed higher IC_{50} values and weaker antioxidant properties in the DPPH, ORAC, and X/XO assays. The lower antioxidant activity of hexane extracts could be due to the lower polarity of the solvent system, indicating that the majority of phenol and flavonoid compounds are present in the methanolic extracts as determined by Folin and TLC analysis. Additionally, all methanolic extracts showed higher extraction yields compared to the corresponding hexane extract, indicating that the majority of phytoconstituents are hydrophilic molecules. ORAC is a method for antioxidant activity widely used in food sciences, but it is not the first time that certain ferns have also been evaluated using this methodology. The main advantage of this methodology is that the use of fluorescence in the ORAC assay avoids interference with the colored samples [49].

In the superoxide method, superoxide radicals are reduced by receiving one electron (SET mechanism). Flavonoids are known to possess antiradical activity by SET mechanisms; for this reason, the reported activity in this method may be due to flavonoids [50]. The successful results with IC_{50} values better than trolox demonstrate an exploitable antioxidant activity in line with previous results for *Polypodium leucotomos* [51].

The isolation of some phytoecdysteroids in certain ferns has been the focus for certain medicinal applications [52]. For example, phytoecdysteroids have already been isolated in *Polypodium vulgare* [53]. The antioxidative properties by singlet oxygen quenching (SET) and the promotion of differentiation of human keratinocytes of these components may be the reason for obtaining certain bioactivities in hexane extracts [54,55]. Contrary to expectation, all hexane extracts exerted antioxidant properties in the different tested methods. The content of terpenoids in ferns could be responsible for these results in relation to the antioxidant activity of the hydrofobic (hexane) extracts [43,56]. Additionally, carotenoids, which are tetraterpenoids, have been reported in a number of fern species [57].

According to tyrosinase inhibition, *Selaginella tamariscina* and *Stenoloma chusanum* are the only ferns that have been described as anti-tyrosinase agents, with flavonoids involved in this activity [42,58,59]. In our study, methanolic extracts also displayed the lowest IC_{50} values, displaying better anti-tyrosinase activity than hexane extracts. This might ascribe the activity to phenolic compounds; nevertheless, the Pearson values dismiss the positive correlation between IC_{50} values in the tyrosinase assay and TPC. Due to the fact that flavonoids have been described in certain species of the *Polypodiaceae* family [60,61], we might assume that anti-tyrosinase activity of methanolic and hexane extracts could be due to flavonoids; however, this is not completely in agreement with our results, and other authors have found that the inhibition of tyrosinase could be also due to cycloartanes derivatives isolated in PV [62–64].

5. Conclusions

Regarding the results obtained, particularly for the methanolic extracts, antioxidant and potential depigmenting activity has been reinforced for the species *Polypodium vulgare*, *Asplenium adiantum-nigrum*, *Asplenium trichomanes*, and *Ceterach officinarum*. Hydrophilic extracts of these species could be of interest to develop pharmaceutical or cosmetic products, but further studies are needed to better understand the properties and safety aspects of these species.

Supplementary Materials: The following are available online at http://www.mdpi.com/1999-4907/10/2/179/s1, Figure S1: TLC analysis of methanol and hexane extracts of the four fern samples.

Author Contributions: Conceptualization, V.L. and A.F.; Methodology, V.L. and A.F.; Software, V.L. and F.L.; Validation, V.L. and F.L.; Formal Analysis, V.L.; Investigation, A.F., G.C. and F.L.; Resources, V.L. and E.M.T.; Data Curation, A.F., F.L. and V.L; Writing—Original Draft Preparation, A.F.; Writing—Review & Editing, V.L.; Visualization, V.L.; Supervision, V.L., E.M.T. and M.M.; Project Administration, V.L.; Funding Acquisition, V.L. and E.M.T.

Funding: This research received no external funding.

Acknowledgments: We acknowledge Universidad San Jorge for technical assistance and laboratory facilities.

Conflicts of Interest: The authors declare that they do not have any conflict of interest.

Abbreviations

AA	*Asplenium adiantum-nigum*
AAPH	2,2′-azobis(2-amidinopropane) dihydrochloride
AT	*Asplenium trichomanes*
CO	*Ceterach officinarum*
DMSO	dimethyl sulfoxide
DPPH	2,2-diphenyl-1-picrylhydrazyl
GAE	gallic acid equivalents
NBT	nitrotetrazolium blue chloride
ORAC	oxygen radical absorbance capacity
PV	*Polypodium vulgare*
TE	trolox equivalents
TLC	thin layer chromatography
TPC	total phenolic content
X/XO	xanthine/xanthine oxidase

References

1. Leiter, U.; Keim, U.; Eigentler, T.; Katalinic, A.; Holleczek, B.; Martus, P.; Garbe, C. Incidence, Mortality, and Trends of Nonmelanoma Skin Cancer in Germany. *J. Investig. Dermatol.* **2017**, *137*, 1860–1867. [CrossRef] [PubMed]

2. Tran, T.N.T.; Schulman, J.; Fisher, D.E. UV and pigmentation: Molecular mechanisms and social controversies. *Pigment Cell Melanoma Res.* **2008**, *21*, 509–516. [CrossRef] [PubMed]

3. Kimlin, M.G.; Guo, Y.M. Assessing the impacts of lifetime sun exposure on skin damage and skin aging using a non-invasive method. *Sci. Total Environ.* **2012**, *425*, 35–41. [CrossRef] [PubMed]

4. Jansen, R.; Wang, S.Q.; Burnett, M.; Osterwalder, U.; Lim, H.W. Photoprotection Part I. Photoprotection by naturally occurring, physical, and systemic agents. *J. Am. Acad. Dermatol.* **2013**, *69*, 12.

5. Smit, N.P.M.; Kolb, R.M.; Lentjes, E.; Noz, K.C.; van der Meulen, H.; Koerten, H.K.; Vermeer, B.J.; Pavel, S. Variations in melanin formation by cultured melanocytes from different skin types. *Arch. Dermatol. Res.* **1998**, *290*, 342–349. [CrossRef]

6. Gloster, H.M.; Neal, K. Skin cancer in skin of color. *J. Am. Acad. Dermatol.* **2006**, *55*, 741–760. [CrossRef] [PubMed]

7. Sklar, L.R.; Almutawa, F.; Lim, H.W.; Hamzavi, I. Effects of ultraviolet radiation, visible light, and infrared radiation on erythema and pigmentation: A review. *Photochem. Photobiol. Sci.* **2013**, *12*, 54–64. [CrossRef]

8. Ortonne, J.P.; Passeron, T. Melanin pigmentary disorders: Treatment update. *Dermatol. Clin.* **2005**, *23*, 209–226. [CrossRef]

9. Gonzalez, S.; Fernandez-Lorente, M.; Gilaberte-Calzada, Y. The latest on skin photoprotection. *Clin. Dermatol.* **2008**, *26*, 614–626. [CrossRef]

10. Jansen, R.; Osterwalder, U.; Wang, S.Q.; Burnett, M.; Lim, H.W. Photoprotection Part II. Sunscreen: Development, efficacy, and controversies. *J. Am. Acad. Dermatol.* **2013**, *69*, 14.

11. Chen, A.C.; Halliday, G.M.; Damian, D.L. Non-melanoma skin cancer: Carcinogenesis and chemoprevention. *Pathology* **2013**, *45*, 331–341. [CrossRef] [PubMed]

12. Saewan, N.; Jimtaisong, A. Natural products as photoprotection. *J. Cosmet. Dermatol.* **2015**, *14*, 47–63. [CrossRef] [PubMed]

13. Nichols, J.A.; Katiyar, S.K. Skin photoprotection by natural polyphenols: Anti-inflammatory, antioxidant and DNA repair mechanisms. *Arch. Dermatol. Res.* **2010**, *302*, 71–83. [CrossRef] [PubMed]

14. Del Rosso, J.Q. Use of *Polypodium leucotomas* Extract in Clinical Practice: A Primer for the Clinician. *J. Clin. Aestheti. Dermatol.* **2016**, *9*, 37–42.

15. Murbach, T.S.; Beres, E.; Vertesi, A.; Glavits, R.; Hirka, G.; Endres, J.R.; Clewell, A.E.; Szakonyine, I.P. A comprehensive toxicological safety assessment of an aqueous extract of *Polypodium leucotomos*, Fernblock, R)). *Food Chem. Toxicol.* **2015**, *86*, 328–341. [CrossRef]

16. Palomino, O.M. Current knowledge in *Polypodium leucotomos* effect on skin protection. *Arch. Dermatol. Res.* **2015**, *307*, 199–209. [CrossRef]

17. Wolf, P.G.; Pryer, K.M.; Smith, A.R.; Hasebe, M. Phylogenetic studies of extant pteridophytes. In *Molecular Systematics of Plants, II: DNA Sequencing*; Soltis, D.E., Soltis, P.S., Doyle, J.J., Eds.; Kluwer Academic Publishers: Dordrecht, The Netherlands, 1998; pp. 541–556.

18. Christenhusz, M.J.M.; Chase, M. Trends and concepts in fern classification. *Ann. Bot.* **2014**, *113*, 571–594. [CrossRef]

19. Berman, B.; Ellis, C.; Elmets, C. *Polypodium Leucotomos*—An Overview of Basic Investigative Findings. *J. Drugs Dermatol.* **2016**, *15*, 224–228.

20. Cao, H.; Chai, T.T.; Wang, X.; Morais-Braga, M.F.B.; Yang, J.H.; Wong, F.C.; Wang, R.B.; Yao, H.K.; Cao, J.G.; Cornara, L.; et al. Phytochemicals from fern species: Potential for medicine applications. *Phytochem. Rev.* **2017**, *16*, 379–440. [CrossRef]

21. Amoroso, V.B.M.; Rainear, A.; Villalobos, A.P. Bringing back the lost value of Philippine edible ferns: Their antioxidant, proteins and utilization. *Int. J. Adv. Res.* **2017**, *5*, 757–770. [CrossRef]

22. Tomsik, P. Ferns and Lycopods—A Potential Treasury of Anticancer Agents but Also a Carcinogenic Hazard. *Phytother. Res.* **2014**, *28*, 798–810. [CrossRef] [PubMed]

23. Nguyen, P.H.; Zhao, B.T.; Ali, M.Y.; Choi, J.S.; Rhyu, D.Y.; Min, B.S.; Woo, M.H. Insulin-Mimetic Selaginellins from *Selaginella tamariscina* with Protein Tyrosine Phosphatase 1B, PTP1B, Inhibitory Activity. *J. Nat. Prod.* **2015**, *78*, 34–42. [CrossRef] [PubMed]

24. Zheng, X.K.; Wang, W.W.; Zhang, L.; Su, C.F.; Wu, Y.Y.; Ke, Y.Y.; Hou, Q.W.; Liu, Z.Y.; Gao, A.S.; Feng, W.S. Antihyperlipidaemic and antioxidant effect of the total flavonoids in *Selaginella tamariscina*, Beauv., Spring in diabetic mice. *J. Pharm. Pharmacol.* **2013**, *65*, 757–766. [CrossRef] [PubMed]

25. Saiz, J.C.M.; Pataro, L.; Sotomayor, S.P. Atlas of the pteridophytes of the Iberian Peninsula and the Balearic Islands. *Acta Bot. Malacit.* **2015**, *40*, 5–55.

26. Do, Q.T.; Bernard, P. Reverse pharmacognosy: A new concept for accelerating natural drug discovery. In *Lead Molecules from Natural Products: Discovery and New Trends*; Khan, M.T.H., Ather, A., Eds.; Elsevier: Amsterdam, The Netherlands, 2006; pp. 1–20.

27. Tanew, A.; Radakovic, S.; Gonzalez, S.; Venturini, M.; Calzavara-Pinton, P. Oral administration of a hydrophilic extract of *Polypodium leucotomos* for the prevention of polymorphic light eruption. *J. Am. Acad. Dermatol.* **2012**, *66*, 58–62. [CrossRef] [PubMed]

28. The Plant List. Available online: http://www.theplantlist.org/ (accessed on 16 November 2016).

29. Banco de Datos de Biodiversidad de Cataluña, Generalidad de Cataluña. Available online: http://biodiver.bio.ub.es/biocat/index.jsp (accessed on 16 November 2016).

30. Bonnier, G.; de Layens, G. *Claves Para la determinación de Plantas Vasculares*, 1st ed.; Ediciones Omega, S.L.: Barcelona, Spain, 1988.

31. Wagner, H.; Bladt, S. *Plant Drug Analysis: A Thin Layer Chromatography Atlas*, 2nd ed.; Springer: Berlin, Germany, 1996.

32. Singleton, V.L. Citation classic–Colorimetry of total phenolics with phosphomolybdic-phosphotungstic acid reagents. *Curr. Contents Agric. Biol. Environ. Sci.* **1985**, *48*, 18.

33. Blois, M.S. Antioxidant determinations by the use of a stable free radical. *Nature* **1958**, *181*, 1199–1200. [CrossRef]

34. Cásedas, G.; Les, F.; Gómez-Serranillos, M.P.; Smith, C.; López, V. Bioactive and functional properties of sour cherry juice (*Prunus cerasus*). *Food Funct.* **2016**, *7*, 4675–4682. [CrossRef]

35. Davalos, A.; Gomez-Cordoves, C.; Bartolome, B. Extending applicability of the oxygen radical absorbance capacity (ORAC-fluorescein, assay. *J. Agric. Food Chem.* **2004**, *52*, 48–54. [CrossRef]

36. Fridovich, I. Quantitative aspects of the production of superoxide anion radical by milk xanthine oxidase. *J. Biol. Chem.* **1970**, *245*, 4053–4057.

37. Thayer, W.S. Superoxide-Dependent and Superoxide-Independent Pathways for Reduction of Nitroblue Tetrazolium in Isolated Rat Cardiac Myocytes. *Arch. Biochem. Biophys.* **1990**, *276*, 139–145. [CrossRef]

38. Senol, F.S.; Orhan, I.E.; Ozgen, U.; Renda, G.; Bulut, G.; Guven, L.; Karaoglan, E.S.; Sevindik, H.G.; Skalicka-Wozniak, K.; Caliskan, U.K.; et al. Memory-vitalizing effect of twenty-five medicinal and edible plants and their isolated compounds. *S. Afr. J. Bot.* **2016**, *102*, 102–109. [CrossRef]

39. Ho, R.; Teai, T.; Bianchini, J.P.; Lafont, R.; Raharivelomanana, P. *Ferns: From Traditional Uses to Pharmaceutical Development, Chemical Identification of Active Principles, Working with Ferns: Issues and Applications*; Springer: New York, NY, USA, 2010; pp. 321–346.

40. European Medicines Agency (EMA). *Assessment Report on Polypodiumvulgare L.*; Rizoma European Medicines Agency: London, UK, 2008; p. 22.

41. Garcia, F.; Pivel, J.P.; Guerrero, A.; Brieva, A.; Martinez-Alcazar, M.; Caamano-Somoza, M.; Gonzalez, S. Phenolic components and antioxidant activity of Fernblock (R), an aqueous extract of the aerial parts of the fern *Polypodium leucotomos*. *Methods Find. Exp. Clin. Pharmacol.* **2006**, *28*, 157–160. [CrossRef] [PubMed]

42. Iwashina, T.; Matsumoto, S. *Flavonoid Properties of Six Asplenium Species in Vanuatu and New Caledonia, and Distribution of Flavonoid and Related Compounds in Asplenium, Bulletin of the National Museum of Nature and Science*; Series B; Botany/National Museum of Nature and Science: Tokyo, Japan, 2011; p. 13.

43. Zhang, M.; Cao, J.G.; Dai, X.L.; Chen, X.F.; Wang, Q.X. Flavonoid Contents and Free Radical Scavenging Activity of Extracts from Leaves, Stems, Rachis and Roots of Dryopteris erythrosora. *Iran. J. Pharm. Res.* **2012**, *11*, 991–997.

44. Prochazkova, D.; Bousova, I.; Wilhelmova, N. Antioxidant and prooxidant properties of flavonoids. *Fitoterapia* **2011**, *82*, 513–523. [CrossRef] [PubMed]

45. Robbins, R.J. Phenolic acids in foods: An overview of analytical methodology. *J. Agric. Food Chem.* **2003**, *51*, 2866–2887. [CrossRef] [PubMed]

46. Carocho, M.; Ferreira, I. A review on antioxidants, prooxidants and related controversy: Natural and synthetic compounds, screening and analysis methodologies and future perspectives. *Food Chem. Toxicol.* **2013**, *51*, 15–25. [CrossRef] [PubMed]

47. Karadag, A.; Ozcelik, B.; Saner, S. Review of Methods to Determine Antioxidant Capacities. *Food Anal. Methods* **2009**, *2*, 41–60. [CrossRef]

48. Roginsky, V.; Lissi, E.A. Review of methods to determine chain-breaking antioxidant activity in food. *Food Chem.* **2005**, *92*, 235–254. [CrossRef]

49. Kamisan, F.H.; Yahya, F.; Mamat, S.S.; Kamarolzaman, M.F.F.; Mohtarrudin, N.; Kek, T.L.; Salleh, M.Z.; Hussain, M.K.; Zakaria, Z.A. Effect of methanol extract of Dicranopteris linearis against carbon tetrachloride-induced acute liver injury in rats. *BMC Complement. Altern. Med.* **2014**, *14*, 10. [CrossRef]

50. Robak, J.; Gryglewski, R.J. Flavonoids are scavengers of superoxide anions. *Biochem. Pharmacol.* **1988**, *37*, 837–841. [CrossRef]

51. Gomes, A.J.; Lunardi, C.N.; Gonzalez, S.; Tedesco, A.C. The antioxidant action of *Polypodium leucotomos* extract and kojic acid: Reactions with reactive oxygen species. *Braz. J. Med Biol. Res.* **2001**, *34*, 1487–1494. [CrossRef]

52. Thiem, B.; Kikowska, M.; Malinski, M.P.; Kruszka, D.; Napierala, M.; Florek, E. Ecdysteroids: Production in plant in vitro cultures. *Phytochem. Rev.* **2017**, *16*, 603–622. [CrossRef]

53. Arai, Y.; Shiojima, K.; Ageta, H. Fern constituents: Cyclopodmenyl acetate, a cycloartanoid having a new 33-carbon skeleton, isolated from *Polypodium vulgare*. *Chem. Pharm. Bull.* **1989**, *37*, 560–562. [CrossRef]

54. Lafont, R.; Dinan, L. Practical uses for ecdysteroids in mammals including humans: And update. *J. Insect Sci.* **2003**, *3*, 30. [CrossRef]

55. Mamadalieva, N.Z. Phytoecdysteroids from *Silene* plants: Distribution, diversity and biological (antitumour, antibacterial and antioxidant, activities. *Bol. Latinoam. Y Del Caribe De Plantas Med. Y Aromat.* **2012**, *11*, 474–497.

56. Graßmann, J. *Terpenoids as Plant Antioxidants*; Gerald, L., Ed.; Vitamins & Hormones, Academic Press: Cambridge, MA, USA, 2005; pp. 505–535.

57. Takaichi, S. Tetraterpenes: Carotenoids. In *Natural Products*; Ramawat, K.G., Mérillon, J.-M., Eds.; Springer: Berlin, Germany, 2013; pp. 3251–3283.

58. Nguyen, P.H.; Ji, D.J.; Han, Y.R.; Choi, J.S.; Rhyu, D.Y.; Min, B.S.; Woo, M.H. Selaginellin and biflavonoids as protein tyrosine phosphatase 1B inhibitors from *Selaginella tamariscina* and their glucose uptake stimulatory effects. *Bioorg. Med. Chem.* **2015**, *23*, 3730–3737. [CrossRef]

59. Wu, S.Q.; Li, J.; Wang, Q.X.; Cao, J.G.; Yu, H.; Cao, H.; Xiao, J.B. Chemical composition, antioxidant and anti-tyrosinase activities of fractions from *Stenoloma chusanum*. *Ind. Crop. Prod.* **2017**, *107*, 539–545. [CrossRef]

60. Su, W.; Li, P.Y.; Huo, L.N.; Wu, C.Y.; Guo, N.N.; Liu, L.Q. Phenolic content and antioxidant activity of Phymatopteris hastate. *J. Serb. Chem. Soc.* **2011**, *76*, 1485–1496. [CrossRef]

61. Chai, T.T.; Quah, Y.; Ooh, K.F.; Ismail, N.I.M.; Ang, Y.V.; Elamparuthi, S.; Yeoh, L.Y.; Ong, H.C.; Wong, F.C. Anti-Proliferative, Antioxidant and Iron-Chelating Properties of the Tropical Highland Fern, *Phymatopteris triloba (Houtt, Pichi Serm (Family Polypodiaceae). Trop. J. Pharm. Res.* **2013**, *12*, 747–753. [CrossRef]

62. Ageta, H.; Arai, Y. Chemotaxonomy of ferns 3. triterpenoids from *Polypodium–polypodioides*. *J. Nat. Prod.* **1990**, *53*, 325–332. [CrossRef]

63. Arai, Y.; Yamaide, M.; Yamazaki, S.; Ageta, H. Fern constituents–Triterpenoids isolated from *Polypodium-vulgare, Polypodium-fauriei* and *Polypodium-virginianum*. *Phytochemistry* **1991**, *30*, 3369–3377. [CrossRef]

64. Khan, M.T.H.; Khan, S.B.; Ather, A. Tyrosinase inhibitory cycloartane type triterpenoids from the methanol extract of the whole plant of *Amberboa ramosa* Jafri and their structure-activity relationship. *Bioorg. Med. Chem.* **2006**, *14*, 938–943. [CrossRef] [PubMed]

 *forests*

Review

Rediscovering the Contributions of Forests and Trees to Transition Global Food Systems

James L. Chamberlain [1,*], Dietrich Darr [2] and Kathrin Meinhold [2]

[1] USDA Forest Service, 1710 Research Center Drive, Blacksburg, VA 24060, USA
[2] Faculty of Life Sciences, Rhine-Waal University of Applied Sciences, Marie-Curie-Str. 1, 47533 Kleve, Germany; dietrich.darr@hochschule-rhein-waal.de (D.D.); kathrin.meinhold@hochschule-rhein-waal.de (K.M.)
* Correspondence: james.l.chamberlain@usda.gov

Received: 7 August 2020; Accepted: 12 October 2020; Published: 16 October 2020

Abstract: The importance of forests to safeguard agricultural production through regulating ecosystem services such as clean water, soil protection, and climate regulation is well documented, yet the contributions of forests and trees to provide food for the nutritional needs of the increasing human population has not been fully realized. Plants, fungi, and animals harvested from forests have long provided multiple benefits—for nutrition, health, income, and cultural purposes. Across the globe, the main element of "forest management" has been industrial wood production. Sourcing food from forests has been not even an afterthought but a subordinate activity that just happens and is largely invisible in official statistics. For many people, forests ensure a secure supply of essential foods and vital nutrients. For others, foraging forests for food offers cultural, recreational, and diversified culinary benefits. Increasingly, these products are perceived by consumers as being more "natural" and healthier than food from agricultural production. Forest-and wild-sourced products increasingly are being used as key ingredients in multiple billion dollar industries due to rising demand for "natural" food production. Consumer trends demonstrate growing interests in forest food gathering that involves biological processes and new forms of culturally embedded interactions with the natural world. Further, intensifying calls to "re-orient" agricultural production provides opportunities to expand the roles of forests in food production; to reset food systems by integrating forests and trees. We use examples of various plants, such as baobab, to explore ways forests and trees provide for food security and nutrition and illustrate elements of a framework to encourage integration of forests and trees. Forests and trees provide innovative opportunities and technological and logistical challenges to expand food systems and transition to a bioeconomy. This shift is essential to meet the expanding demand for secure and nutritious food, while conserving forest biodiversity.

Keywords: bioeconomy; food and nutrition security; forests and trees; forest foods; wild harvesting

---

1. Introduction

Global demand for food is projected to increase as much as threefold by 2050 [1]. By some estimates, food production will need to increase more than 70 percent over the next 40 years [2]. Supplying food to the growing world's population will put increasing demands on agricultural systems and land. With significant challenges to sustain past rates of yield increase for major agricultural crops in the future [3,4], the pressure to convert natural ecosystems and forest land to agricultural production will likely intensify [5], a practice that has led to tremendous loss of biodiversity. The conventional approach to achieving food security has resulted in a loss of forest cover that directly and indirectly supports the food security and nutritional needs of hundreds of millions of people, especially in developing countries. Hence, one of the greatest challenges of the 21st century is increasing food production to improve food security, without reducing forest area, or biodiversity, and at the same time not just feeding but nourishing people [6].

More than a billion people around the world depend on forests and trees for their livelihood, and forest foods for nutrition [7,8]. Rural people living proximal to forests derive a diverse set of foodstuffs from forests [9], which is particularly important for poor populations [10–12]. At the same time, wild foods are integral to the diets of vulnerable and higher-income urban populations in many parts of the world [13–17].

According to the high-level panel of experts on food security and nutrition of the Committee on World Food Security [18], forests contribute to food security and nutrition in several ways. Direct provision of food from forests may represents only less than one percent of global food supply, but the contribution to dietary quality and diversity is crucial to family health and wellbeing. Forest foods, especially fruits and vegetables are rich in the micronutrients that impoverished people often lack [19]. In 1997, approximately 390 million metric tons of food was produced by trees in developing countries [8]. On average, more than 258 metric tons of forest foods are reportedly harvested from public lands in the United States [20] each year. A recent study of households across Europe that collect forest foods found the mean to be about 60 kg per year, with a median of 20 kg [21].

Forests provide energy to process agricultural and forest foods for consumption. In general, wood fuel contributes about 6 percent of total primary energy supply, yet in some regions of Africa, wood fuel contributes upwards of 25 percent of energy needs [18]. In addition to the use of wood fuel for household subsistence, its sale and conversion into charcoal to supply urban centers provides employment and income opportunities in many rural areas [22]. In many developing countries, charcoal is the main energy source for urban households [23]. At the same time, fuelwood scarcity may negatively affect household food security and health, as typical household coping strategies include selling or bartering food to procure fuelwood, the use of lower-quality fuelwood substitutes such as dung, or reducing the amount of fuelwood use by eating fewer meals, undercooking food, or boiling water insufficiently to save fuelwood [24]. While the impact of fuelwood harvest on forests is highly context-specific, evidence suggests that harvest volumes can exceed sustainable levels [25,26]. Such negative environmental outcomes are partially attributed to the largely informal and unregulated wood fuel sector and wood fuel value chains with weak enforcement and governance mechanisms [27].

According to the FAO [6], commercialization of forest foods—and other forest products such as fuelwood—account for about 20 percent of income for rural households in developing countries. The formal forest sector employs more than 13 million people, providing income that allows for the purchase of food and other necessities [18]. Total income from production of non-wood forest products (a subset of non-timber forest products (NTFP)) is estimated to be more than 88 billion USD, although estimates are acknowledged widely to be lower than actual [28] (p. 25). Animal products (e.g., bushmeat, game) generate another 10.5 billion USD in income, and collection of medicinal and aromatic plants some 700 million USD [28]. Estimates of forest sector employment do not include the innumerable number of people worldwide that function in the informal forest economy, gathering for direct consumption, or bartering and trading products with no transaction records. By some estimates, more than 300 million people earn part of their annual livelihood and food from forests [8]. In some situations, food gathered from forests can reduce the family food budget as much as 60 percent [29].

Forests provide many products and services that are consumed by people everywhere, or used to serve their needs. A tremendous diversity of forest plants, fungi, and animals are harvested to provide food, medicine, and other essential and luxury items. While accounts of the roles forests have in human nutrition and food security via provision of food, energy, and income opportunities as illustrated are not inaccurate, we contend that they tend to neglect important and critical aspects—most importantly due to the informal nature of markets and consumption of forest products that is unaccounted for in official global statistics. Forests can indeed contribute to food systems at the local and global level in many further ways, as we demonstrate in the next section. While steps are being taken to improve the situation [30], more can be done. Recognizing these contributions is necessary to more fully account for and realize the forests' potential for sustainable human nutrition.

In this narrative, we focus on forest plants and fungi used as food, and forests and trees as provisioning units for food systems and the bioeconomy, by reviewing and synthesizing relevant literature and subsequently integrating the concepts. We start with an examination of the value of forests and trees to traditional and contemporary cultures, growing consumer demand for sustainable products, and potential public health risks. The myriad of production systems, from wild harvesting to single species plantations, are discussed. We then reflect on elements of a framework to encourage and support transition to a bioeconomy that integrates forests and trees to provide for food security and nutrition needs.

2. Contributions of Forests and Trees to People in Food Systems

2.1. Cultural Identities and Nature-Related Values

Indigenous people living in tropical and temperate regions have strong cultural ties to forest foods. The collection and use of forest plants and fungi to meet the nutritional needs of native peoples is deeply embedded in their cultures, with knowledge transferred through generations. For example, forest foods have contributed to the cultural identity of indigenous people across the temperate and boreal forest regions [31], as evidenced by the hundreds of edible plants documented contributing to diverse diets. Similarly, Native Americans have a long history of harvesting mushrooms for food and other purposes [32]. Likewise, traditional hunter—gatherers in the Congo basin subsisted for extended periods by foraging wild forest food, such as wild yams [33]. Contrary to native populations in Canada and Mexico, Amazonian Indians tended to consume fewer wild greens but preferred wild fruits and tubers [34]. The sharing of knowledge about these forest foods adds to the edification of next generations and can provide valuable insights into managing forests for these resources.

In addition to the cultures of native peoples, forest foods embody significant importance to the cultural identity, lifestyles, and intangible values of urban populations. Foraging landscapes for food is very much a part of urban cultures around the world and has increased in popularity over the last two decades. A simple search of the social media platform, *Facebook*, reveals many groups formed around foraging food from natural habitats. The global membership of foragers on Facebook have commonality in the culture of food plants and share information and knowledge to promote their practices. In Berlin, urban residents commonly use and consume wild plants for food, personal joy, and medicinal uses [35]. The collection of wild onions, also known as ramps or leeks (*Allium tricoccum* Ait.) from the forests of Appalachia, in the US, is deeply embedded in the region's culture and has grown in popularity among urban consumers over the last 25 years [36]. With the collection of wild mushrooms and other forest foods gaining popularity among urban middle class and hipster populations in Sweden since the 20th century, a vivid community has developed providing related educational and support services including evening classes, study circles, clubs, and exhibitions [37]. Foraging for forest foods is supporting a renaissance among urban cultures around the world and may have significant impacts on forest-based economies and ecologies [38]. The demographics of contemporary foragers suggest that verified sustainable sourcing of forest foods would be important to these consumers and advocates.

Studies have highlighted the roles of formal education systems in mainstreaming the contribution of wild edible plants in human diets. For example, a progressive school in Berlin, Germany, that offered regionally produced food provided opportunities to grow fruit and vegetables in the school's garden and foraging wild edible plants from a neighboring vacant plot; it was involved in the collaborative planning and management of their food production and found that students were better educated about nature and food production and had improved their diets [39]. Some Native American tribes, such as the Eastern Band of the Cherokee Indians, are integrating traditional foods into their educational programs. Conversely, formal education systems in some Chilean traditional communities largely neglect local knowledge of traditional food plants and their uses, contributing to a process of biocultural homogenization that leads to reduced use of biodiversity for household nutrition [40].

2.2. Satisfying Consumer Demand for Healthy, Sustainably Produced Natural Foods

Undeniably, forest foods provide the calories and macro- and micronutrients essential for human nourishment. More broadly, they contribute to meeting the diversifying culinary demands of a changing global population. Economic advancement, increasing levels of income and living standards, and the emergence of an urban middle class in many countries have prompted significant transitions of diets, eating habits, and consumer preferences related to food products. Changing socioeconomic conditions and the modernization of lifestyles lead to a growing diversity of consumer tastes and preferences [41–43]. Nevertheless, among the major food-related consumer trends are the increasing demand for healthy, sustainably produced, and convenience food products [44], the latter denoting increasing demand for food products that aim to reduce the time and effort required by consumers to buy, store, prepare, and consume their food. This is largely triggered by sociodemographic changes such as the increasing number of single households, declining culinary skills, and more stressful lifestyles and has led to a diversity of fresh, ready-to-eat, single-portion packaged, snack, or pre-prepared food products [44]. Forest foods can contribute to satisfying these demands. Wiersum [38] describes the increasing appreciation of forest foods in Europe as driven by a growing interest in heritage-inspired and more natural forest production systems and experimentation with new biocultural practices.

While forest foods have been portrayed as old fashioned and reserved for the poor [45,46], there is evidence that wild edible plants are perceived positively by more affluent consumers [13,47] and some have been integrated into mainstream markets. Consequently, there is increasing potential for wild foods to contribute to bioeconomies as sources of sustainable, healthy raw material [48]. For example, wild edible fruits and other forest foods often are perceived by consumers as being healthy and nutritious and linked to cultural identity [16,36,49,50]. Aworh [13] reports that some African traditional leafy vegetables, including tree leaves, are increasingly being offered in fine dining restaurants and supermarket chains in urban centers of Kenya and other African countries. Other studies provide evidence that, while traditionally linked to consumption by rural poor and as a safety net in times of food shortage, some forest foods, such as baobab (*Adansonia digitata* L.), are increasingly processed into high-value food products and sold to high-income consumer segments through local urban specialty and delicacy stores in Malawi [51] or in international markets [52]. Similar reports also exist for shea (*Vitellaria paradoxa* C.F. Gaertn), acai (*Euterpe oleracea* Mart.), pine nuts, or edible insects (e.g., *Ruspolia differens* Scopoli). At the same time, technological advancements facilitate the use of wood products for processing into food ingredients for mainstream markets, as demonstrated by the importance of xylitol produced from birch wood (*Betula pendula* Roth.) in Finland as a low-calorie sweetener possessing dental health benefits [53]. Hence, forests have significant potential to supply healthy, natural, and sustainably produced food products that health-conscious and sustainability-oriented consumer would buy and are beginning to be considered in discussions about "future smart food". Transitioning to a bioeconomy may spur this shift more rapidly.

2.3. Public Health—Contributions and Risks

While forest and trees contribute to improving public health, there also are potential risks to consuming forest foods. Studies have confirmed the antioxidant, antihypercholesterolemic, antidiabetic, anti-inflammatory, anti-amyloidogenic, antimutagenic, antiviral, or antimicrobial activity of phytochemical compounds contained in wild edible plants [54]. Integration of forest foods into food systems, by altering consumption habits and preferences, production, harvesting and processing practices, and the structure and governance mechanisms of food value chains, could contribute to reducing malnutrition [55]. Food production systems that include forest foods could appropriately address undernourishment, food insecurity, and micronutrient deficiency, problems that affect more than 800 million people, particularly in the Global South [6]. In addition, integrated food systems can contribute to halting and reversing the prevalence and severity of diabetes [56], cancer [57], cardiovascular [58], and other non-communicable diseases related to dietary excess and physical inactivity.

Conversely, consumption of forest foods can increase exposure to human health risks. For example, due to the increasing popularity of collecting mushrooms and wild plants by city dwellers, accidental poisoning may occur more frequently if they are mixed with similar-looking inedible or even poisonous plants. For example, confusing self-collected wild garlic (*Allium ursinum* L.) with lily of the valley (*Convallaria majalis* L.) has tremendous risks as the latter has high content of digitalis glycosides [59]. Furthermore, the hunting, handling, processing, and consumption of bushmeat can increase the occurrence of wildlife zoonotic diseases, the emergence and spread of which has been amplified by changes in climate, land use, and biodiversity [60]. In many cases, disadvantaged social groups, particularly, are most affected by these risks. In Cameroon, marginalized groups of young, poorly educated, jobless hunters and women preparing bat meat for consumption were exposed to the risk of Ebola virus infections [61]. Likewise, households with a lower food security status hunted and consumed a larger diversity of bushmeat species, which along with the potential increased negative impacts on wildlife conservation also increased their risk of infections with zoonotic diseases [62]. Moves to ban wildlife consumption, particularly in view of the Covid-19 pandemic, may have unintended consequences particularly on those who depend on wild meat for nutrition. It is important to not demonize foraged foods, but instead increase awareness on proper handling, harvesting, and consumption. Aversions to risks such as these may be overshadowed by the urgent need to feed families.

3. Forests and Tree Food Production

Forests are an assemblage of fauna and flora that form a system of mutual benefits. The plants that make up forests include forbs, grasses, and herbaceous plants that cover the forest floor, shrubs, and other woody species that inhabit an understory created by trees that dominate overhead. Traditionally, forests have been perceived as large expanses of land dominated by trees. A more contemporary perspective on "forests" include trees in all of their settings: natural and manipulated habitats, agroforests, single-species plantings, and on lands where they are not the dominant growth habit, such as fields, roadsides and urban landscapes. Natural forests provide tremendous biodiversity harvested for food for local and nonlocal consumption. Agroforests that integrate trees and crops are designed to provide food products directly and protective services (e.g., soil erosion, habitat for pollinators, mitigating microclimate conditions) to improve production. Secondary forests and single-species plantations provide many food products for self-consumption and for commodity markets. Trees outside of these habitats provide food valued by urban and rural households, as well as others. Fields and urban landscapes often have trees that produce food and other important products. Figure 1 illustrates contributions of forest and trees to food systems.

Production of forest foods range from unmanaged wild foraging to highly managed single-species plantations [29]. Some production systems appear to be random accumulation of plants with few inputs, while others are very organized with high human interventions. The scale of production depends greatly on the purpose and demand for the products. Conceivably, the amount of forest foods needed for direct household consumption are less than amounts wanted for income generation. In many countries, harvesting forest foods is embedded in the culture, and access is uninhibited because the products are considered public resources.

Forests have been altered by people living in and near them for millennia [63]. Evidence indicates that people have "managed" forests to increase food production for tens of thousands of years, and that most forests have been changed significantly over time to provide for the direct nutritional needs of families [64–66]. As Parrotta et al. [63] indicate, many forms of traditional forest management that include food production—such as multi-storied agroforests, home gardens, and shifting cultivation—remain but are undervalued for their contributions.

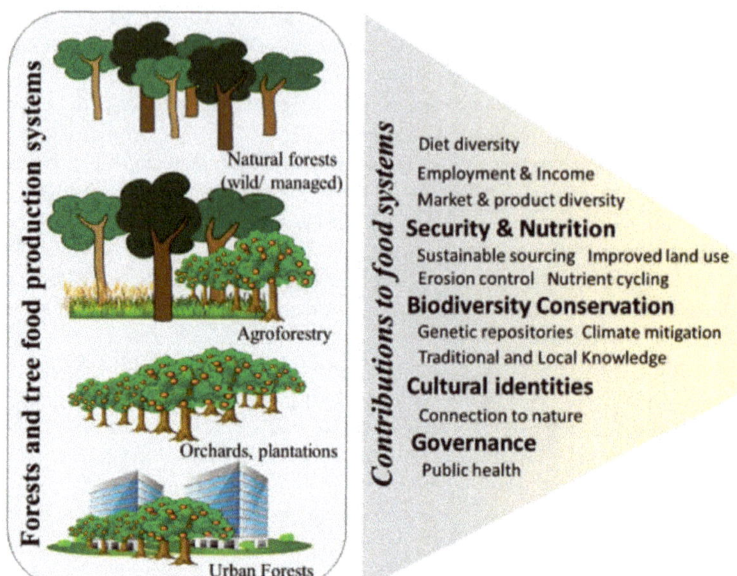

Figure 1. Contributions of forest and trees to food systems.

3.1. Wild Harvesting/Foraging

People have been foraging food from forests since the beginning of time. Wild-harvesting forest foods supported families and communities since before agricultural technologies made it possible to grow food in fields. Long before the technologies existed to grow food, indigenous people were harvesting food from forests for direct consumption. With technological developments that allowed for food production in fields, demand for forest foods may have decreased, but forests continue to supply critical nutritional needs to families and communities. It is fundamental to the identities of cultures around the world, and many forest food plants are considered cultural key stone species [67] that identify the people who depend on them. The loss of these species or access to the plants would drastically affect the people and their cultures.

Given that foraging often takes place on open-access or communal lands and involves some competition over the resource between foragers, some communities of foragers have developed governance and management practices to ensure production of these forest foods. In northeastern Brazil, harvesters of pequi (*Caryocar coriaceum* Wittn) have rules on how to harvest the fruit to maximize quality without detriment to future harvests or the vitality of the tree [68]. Management by Native Americans by pruning or coppicing has been shown to increase biomass production in some forest plants [69]. Some Native American tribes, such as the Eastern Band of the Cherokee Indians, have formal forest management plans with explicit objectives for production of traditional foods [70]. Formal regulations have been established by many US state governments to manage the harvest of American ginseng roots to encourage conservation [71]. Traditional and local knowledge can inform better management of forests for wild foods. However, the effectiveness of governance systems may strongly differ or lead to unintended consequences [72,73].

Wild harvesting of forest food often is perceived to be done with little or no science-based management [74]. Many contemporary foragers have developed management strategies to ensure long-term sourcing of the products. Longtime harvesters of *Allium tricoccum* (ramps, leeks) in the US may rotate harvest sites and select larger leafed plants to help manage the resource [36]. In some situations, such as berry harvesting in northern European countries, cultural harvesting practices have integrated with science-based knowledge to ensure management of the resources to supply contemporary demands for forest foods.

3.2. Managed Forests

Traditional forestry organizations, in general have not managed forests for food, although silvicultural practices, used for timber production, may provide opportunities to support food production needs [31,70,74]. The inclusion of food bearing plants in forest management can improve biodiversity conservation while addressing food security and nutrition. For example, spring ephemeral forest herbs, such as *Allium* spp., are harvested for food and contribute to nitrogen cycling; improving forest health while providing sustenance to humans. Managing forests for American ginseng, a medicinal plant, would require extending the rotation length for timber, reducing pressures from nonnative invasive plants, and encouraging growth of native understory plants [70,71]. Thinning forest stands may improve conditions for some food plants, such as berries, but may decrease production of others, such as mushrooms [75]. Managing forests for forest foods requires consideration of the impacts silvicultural treatments have on associated resources.

One of the rare examples of natural forests that are being managed primarily for forest foods are the walnut (*Juglans regia* L.) fruit forests in Kyrgyzstan. Due to the high number and genetic diversity of crop wild relatives that these forests harbor and their environmental importance for slope stabilization and equalization of water flows across seasons for downstream agricultural irrigation areas, these forests have been classified as protected forests since during Soviet times [76]. The production of walnut and other wild edible plants such as apples, plums, barberry, and rosehip were—and still are—important forest management objectives. In addition, limited quantities of timber and fuelwood were to be harvested through sanitary felling, regular thinning, and felling to induce forest rejuvenation according to management plans. It has been shown that careful silvicultural interventions can improve the stability and quality of stands and the production of walnut fruits [77]. However, due to weak enforcement of legal norms, and tremendous economic pressures, these forests have been continuously affected by uncontrolled and partially illegal timber exploitation, ineffective forest restoration measures, extraction of unsustainable amounts of fuelwood and commercial NTFP, and overgrazing [72]. This negatively affects the production of walnut and other forest foods.

Silvicultural prescriptions for forest foods require consideration of the impacts on resource availability [78,79]. Changing light and moisture regimes can have beneficial and detrimental impacts to forest plants desired for food. Creating gaps in the forest canopies can improve habitat for berries and other plants that like lots of sun. Conversely, gaps can increase the light getting to the forest floor to the detriment of shade loving plants. Some forest plants, such as hazelnuts (*Corylus* spp.), survive in the shade of forest canopies but need more light to produce fruit [31]. As production of maple syrup is directly related to the amount of foliage, trees managed for sap often are thinned and widely spaced to encourage large-crowned trees [80]. Although this may not be conducive to co-management for timber, it may present opportunities to grow understory food-producing plants, such as ramps in agroforestry.

3.3. Agroforests

A significant amount of wild foods come from agricultural lands that have some tree cover [6,29]. Agroforestry, a collection of food production systems, integrates woody perennial plants with agricultural production in spatial and temporal combinations with the intent of improving productivity, encouraging positive interactions within the system, without compromising future production potential [63]. Zomer et al. [81] estimated that over a billion ha worldwide of all agricultural lands were under some sort of agroforestry (i.e., agricultural lands with more than 10 percent tree cover) and ranging from the purposeful retention of naturally occurring trees on farms [82] to the establishment of multi-layered home gardens composed of annual and perennial herbs, shrubs, trees, and vines [83]. Zomer et al. [81] estimated that more than 96 percent of the agricultural lands in Central America were classified as agroforestry, while that proportion was over 80 percent in Southeast Asia and South America. Over the decade ending in 2010, Zomer et al. [81] estimated that agroforestry lands increased 3 percentage points, more than 82 million ha.

Agroforestry systems support food security and nutrition directly and indirectly through a number of ways. For one, they provide ecosystem services and positive livelihood impacts [82,84]. By integrating native trees in their production system farmers can generate additional income, directly benefit from nutritious food products, and increase their resilience to market or climatic shocks [85,86]. Furthermore, agroforestry can help maintain tree and associated biodiversity [87], reduce soil erosion, and improve soil characteristics, which can increase crop yield and household food availability year-round [88,89]. As an example, traditional agroforestry systems in Indonesia are associated with more frequent consumption of healthy foods than natural forests or tree crop plantations, as they are known for the diversity of crops and management of wild foods [19]. Home gardens in Uganda significantly contribute to household nutrition security, compared to anticipated shifts in industrial agricultural systems envisioned by governmental strategies [90].

Agroforestry systems with native fruit trees and traditional forest foods can be of particular interest to achieve nutrition and food security objectives given that many forest fruits and plants are important sources of macro- and micronutrients. However, bottlenecks limit the benefits from indigenous fruits produced in traditional agroforestry systems [91], among them market insufficiencies and failures such as limited demand, inadequate supply, and marketing channels or supply control mechanisms [92–94]. One strategy to realize the potential of native fruit tree species in traditional agroforestry systems to enhance food security, livelihoods, and resilience for future challenges, is to stimulate emergence of markets and development and promotion of businesses and innovations to meet the arising market demand [95].

3.4. Orchards and Plantations

Single species plantings of food-producing trees provide significantly to security and nutritional needs. Hundreds of tree species are cultivated by households for direct consumption and for sale. Smith et al. [96] estimated that more than 170 tropical and subtropical tree species are under cultivation. A subset of these are grown for international commodities markets, and a few are being grown for these markets on a large scale by small landholders [97].

Many non-timber forest species undergo a "dynamic process of domestication" that transgresses their collection from the wild and often includes their simultaneous cultivation on farms [98]. Certainly, there is a progression of sourcing forest foods that originate in natural habitats of gathering fruits and proceeds to growing fruit around homes. When production of home grown fruit can no longer supply peoples' desires, trees would be planted in clusters, or orchards and plantations. For example, for baobab, it has now been shown that its density is often higher in villages and fields in contrast to natural plains [99]. While a number of NTFP of high commercial value are grown in intensively managed plantations (e.g., rubber, oil palm, walnuts) and, hence, commonly considered as agricultural crops rather than forest products, some of these products are predominantly still being produced by small-scale and/or non-industrial producers. For example, in some parts of China, approximately 90 percent of the walnut plantations were managed by smallholder farmers [100]. With a total production volume of 1.6 million tons of walnuts with shell harvested from approximately 390 thousand ha of walnut plantations in 2018 [101], China has displaced the US as the leading global walnut producer already since the 1990s. The cultivation of walnuts in smallholder orchards and plantations provides important livelihoods, contributes to household incomes, and is significant in terms of food production in the Central Asian highlands and many of the countries along the Silk Road [102]. There is great potential to increase the contribution of such smallholder-managed plantations to global food production by improving management for this and other food bearing trees. For example, the largely non-industrial walnut plantations in India yielded only 0.84 t/ha [103]; in China, yields were about 2.3 t/ha, and in Iran, yields were about 2.5 t/ha [104], while well-managed industrial walnut plantations in California, U.S., can attain average yields of 4.2 t/ha. This yield gap is typically caused by a lack of advanced propagation techniques, root stocks, and cultivars, as well as lack or sub-optimal use of agronomic inputs such as irrigation, fertilizer, or pesticides.

3.5. Urban and Peri-Urban

A small portion of the estimated 600,000 km^2 [105] of urban and peri-urban land in the world produces over 15 percent of the world's food [106]. In the US alone, more than 6400 km^2 of urban land could be planted with food bearing trees and contribute to food production systems [106]. A diversity of urban habitats, such as city parks, campuses, street trees, urban woodlots, cemeteries, residential yards, and allotment gardens, provide plant material and fungi to urban foragers [107]. While urban foragers may depend on these food resources to varying degrees, from occasional recreational use to regular livelihood supplement, collection of foods from urban environments embodies important aspects of urban well-being [107]. Estimates suggest that daily recommended minimum intake of fruit could be achieved via planting fruit trees in publicly accessible open spaces [106]. Despite its importance, the foraging of food products on public lands is often neglected, heavily regulated, or even prohibited in many places [107–109].

At the same time that urban lands could be sources of tree foods, cities are major pollution hotspots, and hence, urban food security aspects need to be balanced with concern for food safety. Of particular relevance is the uptake and translocation of pollutants such as heavy metals from contaminated soils or airborne metal emissions to plant edible parts [110]. For example, wild mushrooms collected from urban habitats in Berlin, Germany, accumulated extremely high amounts of lead and cadmium and did not meet the EU standards for these trace metals in 86 percent and 54 percent of the cases [111]. Yet, circumstantial evidence from the San Francisco East Bay [112], in the US, indicated that wild edible plants grown in soils with elevated heavy metal concentrations in high-traffic industrial urban areas were safe to eat after rinsing in tap water. Yet, such concerns are not limited to urban and peri-urban areas as contamination of wild fish, meat, and plants with heavy metals can occur in natural ecosystems, whence these resources inhabit due to geological conditions, such as the Kamchatka region in the Russia Far East [113].

4. Realigning Food Systems for Forests and Trees

Clearly, forests contribute much to the provisioning of food, for direct consumption, and indirectly for income generation to purchase food. Markets for forests foods, such as baobab, range from local-based that supply households or communities, to global and corporate-based, serving multiple countries. Unlike agricultural and timber-based commodities, the value chains for forest foods are less developed or understood [114]. They may supply segments of industries (e.g., pharmaceuticals, nutraceuticals, natural foods, health foods) that are enigmatic to agricultural or forestry institutions that set policies, track production, and promote innovation and market opportunities. As such, many of these products remain invisible in national and international governance and lacking in resource management and production. [30]. Integration into, and acceptance by, appropriate institutions could increase visibility of their importance. Approval of baobab fruit pulp as a safe novel food would open European markets to the export of this African-grown product [115]. International discussions such as on the acceptance and use of "novel foods", or the related access rights and benefit sharing mechanisms with local communities under the Nagoya protocol can contribute [116,117] to increasing the attention on the importance of forests for these "other" resources. Fostering a transition to an integrated food system that includes forests and trees, involves innovative new business models and products, and markets that are driven by progressive consumer awareness and demand, with governance by stakeholders at all levels, and evidence based sustainable forest management and production (Figure 2). In essence, these elements support a framework for transitioning to a bioeconomy that can meet increasing global demand for healthy and nutritious food and increasing concerns for biodiversity conservation.

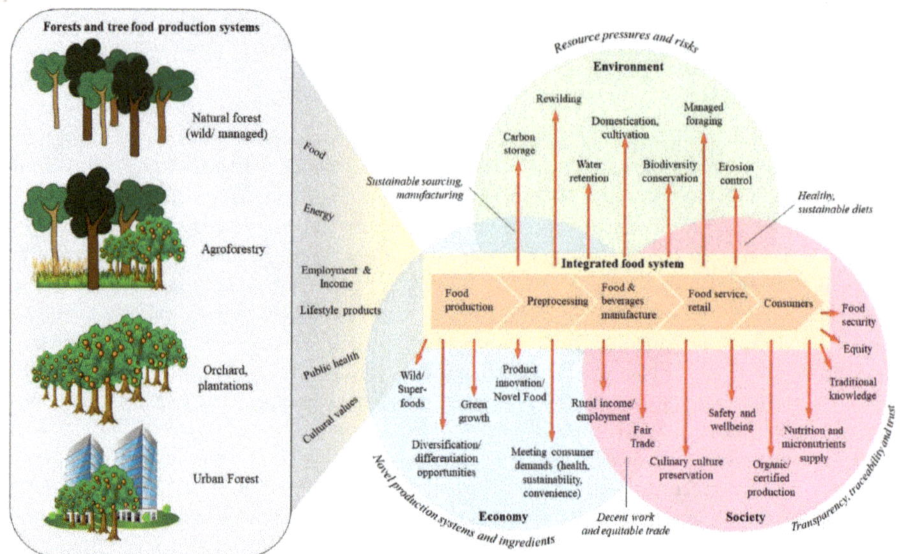

Figure 2. Integration of forests and trees into food systems has many elements that lead to a framework for transitioning to a bioeconomy. Source: modified from food system framework as presented by the Institute of Food Science and Technology [118].

4.1. Shift to a Bioeconomy

Realignment of forest production systems and management to integrate into food systems will require a fundamental change in priorities to include edible plants and fungi (and animals). In a bioeconomy, the intrinsic values of forests to supply healthy and nutritious natural foods are recognized, meaning that even though the economic value of forest foods may be less than for other resources, the contribution of forests to food security and nutrition are integrated into economic planning and resource management. A fundamental element to support transitioning to a bioeconomy is full recognition that edible forest resources are natural resources that require management efforts similar to other forest products. With this awareness, a pathway from status quo to a more progressive bio-based economy can be charted.

The bioeconomy is emerging as a major strategic economic movement of the 21st century [119–122]. The motivation driving this was originally oriented on biotechnology to replace fossil fuels with biofuels, particularly wood fuels, and other alternative energy. Ensuring and producing healthy and safe food are viewed by some as central objectives of a bioeconomy [123,124]. The path to a bioeconomy will touch all levels of society, from consumers to producers. It is based on sustainably and equitably sourcing products, with little or no negative environmental impacts, and supported with strong science-based knowledge. Figure 2 illustrates elements to the three foundational pillars (i.e., economy, society, ecology) of sustainable bio-based economies, and the contributions from forests and trees in integrated food systems.

Although, they have been, and continue to be, a major part of the global economy, edible forest products remain under-valued in the forest-based economy. For the most part, this economic model includes primarily wood-based products. However, a great deal of forest biodiversity is harvested for non-wood products that contribute to bio-based economies. Forest foods, such as from baobab, provide opportunities for transitioning to a bioeconomy that serves diverse markets, innovates to supply sustainably sourced products, and incorporates arrangements that promote Fair Trade and other social-equity systems. Investment in "green" market infrastructure to improve production, processing, and distribution may help to facilitate a transition to such a model.

4.2. Forests and Trees as Provisioning Units

Forest production systems and their management to include foods will expand beyond "traditional" objectives in transitioning to a bioeconomy. Management decisions, analyses, plans, and actions will expand to include food production possibilities. The general concept of multiple-use forest management [125] would expand to include food products and associated services. The ecosystem management paradigm [126,127] would include the manipulation of forests for production of food. In a transition to a bioeconomy, forest production systems expand to include societal costs and benefits related to more healthy food consumption, optimizing timber production, while providing environmental services to society.

In a bioeconomy, forest foods are produced in a variety of systems; forests that are open to wild harvesting, forests managed for food, agroforests, urban and peri-urban forested landscapes, and where appropriate, single-species plantings. They blend into diverse mosaics of habitats across landscapes. This would include trees outside of traditionally perceived forests, across a spectrum of land uses to natural forest settings that may serve as refugia for biodiversity. For most efficient production, single-species plantings would be strategically located, while agroforests would blend into the landscape, providing food and other products for personal consumption and to supplement incomes. Urban trees would be managed by the people and communities where they live.

While sustainably harvested for local consumption over generations, more intensive utilization of some forest food species may be related to significant negative impacts on forest ecosystems and biodiversity. For this reason, a transition to a bioeconomy may entail domestication and cultivation in more intensively managed production systems outside natural forests. In a bioeconomy these and other systems would adhere to evidence-based production practices to supply sustainably sourced products.

4.3. Consumer Demand Pushing Innovative Business Models

Ultimately, consumer awareness, demand, and support will drive a shift to a bioeconomy. The growing population of urban consumers who want sustainably harvested, all-natural, locally, and justly produced products are creating the demand for such a shift [38,44,128,129]. These consumers are more likely to demand Fair Trade products that are certified sustainably and organically sourced. Consumer demand for forest foods that are socially, economically, and ecologically respectful as well as increasing regulatory push will spur industry to supply such products. Social marketing [130,131] to this demographic could generate energy to transition from the status quo. The more that consumers demand these kinds of products, the more motivation the industry has to innovate.

Policy actions at the global and national levels could encourage industry shifts more expeditiously. A transition to a bioeconomy may spur new and creative business models (Figure 2) and investment in product innovation and technology development to satisfy growing demand for healthy, sustainable, and fairly sourced foods. Identification, domestication, and cultivation of promising species, planted in appropriate systems to satisfy the growing demand for healthy, sustainable, and fairly sourced foods are elemental for a transition. Further, innovation in processing, distribution and marketing, and other elements of value chains may stimulate transitions in local and global economies.

4.4. Innovation Opportunities

A diverse range of sectors (e.g., pharmaceuticals, nutraceuticals) in the functional food industry, companies that manufacture natural and homeopathic preparations, are constantly searching for new, as yet uncharacterized, plant-based raw materials. They are looking for natural substances that can replace artificial ingredients to extend product shelf life, improve technological properties, or have other properties that are valued by consumers (e.g., taste), which do not have to be declared on food labels due to their natural origin, which will help to expand their product portfolios and retain competitive advantage. Phytochemicals contained in trees and forest plants provide significant opportunities for development of new food products [128,129]. In view of the trend towards healthy nutrition and the growing demand for food products containing natural ingredients, new, natural food colorants, flavors, and functional ingredients are gaining importance. Natural ingredients of plant origin, which are much better accepted, or perceived as adding value, by critical and increasingly well-informed consumers, are of particular interest to a progressive food industry [44,57]. For example, natural colorants derived from fruits, such as anthocyanins, betalains, carotenoids, and chlorophylls possess bioactive properties in addition to their coloring ability [132]. In contrast, synthetic additives are sometimes suspected of causing allergic reactions or triggering ADHD syndrome in children [133]. Many wild products however are hidden ingredients in food, pharmacological, and cosmetic industries and are not sustainably or ethically sourced and, in some cases, contributing to the depletion of valuable species.

Forest biodiversity provides an unlimited reservoir of potentially useful substances, as the phytochemical composition of most forest plant species has yet to be thoroughly characterized. As an example, extracts from the male flower of sweet chestnut (*Castanea sativa* Mill.) used as an alternative to potassium sorbate in pastry products increased contents of reducing agents and radical scavengers, while maintaining the nutritional and chemical profiles of the pastry [134]. So-called superfruits such as Acai (*Euterpe oleracea* Martius), Acerola (*Malphigia emarginata* D.C.), Camu-camu (*Myrciaria dubia* (Kunith) McVaugh), Goji berry (*Lycium barbarum* L.), Jaboticaba (*Myrciaria cauliflora* (Mart.) O. Berg), Jambolão (*Syzygium cumini* (L.) Skeels)), Maqui (*Aristotelia chilensis* (Molina) Stuntz), Noni fruit (*Morinda citrifolia* L.), and Pitanga (*Eugenia uniflora* L.) contain a variety of phytochemicals (e.g., phenolic acids, flavonoids, proanthocyanidins, iridoids, coumarins, hydrolysable tannins, carotenoids, and anthocyanins) that have promising health benefits to humans [135]. Clinical trials have confirmed many of these effects in vitro and in vivo. For example, human intervention studies suggest that the juice of the noni berries from French Polynesia may provide a number of health benefits, including protection against tobacco smoke-induced DNA damage, blood lipid and homocysteine elevation, and systemic inflammation [136]. Baobab-enriched bread has been shown to reduce starch digestion and glycemic response in humans [137].

A re-emerging trend shows growing interest in foraging forest foods, described by Wiersum [38] as a relatively new multidimensional phenomenon, which involves biological processes of "re-wildering" and new culturally embedded interactions between people and nature. Some key examples of ingredients from forests and tress include baobab, moringa (*Moringa oleifera* L.), shea butter, argan (*Argania spinose* L.), and pygeum (*Prunus africana* (Hook. F.) Kalkman). All are sourced to supply multiple, billion-dollar industries that serve growing demand for "natural" food production. A transition to an integrated food system would leverage the true potential of forest foods to satisfy global food production, including changing food-related preferences and behaviors of consumers. It would acknowledge the importance of forest foods vis-à-vis agricultural food production; the structure and governance of forest food value chains; and quality and food safety aspects of provisioning forest food.

4.5. Processing, Marketing, Distribution

Transitioning to a bioeconomy may involve changes to the processing, marketing, and distribution of forest foods. Distribution channels range from traditional, local markets to highly sophisticated global supply chains. Forest foods such as maple syrup or acai and baobab powder are processed as ingredients in a wide range of food products and can be found in global markets. Most forest foods, however, are still sold unprocessed and undergo limited value-added processing. More local processing to add value to products could, however, have positive effects on income and profits [138,139]. The distribution of benefits, however, depends greatly on the value chain structure and governance, as middlemen or elite traders may inhibit smallholder producers from realizing more benefits. Processing may contribute to preserving forests foods, so that they can be distributed to more distant markets, although processing may impact the nutritional qualities of the products. Many forests foods are available only in local or regional markets with relatively short distribution chains. Reaching high-value export markets may be a worthwhile strategy for some forest foods, but these markets may be difficult for smallholder producers to enter and may be very risky for less affluent producers. A focus on local markets is linked to consumer preference for locally produced food products traded through short supply chains [140]. Hence, local market development that benefits small producers would be an important element for a bioeconomy strategy [140].

Different approaches have been used, such as Fair Trade, Fair Wild, and organic certification, to ensure higher sustainability in the production and equitable sharing of benefits from utilization of natural resource. Studies show that consumers are willing to purchase and to pay premium prices for food products with attributes such as local production, organic or sustainable production, social fairness, or geographical indication [42,141]. The provision of health information on packaging labels can increase consumers' willingness to pay [42,142]. The comparatively high costs involved in obtaining certification, however, hinders their wider acceptance, particularly for small-scale producers and companies. To encourage participation and transition to a bioeconomy, some organizations are providing cost-share programs and other incentives to reduce financial burdens of certification.

Special considerations for food quality and safety are needed to target some more advanced markets. To integrate forest foods into agricultural-based food systems will require similar quality and safety standards. A few forest foods, such as shea and Arabic gum, have associated standards under the Codex Alimentarius International Food Standards [143]. Establishing standards may be complicated by high intra-species variety within wild forest foods and inferior product attributes (e.g., size or sugar content) compared to cultivated products [144]. Forests foods typically do not undergo quality checks for local markets similar to agricultural production practices, which may raise concerns considering food safety (e.g., microbiological contamination due to insufficient post-harvest handling, risk of zoonosis in the case of bushmeat), and may limit access to export markets. Such issues may counteract positive effects from nutritional composition of forest foods. As noted, consumer awareness and preferences for healthy, natural, and sustainably sourced food, quality, and safety of forest foods could be a major driver in transitioning to a bio-based economy

4.6. Knowledge and Governance

Integrated food systems with forests and trees recognize the contributions of traditional and local knowledge and governance by the array of stakeholders. A bioeconomy strategy incorporates knowledge systems with science and policy to the design and implementation of sustainable food systems that includes stakeholder knowledge from all levels. Many food systems are based on traditional land-use practices, and some have been the subject of a great deal of scientific analyses. The combined knowledge can lead to better management for improved local conditions.

In general, forest governance in a bioeconomy strategy would be inclusive of all stakeholders. Stakeholders at all levels of society would be involved in conversations and dialogue about managing all forest resources. Management of the forest bioeconomy would be vertically integrated to include value chains with fair and equitable distribution of benefits. State agencies would partner with local communities and private entities to manage forests for the benefit of all. Governance of distribution channels and standards would ensure that products achieved an acceptable level of certification.

Finally, transitioning to a bioeconomy that integrates forest foods should be approached from a global justice perspective because a significant share of forest food species (current and potential) originate from tropical and subtropical ecosystems. Many consuming countries are economically advanced, yet source products from the Global South. The global nature of food systems obligates consuming countries wanting to transition to a bioeconomy to consider and mitigate the impacts of consumption of products from developing countries. Considering global inequality, this view proposes the pursuit of global egalitarian values rooted in humanitarianism and cosmopolitanism [145]. High-valued markets for healthy, sustainably produced lifestyle food products represent opportunities for entrepreneurial producers, although there is risk that a fad for such superfoods could end as another example of exploiting resources for short-lived consumer demands. Potential safeguards include access and benefit sharing agreements as stipulated by the Nagoya Protocol that aim to protect the customary knowledge and intellectual property rights of local communities in their resources [146]. Likewise, increasing the share of value-adding activities in the countries of origin notwithstanding the variety of logistical and technological challenges in current food supply chains can ensure that local producers and communities obtain a fair share of the benefits. Equally important will be to foster the use of locally produced forest food by local populations, which can contribute to the localization of food production.

5. Conclusions

Forests and trees are important sources of products for people's food security, nutrition, sovereignty, income/livelihoods, and culinary cultural identity. They provide fruit, nuts, vegetables, medicinal and aromatic plants, insects, and wild meat, rich in micro and macronutrients that many people lack. They are important to the basic sustenance of rural and urban, poor and affluent people, worldwide, who directly and indirectly consume non-timber forest products. Forests and trees provide "security nets" for food products during droughts and other environmentally stressful times. They enrich the culinary culture of people throughout. The diversity of spatial and temporal factors that affect forests and trees in providing food products needs full consideration to reintegrate and expand these resources into food systems.

Foraging forest foods is expanding globally, and this trend is expected to continue as populations increase and consumer demand evolves. This use of forests increases pressures on biodiversity, putting excess strain on plant populations that are impacted by gathering products. While the harvest of some plant organs (e.g., fruits and nuts) may not impact populations, the harvest of other organs (e.g., roots, tubers) can have significant impact on sustainability.

At the same time, foraging is supporting a renaissance among urbanites who are a driving force in global markets for sustainably sourced forest foods, which requires scrutiny and verification through appropriate certification. With continued education of an expanding demographic worldwide demand for sustainably produced forest foods can be expected to grow. Unfortunately, there are public health concerns, such as transmission of viruses and disease from animals to humans, associated with consumption of forests foods.

Forest foods are produced in myriad of systems, from wild-harvesting natural populations to intense single-species plantings. Over generations, people living near forests, and dependent on them for sustenance, have developed formal and informal management practices that could support more prescription management strategies of open-access or communal forest lands to support sustainable sourcing of much needed foods. There are opportunities across rural and urban landscapes to integrate trees into food systems and support transition to a bioeconomy. Agroforestry systems with native fruit trees and traditional forest foods can contribute important macro- and micronutrients needed to achieve nutrition and food security objectives. Orchards and plantations are important sources of large volumes of nutritious foods, although loss of biodiversity by converting natural forests to single plantings must be fully considered and avoided. The spectrum of production systems will be required to achieve food security and nutrition.

Realigning forest production systems to a bioeconomy model is a process that may take decades. In some perspectives, that process has been underway for many years, as is evident by efforts to ensure Fair Trade, certification of production, and sustainability of sourcing. It will require involvement of all stakeholders, from individual farmers to international level policy makers. Providing forest foods to local and external markets will require integration of forests and trees across a mosaic of production landscapes. Innovations and technology development are essential for a transition to a bioeconomy with forest foods. Ultimately, transition to a bioeconomy where forests and trees are integral to food systems will be driven by consumer demand, which can be influenced through social marketing and education that is supported by research on such matters as phytochemistry, traditional practices, and sustainable production methods.

Author Contributions: All authors contributed to the preparation of this manuscript. J.L.C. conceived the concept for the article and invited the other authors to contribute because of their international expertise. J.L.C., D.D., and K.M., contributed to the initial draft of the manuscript, and all subsequent drafts. All authors have read and agreed to the published version of the manuscript.

Funding: This research received no external funding.

Acknowledgments: At the 2019 World Forest Congress in Curitiba, Brazil, the Board of the International Union of Forestry Research Organizations (IUFRO) commissioned the global task force *"Unlocking the Bioeconomy and Nontimber Forest Products"*. The authors on this manuscript are members of the Task Force and represent two of the more than 25 institutions from over 23 countries involved in this effort. The expressed views of the authors do not necessarily represent that of their respective institutions. The task force is investigating how non-timber forest products can be integrated into global and national efforts to transition to a bioeconomy. The authors acknowledge the contribution of Giulia Muir, a Task Force member with UN FAO, to the earliest version of this manuscript, who due to COVID-19 was unable to continue adding to further development of this paper.

Conflicts of Interest: No conflicts of interest have been identified.

References

1. Green, R.E.; Cornell, S.J.; Scharlemann, J.P.W.; Balmford, A. Farming and the Fate of Wild Nature. *Science* **2005**, *307*, 550–555. [CrossRef] [PubMed]
2. FAO. *Biodiversity for Food and Agriculture: Contributing to Food Security and Sustainability in a Changing World*; PAR Platform, FAO: Rome, Italy, 2011.
3. Jaggard, K.W.; Qi, A.; Ober, E.S. Possible changes to arable crop yields by 2050. *Philos. Trans. R. Soc. B Biol. Sci.* **2010**, *365*, 2835–2851. [CrossRef] [PubMed]
4. Iizumi, T.; Furuya, J.; Shen, Z.; Kim, W.; Okada, M.; Fujimori, S.; Hasegawa, S.; Nishimori, M. Responses of crop yield growth to global temperature and socioeconomic changes. *Sci. Rep.* **2017**, *7*, 7800. [CrossRef] [PubMed]
5. Bahar, N.H.; Lo, M.; Sanjaya, M.; Van Vianen, J.; Alexander, P.; Ickowitz, A.; Sunderland, T. Meeting the food security challenge for nine billion people in 2050: What impact on forests? *Glob. Environ. Chang.* **2020**, *62*, 102056. [CrossRef]
6. FAO. *The State of the World's Forests 2018: Forest Pathways to Sustainable Development*; FAO: Rome, Italy, 2018.
7. Angelsen, A.; Jagger, P.; Babigumira, R.; Belcher, B.; Hogarth, N.J.; Bauch, S.; Börnder, J.; Smith-Hall, C.; Wunder, S. Environmental Income and Rural Livelihoods: A Global-Comparative Analysis. *World Dev.* **2014**, *64*, S12–S28. [CrossRef]
8. Pimentel, D.; McNair, M.; Duck, L.; Pimentel, M.; Kamil, J. The value of forests to world food security. *Hum. Ecol.* **1997**, *25*, 91–120. [CrossRef]
9. Penafiel, D.; Lachat, C.; Espinel, R.; van Damme, P.; Kolsteren, P. A systematic review on the contributions of edible plant and animal biodiversity to human diets. *EcoHealth* **2011**, *8*, 381–399. [CrossRef] [PubMed]
10. Ickowitz, A.; Powell, B.; Salim, M.A.; Sunderland, T.C. Dietary quality and tree cover in Africa. *Glob. Environ. Chang.* **2014**, *24*, 287–294. [CrossRef]
11. Nykänen, E.A.; Dunning, H.E.; Aryeetey, R.N.O.; Robertson, A.; Parlesak, A. Nutritionally optimized, culturally acceptable, cost-minimized diets for low income Ghanaian families using linear programming. *Nutrients* **2018**, *10*, 461. [CrossRef] [PubMed]
12. Mollee, E.; Pouliot, M.; McDonald, M.A. Into the urban wild: Collection of wild urban plants for food and medicine in Kampala, Uganda. *Land Use Policy* **2017**, *63*, 67–77. [CrossRef]
13. Aworh, O.C. From lesser-known to super vegetables: The growing profile of African traditional leafy vegetables in promoting food security and wellness. *J. Sci. Food Agric.* **2018**, *98*, 3609–3613. [CrossRef] [PubMed]
14. Nero, B.F.; Kwapong, N.A.; Jatta, R.; Fatunbi, O. Tree Species Diversity and Socioeconomic Perspectives of the Urban (Food) Forest of Accra, Ghana. *Sustainability* **2018**, *10*, 3417. [CrossRef]
15. Sneyd, L.Q. Wild food, prices, diets and development: Sustainability and food security in urban Cameroon. *Sustainability* **2013**, *5*, 4728–4759. [CrossRef]
16. Schlesinger, J.; Drescher, A.; Shackleton, C.M. Socio-spatial dynamics in the use of wild natural resources: Evidence from six rapidly growing medium-sized cities in Africa. *Appl. Geogr.* **2015**, *56*, 107–115. [CrossRef]

17. Cloete, P.C.; Idsardi, E.F. Consumption of indigenous and traditional food crops: Perceptions and realities from South Africa. *Agroecol. Sustain. Food Syst.* **2013**, *37*, 902–914. [CrossRef]

18. HLPE. *Sustainable Forestry for Food Security and Nutrition: A Report by the High Level Panel of Experts on Food Security and Nutrition of the Committee on World Food Security*; FAO: Rome, Italy, 2017.

19. Ickowitz, A.; Rowland, D.; Powell, B.; Salim, M.A.; Sunderland, T. Forests, trees, and micronutrient-rich food consumption in Indonesia. *PLoS ONE* **2016**, *11*, e0154139. [CrossRef]

20. Chamberlain, J.; Teets, A.; Kruger, S. *Nontimber Forest Products in the United States: An Analysis for the 2015 National Sustainability Report*; General Technical Report SRS-229; U.S. Department of Agriculture, Forest Service, Southern Research Station: Asheville, NC, USA, 2018; 36p.

21. Lovrić, M.; Da Re, R.; Vidale, E.; Prokofieva, I.; Wong, J.; Pettenella, D.; Verkerk, P.J.; Mavsar, R. Non-wood forest products in Europe–A quantitative overview. *For. Policy Econ.* **2020**, *116*, 102175. [CrossRef]

22. Mwampamba, T.H.; Ghilardi, A.; Sander, K.; Chaix, K.J. Dispelling common misconceptions to improve attitudes and policy outlook on charcoal in developing countries. *Energy Sustain. Dev.* **2013**, *17*, 75–85. [CrossRef]

23. Brobbey, L.K.; Hansen, C.P.; Kyereh, B.; Pouliot, M. The economic importance of charcoal to rural livelihoods: Evidence from a key charcoal-producing area in Ghana. *For. Policy Econ.* **2019**, *101*, 19–31. [CrossRef]

24. Scheid, A.; Hafner, J.; Hoffmann, H.; Kächele, H.; Sieber, S.; Rybak, C. Fuelwood scarcity and its adaptation measures: An assessment of coping strategies applied by small-scale farmers in Dodoma region, *Tanzania*. *Environ. Res. Lett.* **2018**, *13*, 095004. [CrossRef]

25. Amare, D.; Mekuria, W.; Wondie, M.; Teketay, D.; Eshete, A.; Darr, D. Wood Extraction among the Households of Zege Peninsula, Northern Ethiopia. *Ecol. Econ.* **2017**, *142*, 177–184. [CrossRef]

26. Rehnus, M.; Nazarek, A.; Mamadzhanov, D.; Venglovsky, B.I.; Sorg, J.P. High demand for fuelwood leads to overuse of walnut-fruit forests in Kyrgyzstan. *J. For. Res.* **2013**, *24*, 797–800. [CrossRef]

27. Sola, P.; Schure, J.; Atyi, R.E.; Gumoo, D.; Okeyo, I.; AwoNo, A. Woodfuel policies and practices in selected countries in Sub-Saharan Africa—A critical review. *Bois For. Trop.* **2019**, *340*, 27–41. [CrossRef]

28. FAO—Food and Agriculture Organization of the United Nations. *State of the World's Forests: Enhancing the Socioeconomic Benefits from Forests*; FAO: Rome, Italy, 2014; 133p, ISBN 978-92-5-108269-0.

29. Powell, B.; Thilsted, S.H.; Ickowitz, A.; Termote, C.; Sunderland, T.; Herforth, A. Improving diets with wild and cultivated biodiversity from across the landscape. *Food Secur.* **2015**, *7*, 535–554. [CrossRef]

30. Sorrenti, S. *Non-Wood Forest Products in International Statistical Systems*; Non-Wood Forest Product Series No. 22; UN FAO: Rome, Italy, 2017.

31. Chamberlain, J.L.; Small, C.J.; Baumflek, M. Sustainable production of temperate and boreal nontimber forest products: Examples from North America. In *Achieving Sustainable Management of Boreal and Temperate Forests*; Stanturf, J.A., Ed.; Burleigh Dodds Science Publishing: Cambridge, UK, 2019. [CrossRef]

32. Anderson, M.K.; Lake, F.K. California Indian Ethnomycology and Associated Forest Management. *J. Ethnobiol.* **2013**, *33*, 33–85. [CrossRef]

33. Yasuoka, H. Long-term foraging expeditions (Molongo) among the baka hunter-gatheres in the northwestern Congo Basin, with special reference to the "wild yam question". *Hum. Ecol.* **2006**, *34*, 275–296. [CrossRef]

34. Katz, E.; Lopez, C.L.; Fleury, M.; Miller, R.P.; Paye, V.; Dias, T.; Silva, F.; Oliveira, Z.; Moreira, E. No greens in the forest? Note on the limited consumption of greens in the Amazon. *Acta Soc. Bot. Pol.* **2012**, *81*, 283–293. [CrossRef]

35. Landor-Yamagata, J.L.; Kowarik, I.; Fischer, L.K. Urban foraging in Berlin: People, plants and practices within the metropolitan green infrastructure. *Sustainability* **2018**, *10*, 1873. [CrossRef]

36. Baumflek, M.; Chamberlain, J.L. Ramps Reporting: What 70 years of popular media tells us about a cultural keystone species. *Southeast. Geogr.* **2019**, *59*, 77–96. [CrossRef]

37. Svanberg, I.; Lindh, H. Mushroom hunting and consumption in twenty-first century post-industrial Sweden. *J. Ethnobiol. Ethnomed.* **2019**, *15*, 42. [CrossRef] [PubMed]

38. Wiersum, K.F. New interest in forest products in Europe as an expression of biocultural dynamics. *Hum. Ecol.* **2017**, *45*, 787–794. [CrossRef] [PubMed]

39. Fischer, L.K.; Brinkmeyer, D.; Karle, S.J.; Cremer, K.; Huttner, E.; Seebauer, M.; Nowikow, U.; Schütze, B.; Voigt, P.; Völker, S.; et al. Biodiverse edible schools: Linking healthy food, school gardens and local urban biodiversity. *Urban For. Urban Green.* **2019**, *40*, 35–43. [CrossRef]

40. Barreau, A.; Ibarra, J.T.; Wyndham, F.S.; Kozak, R.A. Shifts in Mapuche Food Systems in Southern Andean Forest Landscapes: Historical Processes and Current Trends of Biocultural Homogenization. *Mt. Res. Dev.* **2019**, *39*, R12–R23. [CrossRef]

41. Ali, J.; Kapoor, S.; Moorthy, J. Buying behavior of consumers for food products in an emerging economy. *Br. Food J.* **2010**, *112*, 109–124. [CrossRef]

42. Moser, R.; Raffaelli, R.; Thilmany-McFadden, D. Consumer Preferences for Fruit and Vegetables with Credence-Based Attributes: A Review. *Int. Food Agribus. Manag. Rev.* **2011**, *14*, 121–142. [CrossRef]

43. Rodrigues, D.M.; Rodrigues, J.F.; Rios de Souza, V.; Souza Carneiro, J.D.; Vilela Borges, S. Consumer preferences for Cerrado fruit preserves: A study using conjoint analysis. *Br. Food J.* **2018**, *120*, 827–838. [CrossRef]

44. Grunert, K.G. (Ed.) *Consumer Trends and New Product Opportunities in the Food Sector*; Wageningen Academic Publishers: Wageningen, The Netherlands, 2017.

45. Sardeshpande, M.; Shackleton, C. Wild edible fruits: A systematic review of an under-researched multifunctional NTFP (non-timber forest product). *Forests* **2019**, *10*, 467. [CrossRef]

46. Kuznesof, S.; Tregear, A.; Moxey, A. Region foods: A consumer perspective. *Br. Food J.* **1997**, *99*, 199–206. [CrossRef]

47. Garekae, H.; Shackleton, C.M. Foraging wild food in urban spaces: The contribution of wild foods to urban dietary diversity in South Africa. *Sustainability* **2020**, *12*, 678. [CrossRef]

48. Lovrić, M.; Lovrić, N.; Mavsar, R. Mapping forest-based bioeconomy research in Europe. *For. Policy Econ.* **2019**, *110*, 101874. [CrossRef]

49. Mungofa, N.; Malongane, F.; Tabit, F.T. An exploration of the consumption cultivation and trading of indigenous leafy vegetables in rural communities in the greater Tubatse local municipality, Limpopo province, South Africa. *J. Consum. Sci.* **2018**, *3*, 53–67.

50. Maroyi, A. Potential role of traditional vegetables in household food security: A case study from Zimbabwe. *Afr. J. Agric. Res.* **2011**, *6*, 5720–5728.

51. Darr, D.; Chopi-Msadala, C.; Namakhwa, C.D.; Meinhold, K.; Munthali, C. Processed baobab (*Adansonia igitate* L.) food products in Malawi: From poor men's to premium-price specialty food? *Forests* **2020**, *11*, 698. [CrossRef]

52. Gebauer, J.; Assem, A.; Busch, E.; Hardtmann, S.; Möckel, D.; Krebs, F.; Ziegler, T.; Wichern, F.; Wiehle, M.; Kehlenbeck, K.D. Baobab (*Adansonia digitata* L.): Wildobst aus Afrika für Deutschland und Europa?! *Erwerbs Obstbau* **2014**, *56*, 9–24. [CrossRef]

53. Holmborn, B. Xylitol: A healthy sweetener from birch wood. *Actual. Chim.* **2002**, *11–12*, 52–53.

54. Lim, T.K. *Edible Medicinal and Non-Medicinal Plants. Volume 1–6, Fruits*; Springer: Dordrecht, The Netherlands, 2012. [CrossRef]

55. Wells, J.C.; Sawaya, A.L.; Wibaek, R.; Mwangome, M.; Poullas, M.S.; Yajnik, C.S.; Demaio, A. The double burden of malnutrition: Aetiological pathways and consequences for health. *Lancet* **2020**, *395*, 75–88. [CrossRef]

56. Popkin, B.M. Nutrition transition and the global diabetes epidemic. *Curr. Diabetes Rep.* **2015**, *15*, 64. [CrossRef]

57. Lachance, J.C.; Radhakrishnan, S.; Madiwale, G.; Guerrier, S.; Vanamala, J.K.P. Targeting hallmarks of cancer with a food-system-based approach. *Nutrition* **2020**, *69*, 110563. [CrossRef]

58. Sammugam, L.; Pasupuleti, V.R. Balanced diets in food systems: Emerging trends and challenges for human health. *Crit. Rev. Food Sci. Nutr.* **2019**, *59*, 2746–2759. [CrossRef]

59. Martens, F. Akute und chronische Vergiftungen mit Digitalisglykosiden. *Notf. Hausarztmedizin* **2005**, *31*, A306–A309. [CrossRef]

60. Ellwanger, J.H.; Kulmann-Leal, B.; Kaminski, V.L.; Valverde-Villegas, J.M.; Da Veiga, A.B.G.; Spilki, F.R.; Fearnside, P.M.; Caesar, L.; Giatti, L.L.; Wallau, G.L.; et al. Beyond diversity loss and climate change: Impacts of Amazon deforestation on infectious diseases and public health. *An. Acad. Bras. Cienc.* **2020**, *92*, e20191375. [CrossRef] [PubMed]

61. Akem, E.S.; Pemunta, N.V. The bat meat chain and perceptions of the risk of contracting Ebola in the Mount Cameroon region. *BMC Public Health* **2020**, *20*, 593. [CrossRef] [PubMed]

62. Friant, S.; Ayambem, W.A.; Alobi, A.O.; Ifebueme, N.M.; Otukpa, O.M.; Ogar, D.A.; Alawa, C.B.I.; Goldberg, T.L.; Jacka, J.K.; Rothman, J.M. Eating Bushmeat Improves Food Security in a Biodiversity and Infectious Disease "Hotspot". *EcoHealth* **2020**, *17*, 125–138. [CrossRef] [PubMed]

63. Parrotta, J.A.; Dey de Pryck, J.; Obiri, B.D.; Padoch, C.; Powell, B.; Sandbrook, C.; Sandbrook, C. Chapter 3: Historical, Environmental and Socio-Economic Context of Forests and Tree-Based Systems for Food Security and Nutrition. In *Forests, Trees and Landscapes for Food Security and Nutrition: A Global Assessment Report*; IUFRO World Series Volume 33; Bhaskar, V., Wildburger, C., Mansourian, S., Eds.; IUFRO: Vienna, Austria, 2015; 172p.

64. Hladik, C.M.; Linares, O.F.; Hladik, A.; Pagezy, H.; Semple, A. Tropical forests, people and food: An overview. In *Tropical Forests, People and Food. Biocultural Interactions and Applications to Development*; Man and Biosphere Ser. 13; Hladik, C.M., Hladik, A., Linares, O.F., Pagezy, H., Semple, A., Hadley, M., Eds.; UNESCO: Paris, France; Parthenon: New York, NY, USA, 1993.

65. Boerboom, J.H.A.; Wiersum, K.F. Human impact on tropical moist forest. In *Man's Impact on Vegetation*; Holzner, W., Werger, M.J.A., Ikusima, I., Eds.; W Junk: The Hague, The Netherlands, 1983; pp. 83–106.

66. Sauer, C.O. *Agricultural Origins and Dispersals*, 2nd ed.; MIT Press: Cambridge, MA, USA; London, UK, 1969.

67. Garibaldi, A.; Turner, N. Cultural keystone species: Implications for ecological conservation and restoration. *Ecol. Soc.* **2004**, *9*. [CrossRef]

68. Silva, R.R.; Gomes, L.J.; Albuquerque, U.P. Plant extractivism in light of game theory: A case study in northeastern Brazil. *J. Ethnobiol. Ethnomed.* **2015**, *11*, 6. [CrossRef]

69. Anderson, K. *Tending the Wild: Native American Knowledge and the Management of California's Natural Resources*; University of California Press: Berkeley, CA, USA, 2005; 588p, ISBN 9780520280434.

70. Chamberlain, J.L.; Emery, M.R.; Patel-Weynand, T. (Eds.) *Assessment of Nontimber Forest Products in the United States under Changing Conditions*; General Technical Report SRS–232; U.S. Department of Agriculture, Forest Service, Southern Research Station: Asheville, NC, USA, 2018; 260p.

71. Schmidt, J.P.; Cruse-Sanders, J.; Chamberlain, J.L.; Ferreira, S.; Young, J.A. Explaining harvests of wild-harvested herbaceous plants: American ginseng as a case study. *Biol. Conserv.* **2019**, *231*, 139–149. [CrossRef]

72. Tieguhong, J.C.; Ingram, V.; Mala, W.A.; Ndoye, O.; Grouwels, S. How governance impacts non-timber forest product value chains in Cameroon. *For. Policy Econ.* **2015**, *61*, 1–10. [CrossRef]

73. Wynberg, R.P.; Laird, S.A. Less is often more: Governance of a non-timber forest product, marula (*Sclerocarya birrea* subsp. caffra) in southern Africa. *Int. For. Rev.* **2007**, *9*, 475–490. [CrossRef]

74. Chamberlain, J.; Small, C.; Baumflek, M. Sustainable Forest Management of Nontimber Forest Products. *Sustainability* **2019**, *11*, 2670. [CrossRef]

75. Pilz, D.; Molina, R.; Mayo, J. Effects of thinning young forests on chanterelle mushroom production. *J. For.* **2006**, *104*, 9–14.

76. Schmidt, M. *Mensch und Umwelt in Kirgistan: Politische Ökologie im postkolonialen und postsozialistischen Kontext*; Erdkundliches Wissen Band 153; Franz Steiner: Stuttgart, Germany, 2013.

77. Sorg, J.P.; Urech, Z.L.; Mamadzhanov, D.; Rehnus, M. Thinning effects on walnut stem and crown diameter growth and fruiting in the walnut-fruit forests of Kyrgyzstan. *J. Mt. Sci.* **2016**, *13*, 1558–1566. [CrossRef]

78. Ellum, D. Demographic Patterns and Disturbance Responses of Understory Vegetation in a Managed Forest of Southern New England: Implications for Sustainable Forestry and Biodiversity. Ph.D. Dissertation, Yale University, New Haven, CT, USA, 2007; 235p.

79. Duguid, M.C.; Frey, B.R.; Ellum, D.S.; Kelty, M.; Ashton, M.S. The influence of ground disturbance and gap position on understory plant diversity in upland forests of southern New England. *For. Ecol. Manag.* **2013**, *303*, 148–159. [CrossRef]

80. Pierce, A. Sugar Maple (*Acer saccharum*). In *Tapping the Green Market: Certification and Management of Non-Timber Forest Products*; Laird, S.A., Guillen, A., Pierce, A.R., Eds.; Earthscan Publications: Sterling, VA, USA, 2002; pp. 162–171.

81. Zomer, R.J.; Trabucco, A.; Coe, R.; Place, F.; van Noordwijk, M.; Xu, J.C. *Trees on Farms: An Update and Reanalysis of Agroforestry's Global Extent and Socio-Ecological Characteristics*; Working Paper 179; World Agroforestry Centre (ICRAF) Southeast Asia Regional Program: Bogor, Indonesia, 2014. [CrossRef]

82. Amare, D.; Wondie, M.; Mekuria, W.; Darr, D. Agroforestry of smallholder farmers in Ethiopia: Practices and benefits. *Small Scale For.* **2019**, *18*, 39–56. [CrossRef]

83. Whitney, C.W.; Luedeling, E.; Tabuti, J.R.S.; Nyamukuru, A.; Hensel, O.; Gebauer, J.; Kehlenbeck, K. Crop diversity in homegardens of southwest Uganda and its importance for rural livelihoods. *Agric. Hum. Values* **2018**, *35*, 399–424. [CrossRef]

84. Assogbadjo, A.E.; Glèlè, K.R.; Vodouhê, F.G.; Djagoun, C.A.M.S.; Codjia, J.T.C.; Sinsin, B. Biodiversity and socioeconomic factors supporting farmers' choice of wild edible trees in the agroforestry systems of Benin (West Africa). *For. Policy Econ.* **2012**, *14*, 41–49. [CrossRef]

85. Leakey, R.; van Damme, P. The role of tree domestication in green market product value chain development. *For. Trees Livelihoods* **2014**, *23*, 116–126. [CrossRef]

86. Reed, J.; van Vianen, J.; Foli, S.; Clendenning, J.; Yang, K.; MacDonald, M.; Petrokofsky, G.; Padoch, C.; Sunderland, T. Trees for life: The ecosystem service contribution of trees to food production and livelihoods in the tropics. *For. Policy Econ.* **2017**, *84*, 62–71. [CrossRef]

87. Fifanou, V.G.; Ousmane, C.; Gauthier, B.; Brice, S. Traditional agroforestry systems and biodiversity conservation in Benin (West Africa). *Agrofor. Syst.* **2011**, *82*, 1–13. [CrossRef]

88. Apuri, I.; Peprah, K.; Achana, G.T.W. Climate change adaptation through agroforestry: The case of Kassena Nankana West District, Ghana. *Environ. Dev.* **2018**, *28*, 32–41. [CrossRef]

89. Félix, G.F.; Diedhiou, I.; Le Garff, M.; Timmermann, C.; Clermont-Dauphin, C.; Cournac, L.; Groot, J.C.J.; Tittonell, P. Use and management of biodiversity by smallholder farmers in semi-arid West Africa. *Glob. Food Secur.* **2018**, *18*, 76–85. [CrossRef]

90. Whitney, C.W.; Tabuti, J.R.S.; Hensel, O.; Yeh, C.H.; Gebauer, J.; Luedeling, E. Homegardens and the future of food and nutrition security in southwest Uganda. *Agric. Syst.* **2017**, *154*, 133–144. [CrossRef]

91. Jamnadass, R.H.; Dawson, I.K.; Franzel, S.; Leakey, R.R.B.; Mithöfer, D.; Akinnifesi, F.K.; Tchoundjeu, Z. Improving livelihoods and nutrition in sub-Saharan Africa through the promotion of indigenous and exotic fruit production in smallholders' agroforestry systems: A review. *Int. For. Rev.* **2011**, *13*, 338–354. [CrossRef]

92. Gruère, G.; Giuliani, A.; Smale, M. *Marketing Underutilized Plant Species for the Benefit of the Poor: A Conceptual Framework*; EPT Discussion Paper #154; Intl Food Policy Res Inst.: Washington, DC, USA, 2006.

93. Leakey, R.R.B.; Tchoundjeu, Z.; Schreckenberg, K.; Shackleton, S.E.; Shackleton, C.M. Agroforestry Tree Products (AFTPs): Targeting Poverty Reduction and Enhanced Livelihoods. *Int. J. Agric. Sustain.* **2005**, *3*, 1–23. [CrossRef]

94. Meinhold, K.; Darr, D. The Processing of Non-Timber Forest Products through Small and Medium Enterprises—A Review of Enabling and Constraining Factors. *Forests* **2019**, *10*, 1026. [CrossRef]

95. Meinhold, K.; Darr, D. Using a multi-stakeholder approach to increase value for traditional agroforestry systems: The case of baobab (*Adansonia digitata* L.) in Kilifi, Kenya. *Agrofor. Syst.* **2020**, resubmitted.

96. Smith, N.J.H.; Williams, J.T.; Plucknett, D.L.; Talbot, J.P. *Tropical Forests and Their Crops*; Cornell University Press: Ithaca, NY, USA, 1992.

97. Watson, G.A. Tree crops and farming systems development in the humid tropics. *Exp. Agric.* **1990**, *26*, 143–159. [CrossRef]

98. Muir, G.F.; Sorrenti, S.; Vantomme, P.; Vidale, E.; Masiero, M. Into the wild: Disentangling non-wood terms and definitions for improved forest statistics. *Int. For. Rev.* **2020**, *22*, 101–119. [CrossRef]

99. Venter, S.M.; Witkowski, E.T.F. Baobab (*Adansonia digitata* L.) density, size-class distribution and population trends between four land-use types in northern Venda, South Africa. *For. Ecol. Manag.* **2010**, *259*, 294–300. [CrossRef]

100. Yan, M.; Terheggen, A.; Mithöfer, D. Who and what set the price of walnut for small scale farmers in Southwest China? *J. Agribus. Dev. Merging Econ.* **2017**, *7*, 135–152. [CrossRef]

101. FAOSTAT. Production Quantity and Area Harvested of Walnuts, with Shell (1961–2018). 2020. Available online: http://faostat.fao.org/site/567/DesktopDefault.aspx?PageID=567 (accessed on 24 June 2020).

102. Shigaeva, J.; Darr, D. On the socio-economic importance of natural and planted walnut (*Juglans regia* L.) forests in the Silk Road countries: A systematic review. *For. Policy Econ.* **2020**, accepted. [CrossRef]

103. Malhotra, S.P. *World Edible Nuts Economy*; Concept Publishing: New Delhi, India, 2008.

104. Vahdati, K.; Hassani, D.; Rezaee, R.; Jafari, M.H.; Sayadi, S.; Khorami, S. Following Walnut Footprint in Iran. In *Following Walnut Footprints (Juglans regia L.): Cultivation and Culture, Folklore and History, Traditions and Uses*; Avanzato, D., McGranahan, G.H., Vahdati, K., Botu, M., Iannamico, L., Assche, J.V., Eds.; Scripta Horticulturae (Scripta Horticulturae 17): Leuven, Belgium, 2014.

105. Angel, S.; Parent, J.; Civco, D.; Blei, A.; Potere, D. *A Planet of Cities: Urban Land Cover Estimates and Projections for All Countries, 2000–2050*; Lincoln Institute of Land Policy Working Paper; Lincoln Institute of Land Policy: Cambridge, MA, USA, 2010; 103p.

106. Clark, K.H.; Nicholas, K.A. Introducing urban food forestry: A multifunctional approach to increase food security and provide ecosystem services. *Landsc. Ecol.* **2013**, *28*, 1649–1669. [CrossRef]

107. Shackleton, C.; Hurley, P.T.; Dahlberg, A.C.; Emery, M.R.; Nagendra, H. Urban foraging: A Ubiquitous Human Practice Overlooked by Urban Planners, Policy, and Research. *Sustainability* **2017**, *9*, 1884. [CrossRef]

108. McLain, R.; Poe, M.; Hurley, P.; Lecompte-Mastenbrook, J.; Emery, M. Producing edible landscapes in Seattle's urban forest. *Urban For. Urban Green.* **2012**, *11*, 187–194. [CrossRef]

109. Poe, M.R.; McLain, R.J.; Emery, M.; Hurley, P.T. Urban forest justice and the rights to wild foods, medicines, and materials in the city. *Hum. Ecol.* **2013**, *41*, 409–422. [CrossRef]

110. Gori, A.; Ferrini, F.; Fini, A. Growing healthy food under heavy metal pollution load: Overview and major challenges of tree based edible landscapes. *Urban For. Urban Green.* **2019**, *38*, 403–406. [CrossRef]

111. Schlecht, M.T.; Säumel, I. Wild growing mushrooms for the Edible City? Cadmium and lead content in edible mushrooms harvested within the urban agglomeration of Berlin, Germany. *Environ. Pollut.* **2015**, *204*, 298–305. [CrossRef] [PubMed]

112. Stark, P.B.; Miller, D.; Carlson, T.J.; de Vasquez, K.R. Open-source food: Nutrition, toxicology, and availability of wild edible greens in the East Bay. *PLoS ONE* **2019**, *14*, e0202450. [CrossRef]

113. Shmatkov, N.; Brigham, T. Non-timber forest products in community development: Lessons from the Russian Far East. *For. Chron.* **2003**, *79*, 113–118. [CrossRef]

114. Ingram, V.; Ewane, M.; Ndumbe, L.N.; Awono, A. Challenges to governing sustainable forest food: *Irvingia* spp. from southern Cameroon. *For. Pol. Econ.* **2017**, *84*, 29–37. [CrossRef]

115. Buchmann, C.; Prehsler, S.; Hartl, A.; Vogl, C.R. The importance of baobab (*Adansonia digitata* L.) in rural West African subsistence—Suggestion of a cautionary approach to international market export of baobab fruits. *Ecol. Food Nutr.* **2010**, *49*, 145–172. [CrossRef]

116. Hermann, M. The impact of the European Novel Food Regulation on trade and food innovation based on traditional plant foods from developing countries. *Food Policy* **2009**, *34*, 499–507. [CrossRef]

117. Heinrich, M.; Scotti, F.; Andrade-Cetto, A.; Berger-Gonzalez, M.; Echeverria, J.; Friso, F.; Garcia-Cardona, F.; Hesketh, A.; Hitziger, M.; Maake, C.; et al. Access and Benefit Sharing Under the Nagoya Protocol—Quo Vadis? Six Latin American Case Studies Assessing Opportunities and Risk. *Front. Pharmacol.* **2020**, *11*, 765. [CrossRef] [PubMed]

118. ISFT. *Food System Framework: A Focus on Food Sustainability*; Institute of Food Science + Technology: London, UK, 2018. Available online: http://www.3keel.com/wp-content/uploads/reports/IFST%20Sustainable%20Food%20System%20Framework_0.pdf (accessed on 17 September 2020).

119. Duchesne, L.; Wetzel, S. The bioeconomy and the forestry sector: Changing markets and new opportunities. *For. Chron.* **2003**, *79*, 860–864. [CrossRef]

120. Dietz, T.; Börner, J.; Förster, J.J.; Braun, J.V. Governance of the Bioeconomy: A Global Comparative Study of National Bioeconomy Strategies. *Sustainability* **2018**, *10*, 3190. [CrossRef]

121. Hetemäki, L.; Hanewinkel, M.; Muys, B.; Ollikainen, M.; Palahí, M.; Trasobares, A. *Leading the Way to a European Circular Bioeconomy Strategy. From Science to Policy 5*; European Forest Institute: Joensuu, Finland, 2017; 52p. [CrossRef]

122. Pülzl, H.; Kleinschmit, D.; Arts, B. Bioeconomy—An emerging meta-discourse affecting forest discourses? Scandinavian. *J. For. Res.* **2014**, *29*, 386–393. [CrossRef]

123. Albrecht, S.; Gottschick, M.; Schorling, M.; Stint, S. Bio-economy at a crossroads. Way forward to sustainable production and consumption or industrialization of biomass? *Gaia Ecol. Perspect. Sci. Soc.* **2012**, *21*, 33–37.

124. Levidow, L.; Birch, K.; Papaioannou, T. Divergent paradigms of European agro-food innovation: The knowledge-based bio-economy (KBBE) as an R&D agenda. *Sci. Technol. Hum. Values* **2013**, *38*, 94–125.

125. Behan, R.W. Multiresource forest management: A paradigmatic challenge to professional forestry. *J. For.* **1990**, *88*, 12–18.

126. Galindo-Leal, C.; Bunnell, F.L. Ecosystem management: Implications and opportunities of a new paradigm. *For. Chron.* **1995**, *71*, 601–606. [CrossRef]

127. Lackey, R.T. Seven pillars of ecosystem management. *Landsc. Urban Plan.* **1998**, *40*, 21–30. [CrossRef]

128. Leakey, R.R.B. Potential for novel food products from agroforestry trees: A review. *Food Chem.* **1999**, *66*, 1–14. [CrossRef]

129. Devappa, R.K.; Rakshit, S.K.; Dekker, R.F.H. Forest biorefinery: Potential of poplar phytochemicals as value-added co-products. *Biotechnol. Adv.* **2015**, *33*, 681–716. [CrossRef] [PubMed]

130. Serrat, O. The future of social marketing. In *Knowledge Solutions*; Springer: Singapore, 2017; pp. 119–128.

131. McKenzie-Mohr, D. *Fostering Sustainable Behavior: An Introduction to Community-Based Social Marketing*; New Society Publishers: Gabriola Island, BC, Canada, 2011; 171p.

132. Albuquerque, B.R.; Oliveira, M.B.P.P.; Barros, L.; Ferreira, I.C.F.R. Could fruits be a reliable source of food colorants? Pros and cons of these natural additives. *Crit. Rev. Food Sci. Nutr.* **2020**. [CrossRef] [PubMed]

133. Classen, H.G.; Elias, P.S.; Hammes, W.P.; Winter, M. *Toxikologisch-Hygienische Beurteilung von Lebensmittelinhaltsstoffen und Zusatzstoffen*; Behr's Verlag: Hamburg, Germany, 2001.

134. Celeja, C.; Barros, L.; Barreira, J.C.M.; Soković, M.; Calhelha, R.C.; Bento, A.; Oliveira, M.B.P.P.; Ferreira, I.C.F.R. Castanea sativa male flower extracts as an alternative additive in the Portuguese pastry delicacy "pastel de nata". *Food Funct.* **2020**, *11*, 2208–2217. [CrossRef] [PubMed]

135. Chang, S.K.; Alasalvar, C.; Shahidi, F. Superfruits: Phytochemicals, antioxidant efficacies, and health effects—A comprehensive review. *Crit. Rev. Food Sci. Nutr.* **2019**, *59*, 1580–1604. [CrossRef] [PubMed]

136. West, B.J.; Deng, S.; Isami, F.; Uwaya, A.; Jensen, C.J. The Potential Health Benefits of Noni Juice: A Review of Human Intervention Studies. *Foods* **2018**, *7*, 58. [CrossRef]

137. Coe, S.A.; Clegg, M.; Armengol, M.; Ryan, L. The polyphenol-rich baobab fruit (*Adansonia digitata* L.) reduces starch digestion and glycemic response in humans. *Nutr. Res.* **2013**, *33*, 888–896. [CrossRef] [PubMed]

138. Aworh, C.O. Promoting food security and enhancing Nigeria's small farmers' income through value-added processing of lesser-known and under-utilized indigenous fruits and vegetables. *Food Res. Int.* **2015**, *76*, 986–991. [CrossRef]

139. Mahapatra, A.K.; Shackleton, C.M. Exploring the relationships between trade in natural products, cash income and livelihoods in tropical forest regions of Eastern India. *Int. For. Rev.* **2012**, *14*, 62–73. [CrossRef]

140. Shackleton, S.; Shanley, P.; Ndoye, O. Invisible but viable: Recognising local markets for non-timber forest products. *Int. For. Rev.* **2007**, *9*, 697–712. [CrossRef]

141. Botelho, A.l.; Dinis, I.; Lourenco-Gomes, L.; Moneira, J.; Pinto, L.C.; Simoes, O. The effect of sequential information on consumers' willingness to pay for credence food attributes. *Appetite* **2017**, *118*, 17–25. [CrossRef] [PubMed]

142. Hoke, O.; Campbell, B.; Brand, M.; Thao, H. Impact of Information on Northeastern U.S. Consumer Willingness to Pay for Aronia Berries. *HortSciences* **2017**, *52*, 395–400. [CrossRef]

143. Codex Alimentarius International Food Standards. 2020. Available online: http://www.fao.org/fao-who-codexalimentarius/codex-texts/list-standards/en/ (accessed on 21 September 2020).

144. Smanalieva, J.; Iskakova, J.; Oskonbaeva, Z.; Wichern, F.; Darr, D. Determination of physicochemical parameters, phenolic content, and antioxidant capacity of wild cherry plum (*Prunus divaricata* Ledeb.) from the walnut-fruit forests of Kyrgyzstan. *Eur. Food Res. Technol.* **2019**, *245*, 2293–2301. [CrossRef]

145. Landesman, B.M. Global Justice. In *Encyclopedia of Global Justice*; Chatterjee, D.K., Ed.; Springer: Dordrecht, The Netherlands, 2011.

146. Teran, M.Y. The Nagoya Protocol and Indigenous Peoples. *Int. Indig. Policy J.* **2016**, *7*. [CrossRef]

Publisher's Note: MDPI stays neutral with regard to jurisdictional claims in published maps and institutional affiliations.

 *forests*

Review

Evaluation of Wild Foods for Responsible Human Consumption and Sustainable Use of Natural Resources

Jeferson Asprilla-Perea [1], José M. Díaz-Puente [2,*] and Susana Martín-Fernández [3]

[1] Departamento de Biología, Universidad Tecnológica del Chocó "Diego Luis Córdoba",
Cra. 22 #18b-10 Neighborhood Nicolás Medrano, Ciutadella Universitaria,
270001 Quibdó, Chocó, Colombia; jeferson.asprilla@utch.edu.co

[2] Agricultural, Food and Biosystems Engineering School, Universidad Politécnica de Madrid,
Avda. Puerta de Hierro 2, 28040 Madrid, Spain

[3] Forestry and Natural Resources Engineering School, Universidad Politécnica de Madrid,
Ciudad Universitaria, 28040 Madrid, Spain; susana.martin@upm.es

* Correspondence: jm.diazpuente@upm.es; Tel.: +34-910-670-932

Received: 18 May 2020; Accepted: 16 June 2020; Published: 18 June 2020

Abstract: Traditional consumption of plants, fungi and wild animals constitutes a reality for the feeding of diverse human groups in different tropical territories of the world. In this regard, there are two views within the academic community: (1) those who defend the importance of the traditional consumption for family food security in rural areas, especially in tropical countries with emerging development; and (2) those who affirm their inconvenience as they are considered vectors of rapidly spreading diseases worldwide. A systematic literature review and an Analytic Hierarchy Process (AHP) with experts were carried out to identify the contributing criteria and dimensions in Science, Technology and Innovation (STI) that help evaluate the potential of wild foods for responsible consumption in terms of human health and nature conservation. Four dimensions were identified. The first three are: (1) importance of food for the community that consumes it (w = 0.31); (2) nutritional value and risks for human health (w = 0.28) and (3) sustainability of the local use of wild food model (w = 0.27). These three obtained similar integrated relative weights, which suggests the possible balanced importance in the formulation of multidisciplinary methods for estimating the potential of wild foods. The fourth identified dimension is: (4) transformation techniques for turning wild foods into products with commercial potential, obtained an integrated relative weight of 0.14, which, although is lower than the other three, still contributes to the potential of this type of food. The study found ten assessment criteria to evaluate the identified dimensions, constituting a starting point to estimate the potential of this type of food.

Keywords: assessment of wild food; dimensions in science; technology and innovation (STI); estimation of potential; food security; tropical forest areas

---

1. Introduction

Wild biological diversity as a source of food resources, contributes to family food security through ancestral practices of exploitation of fungi, plants and animals, especially in tropical forest areas [1]. These contributions are reflected at different levels of importance regarding their use. In many rural areas they are the main option for the consumption of animal protein, cereals, tubers, vegetables and fruits. In urban areas they constitute an alternative that complements the supply of non-wild and commercial food within the usual family basket of the territory [2–9]. Wild foods also contribute to the family economy through the generation of income from activities such as hunting, gathering, planting or raising plants, fungi and animals. Usually, the generation of economic income is derived from the occasional or regular sale of said products and the economic resources obtained are used for family subsistence [1].

The consumption of wild food is thus a reality for populations in different tropical territories of the world. This reality deserves a deep analysis, as there are currently different studies that attribute to zoonosis the origin of rapidly spreading diseases worldwide [10–13], and at the same time recognize the contributions of wild foods to the family food security of these territories [1]. One way of solving this dilemma is to generate empirical studies and arguments that enable responsible consumption of this type of food, understanding the processes of manipulation and ingestion of wild foods as responsible consumption that minimizes the risks to human health or biological and/or ecosystem conservation.

These studies are especially necessary in tropical areas that currently occupy some 4 billion hectares and account for about 31% of the world's land surface [14]. Some 800 million people live in these areas [15,16] of whom 38% are undernourished [17]. Tropical forests are distributed in countries in Africa, Asia and Latin America and the Caribbean.

The academic community has been making efforts in Science, Technology and Innovation (STI) to address this need. Disciplinary studies define ethnobiological aspects to describe the wild species of plants, fungi or animals consumed, the frequency and forms of this consumption, as well as the social groups that consume them. This knowledge base serves to analyze the importance of these wild foods in the food security of families living in specific territories [2,8,18–22]. This type of analysis evaluates the nutritional value and risks to human health from the consumption of certain foods (bromatological studies); evaluates possible techniques to develop different products with more commercial potential [23–36]; and describes and/or enhances ancestral planting or breeding practices [37–42]. All this information allows us to understand the importance of wild foods for family food security in tropical areas. Even so, a knowledge gap continues to exist that enables identification of wild foods that can be consumed and/or manipulated without risks to human health or the conservation of natural resources. Furthermore, there is no known theoretical approach that allows the multidisciplinary estimation of the potential of wild foods from its advances in STI.

Therefore, the present study sought to contribute to the identification and definition of evaluation alternatives of those dimensions in STI that define the potential of wild foods for responsible consumption in territories associated with tropical forests. Once the dimensions are identified, a methodological process is developed to weigh and rank them according to their importance in explaining the potential of wild food. Prioritizing the dimensions according to their importance could help define a logical order in evaluating the potential of a wild food. By evaluating the most important dimension first, the non-positive results might indicate that the evaluation is not worth continuing, unless these results can be reversed through STI processes. Finally, criteria are defined to guide the practical measurement of each dimension. In order to meet these objectives, the following research questions were addressed: (1) what are the dimensions in STI that define the potential of wild foods for responsible consumption in tropical forest territories? (2) what is the relative importance of these dimensions in defining the potential of a wild food? and (3) what evaluation criteria could be defined to measure the identified dimensions?

The results generated in the present study—dimensions and evaluation criteria—will contribute to the formulation of multidisciplinary methods that allow estimating the potential of wild foods; an estimate that allows these foods to be analyzed as alternatives in planning food and nutrition security in tropical forest areas where their traditional consumption is a reality today.

2. Methods

The methodological process conducted in the present study was developed under three stages (see Figure 1). In the first stage, a systematic literature review was carried out, through which the main dimensions in STI that define the potential of wild foods for responsible consumption in tropical forest areas were identified. The second stage involved prioritizing these dimensions according to their importance in determining the potential of these foods. For the development of this stage, 15 experts were consulted. Experts were from different knowledge areas related to the use of wild foods by human populations in tropical forest areas. One of the most frequent methods applied in participatory processes in sustainable forest management [43–50] is the Analytic Hierarchy Process (AHP). This method is the most popular multiple criteria decision method used in public participation processes regarding nature resources management. In fact, according to Esmail and Geneletti [51] and Díaz-Balteiro and Romero [52] over 98 applications of this tool have been recorded in nature conservation and forest management. We applied this method to rank the dimensions in STI identified. AHP is based on pairwise comparisons between criteria and alternatives, which are compiled into square matrices whose coefficients are numerical values assigned to the preferences indicated by the participants. Finally, in the third stage the evaluation criteria were defined as a proposal to value wild foods in each of the identified dimensions.

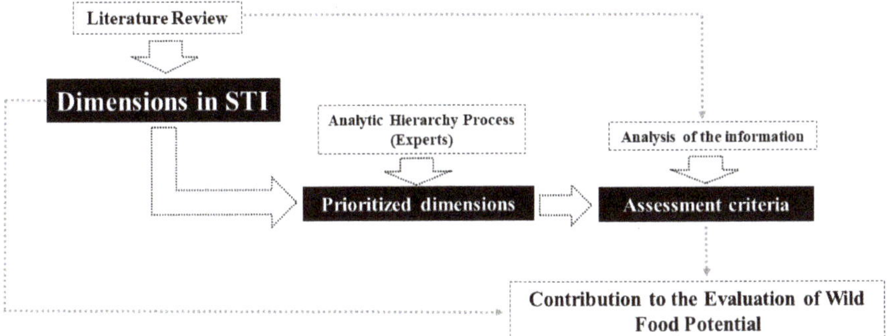

Figure 1. Methodological steps of the study.

2.1. Identify the Dimensions (Literature Review)

To identify the dimensions, it was necessary to previously recognize the importance of this type of food for people who use it in communities associated with tropical forests as well as the main challenges and needs in STI for responsible consumption. The recognition of the importance of wild foods and their main challenges and needs in STI was carried out through a systematic literature review, for which a methodological process of three stages was developed. That process includes: (a) identifying and obtaining documents; (b) reviewing and selecting the most relevant documents and (c) analyzing the information and structuring the results [53,54].

2.1.1. Identification and Retrieval of Documents

A total of 182 documents were identified among books, book chapters and scientific papers. To obtain these documents, consultations were made on scientific websites: Web of ScienceTM Core Collection, BIOSIS Citation IndexSM, BIOSIS Previews®, Current Contents Connect®, Derwent Innovations IndexSM, Inspec®, MEDLINE® and SciELO Citation Index (all linked to WEB OF SCIENCE), as well as queries on Google Academic. Searches were conducted in English and Spanish using keywords in the document titles and were guided by the terms: wild foods, wild vegetables, indigenous vegetables, wild edible plants, wild meat, edible wild fruit and bushmeat. Each of the words was also searched in combination with the terms of food security and challenge. No language restrictions, years of publication, or area of knowledge were programmed.

2.1.2. Review and Selection of Documents

Only documents published in peer-reviewed journals or books published by renowned publishers were included. For the inclusion of a document, in addition to the above characteristics, it was verified that its content contributed to answering the research questions raised in the present study and its area of study corresponded to tropical forest areas. After the review and analysis process of the 182 identified documents, only 45 met the defined criteria and were included (8 books and 37 scientific articles). These 45 documents are cited in the reference section of this article and the 137 documents discarded are listed in Table A1.

2.1.3. Analysis of Information

Through a detailed review of the collected documents, the importance of wild foods for family food security in tropical forest areas and the main challenges and needs in STI were identified, so that this type of food could be responsibly consumed in these territories. During the analysis process for the identification of these elements, and when necessary, contributions were made to the results based on empirical experiences of the authors and/or knowledge generated during technical discussions in regular meetings.

Finally, the identification of the dimensions of technological and/or scientific knowledge that define the potential of wild foods was carried out through analyzing the main challenges and needs in STI so that this type of food could be considered as a viable alternative in food and nutritional security planning. The main arguments for the recognition of each dimension were the identification of its contribution to understanding the importance of the food for the communities that consume it or its possible contributions to the solution of the challenges and/or needs in STI.

2.2. Prioritize the Dimensions (Analytic Hierarchy Process)

To identify the preferences of each expert, the multicriteria decision making method Analytic Hierarchy Process (AHP) was used [55] during the last three months of 2018. According to Saaty [56]: "The purpose of the method is to enable the decision maker to structure a multi-criteria problem by building a hierarchical model". In this case the structure had two levels: objective and criteria. The objective of the model in our study was: ordering dimensions according to their capacity to define the suitability of wild food to be consumed in tropical forest areas, broken down into four criteria corresponding to the identified dimensions. The schema of this methodology is shown in Figure 2.

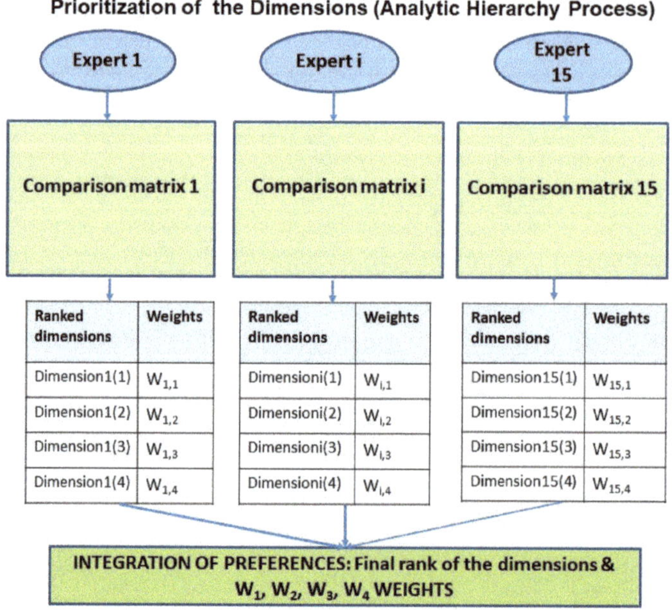

Figure 2. Structure of the dimension prioritization process.

Pairwise comparisons were then performed between the dimensions. A pairwise comparison matrix allows subjective assessments to be converted into relatively important global scores or weights. The comparisons are done by asking the following question: How important is the Ci criterion (dimension i) in relation to the Cj criterion (dimension j)?

A comparison matrix, C4x4, is built for every participant. Every entry $C_{i,j}$ is a number representing the comparison between criteria i and j, according to the scale used. A comparison matrix, C, has three basic properties, namely positivity ($C_{ij} > 0$, for all i, j); homogeneity ($C_{ij} = 1$, if criteria i and j are considered equally important: specifically $C_{ii} = 1$ for each i) and reciprocity ($C_{ji} = 1/C_{ij}$ for all i, j). From this perspective, only 1/2m (m − 1) comparisons need to be made in the comparison matrix.

The AHP pairwise comparison scale was adopted from Saaty [57]:

1 = Both criteria equally important.

3 = Very slight importance of one criterion over the other.

5 = Moderate importance of one criterion over the other.

7 = Demonstrated importance of one criterion over the other.

9 = Extreme of absolute importance of one criterion over the other.

Values express intermediate preference between the two contiguous odd values.

Once all the participants' matrices were obtained, we applied the eigenvector method proposed by Saaty [57] to obtain the weights of the criteria from each matrix. As a measure of consistency of the preference (if A is preferable or indifferent to B, and B is preferable or indifferent to C, then A is preferable or indifferent to C) reflected in every matrix. We applied the Consistency Ratio CR = CI/RI, to measure this transitivity, where CI, the Consistence Index is CI = (lmax − n)/(n − 1) and RI, the Random Index, is RI = 1.98(n − 2)/n, lmax is the maximum eigenvalue of the matrix and n the number of rows of the matrix. If CR is less than 0.1, then the preference is consistent and the estimate is accepted.

The comparisons were done individually by 15 research experts from different tropical countries with training profiles and/or research experience in different areas of knowledge. These areas include: (1) ethnobiology, (2) food and nutritional security, (3) agronomy and/or zootechnics, (4) biology and/or ecology, (5) agricultural and/or natural resources economics and (6) agroindustry. The quality requirements defined for the selection of the experts were: Doctor's degree (PhD) with scientific publications in high-quality indexed journals and having participated in research projects in their area of knowledge.

The results obtained by each expert were averaged to obtain the final relative weight of each dimension by the area of knowledge (see Figure 2). For each area of knowledge, the dimensions were ordered by their importance and a hierarchy of importance was drawn that defined the potential of wild foods for responsible consumption in territories associated with tropical forests.

In addition, three analyses of variance were carried out to see the relationship among the factors "Area of knowledge" and "Dimension", and the quantitative variable "Weight" obtained with the AHP method. These three analyses were: (1) analysis of the influence of the "Area of knowledge" on the average behavior of the variable "Weight" of the expert; (2) analysis of the influence of the "Dimension" on the average behavior of the variable "Weight" of the expert and (3) if the factor "Dimension" influenced the integrated value of the "Weights by area of knowledge".

2.3. Define the Criteria

Once the dimensions were obtained and prioritized, a set of evaluation criteria was defined based on the needs identified in STI during the literature review. The evaluation criteria were obtained from a rigorous analysis process on the STI needs related to each of the identified dimensions. From this perspective, a matrix was built in which each evaluation criterion responds to a mechanism to solve a need in STI. The evaluation criteria constitute the minimum conditions a wild food must meet for responsible consumption or for being considered as a viable alternative in the planning of food and nutritional security for tropical forest territories. Based on the results, a set of dimensions and their respective evaluation criteria is proposed to check the minimum conditions of wild food with respect to its advances in STI.

3. Results

3.1. Key Challenges in STI of Wild Foods

Two main challenges in STI were identified in the literature review. They are the challenges for wild food to be consumed in a responsible way or considered as a viable alternative in the planning of food and nutrition security in tropical forest areas. They are related to (1) the negative effects on biodiversity conservation as a result of unplanned extractive harvesting practices and (2) the possible risks to human health due to the lack of assessment of their nutritional and health quality.

With regard to the first challenge, some studies conducted in different areas of tropical forest in Africa and Latin America show concerns about the frequent and extractive use of wild plants, fungi and animals. The most negative effects in the medium and long term are the reduction of populations of vulnerable species, local extinction and habitat fragmentation, with consequences on the functioning of ecosystems and people's lives [1,9,58–62].

Concerning the second challenge, food products whose nutrition has not been studied, can generate food imbalances for individuals who consume them instead of providing nutrients. This would be counterproductive for food security [2,5,63,64]. In addition, more than 35 new infectious diseases have emerged in humans [53] in recent decades, many of which are attributed to the handling and consumption of plants, fungi or jungle animals [2,5,11–13,63–67]. Nevertheless, the consumption of plants, fungi and wild animals in tropical forest areas as food is not commonly related to government food security policies. On the contrary, this consumption is related to the ancestral traditions and socio-cultural customs of local communities.

On a higher-detailed scale, these challenges could be expressed through the following needs and dimensions in STI identified in the literature review.

3.2. Needs and Dimensions in STI That Define the Potential of Wild Foods

Ten needs in STI were identified in the literature review. They were analyzed and conceptualized in four technological and/or scientific knowledge dimensions. These dimensions in STI constitute the structural axes for estimating the potential of wild foods for responsible consumption as an alternative resource in the planning of interventions in favor of food and nutrition security for tropical forest territories. These dimensions are: (a) importance of food for the community that consumes it; (b) sustainability of the local model of wild food use; (c) nutritional value and risks to human health and (d) processing techniques into products with commercial potential. By its nature, the understanding of these dimensions must be addressed from a multidisciplinary approach since knowledge is required from different disciplines for the study such as sociology, anthropology, biology, ecology, economics, agronomy, zootechnics, veterinary sciences and laws, among others. Each dimension and its respective needs in STI are detailed below.

3.2.1. Importance of Food for the Community That Consumes It

This dimension defines the importance of wild foods based on traditional uses and other aspects that shape the use of a food in a specific community. The food whose pattern of use (obtained through scientific research) demonstrates the importance of this edible food's usage for the territory will have greater potential. Below are the needs for advances in STI related to the dimension: (1) recognition and documentation of the traditional use of wild species in the feeding of communities living in tropical forest territories to include lists of species used as food and studies with different types of use where food is included, etc.; (2) recognition of the pattern of use of wildlife in the feeding of communities in tropical forest areas, with special emphasis on the species/culture/territory relationship and parts of the plant, fungus or animal used, ways of use, frequency of use, economic assessment of the contribution of these products to food and nutritional security, etc., and (3) identification of drivers for the consumption of wild foods including preferences between wild and non-wild foods, limitations to obtaining other alternatives, effects of culture on the consumption of wild foods, etc.

3.2.2. Sustainability of the Local Model of Wild Food Use

This dimension is based on the sustainability of the source from which wild food could be obtained for family consumption and contribution to food and nutritional security in the territory. The wild food will have greater potential if it has at least a sustainable harvesting mechanism tested in the territory. The needs for advances in STI related to the dimension are: (1) recognition of supply sources of the wild food in the territory whether extractive and/or non-extractive use; (2) generation of technological and/or scientific knowledge on mechanisms for the sustainable use of wild food and (3) cost/benefit analysis of such mechanisms.

3.2.3. Nutritional Value and Risks to Human Health

This dimension is based on technological and/or scientific knowledge about the nutritional value of wild food products and the assessment of possible risks to human health due to their intake. The wild food with known nutritional value whose result is similar or better than commercial non-wild foods belonging to the same group (fruit vs. fruit, meat vs. meat, vegetables vs. vegetables, etc.) will have greater potential. Food with studies showing that its consumption does not create risks to human health will also have greater potential. Below are the needs for advances in STI related to this dimension: (1) identification of the nutritional value of wild food products being consumed by human populations in the territory studied; (2) identification of the biological assimilation of these wild food products and (3) assessment of food quality with regard to risks to human health due to its consumption.

3.2.4. Processing Techniques into Products with Commercial Potential

This dimension values the technological and/or scientific knowledge advances about the transformation of wild foods into products with potential economic importance for human communities. Food with advances in technological developments that allow transformation into products with commercial potential will have greater overall potential. There is one need for advances in STI related to this dimension and that is: the generation of scientific knowledge, experimental developments and/or technological developments for the transformation of raw materials into products with commercial potential.

3.3. Sorting of the Dimensions According to Experts' Preferences

Once the dimensions were identified, each expert using the AHP method arranged them in order. The results of this consultation with the fifteen experts and the application of the AHP method can be found in Table 1.

As a result of the application of the AHP method by each expert, the importance of each dimension for each expert was obtained. This importance was expressed by the weights shown in Table 1. These were the weights obtained for each expert who was using the AHP process for each dimension. In Table 1, the experts were grouped according to their knowledge area to make it easier to compare results.

Table 1 also shows that the consistency of preference for all experts was acceptable since no CR value had exceeded the limit value of 0.1 [57].

Table 1. Analytic Hierarchy Process (AHP) weights for each expert Ei of each knowledge area, for each dimension and Consistency Ratio (CR).

Dimensions/Knowledge Area	Agronomy and/or Zootechnics			Agroindustry		Biology and/or Ecology			Natural Resources Economics		Ethnobiology		Food and Nutritional Security		
	E1	E2	E3	E4	E5	E6	E7	E8	E9	E10	E11	E12	E13	E14	E15
Importance of food for the community that consumes it	0.15	0.43	0.25	0.59	0.61	0.43	0.28	0.04	0.31	0.22	0.14	0.23	0.20	0.09	0.33
Sustainability of the local model of wild food use	0.64	0.04	0.25	0.12	0.05	0.09	0.57	0.10	0.56	0.15	0.14	0.29	0.59	0.25	0.49
Nutritional value and risks to human health	0.15	0.42	0.25	0.20	0.09	0.05	0.10	0.60	0.08	0.47	0.48	0.37	0.12	0.63	0.13
Processing techniques into products with commercial potential.	0.05	0.11	0.25	0.09	0.25	0.43	0.04	0.25	0.05	0.15	0.23	0.10	0.09	0.04	0.05
Consistency Rate (CR)	0.04	0.09	0	0.07	0.07	0.02	0.08	0.03	0.04	0	0	0	0.04	0	0.08

Once the results from the AHP method were obtained, they were analyzed using an ANOVA if the knowledge area to which each expert belonged influenced the AHP results. The *p*-value of this test was 1, so we could not reject that the knowledge area did not influence the average behavior of the variable AHP-weight. The residuals met the requirements of normality, (*p*-value of the Kolmogorov–Smirnov test = 0.084217), homoscedasticity (Levene's test *p*-value = 0.4972), independence (autocorrelation coefficient for lag = 1 was −0.23514, which did not belong to the 95% Confidence Interval (−0.250031; 0.253031)), and the residuals average was 0.

These findings allowed us to analyze the AHP weights as a whole and not by the knowledge area. We ranked for every expert the dimensions from the most preferred to the least according to their AHP weight value and also integrated the AHP weights of the experts calculating the geometric mean. Table 2 shows the global results of this ranking.

Table 2. Global preference order of the dimensions according to the experts' individual rankings.

Dimension	First Place	Second Place	Third Place	Fourth Place	Total	Weight Integrated Value	Standard Deviation	Final Ranking
Importance of food for the community that consumes it	5	6	3	1	15	0.31	0.17	1
Sustainability of the local model of wild food use	6	2	4	3	15	0.27	0.22	3
Nutritional value and risks to human health	6	3	5	1	15	0.28	0.2	2
Processing techniques into products with commercial potential	1	4	2	8	15	0.14	0.11	4

The first columns of Table 2 show how many times each dimension has been chosen from first place to the last one for the 15 experts together. These data have been obtained from Table 1 by ordering the dimensions from the most preferred to the last for each expert and then counting how many times each dimension is ranked in first, second, third or fourth place. The dimensions appear in the table in order of preference. The result from the consultation with experts strictly analyzed by the relative weight values suggest the dimension "Importance of food for the community that consumes it" as the most important in defining the potential of wild foods for responsible consumption. A total of 11 experts out of 15 considered it as the first or second most important dimension. The following dimensions in order of importance would be: "Nutritional value and risks to human health" and "Sustainability of the local model of wild food use" with practically equal integrated weights: 0.28 and 0.27 respectively. The number of experts who consider them as the most important dimension is the same, and they are the dimensions in which the standard deviation has been highest. Finally, the dimension considered to be in the 4th position, by more than 50% of the experts, was "Processing techniques into products with commercial potential".

3.4. Criteria for Assessing the Potential of Wild Foods

Based on the literature review, ten assessment criteria are proposed according to the needs in STI identified in each dimension. These criteria include the minimum conditions that a wild food must meet to be responsibly considered for its consumption in territories associated with tropical forests. Figure 3 shows the assessment criteria for each of the dimensions.

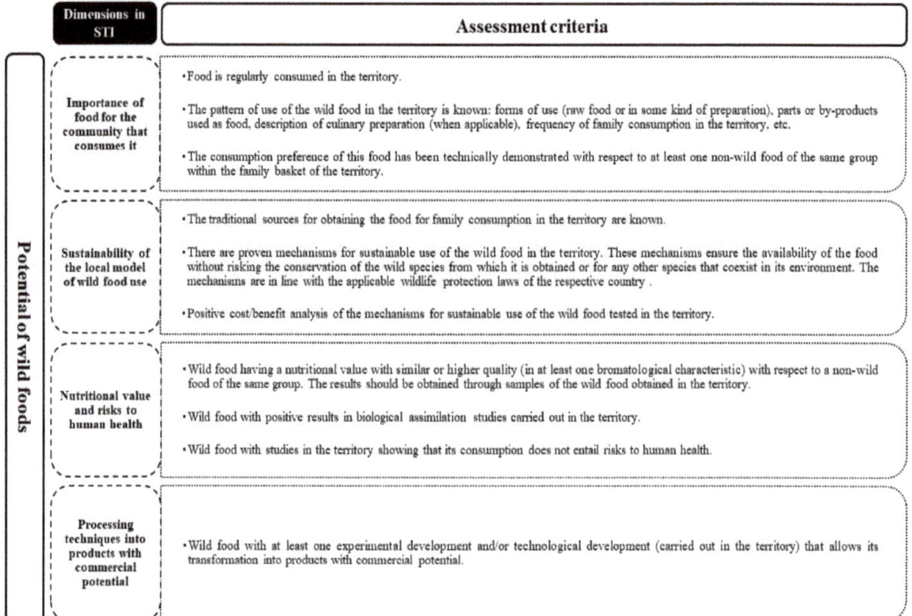

Figure 3. Evaluation criteria for each of the Science, Technology and Innovation (STI) dimensions that define the potential of wild foods.

4. Discussion and Conclusions

The dimensions identified in this study were related to the approach of Arenas and Scarpa [68] where it is stated that the recognition of the potential for human consumption of wild fruits depends on three aspects: cultural acceptance, the abundance of the species in their natural environment and fruit quality. The present research built upon this and proposed four dimensions in STI that define the potential of wild foods as an alternative resource in the planning of food and nutritional security for tropical forest territories. These dimensions are: (a) importance of the food for the community that consumes it; (b) sustainability of the utilization model, (c) nutritional value and risks to human health and (d) processing techniques into products with commercial potential.

The integration of the opinion of the experts has allowed ordering the dimensions by their importance. The consistency of the results makes this ordering applicable in the decision-making process. The results of AHP indicate the unanimity when considering the dimension "Importance of food for the community that consumes it" the most important one. A diet should not be imposed if there is no local tradition in its consumption. It is assumed that initiatives concerning food and nutritional security that include traditional food in the territory could be more relevant and sustainable.

The AHP results also show that the nutritional quality of food consumed by the population should be explored and considered first, as well as the management of sustainable use mechanisms, before considering the processing of wild food and its exploitation for sale in local markets or other places. On the other hand, the differences between the integrated weights of the first three dimensions (a, b and c) vary between 8 and 10%, lower than the variability of the weights of these dimensions (see Table 2). This indicates, for practical purposes, the need to consider the dimensions with equal demand when deciding whether a food should be incorporated into the diet of a population or not.

Knowing the "Importance of food for the community that consumes it", (a) how often it is consumed and what people in the community consume it, is decisive in managing the foods, since cultural preferences and traditional practices tend to maintain them and their modification is difficult [69]. Failure to apply sustainable wild food management (b) can lead to over-exploitation of these species. In the case of forest species there is frequently multiple use of these items as food, firewood, construction or fencing [69]. The dimension of "Nutritional value and risks to human health" (c) will allow measuring whether the wild food consumed by a community meets the requirements of an adequate diet for the population, as demonstrated by the work of Fungo et al. [70] in Nigeria, Gabon and Congo, where, due to wild berries, they are able to get an adequate diet. Knowing the nutritional value of food determines its safety for the health of the community. The lack of studies on the risks consumption entails for human health can cause serious stomach problems, pains in the chest or even the death in these tropical areas [71]; or complex problems such as the COVID-19 pandemic [11–13], which is attributed by some experts to the unplanned management and consumption of wildlife.

The results of the fourth dimension on processing techniques (d) also show their contribution to estimating the potential of wild foods, but the other three dimensions should be addressed as a priority and are considered by experts to be more important.

The ten criteria defined for the evaluation of the four dimensions form the basis to create mechanism that allow the formulation of multidisciplinary methods (qualitative and/or quantitative) for estimating the potential of wild foods for responsible consumption or as an feasible alternative in planning food and nutrition security for territories associated with tropical forests.

The dimensions and criteria presented are not intended to be suggested as absolute variables for evaluating the potential of wild foods for responsible consumption in tropical forest areas. On the contrary, this study seeks to contribute to discussions on methods for multidisciplinary estimation of the potential of this type of food that, although not officially included in the planning of food security policies, does constitute a reality for the food of various human groups in different tropical territories on this planet.

Author Contributions: Conceptualization, J.A.-P. and J.M.D.-P.; methodology, J.A.-P., J.M.D.-P. and S.M.-F.; software, S.M.-F.; validation, J.A.-P., J.M.D.-P. and S.M.-F.; formal analysis, J.A.-P., J.M.D.-P. and S.M.-F.; investigation, J.A.-P., J.M.D.-P. and S.M.-F.; resources, J.A.-P. and J.M.D.; data curation, S.M.; writing—original draft preparation, J.A.-P., J.M.D. and S.M.; writing—review and editing, J.A.-P., J.M.D. and S.M.; visualization, J.A.-P. and S.M.-F.; supervision, J.A.-P., J.M.D.-P. and S.M.-F.; project administration, J.A.-P. and J.M.D.-P. All authors have read and agreed to the published version of the manuscript.

Funding: This research received no external funding.

Acknowledgments: The authors thank Ramón Rosales, Clara María Mejía Doria, Francisco Javier Castellanos Galeano, Israel Ríos Castillo, Hamleth Valois Cuesta, Rubén Cornelio Montes Pérez, Hugo Fernando López Arévalo, Oscar Vergara Garay, Oscar Gustavo Retana Guiascón, Julio Ricardo Sanabria Botero, William Narváez-Solarte, Alex Mauricio Jiménez Ortega, Mélida Martínez Guardia, Ana Afonso Gallegos and Harley Quinto Mosquera for their valuable technical contribution in the application of AHP. They also wish to express their gratitude to the Technological University of Chocó and the Technical University of Madrid for their institutional support.

Conflicts of Interest: The authors declare no conflict of interest.

Appendix A

Table A1. Documents discarded in the literature review and not listed in the reference section.

Geographical Area	Journal Name	Document Type	Document Title	Language	Year of Publication
Africa	Energy Policy	Paper	Biofuels and food security: Micro-evidence from Ethiopia	English	2013
Africa	Energy Policy	Paper	Potential impacts of biofuel development on food security in Botswana: A contribution to energy policy	English	2012
Africa	African Affairs	Paper	Biofuels, food security, and Africa	English	2010
Africa	International Food and Agribusiness Management Review	Paper	The Impact of biofuel production on food security: A briefing paper with a particular emphasis on maize-to-ethanol production	English	2008
Africa	Economics-The Open Access Open-Assessment E-Journal	Paper	South African food security and climate change: Agriculture futures	English	2013
Africa	Review of Development Economics	Paper	Climate change, agriculture and food security in Tanzania	English	2012
Africa	Climate and Development	Paper	Climate variability, yield instability and global recession: The multi-stressor to food security in Botswana	English	2012
Africa	Climatic Change	Paper	East African food security as influenced by future climate change and land use change at local to regional scales	English	2012
Africa	International Journal of Pest Management	Paper	Food security, politics and perceptions of wildlife damage in Western Ethiopia	English	2012
Africa	International Journal of Climate Change Strategies and Management	Paper	Gendered response and risk-coping capacity to climate variability for sustained food security in Northern Cameroon	English	2012
Africa	Environment Development and Sustainability	Review paper	Climate change and variability in Sub-Saharan Africa: A review of current and future trends and impacts on agriculture and food security	English	2011
Africa	American Journal of Agricultural Economics	Paper	Does adaptation to climate change provide food security? A micro-perspective from Ethiopia	English	2011

Table A1. *Cont.*

Geographical Area	Journal Name	Document Type	Document Title	Language	Year of Publication
Africa	Mitigation and Adaptation Strategies for Global Change	Paper	Food security and climate change in drought-sensitive savanna zones of Ghana	English	2011
Africa	Area	Paper	Farming flexibility and food security under climatic uncertainty: Manang, Nepal Himalaya	English	2010
Africa	Sustainability Science	Paper	Food security and seasonal climate information: Kenyan challenges	English	2010
Africa	Journal of Human Ecology	Paper	Climate variability, environment change and food security nexus in Nigeria	English	2009
Africa	Environmental Science and Technology	Paper	Markets, climate change, and food security in West Africa	English	2009
Africa	Erdkunde	Paper	Climate change and food security in tropical West Africa—A dynamic-statistical modeling approach	English	2008
Africa	Journal of Energy in Southern Africa	Paper	The impacts of climate change on food security and health in Southern Africa	English	2008
Africa	Proceedings of the National Academy of Sciences of the United States of America	Paper	Warming of the Indian Ocean threatens eastern and southern African food security but could be mitigated by agricultural development	English	2008
Africa	Climatic Change	Paper	The economic and food security implications of climate change in Mali	English	2005
Africa	Philosophical Transactions of the Royal Society B-biological Sciences	Paper	Weather patterns, food security and humanitarian response in sub-Saharan Africa	English	2005
Africa	Agricultural Water Management	Paper	Adaptation to climate change to enhance food security and preserve environmental quality: example for southern Sri Lanka	English	2004
Africa	Food and Nutrition Bulletin	Paper	Food security indicators after humanitarian interventions including food aid in Zimbabwe	English	2010
Africa	Agrekon	Paper	The status of household food security targets in South Africa	English	2009

Table A1. *Cont.*

Geographical Area	Journal Name	Document Type	Document Title	Language	Year of Publication
Africa	Disasters	Paper	Measuring populations' vulnerabilities for famine and food security interventions: the case of Ethiopia's Chronic Vulnerability Index	English	2008
Africa	Disasters	Paper	The underutilization of street markets as a source of food security indicators in Famine Early Warning Systems: a case study of Ethiopia	English	2008
Africa	Food Culture and Society	Paper	Toward improved understanding of food security a methodological examination based in rural South Africa	English	2013
Africa	Systems Research and Behavioral Science	Paper	Designing sustainable food security policies in sub-Saharan African countries: How social dynamics over-ride utility evaluations for good and bad	English	2012
North America	Economics-The Open Access Open-Assessment E-Journal	Paper	US food security and climate change: Agricultural futures	English	2013
North America	Ecohealth	Paper	Adapting to the impacts of climate change on food security among Inuit in the Western Canadian Arctic	English	2010
North America	International Journal of Circumpolar Health	Paper	Local observations of climate change and impacts on traditional food security in two northern Aboriginal communities	English	2006
North America	Ambio	Paper	A call for urgent monitoring of food and water security based on relevant indicators for the Arctic	English	2013
North America	Journal of Nutrition	Paper	Food security of older adults requesting older Americans act nutrition program in Georgia can be validly measured using a short form of the U.S. household food security survey module	English	2011
North America	Public Health Nutrition	Paper	Does interview mode matter for food security measurement? Telephone versus in-person interviews in the current population survey food security supplement	English	2007

Table A1. *Cont.*

Geographical Area	Journal Name	Document Type	Document Title	Language	Year of Publication
North America	Public Health Nutrition	Paper	The assessment of food security in homeless individuals: a comparison of the Food Security Survey Module and the Household Food Insecurity Access Scale	English	2011
North America	Journal of Nutrition	Paper	Recent advances provide improved tools for measuring children's food security	English	2007
Arctic	International Journal of Circumpolar Health	Paper	Indicators of food and water security in an Arctic health context—results from an international workshop discussion	English	2013
Asia	Renewable and Sustainable Energy Reviews	Review paper	Biofuel and food security in China and Japan	English	2013
Asia	Mountain Research and Development	Paper	Solar greenhouse technology for food security: A case study from Humla District, NW Nepal	English	2012
Asia	Renewable Energy	Paper	Will the development of bioenergy in China create a food security problem? Modeling with fuel ethanol as an example	English	2012
Asia	Journal of Environmental Quality	Paper	Biofuel development, food security and the use of marginal land in China	English	2011
Asia	Energy for Sustainable Development	Paper	Bioenergy and food security: Indian context	English	2009
Asia	Philippine Agricultural Scientist	Paper	Sugar as a biofuel: Implications for Philippine agriculture and food security	English	2008
Asia	Agronomy for Sustainable Development	Paper	Climate change impact on China food security in 2050	English	2013
Asia	Journal of Food Agriculture and Environment	Paper	Climate change, agriculture and food security issues: Malaysian perspective	English	2013
Asia	Sustainability	Paper	Impact of climate and land use changes on water and food security in Jordan: Implications for transcending "The tragedy of the commons"	English	2013

Table A1. *Cont.*

Geographical Area	Journal Name	Document Type	Document Title	Language	Year of Publication
Asia	Environmental Modeling and Assessment	Paper	Modeling of ecological footprint and climate change impacts on food security of the hill tracts of Chittagong in Bangladesh	English	2013
Asia	Chinese Geographical Science	Paper	Soil degradation and food security coupled with global climate change in northeastern China	English	2013
Asia	Ambio	Review paper	Climate change and population growth in Timor Leste: Implications for food security	English	2012
Asia	Food Security	Paper	Adaptation to climate change for food security in the lower Mekong Basin	English	2011
Asia	Journal of Developments in Sustainable Agriculture	Paper	Enhancing food security in the context of climate change and the role of higher education institutions in the Philippines	English	2010
Asia	Asia Pacific Journal of Clinical Nutrition	Review paper	Climate change and food security in East Asia	English	2009
Asia	Food Security	Paper	Climate change, flooding and food security in south Asia	English	2009
Asia	Climatic Change	Paper	Climate change, land use change, and China's food security in the twenty-first century: An integrated perspective	English	2009
Asia	Asia Pacific Journal of Clinical Nutrition	Review paper	Food security in the Asia-Pacific: Climate change, phosphorus, ozone and other environmental challenges	English	2009
Asia	Environmental Management	Paper	Food security in the face of climate change, population growth, and resource constraints: Implications for Bangladesh	English	2004
Asia	Bulletin of Indonesian Economic Studies	Paper	Using climate models to improve Indonesian food security	English	2004
Asia	The Journal of Nutrition	Journal Article	Development and validation of an Arab family food security scale	English	2014

Table A1. *Cont.*

Geographical Area	Journal Name	Document Type	Document Title	Language	Year of Publication
Asia	Advanced Materials Research	Paper	Established the evaluation index system of food security in state farms of Heilongjiang province	English	2011
Asia	Public Health Nutrition	Paper	An empirical study of Taiwan's food security index	English	2010
Asia	Journal of Preventive Medicine and Public Health = Yebang Uihakhoe Chi	Review paper	The concept and measurement of food security	Korean	2008
Asia	Asia Pacific Journal of Clinical Nutrition	Paper	Household food security status measured by the US-household food security/hunger survey module (US-FSSM) is in line with coping strategy indicators found in urban and rural Indonesia	English	2007
Asia	Food Security	Paper	How disaggregated should information be for a sound food security policy?	English	2013
Europe	Food Policy	Paper	Disentangling the consensus frame of food security: The case of the EU common agricultural policy reform debate	English	2014
Europe	Renewable and Sustainable Energy Reviews	Review paper	Biomass energy in Vojvodina: Market conditions, environment and food security	English	2010
Europe	Economics-The Open Access Open-Assessment E-Journal	Paper	Russia's food security and climate change: Looking into the future	English	2013
Europe	Applied Artificial Intelligence	Paper	Food security risk level assessment: a fuzzy logic-based approach	English	2013
Europe	Cybernetics and Systems Analysis	Paper	Integrated modeling of food security management in Ukraine. I. Model for management of the economic availability of food	English	2013
Global	Food Policy	Paper	Policy coherence and food security: The effects of OECD countries' agricultural policies	English	2014

Table A1. *Cont.*

Geographical Area	Journal Name	Document Type	Document Title	Language	Year of Publication
Global	Pastos y Forrajes	Paper	Producción de agroenergía a partir de biomasa en sistemas agroforestales integrados: una alternativa para lograr la seguridad alimentaria y la protección ambiental	Spanish	2010
Global	Revista de Salud Pública (Bogota, Colombia)	Paper	Biocombustibles, seguridad alimentaria y cultivos transgenicos.	Spanish	2009
Global	California Agriculture	Paper	Biofuel policy must evaluate environmental, food security and energy goals to maximize net benefits	English	2009
Global	Renewable and Sustainable Energy Reviews	Review paper	Biofuels: Environment, technology and food security	English	2009
Global	Chemical Engineering Research and Design	Paper	Relation of biofuel to bioelectricity and agriculture: Food security, fuel security, and reducing greenhouse emissions	English	2009
Global	Biofuels Biofuels Bioproducts and Biorefining-Biofpr	Paper	Is the expansion of biofuels at odds with the food security of developing countries?	English	2007
Global	Environment	Paper	The ripple effect: Biofuels, food security and the environment	English	2007
Global	Berichte Uber Landwirtschaft	Paper	Challenges of global change for agricultural development and world food security	German	2013
Global	Current History	Paper	Climate change and food security	English	2013
Global	Science	Review paper	Climate change impacts on global food security	English	2013
Global	Journal of Crop Improvement	Paper	Declining agricultural productivity and global food security	English	2013
Global	Ecohydrology and Hydrobiology	Paper	Food security in a changing climate	English	2013
Global	Global Change Biology	Review paper	How much land-based greenhouse gas mitigation can be achieved without compromising food security and environmental goals?	English	2013

Table A1. *Cont.*

Geographical Area	Journal Name	Document Type	Document Title	Language	Year of Publication
Global	Journal of Renewable and Sustainable Energy	Paper	Achieving food security while switching to low carbon agriculture	English	2012
Global	Bulgarian Journal of Agricultural Science	Paper	Agricultural research in 21st century: Challenges facing the food security under the impacts of climate change	English	2012
Global	Environmental Health Perspectives	Review paper	Climate change and food security: Health impacts in developed countries	English	2012
Global	Current Opinion in Environmental Sustainability	Paper	Climate change, agriculture and food security: A global partnership to link research and action for low-income agricultural producers and consumers	English	2012
Global	Irrigation and Drainage	Paper	Deltas The new challenges to food security under climate change	English	2012
Global	Global Environmental Change-Human and Policy Dimensions	Paper	Farmers and perverse outcomes: The quest for food and energy security, emissions reduction and climate adaptation	English	2012
Global	Proceedings of the Royal Society B-Biological Sciences	Review paper	Food security and climate change: On the potential to adapt global crop production by active selection to rising atmospheric carbon dioxide	English	2012
Global	Isle-Interdisciplinary Studies in Literature and Environment	Paper	Framing Emerson's "farming": Climate change, peak oil, and the rhetoric of food security in the twenty-first century	English	2012
Global	Environmental Science and Policy	Review paper	Options for support to agriculture and food security under climate change	English	2012
Global	Euphytica	Review paper	Implications of climate change for diseases, crop yields and food security	English	2011
Global	Journal Fur Verbraucherschutz Und Lebensmittelsicherheit (Journal of Consumer Protection and Food Safety)	Paper	Threat of the "food security" by the effects of climate change	German	2011

Table A1. *Cont.*

Geographical Area	Journal Name	Document Type	Document Title	Language	Year of Publication
Global	Wiley Interdisciplinary Reviews-Climate Change	Review paper	Adapting to climate change to sustain food security	English	2010
Global	Food Security	Paper	Beyond Copenhagen: Mitigating climate change and achieving food security through soil carbon sequestration	English	2010
Global	Food Policy	Paper	Global water crisis and future food security in an era of climate change	English	2010
Global	Progress in Natural Science-Materials International	Review paper	Climate change impacts on crop yield, crop water productivity and food security—a review	English	2009
Global	Food Security	Paper	Declining global per capita agricultural production and warming oceans threaten food security	English	2009
Global	Science	Paper	Prioritizing climate change adaptation needs for food security in 2030	English	2008
Global	Proceedings of the National Academy of Sciences of the United States of America	Paper	Global food security under climate change	English	2007
Global	Climate Change and Global Food Security	Paper	Assessing the consequences of climate change for food security: A view from the intergovernmental panel on climate change	English	2005
Global	Philosophical Transactions of the Royal Society B-Biological Sciences	Paper	Climate change and food security	English	2005
Global	Climate Change and Global Food Security	Paper	Climate change and tropical agriculture: Implications for social vulnerability and food security	English	2005
Global	Climate Change and Global Food Security	Paper	Greenhouse gases and food security in low-income countries	English	2005
Global	Ids Bulletin-Institute of Development Studies	Paper	Climate change and food security	English	2004
Global	Proceedings of the National Academy of Sciences of the United States of America	Paper	A systems science perspective and transdisciplinary models for food and nutrition security	English	2012

Table A1 *Cont.*

Geographical Area	Journal Name	Document Type	Document Title	Language	Year of Publication
Global	Grey Systems: Theory and Application	Paper	Grey prediction model-based food security early warning prediction	English	2012
Global	Food Security	Paper	A food systems approach to researching food security and its interactions with global environmental change	English	2011
Global	Ciencia and Saude Coletiva	Paper	Conceptualizing and measuring food and nutrition security	Portuguese	2011
Global	Nutrition	Review paper	Food security measurement in cultural pluralism: Missing the point or conceptual misunderstanding?	English	2010
Global	Food Security	Paper	Food security: Definition and measurement	English	2009
Global	Journal of Physical Therapy Science	Review paper	Measuring household food security: The global experience	English	2008
Global	Journal of Tsinghua University (Science and Technology)	Paper	Optical immuno-sensors chip for environmental monitoring and food security detection	Chinese	2007
Global	Applied Geography	Paper	A city and national metric measuring isolation from the global market for food security assessment	English	2013
Global	Chinese Journal of Chromatography	Paper	Development of sample pretreatment techniques-rapid detection coupling methods for food security analysis	Chinese	2013
Global	Food Security	Paper	Rethinking the measurement of food security: From first principles to best practice	English	2013
Global	Advances in Nutrition	Review paper	What are we assessing when we measure food security? A compendium and review of current metrics	English	2013
Global	Environmental Science and Policy	Paper	A framework to assess national level vulnerability from the perspective of food security: The case of coral reef fisheries	English	2012
Latin America and the Caribbean	Energy for Sustainable Development	Paper	Bioenergy and sustainable development: The dilemma of food security and climate change in the Brazilian savannah	English	2010

Table A1. *Cont.*

Geographical Area	Journal Name	Document Type	Document Title	Language	Year of Publication
Latin America and the Caribbean	China Agricultural Economic Review	Paper	Biofuels, food security and compensatory subsidies	English	2010
Latin America and the Caribbean	Mitigation and Adaptation Strategies for Global Change	Paper	Implications of a changing climate on food security and smallholders' livelihoods in Bogota, Colombia	English	2014
Latin America and the Caribbean	Climatic Change	Paper	Climate change and critical thresholds in China's food security	English	2007
Latin America and the Caribbean	Ciencia and Saude Coletiva	Paper	Metodos de analise em programas de seguranca alimentar e nutricional: Uma experiencia no Brasil.	Portuguese	2013
Latin America and the Caribbean	Revista Chilena de Nutrición	Paper	Escalas para medir la seguridad alimentaria en Colombia: ¿Son válidas?	Spanish	2012
Latin America and the Caribbean	Revista Chilena de Nutrición	Paper	Seguridad alimentaria en Colombia y modelo rasch	Spanish	2012
Latin America and the Caribbean	Revista de Salud Pública (Bogota, Colombia)	Paper	Seguridad alimentaria: Un método alterno frente a uno clásico	Spanish	2010
Latin America and the Caribbean	Revista Chilena de Nutrición	Paper	Validéz factorial, consistencia interna y reproductibilidad de la escala de seguridad alimentaria en hogares de Bucaramanga, Colombia	Spanish	2009
Latin America and the Caribbean	Bmc Public Health	Paper	Internal validity of a household food security scale is consistent among diverse populations participating in a food supplement program in Colombia	English	2008
Latin America and the Caribbean	European Journal of Clinical Nutrition	Paper	Psychometric properties of a modified US-household food security survey module in Campinas, Brazil	English	2008
Latin America and the Caribbean	Revista de Nutrição	Paper	Segurança alimentar e nutricional: Desenvolvimento de indicadores e experimentação em um município da Bahia, Brasil	Portuguese	2008
Latin America and the Caribbean	Bmc Public Health	Paper	The 18 household Food Security Survey items provide valid food security classifications for adults and children in the Caribbean	English	2006

Table A1. *Cont.*

Geographical Area	Journal Name	Document Type	Document Title	Language	Year of Publication
Latin America and the Caribbear	Public Health Nutrition	Paper	Self-administration of a food security scale by adolescents: Item functioning, socio-economic position and food intakes	English	2005
Latin America and the Caribbear	Cadernos de Saude Publica	Paper	Validación de escalas de seguridad alimentaria y de apoyo social en una población afro-colombiana: Aplicación en el estudio de prevalencia del estado nutricional en niños de 6 a 18 meses.	Spanish	2005
Latin America and the Caribbear	Archivos Latinoamericanos de Nutrición	Paper	Valoración de informadores clave sobre el plan de acción de las políticas de seguridad alimentaria en Colombia.	Spanish	2005
Oceania	Food Policy	Paper	Impact of water scarcity in Australia on global food security in an era of climate change	English	2013
Oceania	Regional Environmental Change	Paper	Dangerous climate change in the Pacific Islands: Food production and food security	English	2011

References

1. Asprilla-Perea, J.; Díaz-Puente, J.M. Importance of wild foods to household food security in tropical forest areas. *Food Secur.* **2019**, *11*, 15–22. [CrossRef]
2. Asprilla-Perea, J.; Mosquera, Y.; Moreno, A. *Proechimys semispinosus* (Ratón de Espinas): Una especie de fauna silvestre con Potencial Promisorio para comunidades negras del departamento del Chocó, Pacífico Colombiano. *Caldasia* **2012**, *34*, 385–396.
3. Martínez-Pérez, A.; López, P.A.; Gil-Muñoz, A.; Cuevas-Sánchez, J.A. Plantas silvestres útiles y prioritarias identificadas en la Mixteca Poblana, México. *Acta Bot. Mex.* **2012**, *98*, 73–98. [CrossRef]
4. Chandra, K.; Nautiyal, B.P.; Nautiyal, M.C. Ethno-botanical resources as supplementary foods and less known wild edible fruits in district Rudraprayag, Uttarakhand, India. *J. Hum. Ecol.* **2013**, *42*, 259–271. [CrossRef]
5. Kamga, R.T.; Kouamé, C.; Atangana, A.R.; Chagomoka, T.; Ndango, R. Nutritional evaluation of five African indigenous vegetables. *J. Hortic. Res.* **2013**, *21*, 99–106. [CrossRef]
6. Cruz, M.P.; Medeiros, P.M.; Sarmiento-Combariza, I.; Peroni, N.; Albuquerque, U.P. "I eat the manofê so it is not forgotten": Local perceptions and consumption of native wild edible plants from seasonal dry forests in Brazil. *J. Ethnobiol. Ethnomed.* **2014**, *10*, 45. [CrossRef]
7. Bortolotto, I.M.; de Mello Amorozo, M.C.; Neto, G.G.; Oldeland, J.; Damasceno-Junior, G.A. Knowledge and use of wild edible plants in rural communities along Paraguay River, Pantanal, Brazil. *J. Ethnobiol. Ethnomed.* **2015**, *11*, 46. [CrossRef]
8. Fa, J.E.; Olivero, J.; Real, R.; Farfán, M.A.; Márquez, A.L.; Vargas, J.M.; Ziegler, S.; Wegmann, M.; Brown, D.; Margetts, B.; et al. Disentangling the relative effects of bushmeat availability on human nutrition in central Africa. *Sci. Rep.* **2015**, *5*, 8168. [CrossRef]
9. Asprilla-Perea, J.; Díaz-Puente, J.M. Uso de alimentos silvestres de origen animal en comunidades rurales asociadas con bosque húmedo tropical al noroeste de Colombia. *Interciencia* **2020**, *45*, 76–83.
10. Monsalve, S.; Mattar, S.; Gonzalez, M. Zoonosis transmitidas por animales silvestres y su impacto en las enfermedades emergentes y reemergentes. *Rev. MVZ Córdoba* **2009**, *14*, 1762–1773. [CrossRef]
11. Bonilla-Aldana, D.; Villamil-Gómez, W.E.; Rabaan, A.A.; Rodríguez-Morales, A.J. Una nueva zoonosis viral de preocupación global: COVID-19, enfermedad por coronavirus 2019. *Iatreia* **2020**, *33*, 107–110.
12. Palacios, C.; Santos, E.; Cervantes, M.V.; Juárez, M.L. COVID-19, una emergencia de salud pública mundial. *Rev. Clín. Esp.* **2020**. [CrossRef]
13. Rothan, H.A.; Byrareddy, S.N. The epidemiology and pathogenesis of coronavirus disease (COVID-19) outbreak. *J. Autoimmun.* **2020**, *109*, 102433. [CrossRef]
14. Programa CE-FAO. *Una Introducción a los Conceptos Básicos de la Seguridad Alimentaria*; FAO: Roma, Italy, 2011.
15. Groom, B.; Palmer, C. REDD+ and rural livelihoods. *Biol. Conserv.* **2012**, *154*, 42–52. [CrossRef]
16. Kashwan, P.; Holahan, R. Nested governance for effective REDD+: Institutional and political arguments. *Int. J. Commons* **2014**, *8*, 554–575. [CrossRef]
17. Food and Agriculture Organization of the United Nations; International Fund for Agricultural Development; World Food Programme. *El Estado de la Inseguridad Alimentaria en el Mundo. Cumplimiento de los Objetivos Internacionales para 2015 en Relación con el Hambre: Balance de los Desiguales Progresos*; CreateSpace Independent Publishing Platform: Roma, Italy, 2015.
18. Robinson, J.G.; Bennett, E.L. *Hunting for Sustainability in Tropical Forests*; Robinson, J., Bennett, E., Eds.; Columbia University Press: Columbia, NY, USA, 2000.
19. Van den Eynden, V.; Cueva, E.; Cabrera, O. Wild foods from Southern Ecuador. *Econ. Bot.* **2003**, *57*, 576–603. [CrossRef]
20. Schulp, C.J.E.; Thuiller, W.; Verburg, P.H. Wild food in Europe: A synthesis of knowledge and data of terrestrial wild food as an ecosystem service. *Ecol. Econ.* **2014**, *105*, 292–305. [CrossRef]
21. Termote, C.; Raneri, J.; Deptford, A.; Cogill, B. Assessing the potential of wild foods to reduce the cost of a nutritionally adequate diet: An example from eastern Baringo District, Kenya. *Food Nutr. Bull.* **2014**, *35*, 458–479. [CrossRef]
22. Erskine, W.; Ximenes, A.; Glazebrook, D.; da Costa, M.; Lopes, M.; Spyckerelle, L.; Williams, R. The role of wild foods in food security: The example of Timor-Leste. *Food Secur.* **2015**, *7*, 55–65. [CrossRef]

23. Leterme, P.; Buldgen, A.; Estrada, F.; Londoño, A.M. Mineral content of tropical fruits and unconventional foods of the Andes and the rain forest of Colombia. *Food Chem.* **2006**, *95*, 644–652. [CrossRef]

24. Tejada, R.; Chao, L.; Gómez, H.; Painter, R.E.L.; Wallace, R.B. Evaluación sobre el uso de la fauna silvestre en la Tierra Comunitaria de Origen Tacana, Bolivia. *Ecol. Boliv.* **2006**, *41*, 138–148.

25. Bustacara, A.; Joya, F.D. *Elaboración de Tres Productos Cárnicos: Chorizo, Longaniza y Hamburguesa, con 100% Carne de Babilla*; Facultad de Zootecnia, Universidad de la Salle: Bogotá, Colombia, 2007.

26. Palomino, C.; Molina, Y.; Pérez, E. Atributos físicos y composición química de harinas y almidones de los tubérculos de *Colocasia esculenta* (L.) Schott y *Xanthosoma sagittifolium* (L.) Schott. *Rev. Fac. Agron.* **2010**, *36*, 58–66.

27. Ramírez-Rivera, J.; Juárez-Barrientos, J.M.; Herrera-Torres, E.; Navarro-Cortez, R.O.; Hernández-Santos, B. Caracterización fisicoquímica, funcional y contenido fenólico de harina de malanga (*Colocasia esculenta*) cultivada en la región de Tuxtepec, Oaxaca, México. *Cienc. Mar* **2011**, *15*, 37–47.

28. Torres-Rapelo, A.L.; Montero-Castillo, P.M.; Julio-González, L.C. Utilización de almidón de Malanga (*Colocasia esculenta* L.) en la elaboración de salchichas tipo Frankfurt. *Biotecnol. Sect. Agropecu. Agroind.* **2014**, *12*, 97–105.

29. Uchôa-Thomaz, A.M.A.; Sousa, A.E.; Carioca, J.O.; de Morais, S.M.; de Lima, A.; Martins, C.; Alexandrino, C.; Ferreira, P.A. Chemical composition, fatty acid profile and bioactive compounds of guava seeds (*Psidium guajava* L.). *Food Sci. Technol.* **2014**, *34*, 485–492. [CrossRef]

30. Serpa, A.M.; Vásquez, D.C.; Castrillón, D.C.; Hincapié, G.A. Comparison of two dehydration techniques for "pear" guava (*Psidium guajava* L.) on the effects of the vitamin C content and on the behavior of the technical and functional properties of the dietary fiber. *Rev. Lasallista Investig.* **2015**, *12*, 10–20.

31. Alvis, A.; Romero, P.; Granados, C.; Torrenegra, M.; Pajaro-Castro, N. Evaluación del color, las propiedades texturales y sensoriales de salchicha elaborada con carne de babilla (*Caiman Crocodilus Fuscus*). *Rev. Chil. Nutr.* **2017**, *44*, 89–94. [CrossRef]

32. Sardeshpande, M.; Shackleton, C. Wild Edible Fruits: A Systematic Review of an Under-Researched Multifunctional NTFP (Non-Timber Forest Product). *Forests* **2019**, *10*, 467. [CrossRef]

33. Rana, Z.H.; Alam, M.K.; Akhtaruzzaman, M. Nutritional Composition, Total Phenolic Content, Antioxidant and α-Amylase Inhibitory Activities of Different Fractions of Selected Wild Edible Plants. *Antioxidants* **2019**, *8*, 203. [CrossRef]

34. Alam, M.K.; Rana, Z.H.; Islam, S.N.; Akhtaruzzaman, M. Total phenolic content and antioxidant activity of methanolic extract of selected wild leafy vegetables grown in Bangladesh: A cheapest source of antioxidants. *Potravin. Slovak J. Food Sci.* **2019**, *13*, 287–293. [CrossRef]

35. Alam, M.K.; Rana, Z.H.; Kabir, N.; Begum, P.; Kawsar, M.; Khatun, M.; Ahsan, M.; Islam, S.N. Total Phenolics, Total Carotenoids and Antioxidant Activity of Selected Unconventional Vegetables Growing in Bangladesh. *Curr. Nutr. Food Sci.* **2019**, *15*, 1. [CrossRef]

36. Alam, M.K.; Rana, Z.H.; Islam, S.N.; Akhtaruzzaman, M. Comparative assessment of nutritional composition, polyphenol profile, antidiabetic and antioxidative properties of selected edible wild plant species of Bangladesh. *Food Chem.* **2020**, *320*, 126646. [CrossRef] [PubMed]

37. Larrazábal, L. Crianza en cautiverio de perezoso de dos dedos (*Choloepus didactylus*). *Edentata* **2004**, *6*, 30–36. [CrossRef]

38. Viloria, H.; Córdova, C. Production system of taro (*Colocasia esculenta* L. Schott) at Manuel Renaud Parish, Antonio Díaz Municipality, Delta Amacuro State, Venezuela. *Rev. Cient. UDO Agríc.* **2008**, *8*, 98–106.

39. Suárez, J.; Camacaro, M.P.; Giménez, A. Efecto de la temperatura y estado de madurez sobre la calidad postcosecha de la fruta de guayaba (*Psidium guajava* L.) procedente de Mercabar, estado Lara, Venezuela. *Rev. Cient. UDO Agríc.* **2009**, *9*, 60–69.

40. Cifuentes, L.; Moreno, F.; Arango, D.A. Fenología reproductiva y productividad de *Oenocarpus bataua* (Mart.) en bosques inundables del Chocó Biogeográfico, Colombia. *Biota Neotrop.* **2010**, *10*, 101–110. [CrossRef]

41. Álvarez, V.M.; Muriel, S.; Osorio, N. Plantas asociadas al turismo y los sistemas tradicionales de manejo en el occidente cercano antioqueño (Colombia). *Ambient. Desarro.* **2015**, *19*, 67–82. [CrossRef]

42. Sicchar-Valdez, L.A.; Acosta-Díaz, A.; Panduro, S.; Panduro, M.; Ramírez, S.; Monge, M.; Villacorta, D. Supervivencia a condiciones extremas en cautiverio de *Caiman crocodilus* (Linnaeus 1578), lagarto blanco, en Iquitos, Perú. *Conoc. Amaz.* **2015**, *2*, 81–85.

43. Sheppard, R.J.; Meitner, M. Using multi-criteria analysis and visualisation for sustainable forest management planning with stakeholder groups. *For. Ecol. Manag.* **2005**, *207*, 171–187. [CrossRef]

44. Díaz-Balteiro, L.; Romero, C. Sustainability of forest management plans: A discrete goal programming approach. *J. Environ. Manag.* **2004**, *71*, 351–359. [CrossRef]

45. Mendoza, G.A.; Prabhu, R. Qualitative multi-criteria approaches to assessing indicators of sustainable forest resource management. *For. Ecol. Manag.* **2003**, *174*, 107–126. [CrossRef]

46. Mendoza, G.A.; Prabhu, R. Combining participatory modeling and multicriteria analysis for community-based management. *For. Ecol. Manag.* **2005**, *207*, 145–156. [CrossRef]

47. Sugimura, K.; Howard, T.E. Incorporating social factors to improve the Japanese forest zoning process. *For. Policy Econ.* **2008**, *10*, 161–173. [CrossRef]

48. Thomson, A.J. How should we manage knowledge ecosystems? Using additive knowledge management. In *Sustainable Forestry from Monitoring and Modelling to Knowledge Management & Policy Science*; Reynolds, K.M., Thomson, A.J., Köhl, M., Shannon, M.A., Ray, D., Rennolls, K., Eds.; CAB International: Wallingford, UK, 2007; pp. 461–480. ISBN 10-1845931742.

49. Mustajoki, J.; Saarikoski, H.; Marttunen, M.; Ahtikoski, A.; Hallikainen, V.; Helle, T.; Ylisirniö, A.L. Use of decision analysis interviews to support the sustainable use of the forests in Finnish Upper Lapland. *J. Environ. Manag.* **2011**, *92*, 1550–1563. [CrossRef] [PubMed]

50. Ocampo-Melgar, A.; Bautista, S.; de Steiguer, J.E.; Orr, B.J. Potential of an outranking multi-criteria approach to support the participatory assessment of land management actions. *J. Environ. Manag.* **2017**, *195*, 70–77. [CrossRef]

51. Esmail, B.A.; Geneletti, D. Multi-criteria decision analysis for nature conservation: A review of 20 years of applications. *Methods Ecol. Evol.* **2017**, 42–53. [CrossRef]

52. Díaz-Balteiro, L.; Romero, C. Making forestry decisions with multiple criteria: A review and an assessment. *For. Ecol. Manag.* **2008**, *255*, 3222–3241. [CrossRef]

53. Labin, S.N. Research synthesis: Toward broad-based evidence. In *Fundamental Issues in Evaluation*; Smith, N., Brandon, P.R., Eds.; Guilford Press: New York, NY, USA, 2007; pp. 89–110.

54. Mavengahama, S.; McLachlan, M.; de Clercq, W. The role of wild vegetable species in household food security in maize based subsistence cropping systems. *Food Secur.* **2013**, *5*, 227–233. [CrossRef]

55. Saaty, T.L. Axiomatic Foundation of the Analytic Hierarchy Process. *Manag. Sci.* **1986**, *32*, 841–855. [CrossRef]

56. Saaty, T.L. A scaling method for priorities in hierarchical structures. *J. Math. Psychol.* **1977**, *15*, 234–281. [CrossRef]

57. Saaty, T.L. *The Analytic Hierarchy Process*; McGraw Hill: New York, NY, USA, 1980; p. 286.

58. Peres, C.A. Synergistic effects of subsistence hunting and habitat fragmentation on Amazonian forest vertebrates. *Conserv. Biol.* **2001**, *15*, 1490–1505. [CrossRef]

59. Fa, J.E.; Currie, D.; Meeuwig, J. Bushmeat and food security in the Congo Basin: Linkages between wildlife and people's future. *Environ. Conserv.* **2003**, *30*, 71–78. [CrossRef]

60. Laurance, W.F.; Croes, B.M.; Tchignoumba, L.; Lahm, S.A.; Alonso, A.; Lee, M.E.; Campbell, P.; Ondzeano, C. Impacts of roads and hunting on Central African rainforest mammals. *Conserv. Biol.* **2006**, *20*, 1251–1261. [CrossRef] [PubMed]

61. Wright, S.J.; Muller-Landau, H.C. The future of tropical forest species. *Biotropica* **2006**, *38*, 287–301. [CrossRef]

62. Fa, J.E.; Brown, D. Impacts of hunting on mammals in African tropical moist forests: A review and synthesis. *Mamm. Rev.* **2009**, *39*, 231–264. [CrossRef]

63. Pandey, M.; Abidi, A.B.; Singh, S.; Singh, R.P. Nutritional evaluation of leafy vegetable paratha. *J. Hum. Ecol.* **2006**, *19*, 155–156. [CrossRef]

64. Keatinge, D. *Vegetables: Less Visible, but Vital for Human Health. Why Nutrient-Dense Indigenous Vegetables Must Be on the Plate for Economic Development, Food Security, and Health*; The World Vegetables Center (AVRDC), News Brief: Shanhua District, Taiwan, 2012.

65. Leroy, E.M.; Rouquet, P.; Formenty, P.; Souquière, S.; Kilbourne, A.; Froment, J.M.; Bermejo, M.; Smit, S.; Karesh, W.; Swanepoel, R.; et al. Multiple Ebola Virus Transmission Events and Rapid Decline of Central African Wildlife. *Science* **2004**, *303*, 387–390. [CrossRef]

66. Bell, D.; Roberton, S.; Hunter, P.R. Animal origins of SARS coronavirus: Possible links with the international trade in small carnivores. *Philos. Trans. R. Soc. Lond. B Biol. Sci.* **2004**, *359*, 1107–1114. [CrossRef]

67. Karesh, W.B.; Cook, R.A.; Bennett, E.L.; Newcomb, J. Wildlife trade and global disease emergence. *Emerg. Infect. Dis.* **2005**, *11*, 1000. [CrossRef]

68. Arenas, P.; Scarpa, G. Edible wild plants of the Chorote Indians, Gran Chaco, Argentina. *Bot. J. Linn. Soc.* **2007**, *153*, 73–85. [CrossRef]

69. Dejene, T.; Agamy, M.S.; Agúndez, D.; Martin-Pinto, P. Ethnobotanical Survey of Wild Edible Fruit Tree Species in Lowland Areas of Ethiopia. *Forests* **2020**, *11*, 177. [CrossRef]

70. Fungo, R.; Muyonga, J.H.; Ngondi, J.L.; Mikolo-Yobo, C.; Iponga, D.M.; Ngoye, A.; Nchuaji, E.; Chupezi, J. Nutrient and Bioactive Composition of Five Gabonese Forest Fruits and Their Potential Contribution to Dietary Reference Intakes of Children Aged 1–3 Years and Women Aged 19–60 Years. *Forests* **2019**, *10*, 86. [CrossRef]

71. Harris, F.M.; Mohammed, S. Relying on Nature: Wild Foods in Northern Nigeria. *AMBIO* **2003**, *32*, 24–29. [CrossRef] [PubMed]

MDPI
St. Alban-Anlage 66
4052 Basel
Switzerland
Tel. +41 61 683 77 34
Fax +41 61 302 89 18
www.mdpi.com

Forests Editorial Office
E-mail: forests@mdpi.com
www.mdpi.com/journal/forests